PRESS ON!

Selected Works of General Donn A. Starry

Volume I

Selected, Edited, Annotated, and with an Introductory Essay by

Lewis Sorley

Combat Studies Institute Press
US Army Combined Arms Center
Fort Leavenworth, Kansas

Library of Congress Cataloging-in-Publication Data

Starry, Donn A. (Donn Albert), 1925-

Selected works of General Donn A. Starry / selected, edited, annotated, and with an introductory essay by Lewis Sorley.

p. cm.

Includes index.

1. Military art and science--United States--History--20th century. 2. Military doctrine--United States. 3. United States--Military policy. I. Sorley, Lewis, 1934- II. Title.

UA23.S675 2009
355.00973--dc22

2009017784

First printing, September 2009.

Cover photo: Colonel Donn A. Starry while commanding the 11th Armored Cavalry Regiment, Republic of Vietnam, 1970. *Starry Family Collection*

Contents

Page

Volume I

Prologue v

A Note on Reading the Text xxi

Acknowledgments xxiii

Articles, Speeches, and Correspondence

1. Airpower 1
2. Armor 11
3. Army Aviation 145
4. The Battlefield 151
5. Combined Arms 227
6. Command and Control 251
7. Doctrine 261
8. Force Structure 459
9. Intelligence 475
10. Leadership 479
11. Logistics 511
12. Management 529
13. Manpower 579
14. Military History 607
15. Modernization 631
16. Planning 695

Accompanying Disk, Text of Starry Book, *Mounted Combat in Vietnam*

Volume II

17. Soldiers 701
18. Strategy 729
19. Training 767
20. Uniforms 853
21. Values 871
22. Varied Topics 905
23. Vietnam War 947

Page

Oral History Interviews

1. Life and Career of General Donn A. Starry 980
2. Reminiscences of General Creighton Abrams 1171
3. Experiences as a Commander 1196
4. *Soldiers* Magazine Interview "This Is a Tough Business . . ." 1218
5. Desert Storm Lessons Learned 1223
6. Development of Doctrine 1256
7. Air Force: AirLand Battle 1274

Glossary of Acronyms and Abbreviations 1293

Index 1311

About the Editor 1341

Accompanying Disk, Key Starry Speeches

Prologue

Early Days and West Point

Military life came naturally to Donn Starry. His father had fought in the fledgling tank corps during World War I, then served long years in the National Guard. At age four, something of a mascot to his father's outfit, Starry was made a brevet first lieutenant in the Kansas National Guard by Governor Clyde Reed. He still has the "commission" to prove it. He grew up riding horses and reading books. Year after year he accompanied his father to summer training at Fort Riley, Kansas.

In August 1943 Starry enlisted at Fort Leavenworth, Kansas, as a private in the US Army. Before long he was assigned to a service unit at Lafayette College preparing soldiers for admission to West Point. The following June he was discharged for the purpose of entering the Military Academy with the wartime class of 1947. (When the war ended that class was split, some graduating in the originally intended three years and others, including Starry, reverting to the traditional four-year curriculum and graduating as the Class of 1948.)

Starry was motivated to opt for the four-year curriculum in part because that would enable him to take flight training and qualify for commissioning in the Army Air Corps. In anticipation of that development, Starry put only Air Corps on his branch preference sheet and, when that did not come through, found himself arbitrarily assigned to, and commissioned in, the Transportation Corps (which was, to him, a revolting development). Fortunately that also involved a two-year detail to Cavalry (Armor) which, after some further adventures (see below), became his permanent branch.

During cadet days Starry dated his long-time sweetheart, Leatrice Hope Gibbs, known as Letty, of Kansas City. In 1947, as Starry began his final year as a cadet, she moved to Highland Falls, just outside the gates of the Military Academy, and took a job with the Army Athletic Association and then later as a secretary at Holy Trinity Catholic Church at West Point. They were married in Kansas City a week after graduation and so began a life together that spanned 60 years in total, 35 of them at a succession of way stations all across the Army. Four children made their entrances during assignments at various posts—Mike, Paul, Melissa, and Melanie—Fort Knox, Heidelberg, Fort Leavenworth, and Fort Holabird.

First Tour in Europe: 63d Tank Battalion

Following the usual schooling at Fort Riley (Ground General School) and Fort Knox (Armor Officer Basic Course), in the summer of 1949 the Starrys arrived in Mannheim, Germany, for assignment to the 63d Heavy Tank Battalion, 1st Infantry Division, at that time the only tank battalion in US forces in postwar Europe. It proved to be a fortuitous assignment. The battalion was commanded by Lieutenant Colonel Creighton Abrams, one of the best-known young armor leaders of the day after his brilliant command of the 37th Tank Battalion, 4th Armored Division, throughout World War II. There Lieutenant Starry worked for several company commanders, one of them First Lieutenant George S. Patton.

As with every successful officer, Starry received a lot of help and encouragement along the way. Unlike some, he remembered and was grateful for it. In later years he often found occasion to recall Sergeant First Class Willard Lucas, his first platoon sergeant when Starry reported to the 3d Platoon, C Company, of the 63d Tank. Lucas had served in a tank battalion of the 5th Armored

Division throughout the war as a tank commander, platoon sergeant, and acting platoon leader. Lucas was old school, asking Starry such things as, "Would the lieutenant care to inspect the platoon now?" The lieutenant would, and did, looking for things on which Sergeant Lucas had previously coached him. Over time, as they looked at first one thing and then another, Lucas turned Starry into a very knowledgeable young officer, at the same time demonstrating to him the value of listening to and learning from his soldiers. Those were lessons Starry never forgot and continued to put to good use all the way up the ladder of command, and he almost always took time to relate with gratitude the story of how Sergeant Lucas had taken him in hand.

After he had been in the 63d Tank for about a year, Starry was nearing the end of his detail to Cavalry and faced the prospect of reverting, permanently, to the Transportation Corps. A series of increasingly frantic letters and applications seeking to escape that fate was having no discernible effect, and only a wonderfully serendipitous friendship saved the day. Beatrice Patton, widow of the late General George S. Patton, frequently came to visit her son George. That led to her meeting Donn and Letty Starry and to their becoming friends. When Mrs. Patton learned of the branch assignment difficulty, she worked some magic and, after Senator Henry Cabot Lodge Jr. expressed an interest, Starry was allowed to stay with Cavalry. Wrote the colonel in Washington who was then Chief of the Cavalry Section, employing considerable tact and understatement, "it was possible for certain reasons to obtain an exception and he [Starry] will be transferred to Cavalry by Department of the Army orders in the near future." It was a narrow escape. Lieutenant Starry had at the time been on orders for assignment to a truck battalion.

The 63d Tank developed, under the tutelage of Lieutenant Colonel Abrams, quite a lot of talent for subsequent leadership of the Army at much higher levels. Abrams and Starry both attained four-star rank, of course, while George Patton and Ennis Whitehead each attained two stars and Don Packard became a brigadier. Evaluating Starry's performance, Colonel Abrams put a firm dark "x" in the block that read "the most outstanding officer I know."

Approval by Abrams continued, even after Abrams had departed this life, to be a touchstone for Starry (as it was for others who had worked closely with Abrams). Many years later, complimenting Bill Livsey (once executive officer to Chief of Staff Abrams and later a four-star in his own right) for his service in command of Fort Benning and the Infantry Center and School, Starry wrote, "I guess the best thing I can say to you is that I know Abe would have been damned pleased and proud. You know that nothing I can think of would be any higher praise."

Starry proved to be a very consistent man, certain and settled in his values and commitment to his profession and the soldiers entrusted to his care. In this he reflected decisively important early influences, beginning with his father and then the experience of West Point and that early service under the incomparable Creighton Abrams.

Fort Knox and Korea

After three years in Germany Starry was in the summer of 1952 reassigned to Fort Knox and the 3d Armored Division, then training tankers for assignment worldwide. As aide-de-camp to Brigadier General John Tupper Cole, the Assistant Division Commander, Starry accompanied him to the Nevada Test Site for one of the earliest troop tests of nuclear weapons. An airburst collapsed the trench in which Cole had been positioned to watch the shot, so Starry dug his general out, then (asking and receiving permission to do so) got into and drove an M4A3E8 tank, which had been positioned on the test site, to demonstrate that it was undamaged by the detonation. This feat was noted with considerable approval by other tankers present, since at that time nuclear weapons were viewed in some quarters as constituting the end of the line for armored vehicles.

After a year in the 3d Armored Starry, newly promoted to captain, went across post to attend the Armor Officers Advanced Course at Knox, then in August 1954 was sent to Korea and an assignment in the G2 section of Eighth Army Forward, where—Starry recalled—he replaced five lieutenant colonels. His job was coordinator of covert and clandestine collection activities in North Korea, China, and Manchuria. Next was a three-and-a-half-year tour, extending to July 1959, as a combat arms and nuclear weapons instructor at the US Army Intelligence School at Fort Holabird, Maryland, including below-the-zone promotion to major in June 1958. Then it was back to school for Starry himself, the US Army Command and General Staff College at Fort Leavenworth, Kansas, graduating in June 1960. During that year he and others organized a mission church (Mission of the Centurion), which eventually became an Episcopal congregation at Fort Leavenworth.

Second Tour in Europe: Command of 1st Battalion, 32d Armor

Returning to Germany in August 1960, Starry served as a combat command (later brigade) S3 in the 3d Armored Division, then executive officer and (following promotion to lieutenant colonel) commander of the 1st Medium Tank Battalion, 32d Armor, in that division. This was, of course, a critical time in the Cold War, with the Berlin Wall having gone up in August 1961 and extensive fortifications (barbed wire fences, minefields, watch towers, plowed strips) built by the Communists along the inter-German border over the next few years.

Here Starry, in addition to innovative maintenance arrangements and top-scoring gunnery seasons, developed and implemented what came to be known as the Battalion Training Management System. It was in its essence an attempt to overcome the highly negative impact on training of a nearly 30-percent turbulence rate in units (in turn the result of the individual replacement system and specified three-year tours in Germany). Those skills necessary to achieve and sustain combat capability were grouped. Newly assigned soldiers were tested on those skills and then, based on cumulative test data, the battalion training program was designed around an evaluation of how frequently those skills needed to be practiced in order to maintain individual and unit proficiency. Later a version of this system was adopted for use Armywide.

Battalion Commander Starry was serious about combat readiness, also adopting what was called the "Who Fights With Me Tonight?" system to ensure that there would be a soldier to fill every fighting seat in combat and combat support vehicles (to include fuel, ammo, and mess). Thus every day at the time of evening retreat the unit would conduct a stand-to, constructing a battle roster of crews to serve should the battalion be alerted during the night. Any roster shortages in regularly assigned crews were filled as necessary by clerks, mechanics, and others normally assigned noncombat duties. Not surprisingly, his division commander rated Starry as the best tank battalion commander in the 3d Armored Division.

In August 1964 the Starrys returned from Germany, headed for Norfolk, where Lieutenant Colonel Starry would attend the Armed Forces Staff College (a five-month course) and stay on as an Army faculty advisor (for another six months), gaining valuable joint knowledge and experience during the latter period as co-author of an updated version of AFSC Publication 1, *The Joint Staff Officer's Guide*. Then it was on to the US Army War College at Carlisle Barracks, Pennsylvania, as a member of the Class of 1966. Studying and writing about counterinsurgency, Starry concluded that in counterinsurgency operations it was quite likely that military forces could be expected to cope with only a part of the totality constituting war and that successful counterinsurgency operations would require a wider spectrum of skills and training—economic, political, and social as well as military—and would require new doctrine (tactics, techniques, and procedures at both the tactical and operational levels of war) and new and different weapons, organization, and other relevant skills. During his student year Starry also earned a Master of Science degree in political science from the George Washington University.

First Vietnam Tour: USARV Staff and MACOV Study

In August 1966 Starry departed for the first of two assignments to Vietnam, a tour of duty during which he was assigned as Chief of the Plans Branch, Plans and Operations Division, on the G3 Staff of US Army, Vietnam (USARV). As things worked out, however, Starry spent much of this assignment traveling across the entirety of South Vietnam as an operations analyst for the Mechanized and Armor Combat Operations (MACOV) Study Group headed by Major General Arthur L. West Jr. This group was charged with evaluating the operating environment and making appropriate recommendations for improvements in organization and equipment of mechanized and armor units operating in a mixed counterinsurgency and conventional combat environment.

Starry also, he proudly recalls, led a successful USARV staff effort to prevent issue of the M551 Sheridan airborne armored reconnaissance assault vehicle (AARAV) to armored cavalry units in Vietnam. The Sheridan's primary weapon system was the Shillelagh, a beam-rider antitank guided missile launched through a 152mm gun barrel. There were at that time in Vietnam, noted Starry, no targets requiring an antitank guided missile. But there were many targets requiring conventional ammunition such as high-explosive and antipersonnel rounds—types that had not yet been developed for the Sheridan. Under those circumstances, thought Starry, "Sheridan was no more than a very expensive, and quite vulnerable, machinegun platform." Nor had the system been adequately tested for employment in a Vietnam-like environment. The Starry-led articulation of these realities staved off deployment of Sheridan to Vietnam for two years (see below).

Following this very busy year in Vietnam Starry drew Pentagon duty, assigned initially as an operations research analyst in the office of the Army's Assistant Vice Chief of Staff. At the end of 1967 he was promoted below the zone to full colonel and, the following March, became military assistant to Dr. Solis Horowitz, a senior civilian official in the Office of the Secretary of Defense.

Second Vietnam Tour:
Redeployment Planner and Commanding Officer, 11th Armored Cavalry

Back in Vietnam early in 1969, Starry was assigned to MACV headquarters and worked directly with and for General Abrams in writing plans to Vietnamize the war and redeploy US forces. In this close-hold and highly sensitive job, he was responsible for determining how to structure and sequence a series of redeployment (withdrawal) increments of US forces as ordered by the Nixon administration. Included in this work was preparation of detailed plans for the first three redeployment increments (beginning in July–August 1969 with withdrawal of the first 25,000 troops). Starry remembers clearly the instructions from General Abrams: "We have been directed to do this. We must do it. But we should do our best to do it in such a way as to leave South Vietnamese forces the maximum opportunity to successfully defend themselves against attack from the North." Some of the back-channel messages containing proposed troop lists, said Starry, were 80 to 90 pages long.

The redeployment effort well underway, Starry was given command of the celebrated 11th Armored Cavalry Regiment by General Abrams. Starry headed the regiment from December 1969 through July 1970, a tenure that enabled him to lead the "Blackhorse" during the 1970 Cambodian incursion. Starry proved himself an authentic warrior-hero, and these credentials lent important weight to his later credibility on a wide range of topics, from combat vehicle characteristics and development, weaponry, and ammunition to leadership, training, maintenance, and field logistics. At one point, rushing to shove another officer away from an enemy grenade, thereby undoubtedly saving that man's life, Starry himself suffered extensive wounds. Soon, though, he had talked his way out of a field hospital (where he learned he'd been nominated for promotion to brigadier general) and resumed his command. Along with the Purple Heart and the Soldier's Medal, his decorations from this combat tour included the Silver Star, the Distinguished Flying Cross, the Bronze Star Medal for Valor, and nine Air Medals.

Sheridan, the flawed armored reconnaissance vehicle whose deployment to Vietnam Starry had successfully opposed during his first tour, now arrived to plague him on his second. In the intervening two years high-explosive and antipersonnel rounds had been developed for the 152mm gun, but both rounds featured caseless ammunition and were found to be not crew safe. When it came to enemy action even a mine blast, if it vented the hull, inevitably resulted in a catastrophic explosion in the crew compartment, as did an RPG (rocket-propelled grenade) round venting the armor anywhere. Observed Starry, "The whole system was in no way crew safe."

Starry brought away from this command assignment even more intense concern for the soldier and his well-being. This outlook manifested itself in passionate efforts to improve battle dress (protective field clothing) for armor crewmen and other lifesaving innovations in equipment, along with his customary stress on effective training and leadership as the primary determinants of success in battle and saving the lives of soldiers.

Nearly three years of Pentagon duty after the second Vietnam tour gave Starry valuable experience in resource management and force structuring, challenging tasks as during 1970–1973 support for the war in Vietnam was ebbing and the resource stream was drying up, even as redeploying units had to be accommodated or, very often, disbanded and their assets redisposed.

Command of Fort Knox and the Armor Center and School

At the beginning of May 1973 Starry received a second star. Soon Army Chief of Staff General Abrams sent him off to command Fort Knox and the Armor Center and School, in the process providing him with some succinct guidance: "Don't screw up the tank program. Just start with the doctrine, describe the equipment requirements, reshape organization. And get the Army off its ass!"

Knox proved to be an ideal assignment for Starry. There he devoted himself to developing armor tactics and doctrine, applied television training simulators and training devices to the resolution of training problems, and implemented master gunnery courses.

As commanding general he became deeply involved in the formulation of doctrine, including personally drafting parts of Field Manual 100-5, the capstone manual on the operations of the Army in the field, at the famous conference of senior officers convened by General William E. DePuy at Camp A. P. Hill in the autumn of 1974. Starry began doctrine development with a series of concept papers. The first, titled "Modern Armor Battle," he wrote himself. It focused on what had been learned in recent war games about current Soviet tactical schemes and from monitoring Soviet maneuver activity, analyzing the new Soviet doctrine of mass, momentum, and continuous land combat, all this preparatory to describing a concept of how to meet and defeat Soviet forces so deployed.

And at Knox Starry was able to give full force to his concern for soldiers' families. With important input from his wife Letty—whose concern for soldiers and their families her husband describes as tenacious—he established what came to be known as a Community Life Program. Occupants of various family housing areas were invited to elect "mayors" for their areas, then the mayors would appoint small staffs and develop statements of the most needed improvements in their areas and present those to General Starry as the installation commander. Available funds would then be allocated to deal with the most pressing needs. In the course of about two and a half years the unfunded real property maintenance backlog, which had been estimated at nearly $40 million, was reduced to near zero.

The new mayors took their responsibilities seriously. Contractors working in family housing areas soon found themselves being bird-dogged and evaluated by a phalanx of well-informed wives, resulting in standards of performance unlike any they had previously been held to. Starry observed that these residents were only asking for what the contract specified, and that all the contractor had to do to get the wives off his back was live up to his commitments. Starry gives a great deal of the credit for these effective innovations to his wife Letty, and deservedly so. "Honest and sincere," the Fort Knox command sergeant major called the "team" of Donn and Letty Starry, traits that, he added, "are found in so very few of today."

In January 1974 General Abrams sent Starry and Brigadier General Bob Baer to Israel to study the recently completed Yom Kippur War. The results of this mission had far-reaching effects, both for the American Army and for Starry personally. General Abrams had asked that the two officers return and tell him what he personally as Chief of Staff needed to know about the war, and that they also determine what analysis of that war implied for the new tank under development in the United States (a tank that eventually became the "Abrams"). Starry and Baer, close personal and professional friends for many years, were in full accord as to what they saw in Israel, where they had visited the battlefields and talked at length with senior Israeli commanders, especially General Musa Peled, Commander of the Israeli Defense Forces (IDF) Armored Corps, and General Israel Tal, who was in the process of developing for Israel what became the Merkava tank.

Starry and Baer walked the battlefields with commanders who had fought there, both on the Golan Heights and in the Sinai, and were briefed on the Israeli analysis of the war, battle by battle. They examined the hulks of destroyed armored vehicles, looking at armor protection and armor penetration, devoting particular attention to what had been wrought by Soviet T-62 tanks.

Starry was gratified to find that what had transpired in the Yom Kippur War fairly closely resembled what he had described in "Modern Armor Battle." Thus "powerful new antitank weapons, swift-moving formations slashing across the battlefield, and interaction between ground formations and the air arm showed how much the world around our Army had changed as we focused on Vietnam," later said General Scott Wallace in describing the Starry-Baer findings. The greatly increased lethality of the battlefield, the high density of armored vehicles arrayed against one another, and a much accelerated tempo of operations called for new thinking, new materiel, and above all new doctrine.

The visit to Israel was, for Starry, the first of dozens, as over succeeding years he developed close relationships with many of the senior Israeli officers and took from their experiences lessons he sought to apply in his own successively more senior assignments.

When Starry left Fort Knox after a command tour lasting some two and a half years, General DePuy was quite direct in evaluating his performance. "General Starry was the best school commandant in TRADOC," he wrote. "He has been the strongest commandant on the tactical side and on the technical side and throughout the scope of combat development activities. . . . He dominated the Armor School with his strong, brilliant mind and a very practical yet technical bent."

Command of V Corps

Command of V Corps in Germany, beginning in February 1976, was for Starry perhaps his happiest assignment. He was back on familiar ground (so familiar, he told a correspondent, that from the air he could identify the various small towns by the shapes and locations of their church steeples) where, as a young officer, he had first begun learning his profession, but now he was in command of an outfit where he could put into practice the views on doctrine and maneuver developed while leading the Armor School at Knox. "War hasn't changed all that much," Starry wrote at this time to a veteran of the World War I tank corps. "The machines have, but mobility is really a state of mind." Now he set about developing that state of mind in subordinate leaders throughout the corps.

His principal technique was the terrain walk, but with an examinatory aspect. The maneuver and combat support battalions in V Corps each had a designated role and sector to be occupied in time of war. Starry soon determined that virtually none of his subordinate commanders had ever gone out to see where they might have to fight and to calculate the best tactical- and operational-level schemes for defending in their assigned sector. Starry set about systematically meeting each of these commanders on their designated piece of ground, there to question them about their planned dispositions of troops and weaponry and how they saw the battle unfolding. This proved to be highly instructive for both parties, Starry imparting key elements of the new doctrine being developed in the Training and Doctrine Command while learning the capabilities (or lack thereof) of these key troop leaders.

Such a vehicle for training and testing was essential, thought Starry, given the realities of the day. The Army in Europe was then, he estimated, about where it had been in 1960 "when we started building after the doldrums of the late 1950s." But, he wrote to a fellow soldier, "tactically we're pretty weak—ammo downloaded and not much emphasis on readiness." Now he set about trying to deal with those deficiencies.

Those were tactical, while at the operational level there were matters of doctrine also requiring urgent attention. Starry's conceptual approach there was to think in terms of what he called the

Corps Battle, then work on elements of that in a systematic way. He brought in some experts to model aspects of the anticipated battle based on, as he put it, "the target servicing capabilities of the forces involved," beginning with consideration of the defensive battle. They had a current threat estimate that scoped the magnitude of the task.

Starry also, with the approval of USAREUR Commander General George Blanchard, arranged to obtain for issue down to battalion level throughout US Army, Europe, some 700 copies of the draft version of a revised Field Manual 100-5, *Operations*, then under development by TRADOC. The new publication embodied the concepts of "Active Defense," and Starry set about testing the new doctrine throughout V Corps, a task he continued throughout his 16 months in command.

Later Starry recalled that the Active Defense doctrine had not been well received in commands other than V Corps. He thought that was largely because there had been no learning experience like the V Corps terrain walks to help those responsible convince themselves that the new doctrine was reasonable and could in fact be successfully employed. Meanwhile the concept paper "Modern Armor Battle" evolved into "The Corps Battle," one in what became a succession of doctrinal expositions conceived and articulated by Starry ("The Central Duel," "The Central Battle," and others followed in the evolutionary chain).

Training and Doctrine Command

When, in the spring of 1977, Starry was named to be the next TRADOC Commander, USAREUR Commander General George Blanchard sent him a very fine note. "I'd like to thank and commend you for the magnificent job you've done commanding V Corps," he said. "In a little over a year you have completely revitalized it and now have it 'leaning forward in the saddle.'"

As TRADOC's second commander Starry followed the famous General William E. DePuy in that post, each of them holding it for a full four years. Starry often commented on how fortuitous that had been, for he viewed himself as in substantial agreement with DePuy on a wide range of matters, thus ensuring that there would be consistency of outlook and direction during the new command's crucial formative years. General DePuy, Starry told historian Brooks Kleber, "had the greatest influence of any Army man of his era." What Starry admired most about DePuy was that he was always "looking over the hill," always trying to discern where the Army should be going.

Dr. Richard M. Swain, editor of a comparable collection of General DePuy's papers published a number of years ago, wrote therein that "with the able assistance of the commandant of the Armor Center, General Donn Starry, General DePuy wrenched the Army from self-pity and recrimination about its defeat in Vietnam into a bruising doctrinal debate that focused the Army's intellectual energies on mechanized warfare against a first-class opponent."

By the time he himself succeeded to the TRADOC post, then, Starry had already been working closely with DePuy for an extended period, beginning with his command of Fort Knox. When Starry went to Europe to command V Corps, he maintained an extensive correspondence with DePuy and, as noted, secured early copies of key doctrinal materials emanating from TRADOC for distribution and implementation in the corps.

While he was proud of the continuity of outlook and effort represented by his service as successor to General DePuy, Starry proved to have qualities and inclinations that in some respects made him more successful in both the formulation and implementation of policy and doctrine. A senior officer who worked closely with both DePuy and Starry later recalled how DePuy had become impatient with working new doctrine through "the system," and how he had decided "it must be a swift coup, executed personally by the CG TRADOC and CSA," meaning himself and the Army Chief of Staff. "Needless to say," observed his former associate, "DePuy was wrong on that count."

When the 1976 version of FM 100-5 was published under DePuy's stewardship, recalled General Scott Wallace, "a firestorm of controversy surrounded that manual" and led to widespread debate. All this served to underscore the importance of doctrine, which was a good thing, but it also demonstrated the difficulty of gaining agreement regarding what that doctrine should be. Starry's approach was, in contrast and perhaps most important, that of an effective consensus builder.

The "Starry" version of FM 100-5, *Operations*, was ready for print when General Starry departed to head the US Readiness Command and was subsequently formally published in September 1982. That manual, observed Lieutenant Colonel (later Lieutenant General) Leonard D. Holder Jr., laid out the principles of AirLand Battle doctrine. In it "the use of the indirect approach, maintenance of the initiative through speed and violence, flexibility and reliance on the initiative of junior leaders, attack of the enemy in depth, and synchronization of all assets, while not revolutionary, are stressed as key to the AirLand Battle." Lieutenant Colonel Holder assessed the tone of the new manual as positive, "win-oriented," and featuring "close integration of operations, intelligence, and logistics planning." The new manual also looked ahead, addressing not only what the Army was then capable of doing but also what it would be able to do when the family of Army 86 systems came on line.

"What came to be called AirLand Battle," Starry later recalled, "was driven both by an understanding of expanded battlespace at tactical level and . . . by longstanding concerns of Generals Abrams, Starry, and many others concerning how best to cope with the follow-on forces problem with Soviet echelons facing the NATO front, all made more acute by more recent developments in Soviet operational-level doctrine—mass, momentum, and continuous land combat with conventional weapons at the outset, but with tactical/theater-level nuclear weapons should the conventional initiative fail." Dr. Roger Spiller has described AirLand Battle in elegant terms: "It is an artistic doctrine; a modern doctrine which capitalizes upon professional and technical knowledge above all; a doctrine which is informed by its professional past even as it anticipates the future."

AirLand Battle doctrine as laid out in the new manual viewed battles as nonlinear, added Holder, enlarging "the geographical area of conflict and stressing unified air and ground operations throughout a theater." At the same time it established "an operational art—the conduct of campaigns and large unit operations—as a discrete level of war separate from strategy and tactics." That level recognized "the nonquantifiable elements of combat power, especially maneuver, which is considered equal and complementary to firepower." The importance of electronic warfare, and of nuclear and chemical weapons, was acknowledged and their effects on operations were articulated. "Most important," concluded Lieutenant Colonel Holder, "the new manual keeps the human element prominently in the foreground."

All these had to be viewed as significant improvements in this capstone doctrinal document. Assessing the earlier (1976) version of 100-5, Colonel (later Brigadier General) Huba Wass de

Czege, while acknowledging many strengths, also noted that it "underrated the key elements of depth, maneuver, and initiative; ignored the operational level of war; and paid insufficient attention to the human element in battle." These key problems had now been addressed. Thus, concluded Colonel Wass de Czege, the new manual presented "a comprehensive view of modern warfare which accommodates our newest technical capabilities and worldwide commitments." It represented "an innovative general approach to fighting at both the tactical and operational levels which provide[s] for the coordinated employment of all arms, all services, and all means of support."

General Starry later told Army historians that in his view the three most important differences between the 1976 and 1982 versions of FM 100-5 were that the later version included a set of operational concepts that put us back on the nuclear and chemical battlefield, thus ensuring that an enemy's surprise or first use of such weaponry would not enable him to win the war thereby; that it recognized and addressed the importance of attacking the enemy's second and following echelons; and that it dealt with the balance between firepower and maneuver.

Another important attribute of the new FM 100-5 was that it had been written at Fort Leavenworth (not, as had the 1976 version, at Fort Monroe or in off-site drafting sessions that more or less rejected or excluded Leavenworth input). This gave the new manual, right from the start, a degree of acceptability that the previous manual never really achieved and made it far easier to develop a consensus in support of the new doctrine. Likewise TRADOC and Starry personally had worked closely with the Germans in developing the new manual, an essential element in ensuring the viability of allied implementation of the new doctrine in the defense of Europe.

If, as Starry believed and insisted, doctrine rules, its effective implementation depends on good people, well led, in well-trained units equipped with good equipment. Starry became concerned with the delta—the gap—between what the best new equipment was capable of doing and what soldiers—such as tank crews—were capable of getting it to do. Turbulence of assignment and the resulting degradation of training were basic causes of the problem, and these became issues to which Starry devoted intense effort in his most influential years, particularly the potential advantages of unit replacement as compared to the Army's longstanding reliance on an individual replacement system.

During the months in Vietnam when he served as close-hold redeployment planner for General Abrams, Starry had witnessed a dramatic exchange between Abrams and the Army Chief of Staff, then General William C. Westmoreland. The issue was whether the elements redeploying from Vietnam should consist of intact units or, alternatively, aggregations of individuals who had accrued the longest service in Vietnam. General Abrams pressed for redeployment by unit. Westmoreland wanted to bring out, in the interests of "fairness" as he viewed it, those who had been in Vietnam the longest. After a long night of messages sent back and forth (Saigon time was offset from Washington by 13 hours) Westmoreland insisted on his solution and the issue was closed.

Abrams turned to Starry: "I probably won't live to see the end of this, but the rest of your career will be dedicated to straightening out the mess this is going to create." Said Starry later, "How right he was." Pulling all the longest-serving troops out of units all across Vietnam so as to aggregate them under the flag of some outfit being redeployed, then replacing them in all those other units with different people who had been in Vietnam for shorter times, guaranteed constant turbulence, destroyed unit cohesion, and contributed greatly to later widespread problems of

indiscipline and destruction of morale. After this dramatic episode Starry developed a permanent and intense interest in developing a unit rather than individual replacement system, something he fought for during the entire remainder of his active service.

The major reorganization of the Army's command structure resulting in the creation of the Training and Doctrine Command also, besides establishing Forces Command, disestablished what had been known as the Combat Developments Command, with responsibility for combat developments taken up by TRADOC. Starry was one of those critical of the former CDC approach in that it relied on such long-range forecasts to shape combat developments that the postulated future combat environment was fuzzy and indistinct, and in consequence of little practical value. Starry went to the intelligence community for an agreed threat forecast, one he could rely on to shape combat developments. What he was told was that such a forecast could be projected out about eight years, a judgment rendered in about 1978, thus yielding in turn the time horizon for Division 86 (1986) and the whole family of related studies conducted under Starry's supervision.

By late summer of 1979, two years into his assignment at TRADOC, Starry was ready to initiate a study known as Corps 86, a companion to the Division 86 study. Describing the problem the study was to address, the implementing directive said this:

> The Army is planning to meet the numerically superior and increasingly sophisticated threat of the next decade through improved tactical concepts and the introduction of advanced materiel systems. An almost complete replacement of the Army's major systems, as well as the introduction of system capabilities not now present, is envisioned by 1986. Speed and continuity of combat will put individuals under stress not previously experienced in warfare. Complexity of new systems will require new training strategies. The magnitude of the changes caused by incorporating these systems into the Army requires that an evaluation and modification of the battlefield organization be conducted to provide a basis for orderly transition from today's units to those of the mid-1980s.

The new organization, it was stressed, would have to be developed within the limits of "real world resource constraints," and since the most dangerous threat was in Europe, the first such organization developed must be the heavy force that would constitute the main US contribution to the NATO Alliance.

If Starry had one immutable principle (and he did), it was that doctrine rules. All else—combat development, organizational structure, training, every key element of the military profession—must derive from and be driven by doctrine. Not everyone liked this, especially those in the various laboratories and in many quarters of the defense industry, where scientists and engineers preferred to pursue their own interests rather than being dictated to on the basis of some doctrinal concept. But Starry was relentless.

At Fort Monroe he continued the technique, utilized earlier at Fort Knox and in V Corps, of drafting concept papers on various key aspects of doctrine and related matters. At the beginning of 1980, for example, he sent two such documents—one addressing Tactical Command-Control and another on Tactical Intelligence—to the Chief of Staff. They were, explained Starry, designed to describe how we must operate on the battlefield of the future and to provide the basis for doctrinal, materiel, organizational, and training developments. Developing a common understanding of key doctrinal concepts throughout the Army proved to be hard work and a never-ending task. As late as the summer of 1980, for example, while visiting US Army, Europe, training facilities in Germany, Starry told his hosts, per a memorandum for record describing the visit, that he "does not believe that all people understand the active defense, [and] is afraid that many believe the active defense is conducted by all units moving continuously. He emphasized that somebody must hold something."

Earlier in the year Starry had commented unfavorably on a set of films the Pentagon proposed to distribute as a means of portraying the Army's family of modernized equipment. They were not doctrinally sound, Starry observed, and would do more harm than good:

> Overall, they fail to explain the operational concepts of how and why we fight as we do. They portray a linear-oriented, attrition-based fighting scheme that is heavily focused on the ability to kill tanks by every weapon on the battlefield. The important concepts of a battle in depth, based on maneuver and force disruption, are never mentioned. If we fight as these films depict, the Lanchester Law will surely apply and we will just as surely lose.

Clearly there was yet more missionary work to be done on behalf of the new doctrine.

Deep battle became a central concern, perhaps *the* central concern, of the new doctrine. And then, because in order to fight deep it was necessary to see deep and to shoot deep, complementary materiel development requirements were derived. And, since the deep battle tasks were so numerous and so challenging that no one system, and indeed no one service, could handle them by itself, greatly increased Army-Air Force cooperation, based on a shared understanding of battlefield realities, became essential. Starry and an Air Force counterpart, General William Creech at the Tactical Air Command, worked hard to resolve the issues, a task made more difficult by the Air Staff's tendency to view all such cooperative planning as threatening the roles and missions balance. But together Starry and Creech laid out requirements for deep surveillance and target acquisition capabilities for the deployed force commander. This work led to development of JSTARS, an airborne system with attendant ground station that made its debut (with spectacular success) during the first Gulf War in 1991.

During these years, too, Starry's close relations and repeated interactions with Israeli leaders, beginning with his 1973 mission to Israel, were replicated in his most senior assignments by close personal and professional relations with the Germans, British, Canadians, and a range of others.

Another of Starry's distinguishing characteristics was that he was, and is, a voracious reader. His close friends can expect to hear from him frequently about books he has discovered and wants them to know about and then discuss with him.

Starry values the role of storytellers, recognizing that it is their version of what happened that is recorded as history. He is himself a gifted storyteller, able to inform and inspire even when he is the bearer of bad news. A famous example took place at an Armor Conference at Fort Knox where Starry was the featured speaker. Starry delivered an impassioned charge to cast aside complacency and get on with reform. He administered what amounted to a thoroughgoing rebuke with such charm and persuasiveness that his audience, having just been informed of their individual and collective shortcomings, got up and cheered. That speech was known forever after as "Kick in the Grill Doors."

Starry also has a passion for military history, not only for its own sake but because of his conviction that knowledge of military history is essential to an officer's professional development and competence. During his tenure as TRADOC Commander Starry caused there to be established at Fort Leavenworth an entity known as the Combat Studies Institute (CSI). Its mission was to coordinate historical analysis within TRADOC and to study military history and produce publications that would make its findings readily available to officers throughout the Army. Colonel Trevor Dupuy, a noted military historian and analyst in his own right, visited CSI early in its existence and wrote approvingly to Starry. "I consider this to be a truly historical development," he said, "since it is the first time—to the best of my knowledge—that the US armed forces have consciously and systematically moved to realization of the old saying (which I believe) that 'military history is the laboratory of the soldier.'" For his part, General Starry told Army historians as he handed over TRADOC, he considered the Combat Studies Institute "a roaring success." Historian Roger Spiller gave Starry much of the credit for that, recalling that "his support did not extend only to the promotion of military history; it reached far beyond, and aimed at nothing less than intellectual rejuvenation of the entire Army through many other avenues as well."

Starry seldom gave a speech in which he did not include historical references and examples. Likewise his correspondence is filled with such material (in which he took an obvious delight). While Starry was always careful to characterize himself as an amateur historian, his views were sophisticated and interesting enough that he was often invited to speak to gatherings of professional historians and was invariably warmly received by them. On his staff at TRADOC, and later likewise at the Readiness Command, Starry established positions for staff historians, then recruited highly qualified professionals as the first incumbents.

Religious convictions were a major influence on Starry throughout his life and career, although unlike some senior officers he did not force these views on his associates. He served for many years as a lay minister of the Episcopal Church, and military chaplains awarded him the Order of Aaron and Hur in recognition of long and faithful service. That service ranged from frequently, as a lay reader, standing in on short notice for the TRADOC chaplain to fashioning with his own hands a prie dieu, or kneeler, for the historic Fort Monroe Chapel.

No assessment of Starry and his influence could fail to mention his personal style. "Starry likes to operate where the problems are," recalled his former chief of staff at TRADOC. "He conducts most of his business elsewhere, and operates outside the staff." Not that the staff did not have plenty to do, and high standards to be met in evaluating what they accomplished, but General Starry also maintained a vast external network of professional associations and communications, ranging from scientists and defense industry leaders through sergeants with whom he had served, current and former Army and Defense Department civilian officials, and public servants in the Congress and elsewhere. Another senior associate recalled Starry's mode of command as "always on the road and dealing directly." Added his former chief, "Starry liked trends, thrusts, and bar graphs." He also liked people, in or out of TRADOC, who could help make those graphic elements move legitimately in the right direction—and do so *right now*.

Near the end of his TRADOC tenure General Starry staged the first commanders conference conducted as a teleconference. Broadcast to 26 viewing locations across the country, it highlighted new technologies for improving training. These included things that are now commonplace but that were, at that time, cutting edge and highly innovative, such as interactive videodisc, video arcade gaming, voice technology, robotics, fiber optics, microcomputers, two-dimensional simulation, and satellite communications. In the course of the presentation various technologies were combined to demonstrate how soldiers could be trained in near-real conditions for a fraction of the cost of then-current training systems. Starry highlighted the currency of the materials demonstrated by beginning the conference, which originated at Fort Eustis, Virginia, from a commercial video game arcade.

Looking back at what had been accomplished during the Starry years at TRADOC, Lieutenant General John W. Woodmansee Jr., who had been an important part of the Starry team, noted a unique achievement: "TRADOC harnessed creativity, focused it on the out years, and then built a bridge into the program and the budget."

When his four years were up at TRADOC, Starry thanked all who had been part of what had been accomplished and, in saying goodbye, returned to a familiar theme. "I am convinced," he told his friends and associates, "that on that complex, most dangerous, and difficult battlefield of the next war the determined, prompt, correct, and calculated battle actions of trained commanders, soldiers, and units is all that will win."

Readiness Command

During 1981–1983, in his final active duty assignment, General Starry filled a joint billet as Commander of the US Readiness Command with headquarters at MacDill Air Force Base in Florida. As he cabled a friend: "We're still pushing 'Starry on War,' but with a new hat." The Readiness Command was the successor headquarters to what had been known as Strike Command, whose primary responsibility had been deployment planning. Accomplishment of that mission was inhibited by lack of centralized control over transportation assets needed to achieve rapid deployment. When the Readiness Command was created it included a Joint Deployment Agency under the same commander. (Later that agency became the Transportation Command, and at the Readiness Command there was also created a Rapid Deployment Joint Task Force.) Now Starry sponsored development of an intracommand planning process to provide unified and specified commands a common planning system with the Readiness Command, using the best available technology to enable contingency planning involving forces to be deployed.

Also at the Readiness Command Starry again partnered with General Bill Creech at the Tactical Air Command, now adding to the mix his successor as TRADOC Commander, General Glenn Otis, to reach and publish a joint agreement on Joint Attack of the Second Echelon. Working around Joint Staff concerns that militated against anyone else becoming involved in formulating joint doctrine, Starry described these necessary efforts as exercises in developing joint tactics, techniques, and procedures. Another important product they developed was a concept for joint suppression of enemy air defenses.

General Donn A. Starry had begun by enlisting as a private soldier, then after graduation from West Point served for 35 years as a commissioned officer, commanding at platoon, company, battalion, and regimental level, the last in combat in Vietnam. Paralleling the successive commands of this early service, when he retired from active duty in June 1983 Starry had spent the last 10 years of his service in continuous command, having led in succession the Armor School at Fort Knox, V Corps in Germany, TRADOC at Fort Monroe, and finally the Readiness Command at MacDill Air Force Base. Late in this sequence a very influential member of the

House of Representatives wrote to tell Starry that it had been "refreshing to meet an intellectual, anti-bureaucratic, fighting general." That pretty much summed him up.

In "retirement" Starry continued working many of the same problems that had concerned him during his Army days, now operating from a series of highly placed executive positions in the defense industry, and as a long-term member of the Defense Science Board. He also served on the boards of the Eisenhower Foundation, the Army Historical Foundation, and the Washington Institute of Foreign Affairs and was for a decade Chairman of the Board of the US Cavalry Association. In perhaps his favorite role he served for many years as Honorary Colonel of the 11th Armored Cavalry Regiment and mentor to a succession of its regimental commanders.

The Selected Works

Publication of selected works of General Donn A. Starry has been undertaken on the premise that they would constitute a rich source of inspiration and intellectual stimulation of great potential value to current and future Army leaders. These materials in my judgment establish conclusively Starry's stature as an officer of powerful intellect, boundless energy, an impressive range of professional interests, absolute integrity and selflessness, and exceptional ability to motivate and communicate.

The materials chosen for inclusion in the selected works involve not only many of General Starry's major speeches and published works but also an extensive sampling of his letters and messages, for he was a prolific, candid, and often passionate correspondent. The collection has been presented topically and, within topics, chronologically, thus depicting both his very wide range of professional interests and his evolving views on some of the most complex, notably those relating to doctrine.

A technique often employed by Starry was to make a presentation to an audience, often working from a stack of slides or viewgraphs, with his aide or another staff officer sitting in the back taking notes of what questions were asked and what issues were raised for discussion. Then, literally on the fly en route to the next presentation, they would make changes and adjustments in the presentation and try that version on the new audience. Thus many of Starry's key speeches, especially those on AirLand Battle and its antecedents, were and remained works in progress. Starry's late wife Letty maintained that her husband really had only one talk, a signature presentation she called "Starry on War," and that she had heard it so many times she herself could have rendered a credible version without benefit of notes.

General Starry for the most part wrote his own speeches and lectures—assisted in his most senior posts by able research assistants—which rendered them far more vivid and authentic than the usual run of staff-produced texts. These materials thus have an unusual freshness, immediacy, and impact. The same was true for much of the correspondence, often written longhand in Starry's distinctive penmanship (an example of which has been included). His authorship is also extensively represented in some of the key doctrinal materials produced under his supervision. For example, Starry told historian Brooks Kleber that the publication of several manuals had been held up because "I rewrite the drafts to ensure conceptual correctness and uniformity of style." Starry also revealed in an oral history that he had personally drafted certain chapters of Field Manual 100-5, *Operations*, and in correspondence told a friend that he had personally taken on the editorial and leadership task for the field manual on leadership.

Two compact disks are included in this collection. One contains the full text of the book-length monograph *Mounted Combat in Vietnam*, written by Starry with staff assistance at Fort Knox and completed in 1978. This constitutes one of a series of some 20 such volumes various senior officers were tasked to produce during the latter years of the Vietnam War. Fortunately General Starry inherited the task from another officer who had not been able to get to it, thus delaying its preparation and rendering it the only one in the collection that covers the entire war, the others being truncated accounts completed before the war ended and thus of significantly less value.

The second compact disk shows Starry in action as a speaker and briefer and includes some of his best-known presentations, including "Sergeants' Business," "Tanks Forever," "Tanks Forever and Ever," and a classic version of "AirLand Battle." The text of versions of most of these talks is also included in print form in the selected works. General Starry was particularly proud of the first TRADOC commanders conference to be conducted by means of video teleconferencing, and his introduction to that conference (also included) was as previously noted by his design broadcast from a commercial video arcade to dramatize the currency of the new approach.

Dr. William G. Robertson, Director of the Combat Studies Institute at Fort Leavenworth, where this volume was produced, suggested *Press On!* as the collection's apt title. That admonition appears repeatedly—in General Starry's own hand in his correspondence, and in the message traffic he generated, perfectly representing the energy and élan with which he went about his own duties and encouraged others in theirs.

The context in which these materials, both written and verbal, were produced by General Starry was of course the post-Vietnam War era, a time when the Army needed to turn its attention to long-neglected matters of doctrine, training, and materiel development, as well as adaptation to the newly mandated all-volunteer composition of the force. Starry was deeply involved in all these aspects and many more. Whatever his duties of the moment, however, his enduring concerns showed through—the soldier, the unit, and how both were trained and led.

In commissioning the project to collect and publish selected works of General Donn A. Starry, the Army reflected a conviction that these materials would, for current and future soldiers, offer much that is both instructive and inspiring. Here's hoping that will indeed prove to be the case. As General Starry would say: "Press On!"

Lewis Sorley
Potomac, Maryland
31 May 2009

A Note on Reading the Text

General Starry's range of interests, as will be apparent to all who come to this collection, was extraordinarily broad. Many topics he addressed repeatedly over long periods of time, his views—and his articulation of those views—evolving over the years. That is particularly the case with respect to doctrinal matters, the tank-antitank calculus, and the preeminent importance of the individual soldier.

Materials chosen for inclusion in this collection accordingly reflect some repetition, especially in matters of long-term interest to General Starry, illustrating how his thinking developed. Starry himself often made intentional use of repetition, as for example in his well-known speech "Tanks Forever and Ever," in which he recapitulated virtually his entire argument from the earlier "Tanks Forever."

To provide some structure to this wide range of material, it was decided to present items in categories and then, within those categories, chronologically. As will be apparent, quite a number of the pieces could just as well have been inserted in other categories, so where to place them was simply a matter of editorial judgment. The real impact of this material is, in any event, in its cumulative depth and range, so it is hoped that many readers will make their way through the entirety of the collection. For those primarily concerned with particular aspects of General Starry's eclectic interests, in addition to categorizing the materials under topical headings, an extensive index has been provided.

Articles and speeches are, for the most part, rendered in their entirety. Where exceptions have been made, those are indicated in the usual manner by the use of ellipses. When it comes to messages and correspondence, however, only internal omissions have been shown and, in the majority of cases, what has been included is only a pertinent excerpt from a longer communication.

Often General Starry annotated papers to show that he had seen them by drawing a small star and inserting within it a capital letter Y (thus Starry). An example may be seen on the book's back cover and at certain other points in the text. A frequent sign-off, often followed by the hand-drawn star, was "Press On!" which has thus been adopted for this collection's main title.

Finally, there will be found, in some categories of materials, considerable discussion of the threat. In the context of the times, of course, that meant primarily the Soviet and Warsaw Pact threat, which also constituted the principal anticipated adversary in the much-discussed Central Battle. While that threat no longer exists, at least to the same degree, it is important to be aware of the Starry conception and depiction of it in order to understand much of what—in terms of doctrine, tactics, equipment, and training—was driven primarily by contemporary appreciation of that threat.

Acknowledgments

General Scott Wallace conceived of this project, appreciated its potential for educating and inspiring soldiers of today and tomorrow, set it in motion, and gave it his unstinting support. General Marty Dempsey generously continued with his own endorsement and saw the task through to completion. Deputy Under Secretary of the Army Tom Kelly played a key role in facilitating the work.

At TRADOC Dr. J. Britt McCarley, Chief Historian, was unfailingly helpful and supportive, as were members of his staff. Administrative staffers at Fort Monroe were all superb, including John Nerger, Karen O'Kelley, Clyde Johnson, Maria Ivey, Ron Haug, and Rick Parker. Special thanks to Shari Worthey, whose cheerful and timely help with travel was absolutely heroic. It was especially fortuitous that Mike Starry, General Starry's elder son, occupies a key role at TRADOC as Deputy Director of Concept Development and Experimentation in the Army Capabilities Integration Center. He frequently provided answers to questions relating to manuscript elements and in many other ways greatly facilitated the project. Francine Allen, Benjamin King, and Karen Lewis helped by translating a number of arcane acronyms.

Research for this volume was conducted in several repositories, beginning with extensive documentary and photographic files still retained by General Starry. At Fort Monroe I was aided by Dr. Steve McGeorge, Deputy TRADOC Historian. At Fort Leavenworth Kathy Buker in the Combined Arms Research Library assisted. And at the Military History Institute at Carlisle Barracks, where extensive Starry materials are held, my work was greatly facilitated by the Director, Dr. Conrad Crane, and by Dr. Richard Sommers, David Keough, Randy Rakers, Rich Baker, John Kurash, Rodney Foytik, Tom Hendrix, Gary Johnson, Molly Bompane, Steve Bye, Melissa Wiford, Pamela Cheney, and Beth Shaffer. Jessica Sheets was especially helpful. Many of these expert staff members have assisted me on earlier projects over the years, and I am most grateful to them all. Kathy Johnson at *Armor* Magazine was helpful in providing clear maps and images to accompany a number of articles reproduced from that fine publication. Others who took a friendly and helpful interest in this project include Dr. Roger Spiller and Dr. George Hofmann.

The Combat Studies Institute at Fort Leavenworth was charged with producing the finished product, a challenging and laborious task, but one it took on with consummate skill and commitment. Special thanks to Dr. William G. Robertson, Director, whose strong support and managerial skills were invaluable, and who in addition suggested the extremely apt title for the book, and to Marilyn Edwards, who did all the processing and formatting of a lengthy manuscript and then transformed that into book format, tasks she accomplished expertly and with great energy and good cheer. Also at CSI Michael G. Brooks, Karen Taylor, Kelsey Reed, and Amy Castillo provided great support in getting the articles, speeches, and correspondence digitized. Robin Kern digitized the graphics that had originally appeared in the articles. And at M/VISC-TV at Fort Leavenworth David Armstrong and Don Otting transferred the Starry speeches onto a disk. We are most grateful to them all.

The superb book cover and disk labels were created by Kimberly Becker of Becker Design, Ayer Massachusetts.

Finally profound thanks to General Starry himself. His personal involvement in making papers and photographs available, discussing key issues, helping translate acronyms from out of the past, and virtually every other aspect of the project contributed enormously to the successful and timely completion of the task and the quality of the product.

1. Airpower

Page

TACAIR Support in Europe....2
Army-Air Force AirLand Battle Issues....4
Offensive Air Support....6
Joint Army-Air Force Issues (Otis)....7
Joint Army-Air Force Issues (Meyer)....8
Army-Air Force Cooperation....9
Management of Air Assets....10

TACAIR Support in Europe

Message to Lieutenant General Edward C. Meyer
Deputy Chief of Staff for Operations and Plans
15 November 1977

1. This responds to your WDC 106332 regarding CSA/CSAF discussion on TACAIR support in Europe.

2. What are the issues? From the ground commander's viewpoint:

a. There is a perception that we have insufficient air resources to provide adequate reconnaissance and surveillance, especially that necessary to find second-echelon divisions of first-echelon armies, and to find second-echelon armies. Other platforms can contribute, but because of types of sensory equipment being employed, manned aircraft remain in many ways the best vehicle.

b. There is also a perception of insufficient counterair resources to enable the air forces to keep Soviet air off the backs of priority targets, and off the backs of ground forces fighting first-echelon armies as well.

c. There is a perception that there are insufficient air resources to attack interdiction targets, and there is some confusion as to whether those targets are bridges, road nets, airfields, the so-called "choke points," or second-echelon forces themselves. Whatever is decided, it's not just attack aircraft that are in question. It's EW and other suppression, CAP for counterair, and so on. These are expensive missions.

d. The most important perception held by the ground commander is that of the insufficiency of air resources to help service the first-echelon army. In some NATO areas it is probably true that the ground forces can cope with target servicing the first echelon; in most areas this is not the case, however. Defeating the first echelon will indeed require lots of close air support. The numbers developed in V Corps after considerable analysis indicated the corps under breakthrough attack might require as many as 400 close air support sorties for the first three or four days of battle, and about 300 a day for the next three or four days—just against the first echelon, in order to balance up the target servicing equation. Now if NATO's eight corps are attacked by four breakthrough attacks, which is the capability we give the Sovs, then there will be four corps commanders screaming for about as much air as the V Corps commander thinks he needs. The question is, can they depend on it? The problem is we have never been able to get the air forces to sit down and talk reasonably about the matter in order to describe whether or not there is any chance the corps commanders' requirements might be met.

e. Without exception the air commanders of our allies do not believe that air support can be provided in the fight against the first echelon. There are a number of reasons for this, which require too much discussion here. The end result is that the ground commander must assume that about all he can expect against the first echelon is support from aircraft that can't fly the other missions—A10s and Fiats. This amounts to just over 200 airframes in all NATO if memory serves me, and that just isn't enough.

3. What are the solutions? The real solution is the one the Israelis have embarked on. They have concluded that, for a host of reasons, they can't provide much if any close air support in the battle against the first echelon. Therefore they have decided to provide their ground forces with the equipment and the force structure necessary to make up the difference. Their target

servicing equations are not much different from ours. They concluded that they needed more firepower on the first echelon; they also concluded that their air forces couldn't provide it; therefore, from a total force standpoint, they decided to beef up the ground forces—two more divisions, several battalions of artillery, some weapons systems that they hope will redress the firepower imbalance. We too need to do such an Army-Air Force analysis. However, I believe that to be impossible in the real Washington world. Service rivalry for the bucks will intervene. In addition there is a considerable group on the Air Staff of the USAF who do not believe in close air support against the first echelon. Most of them are SAC pilots, but they are influential, and they all believe Dixon and DePuy, now Dixon and Starry, are a couple of nuts. It was largely for that reason that Dixon and DePuy steered around doctrinal issues from the outset. I too have pursued that course, but it is increasingly apparent that sooner or later we must address ourselves to the doctrine. Probably we should delay that as long as possible, for if we start I'm afraid the resulting debate will be very long, very acrimonious, and in the end probably counterproductive.

4. Utility of such a meeting? Depends on where Jones stands with regard to the arguments outlined above. So far he's been supportive of Dixon and DePuy, and of Dixon and me. But he has not been put to the test—someone else was carrying the mail. The Chief will have to judge this one. Is it worthwhile to try and smoke this out at that level, or to continue at Dixon-Starry level? . . .

5. Smile.

Army-Air Force AirLand Battle Issues

Letter to Lieutenant General E. C. Meyer
Deputy Chief of Staff for Operations and Plans
28 December 1978

A belated response to your letter of 30 November concerning the future of our work on the air/land battle.

Two areas you singled out for improvement, allocation procedures and mission priority assignment, TAC and TRADOC have not tackled at all yet. Not that they don't need improvement, it's just that every move toward revised allocation procedures or priority assignment based on the advent of a host of single mission aircraft is immediately viewed by the Air Staff in Washington as an attempt to crack the roles and missions egg. So sensitive is the Air Force to this matter that DePuy and Dixon avoided it like the plague. I have been waiting until I had a better measure of Bill Creech. I now believe that with him we can find a way to address ourselves to this important problem. I need not remind you, however, that whatever we do will fly directly in the face of everything NATO is doing—the centralized control of air by the ATAFs, essentially driven by the bomber mentality, the central fragging of air sorties from BOERFINK, and so on.

Indeed, the NATO air allocation system and BOERFINK itself are simply reincarnations of the command-control and allocation arrangements in the Battle of Britain. BOERFINK itself is a reincarnation of the 11 Group central control headquarters on the southeast coast of England in WW II. Doctrinal revision of allocation procedures and priority assignments will, in the face of that deep a bias, be a formidable task indeed. I suspect that in the end we will have to seek a JCS decision taken in concert with the NATO Military Committee. TAC and TRADOC can do the spade work and bring in a recommendation; but for the life of me I can't see how we will ever resolve the issue without raising it to the highest levels—simply too much ingrained inertia on both sides of the ocean.

You cited four other areas where improvement is required: target acquisition, front line trace ident, reliable interface amongst air ground systems, and tactical suppression of enemy air defenses. In some way we are working on all of these as between TAC and TRADOC.

Progress is slow—much too slow to suit me, and I'm looking for ways to speed things up, even at the risk of having to confront some of the hard issues head on. As you know, we have produced a JMENS on reconnaissance, target acquisition and surveillance. It is far too broad and general to force us to come to grips with the real problems we face. And it is only the beginning. Bill Creech and I have talked about this, and are in the process of drawing up a concept paper for looking at the total problem with some more specificity—sensors, collectors, redundancy, coverage, ELINT, SIGINT, PHOTINT et al., to determine first if we have the right numbers of the right kinds of sensors on the right kinds of vehicles. With that in hand, then it's up to the ground battle commander to say what must be done to those things he asked for information about. We are pulling this effort together at Leavenworth, and in the next few months I will be able to state fairly precisely what we think we need to do back there in the second echelon world, and what kind of information we need to do it. On that basis then we can set to work to define the total sensor problem and the weapon delivery system problem. Apparently it's not been done before in a coherent way, and pulling it together is a real bear.

The latter four of your problem areas are all in some way subsumed by what I have in mind by way of outcomes from our studies.

. . .

In sum, we are a long way from resolving any of the issues and concerns you properly cite. Work is in progress—not enough, not fast enough, with not enough resources applied to problem resolution. But then nobody ever suggested this would be easy. So we'll press on.

Offensive Air Support

Message to General E. C. Meyer
Army Chief of Staff
22 February 1980

. . .

2. You are aware of my problems with the 22 Dec Air Force position paper on apportionment and allocation of offensive air support. This morning I discussed the matter with Bill Creech [Air Force General Wilbur L. Creech, Commander, Tactical Air Command]. He too is concerned about the paper's thrust and its ambiguities, especially with regard to battlefield air interdiction. He intends to ask the Air Staff to publish a clarification. Since resolution of this issue is so critical to how we fight the deep battle, we set our staffs working on a solution. We will review progress on that effort in about a month. If satisfactory results cannot be achieved at TAC/TRADOC, service resolution will be required by you and Lew Allen [Air Force Chief of Staff General Lew Allen Jr.].

3. The three TAC/TRADOC studies underway, Joint Second Echelon Interdiction (JSEI), Joint Suppression of Enemy Air Defenses (J-SEAD), and Joint Counter Air-Air Defense (JCAAD), are not moving as fast as they should—but some progress. Bill Creech has agreed that these studies should permit development of joint prioritized system requirements. Some of the lack of progress can be attributed to failure to resolve the interface problem. We will try to resolve that in the next few weeks so the studies can reach meaningful conclusions. Meantime, as important as these studies are, we do not plan to let them go anywhere until TAC and TRADOC have reached agreement on the interface issue.

4. Bob Dixon [Air Force General Robert Dixon, former Commander, Tactical Air Command] and Bill DePuy [former TRADOC Commander General William E. DePuy] had significant correspondence on the interface problem—between the two of them and with their respective chiefs. Bill Creech and I intend to reaffirm the Dixon-DePuy position. However, it was the failure of the Army and Air Staffs to agree on the TAC-TRADOC interface position and to incorporate that into JCS Pub 2 that led to our current dilemma. We plan to send a recommended Army/Air Force position that can be jointly agreed to at service level and incorporated as a change into JCS Pub 2.

Joint Army-Air Force Issues

Message to Lieutenant General Glenn Otis
Deputy Chief of Staff for Operations
21 April 1980

1. I have in hand a 28 March Air Staff paper which nonconcurs in the CSWS MENS on a roles and missions basis.

2. At the same time we are having extreme difficulty getting anything at all done with TAC. Despite Lew Allen's instruction to Creech to proceed in the requirements business with TRADOC, the TAC staff is apparently completely cowed by the Air Staff. In fact, since the meeting with Shy [Army Chief of Staff General Edward C. Meyer], Lew [Allen], Bill Creech, and me in which we agreed to work out together all these joint problems, nothing much has been done. The 22 December Air Staff memo on BAI is ample evidence. The Air Staff has likewise intimidated the USAFE staff to the extent that Creech and I can't work issues realistically; everywhere we turn it's a roles and missions issue.

3. I've been on the verge of reporting to Shy that the TAC/TRADOC interchange is drying up, not because Creech and I can't agree, but because we are both stymied at every turn by the iron colonels of the Air Staff. This latest problem tipped the scales as far as I'm concerned. Our work with TAC is not inexpensive in resources—I have plenty of other work to do. If progress is to be frustrated again and again by the mental constipation of some Air Force turf protector, then it's probably time for us to reevaluate the utility of our whole program.

[4.] What shall I do next? The CSWS will be part of our JSEI requirement in the TAC/TRADOC study. We are obviously wasting our time.

Joint Army-Air Force Issues

Message to General E. C. Meyer
Army Chief of Staff
21 July 1980

1. Wednesday Bill Creech and I met to resolve a couple of issues which had stymied our staffs, as well as to review the progress on our joint studies. The meeting was fruitful and I am encouraged to believe that in the AirLand business we are on some fertile ground.

2. In February I reported that we were working a response to the 22 December Air Force position paper on apportionment and allocation of offensive air support. We solidified a response which TAC will submit unilaterally to the Air Staff. Bill feels that is the best tactic. While purely an Air Force document, it corrects the essential differences we had with the original paper. It now enables the corps commander to determine the targeting priorities for BAI while allowing the Air Force to make the most effective application of their resources in this effort. The settlement of this issue should allow rapid progress on the AirLand forces interface (ALFI) concept. Both our headquarters are anxious to attack this subsequent priority issue and I anticipate resolution soon. After appropriate staffing, we will forward the ALFI concept as a TAC/TRADOC pamphlet to the Army and Air Staffs. At that point Glenn Otis will have to work out the Air Staff concurrence and subsequent inclusion of the concept in JCS Publication 2.

3. The second major issue discussed centered on what the ultimate product of the joint studies would be. At our 11 October 1979 meeting with you, Lew Allen, and Jack Vessey [Army Vice Chief of Staff General John W. Vessey Jr.], we determined to get TAC/TRADOC into materiel and to make an ultimate product of the joint studies to be a prioritized list of system requirements. Bill Creech has problems with this and for many reasons is not prepared to produce such a list. We can, however, agree on and continue to make meaningful joint progress by focusing our efforts on jointly agreed-to needs. We can then compare current and developmental systems with these needs and determine unfilled needs. TRADOC can then separately prioritize the Army systems which may satisfy a need. Making this our goal allows us to proceed without getting unnecessarily bogged down. This is about the best I can get here. Unless you want to pursue that with Lew Allen, we will proceed on the basis of that settlement.

Army-Air Force Cooperation

Message to General E. C. Meyer
Army Chief of Staff
30 March 1981

1. Last Friday Bill Creech and I met to review several issues relating to the AirLand Battle and to be updated on progress in our joint studies and other related programs.

2. We provided briefings on our concept-based requirements strategy, i.e., the philosophy of future Army requirements being driven by concept-based mission area analyses. We stressed that, as the Army's requirements are driven by a concept that extends the battlefield in depth, it becomes even more important that we consider joint capabilities, lest the Army pursue an unwise/unneeded unilateral solution. Bill was very receptive to our approach. He agreed to provide various levels of TAC support, including TAC participation in the development of our AirLand Battle 2000 concept and membership on the study advisory groups (SAG) for each of our 12 mission area analyses.

3. Our extended battlefield contact team (EBCT) briefed the results of efforts to date, highlighting the AirLand concept and results of the Phase I worldwide tour, to include perceptions received from MACOM commanders and their staffs. Again he was extremely supportive of our efforts to implement this concept now and does not feel we should have any roles and missions problems. In short, he welcomes the expansion of the ground commander's vision of the battlefield and pledged his total support in seeing the effort through.

4. We also ran through a short review of the scope of our proposed joint attack of second echelon study. The objective of this work is to derive a set of how-to procedures that will permit us to implement the AirLand Battle concept today. We also hope to gain an insight into materiel requirements of each service to more effectively carry out the concept in 1986/1987. Again Bill pledged his full support of this important effort.

5. I reported to you in July that we had resolved the Air Force OAS (BAI) policy problems. As a result, General Creech took the lead and sent a revised draft OAS policy agreement to the Air Staff in September 1980. Apparently the Air Staff has about completed its in-house review and is ready to send the document to the Army Staff for coordination. It would be most helpful to us if we can get expeditious support of that document once it is received.

6. Finally, Bill and I agreed to get on with our overdue report to you and General Lew Allen on J-CAAD. Since he has not seen the briefing for several months, he wants one more review. We then plan to forward it through several Air and Army Staff reviews prior to going jointly to you and General Allen.

7. In summary, this may have been the most positive and beneficial of all the meetings I've had so far. I am especially pleased with Bill Creech's endorsement of our AirLand concept and his willingness to work with us for our mutual benefit. We are at the leading edge of a new set of issues to solve as we concentrate our efforts to implement now the AirLand Battle of simultaneous battles at the FLOT and against follow-on echelons. I am confident we can make good progress in these and thereby increase our warfighting potential.

Management of Air Assets

Letter to General Frederick J. Kroesen
Commander in Chief, US Army Europe and Seventh Army
23 July 1981

As you recall, sometime ago we had some problems getting USAF agreement to Battlefield Air Interdiction procedures that would satisfy our ground commanders' needs in fighting the second echelon battle. As a direct result of your work with USAFE, the USAF has taken a position on the BAI issue which I think will improve the synchronization in future AirLand battles. A copy of the recently signed USA and USAF agreement on Apportionment and Allocation of Offensive Air Support is attached.

This agreement reconciles Army and Air Force positions on management of air assets. It provides a description of Air Force management of air resources, while accommodating Army concerns for joint planning and the ground commander's influence on targeting priorities. The agreement states that Close Air Support (CAS) is normally distributed down to corps level, while Battlefield Air Interdiction (BAI) is managed at the air component level. The joint interface occurs at the corps level for CAS and the Allied Tactical Air Force/Army group level.

Your work with USAFE served as a valuable catalyst in getting the USAF to agree to standard procedures for coordinating Offensive Air Support requirements. I am appreciative of your help and confident we can continue to overcome obstacles to more efficient AirLand operations.

2. Armor

Page

Reflections 13

Operations Analysis for Armor 35

Sheridan Critique I 44

The Role of Cavalry 45

Sheridan Critique II 48

Tanks Forever 49

Tank Platoon Organization 54

Tank Unit Organization 55

Tank Platoon Organization and Training 56

Tank Duel Analysis 58

New Main Battle Tank 60

Tank Development 61

Chief of Armor 62

Tank Fire Control Development 63

Tank Suspension Systems 65

Tank Suspension Mine Damage 66

Cavalry Element Tanks 67

Tank Gun Selection 68

Light Armored Vehicle Losses in Battle 69

XM1 Tank Testing 70

Wheeled Reconnaissance Vehicles 71

M60A2 Tank Assessment 74

Mine Detection 75

XM1 Tank-Mounted Barrier Equipment 76

Armor for Armored Vehicles 77

M551 Evaluation 78

Alternative Engine Programs for the XM1 80

	Page
Light Tanks	81
Tank Design	84
Kick in the Grill Doors	85
Tank Main Gun Service Ammunition	88
Mine-Clearing Issues	89
Cavalry Vehicles	91
Tank Track	92
Winter Track Pads	93
Tank Electrical Power	95
Impact of the M1 Tank I	96
Impact of the M1 Tank II	98
Implications of M1 Tank Capabilities	99
M1 Tank Assessment	100
Tanks Forever and Ever	101
Antitank Missiles	106
Evolution of Doctrine: The Armored Force Example	107
Combat Vehicles	113
The Threat and Armor Development	120
Armor/Antiarmor in the Future of Land Combat	125
Light Combat Vehicles	135
Armor on the Modern Battlefield: The Way Ahead	140

Reflections

Annually, April marks the anniversary of the 1917 arrival in France of the first elements of the AEF, the United States's contribution to the Allied defeat of Imperial Germany in the 1914–18 world war.

Subsequent deployments to the AEF included a fledgling group known as the Tank Corps. Tanks came to battle in that war as a means to counter the devastating effects of massed artillery and machine-gun fire on infantry. Some visionary tank persons of the day even foresaw a larger role for tanks—independent of mud, trenches, and massed infantry in collision along the static western front. At the operational level, tanks would strike deep, disrupting command and control, reserve forces, and support infrastructure, and turning forward-deployed enemy forces out of their fixed entrenchments. The battle would be won by encirclement, envelopment, and maneuver. It was a vision far beyond the capabilities of the machinery of the day. Indeed, tactical close action in support of infantry, despite some striking successes, was fraught with substantial mechanical challenges for the fragile machines of the time.

My father, Don Albert Starry, enlisted in the Tank Corps out of college. The Tank Corps and the Air Service were the premier branches of the time. Recruiters from both services worked college campuses of the nation, seeking to enlist the brightest and most active young men into these elite organizations rather than rely on conscript forces. They also sought—at least in places like rural Iowa, where my Dad went to college—young men from the farms, men who had at least some experience with engines and the running gear of machinery. Some tankers of the day enjoyed basic soldier training at Camp Colt, a site now buried in the town or on the battlefield at Gettysburg, Pennsylvania. Cpl. Starry's promotion to sergeant was signed by Capt. Dwight D. Eisenhower, the camp Commander. Since but a single tank was available for training at Camp Colt, Sergeant Starry and some of his buddies were trained with their tanks—French-made Renaults—at the Tank Corps School at Langres, France. Lt. Col. George S. Patton Jr. was the Commandant there. Patton later accompanied his Renault tanks on foot through the wire and across the trenches in the Saint-Mihiel and Meuse-Argonne offensives in September 1918. He took a round through the leg and buttocks on the first day of the latter action—an event the tank was designed to prevent (for those inside), but of which he was, nonetheless, forever proud. Later, of course, Captain Eisenhower would become a General and then President, and the U.S. Army would name a couple of generations of tanks (the M46, M47, M48, and M60) after General Patton, who became the premier U.S. armor Commander of World War II.

World War I was soon finished for the AEF and its Tank Corps. Without a decent requiem for either, both just went away. It was called demobilization—the logical antithesis of mobilization. It was a process with considerable historical precedent in U.S. military affairs. Its political genesis was aggravated by a can-do willingness on the part of the military to simply do the best it could at whatever its civilian masters demanded of it.

If mobilization had seemed frenetic and helter-skelter, demobilization put its predecessor to shame. In 1918 demobilization took the U.S. Army quite by surprise. Numbers and time lines

---

From *Camp Colt to Desert Storm: The History of U.S. Armored Forces*, ed. George F. Hofmann and Donn A. Starry (Lexington: University Press of Kentucky, 1999), 531–61.

are instructive. For example, the Army's rolls on 1 April 1917 included fewer than 130,000 soldiers. In the succeeding nineteen months well over three million Soldiers enlisted or were conscripted. Then the war abruptly ended. Soon it was ruled that draftees and enlistees alike were eligible for immediate discharge. The "war to end all wars" was truly over. While a large staff section had been charged with mobilization planning, one lone Colonel, C. H. Conrad Jr., was charged with planning demobilization. Appointed to the task just a few weeks before 11 November 1918, he was sworn to secrecy lest word that demobilization was even being considered would be condemned as "peace propaganda." Eleven days after the Armistice, Colonel Conrad's demobilization recommendations were forwarded to Chief of Staff Peyton C. March by the War Plans Division of the General Staff. After several false starts, much confusion, and considerable meddling by the press and Congress, legislation governing demobilization was passed on 28 February 1919. By November, one year after the Armistice, 3,416,066 soldiers had been mustered out. Army strength on 30 June 1920 was reported at 209,901—only some 70,000 more than had been in uniform in April 1917. Clearly it was a case of get them all out and home—with back pay and a $60 bonus. The latter was the token reward of a grateful nation for helping make the world safe for democracy and for ending all wars.

So it was that Sergeant Starry and his young Tank Corps buddies—after rushing to the colors, undergoing partial training in the United States and France, and hustling off to combat—sojourned on leave at Monaco's casinos before returning to Camp Meade, Maryland. After a brief stint there, they then made their way back to the hinterlands, to the villages and farms from whence, scarcely a year earlier, they had sallied forth to war. Whatever other benefits of military experience he may have enjoyed, Sergeant Starry took with him back pay and a bonus totaling $81.53.

Many of them would go to war again. But in 1919 such a possibility was so remote as to be unthinkable to the men of C Company, 329th Battalion, Tank Corps, AEF.

While the mobilization and demobilization of personnel visibly occupied center stage during and after the conflict, the procurement of arms and equipment for an Army grown some thirteen times its prewar size in a matter of just over eighteen months was even more dysfunctional.

Traditionally arms, from rifles to artillery pieces, were the responsibility of the U.S. Army Ordnance Department, which designed, engineered, developed, and manufactured weapons and munitions within the bosom of the Army's arsenal system. It was a system developed in the eighteenth century, a time when there was virtually no industrial base in North America. Absent either a robust heavy machinery or arms industry, it was both necessary and expedient to create one internally. However, making rifles, pistols, and even artillery pieces was one thing; tanks were quite another matter. So it was that the AEF Tank Corps fought with French light tanks and British heavies while U.S. industry was consumed by start-up problems ranging from translating millimeters to inches to how to mount what cannon on which tractor-like chassis.

Conversely, tentage, uniforms, food, lumber, tar paper siding for barracks, and additional sinews of war flowed in fair order from a rapidly expanding civilian production base, albeit not without considerable difficulty. Demobilization struck all that preparation with a hurricane-like fury. Clearly most of it—from the design and development of arms to procurement contracts for more wrap leggings—was now a candidate for the dustbin. Whatever the status, it was all consumed by demobilization—conscript and nonregular enlistees sent home, units disbanded, production lines shut down, procurement accounts closed out, everything back to normal.

Furthermore, demobilization left the military with a research, development, and acquisition system quite out of tune with the demands of mechanization.

Despite some obvious opportunities demonstrated by tanks in the Great War just finished, the National Defense Act of 1920 set the policy azimuth for the next two decades. It relegated tank matters to the Chief of Infantry, who, with the other branch chiefs, enjoyed enhanced status in the Army bureaucracy under the new law. The 1923 *Field Service Regulations* proclaimed that: "The coordinating principle which underlies the employment of the combined arms is that the mission of the infantry is the general mission of the entire force."

The stage was set.

As the first post-World War I decade unfolded, many of that war's participants were moved to establish links with their military past, however brief it may have been. So it was that Sergeant Starry—formerly of C Company, 329th Battalion, Tank Corps—became First Lieutenant Starry, commander of Headquarters Company, 2d Battalion, 137th Infantry, 35th Division, Kansas National Guard. He had enjoyed military service. He liked being a soldier. Why he did not take advantage of opportunities to become a professional soldier at the time was never clear. In his civilian employment in Kansas City, Kansas, some of his business associates were National Guardsmen; it was likely they who prevailed on him to accept a commission.

Of course there was no longer a Tank Corps. While a troop of horse cavalry was part of the Kansas City garrison, for reasons now lost he elected to join the infantry. Shortly after his appointment to command Headquarters Company there arrived my own appointment as Brevet First Lieutenant, Kansas National Guard. It was to be the beginning of a long military career. The year was 1929, and I was four years and some months of age when Gov. Clyde M. Reed assigned me to my father for quarters, rations, discipline, and for such other duties as might be assigned by the company commander. Those included, as it turned out, periodic drills at the local armory—first located in an abandoned movie house and later a more substantial building—and attendance at all or part of an annual two-week summer camp at nearby Fort Riley.

In infantry battalions of the time, Headquarters Companies provided what is now called command and control. All battalion telephones, radios, and other electronic gear was assigned to Headquarters Company. Signal flags and other more primitive devices were more widely distributed. Headquarters Company operated the battalion message center and the Headquarters Company commander was the battalion adjutant—not the S1 but the adjutant, a sort of information staff officer on the model of staffs of earlier years.

It was a terribly lean Army. The National Guard was pretty much a mirror image of the Regular Army. Although Headquarters Company was authorized several high frequency radio sets, there was but one on hand. Radio operators and crews took turns operating this lone radio. Ammunition boxes salvaged from summer camp were painted to look like the real thing, complete with wooden knobs and dials and hand-painted scales. Operators and crews for whom there were not enough radios would go through the motions on their wooden mock-ups as the crew picked to operate the real radio practiced.

One older model Ford stake-and-platform truck was assigned to the Kansas City garrison. Companies assigned took turns using it for weekend field exercises. At night it was necessary to park the truck headed downhill for an easy, clutch-assisted start in the morning. It was easier than pushing.

Thus was the condition of the country's defense preparedness for whatever national security challenges might come next. It assured that there would be a reiteration of the mobilization frenzy of World War I, already noted, now further complicated by a host of new technology challenges either to be countered or taken advantage of.

Elsewhere there were new ideas—especially regarding mechanization. The story of Sir Ernest Swinton's invention, the tank—along with the history of the development of concepts for mechanization and mobile, all-arms warfare—began in World War I. Born independently in both the British and French Armies, tanks became the subject of considerable debate regarding design, development, and employment. In the United Kingdom a coterie of single-minded tank and mobility enthusiasts persisted in developing concepts for mobile, all-arms warfare built around tank-led striking forces. In France, Col. Jean Estienne, with the backing of industrialist Louis Renault, was able finally to convince the General Staff of the potential worth of light tanks employed in mass to break the trench-bound stalemate and restore maneuver to the battlefield. However, especially in the United Kingdom, these innovators struggled in the face of stubborn opposition by their less imaginative peers and, worse yet, superiors. They were forced to work around an organizational system dominated by foot infantry and horse cavalry, both of whose leaders abhorred change. Frustrated, many went public with their arguments, and by doing so incurred sufficient enmity from their superiors to bring on early retirement or relegation to remote and inconsequential postings.

Notwithstanding, field trials of a mechanized force were held on England's Salisbury Plain beginning in the late 1920s. These trials were designed to demonstrate new tentative tactics, equipment, and organization. Unable to reach post-trial agreement about what had been learned and what to do about it, the British did not much of anything. Thus it was that as war came to Europe in 1939 the British Army found itself absent agreed upon concepts for all-arms mechanized combat based on armored fighting vehicles. There were inadequate operational and tactical level concepts, structural and organizational alternatives, equipment requirement definitions, and training concepts to implement an idea of warfare they themselves had invented.

The Germans, meanwhile, took mechanization seriously. Armed with the writings of B. H. Liddell Hart and J.F.C. Fuller in the United Kingdom, and having studied reports of the Salisbury Plain trials, Heinz Guderian demonstrated what became the blitzkrieg concept to Adolf Hitler at the Kummersdorf test ground in 1934. With Hitler's approval, Guderian, in just eighteen short months, produced an all-arms panzer division. This division had a fairly well spelled out doctrinal framework. It included operational concepts for mobile warfare at the tactical and operational levels, force structure and organizational schemes, a preliminary array of the types of equipment that would be needed, and some carefully thought out ideas about how to train soldiers and, most importantly, units for mobile warfare. By 1939 the Wehrmacht had further developed Liddell Hart's operational concept of the "Expanding Torrent," foreseeing an all-arms mechanized force supported by tactical airpower capable of cutting deep into the enemy's rear. The mobile, mechanized, all-arms panzer divisions with which the Wehrmacht spearheaded its invasions of Poland, France, the Balkans and the Soviet Union were spawned from this beginning. Despite the predominance of infantry divisions in the Wehrmacht structure, as well as the support of millions of horses for many transport tasks, it was the awesome power of those mechanized spearheads, supported by a fleet of Stuka dive-bombers, that made possible the Wehrmacht's most striking successes. This was especially so on the eastern front.

Some participants in the operations there later reflected that, operationally and tactically, the skillful employment of those all-arms mechanized forces prolonged the war in the east by at least three years.

Meanwhile, scarcely anyone in the United States took mechanization seriously. Pioneers in thought and action were few in number, and the institution proved far more resistant to change than did even the British Army. The means for executing national strategy, if one existed, was restricted by America's abhorrence of large standing armies in peacetime. The Chief of Infantry clung stubbornly to a vision of the dismounted rifleman as the key actor in ground warfare. The Chief of Cavalry opposed mechanization for quite different reasons: he feared it would supersede his beloved horses. Basically it was internecine conflict over the extremely scarce resources provided by a pinchpenny Congress, further exacerbated by an abysmal lack of enlightened thinking by a pride of senior lions. While serving as Chief of Staff in the early 1930s, Gen. Douglas MacArthur testified before the Congress that the Army should not buy too many tanks because they were expensive and became quickly obsolete. Having issued that pronouncement, he decreed that each combat arms branch would pursue mechanization independently of the other branches. This single rationalization created, indeed invited, acrimonious and counterproductive branch contention that lasted well beyond World War II.

There were only two heroes in this drama: Lt. Gen. Daniel Van Voorhis and Maj. Gen. Adna R. Chaffee Jr. Without Chaffee the Army quite likely would have had no tank units at all in 1940. Using his key position on the Army staff in Washington to advantage, Chaffee squirreled away money in the procurement account to provide enough tanks for the equivalent of about three tank battalions. To keep them out of the grasp of the Chief of Infantry, they were designated GHQ battalions. Without Van Voorhis, who commanded the mechanized elements during most of the experimental armored force trials at Fort Knox in the 1930s, there would have been not even tentative operational or tactical level concepts for armored force employment. As it was, the Armored Force—first under General Chaffee and then, following his untimely death, under Maj. Gen. Jacob Devers—struggled to produce relevant doctrine. In the end, each armored division largely provided its own doctrine based on the study and experience of its leaders and Soldiers.

This was the bitter harvest of the long-standing branch impasse between infantry and cavalry over mechanization. The Chief of Infantry clung to his sacred charter under the 1920 National Defense Act, which had assigned tanks to the infantry and, by inference, to no one else. The Chief of Cavalry clung to his horse-mounted troops, but in time accepted the "combat car" as the cavalryman's tank. It is also fair to charge that the search for an adequate tank for either branch was severely inhibited by the Ordnance Department's lack of resolution and constant head butting with maverick tank designer J. Walter Christie. An adequate tank for any employment was simply not available.

Some years ago, historian Edward Katzenbach laid the blame squarely on the horse cavalry in his fascinating paper, "The Horse Cavalry in the Twentieth Century." He alleged that the Army of the most technically advanced nation on earth came to the threshold of World War II firmly wedded to strategy, operational concepts, and tactics deeply rooted in the nineteenth century. However, he erred on several counts. First, the United States on the eve of World War II was by no means the most technically advanced nation on earth. Second, by singling out the horse cavalry as the sole culprit in a bureaucracy whose most noteworthy characteristic was the intellectual inability to cope with both the need and opportunity for change, Katzenbach ignored what is likely the most persistent shortcoming in the history of American military thought: a *blinding* fixation on the infantryman as the centerpiece of all military action. It is a bias that ignores the truth that the mechanization of warfare was simply a means of providing more combat power with far fewer Soldiers. If Soldiers as human beings and as a national resource are precious national commodities, why then did America not wholeheartedly embrace the Soldier-saving capabilities of modern mechanized technology? One reason may be that there has been a persistent notion that manpower in conscript armies is a free resource. How many Soldiers, Sailors, Marines, and Airmen may have died unnecessarily as a result of this fixation is both hard to judge and frightening to contemplate.

Despite increasingly ominous developments in Europe in the second interwar decade, U.S. Army strength on 30 June 1939 stood at 187,893, with an additional 199,491 in the National Guard. It was a force somewhat—but only marginally—larger than it had been on 1 April 1917. The Regular Army alone was short nearly 100,000 Soldiers of the number authorized.

On 31 May 1940 President Roosevelt asked the Congress for authority to federalize the National Guard. Shortly after that a new draft law was passed and conscription began.

In armories everywhere the National Guard stood to arms. Headquarters Company, 2d Battalion, 137th Infantry, joined a host of others as it began readying itself for war. Fortunately for the war effort, Headquarters Company's Brevet First Lieutenant was deemed too young for mobilization. Without his services, the 35th Division moved to Camp Robinson, near Little Rock, Arkansas, for training. After some months equipment began to appear; industrial mobilization had begun. Extension of the law mobilizing the National Guard survived in the Congress by a single vote in 1941, just a few short months before the Japanese attacked Pearl Harbor. While some consideration was given to deploying divisions to the Philippines, the campaign and the force deployed there were lost before that could transpire. Meanwhile, mobilized National Guard units had sent off a number of officers and noncommissioned officers to serve as cadre in new divisions. Guard officers attended appropriate branch schools—infantry officers, for example, went to Fort Benning. Promotions flowed freely as the force expanded, and Capt. Don A. Starry soon left the 137th Infantry's Headquarters Company, 2d Battalion, to command L Company, 3d Battalion. Not long after that, he and other older Guard officers (he was forty-four) were replaced by younger officers and posted to other assignments. So, despite years of training and study and a desire to serve in the infantry, he fought as a headquarters warrior, finishing his war service in 1946 as a Colonel. Not at all new, the idea of using National Guardsmen as individual replacements would be repeated in subsequent wars. This led some thinking Soldiers to wonder why the Army should devote all the time, effort, and expense needed to build National Guard units in peacetime if all the system required was individual replacements in time of war.

Meanwhile, all facets of the Army's research, development, and acquisition process severely lagged behind demand, a demand for equipment for fighting units as well as for training units. The 1st Armored Division deployed to the United Kingdom in April 1942 equipped with obsolete M3 medium tanks, no antitank weapons, and short of other essentials as varied as binoculars, observation aircraft, and training ammunition.

By the summer of 1945 more than eight million Soldiers were under arms when demobilization of a sort began. The post-World War II demobilization did not elicit the same openmouthed astonishment as was the case after World War I. However, while poor Colonel Conrad's dilemma had been studied at length and considerable demobilization planning undertaken, things did not go well. So poorly did they go that the author of an Army study published in July 1952 concluded: "when future scholars evaluate the history of the United States in the first half of the twentieth century they will list the World War II demobilization as one of the cardinal mistakes."

However, with the war in Europe concluded, it was necessary to turn full attention to operations against Japan. Simultaneous establishment of a point system for releasing long-serving individual Soldiers and the need to beef up the Pacific-based force structure for the planned invasion of Japan's home islands were clearly at odds with one another. In the ensuing imbroglio between individual and unit deployments, unit effectiveness and cohesion were destroyed. Vocal dissent was rife, in some cases reaching near-riot proportions. The news media entered the debate to include, interestingly, the *Stars and Stripes*. Never able to refuse an opportunity to be heard, Congress joined the chorus. Fortuitously, nuclear weapons dropped on Hiroshima and Nagasaki not only induced Japan to surrender, they short-stopped what promised to be a most difficult national debate over the endgame in the Pacific.

Unlike Soldiers, tanks and guns are uncomplaining partners to whatever is undertaken. Divisions redeploying from Europe, in whatever combination of individual and unit deployment, simply marched their equipment into designated open areas, dismounted, and departed. Ammunition, weapons, and communications gear were removed; however, fuel, batteries, sighting and fire control equipment, and other impedimenta remained. As supplies and repair parts became scarce again in the postwar years, units stationed in Europe found these divisional parks to be welcome sources of everything from jeep windshields to tank power trains. Many of the removed components had been given over to a German government-run supply firm. Sold to whomever, they helped prime the postwar economic pump. Shortages of unit and individual equipment not available in the supply system were all too frequently made up at Soldier expense from these sources. Equipment readiness rates were sustained by units whose officers and Soldiers stood ready to spend their own scarce money buying back repair parts, major assemblies, and, in some cases, whole vehicles. Lean years had returned.

Back in the enclaves of visionary military thinking, revisions to doctrine, equipment, force structure and organization, and training for individuals and units were being drawn up. They were derived largely from the experience of U.S. forces, especially in the war in Europe, from studies of Soviet operations on the eastern front, and from considerable study of Wehrmacht operations. The changes that resulted struck deeply at some of the most dearly held underpinnings of traditional American military thought.

Possibly the foremost difficulty military forces face in the United States is a historic antipathy toward standing armies in peacetime. Born of America's pre-Revolution experience with the British Army and encouraged by the relative isolation of North America from the rest of the world, it led to a conviction that there would always be time to raise an Army, should the need arise. Politically, a two-part solution was adopted. The national military strategy was founded on strong naval forces. Deployed in far-flung battle lines, this "first line of defense" would buy time for raising, equipping, and training large ground forces. Secretary of War John C. Calhoun proposed the second part of the solution circa 1818: token ground forces in peacetime expanded by a flood of militiamen in wartime. Called the "expansible Army," the idea would last more than a century and a quarter, with Secretary Calhoun's militia reinforcements expanded to include conscripts.

These two fundamental precepts were reinforced by Napoleon's ideas on war, as reported by nineteenth century military theoretician Antoine Henri de Jomini: masses of men and fire (artillery) are necessary to win; one always wins by attacking. In addition, processes had been developed during the Industrial Revolution to turn masses of raw material into finished products quickly and efficiently. In sum, the expansible Army had to outnumber its foes to win, and to make the expansible Army possible one had only to invoke the processes of the Industrial Revolution. Later, modern technology in several forms would further enhance both the destructive power of armed forces and the productive power of factories—training factories and materiel factories.

By the end of World War II, one could detect signals that all this might be coming unraveled. First, in Europe especially, the Soviet Union's substantial postwar conventional military strength revealed the unpleasant truth that no longer could the United States be guaranteed numerical superiority, even with the aid of allies. Therefore, it would henceforth be prudently necessary to maintain substantial standing military forces in peacetime. Faced with a potentially overwhelming Soviet force—mostly mechanized—across the inter-German border, and with limited battle space in which to maneuver in defense of western Europe, a military decision would likely be reached between the deployed forces before mobilization and ensuing deployments could provide the means to salvage a difficult situation.

The expansible Army idea had further engendered the conviction that the U.S. Army could afford to lose the first few battles of its next war because mobilization would subsequently raise masses of Soldiers and materiel. Despite any early losses, the war would, in the long run, be won by sheer weight of arms and men. This arrangement clearly was no longer feasible.

Reinforcing this fracturing of sacred convictions was the advent of nuclear weapons—especially once it was understood that they could be employed at all levels of war: strategic, operational, and tactical. Nuclear weapons offered a relatively inexpensive way to invoke the gods of modern technology by substituting their mass killing power for that of masses of Soldiers and units, thus avoiding the expense and time consumed in mobilization. However, despite this perceived advantage, the politicization of nuclear weapons only aggravated the dilemma of how to fight and win the first and succeeding battles of the next war—with or without nuclear weapons—thereby avoiding early defeat of deployed forces.

The NATO dilemma is particularly instructive. Early on, General Eisenhower, as SACEUR, established a requirement for ninety-six divisions and nine thousand tactical fighter aircraft to defend NATO Europe against the Soviet mechanized threat from the east. However, the NATO member countries—including his own—turned thumbs down to his proposal. They were either unable or unwilling to provide the staggering resources required. Later, as president, Eisenhower settled for twenty-six divisions—twelve of them from the Bundeswehr (the West German Army)—fourteen hundred fighter aircraft, and fifteen thousand operational and tactical level nuclear weapons. Some seven thousand of the latter were deployed, eventually to be removed when the Warsaw Pact collapsed on the eve of Operation Desert Storm in 1991.

Then—interacting in complex ways with traditional concepts of war in U.S. military thinking—came North Korea's surprise invasion of South Korea on a fateful Sunday morning in June 1950. World War II demobilization had left overseas-deployed U.S. Army units with a mechanized capability only at division and regimental level. There was a tank battalion per division, a tank company per regiment, and an armored cavalry troop, all in various states of equipage and with serious personnel shortages and individual and unit training deficiencies. Two armored divisions remained—both at Fort Hood, Texas—considerably distant from the inter-German border where they were destined to fight. Fearful of some not quite understood connection between the North Korean attack and growing Soviet deployments to Europe, partial mobilization was ordered in the United States. Conscription was reinstated. Four divisions were called up, with two deployed to Europe and two to Korea. The Europe deployments included armored divisions. They became the basis for a two-corps mechanized force that included armored cavalry regiments at corps, and tank and armored infantry battalions, self-propelled artillery battalions, and armored cavalry squadrons in the divisions. A host of mechanized support units at both division and corps were envisioned, including separate tank and armored infantry battalions assigned to armor groups. The deployed force would initially be reinforced by active duty divisions deployed from the United States, and later reinforced by a host of National Guard divisions and U.S. Army Reserve support units to be mobilized. The deployment of heavy divisions would remain a complex and controversial transportation problem, as would the readiness of forces to be mobilized. Another political hot potato was the division of mobilization authority between the executive and legislative branches. Notwithstanding, U.S. mechanized force deployments and the designation of reserves to reinforce deployed forces in Europe continued for nearly forty years. Along with allied mechanized strength, it remained a strong and reasonably ready capability until the collapse of the Warsaw Pact nearly four decades later.

Mechanized or not, there can be no greater victory for a military force than to accomplish its mission without having to fight the campaigns and battles for which it so carefully prepared for so long a time.

The Korean War, however, was a sort of field trial in a new era of military affairs. It exposed the bankruptcy of traditional U.S. military thought just described, a bankruptcy best reflected in the popular characterization of Korea as a "limited war." It suggested that all-out mobilization in order to win, after losing the first few battles of a war, was no longer a relevant concept. It brought again into stark focus the price of unpreparedness. Personnel turbulence and resulting unit unreadiness, as well as materiel shortfalls of all kinds, was reflected most tragically in the early losses of the nation's most precious resource: the Soldiers, Sailors, Marines, and Airmen called upon to fight those first battles.

American involvement in the Indochina War began as the Korean War ended. It began with U.S. naval forces deployed to assist with the evacuation of more than a million North Vietnamese into South Vietnam consistent with the 1954 Geneva Accords.

It is frequently said, yet as frequently ignored, that no two wars are ever the same. There is, however, a tragic sameness between Korea and Vietnam. It is a sameness that emphasizes the difficulty of defining a clear-cut connection between political goals and military actions ostensibly undertaken in pursuit thereof. With the absence of a crusade-like framework of total war, consensus over political goals becomes elusive, and on military action even more so—especially as field operations in progress are being beamed into the nation's homes with accompanying commentary that is sometimes relevant and too frequently not.

Mobilization for war in Vietnam was nominal at best. The largest single unit mobilized was an infantry brigade from the Kansas National Guard. The brigade never deployed. Some personnel deployed as individual replacements, reminiscent of earlier practice. Army Chief of Staff Harold K. Johnson publicly recalled making four or five trips to the White House to ask for mobilization, only to be turned away by a president paranoid about anything, including the Vietnam War, which militated against public and budget support for his Great Society. General Johnson's requests were based not so much on an urgent need for numbers of soldiers and units, but more on the likely effect of nonmobilization on the postwar Army. As we shall see, his fears were well founded. Thus, instead of traditional mobilization, structure was added to the Army by simply activating new units and filling them from the conscript replacement stream. Three additional divisions were added to the force structure in this fashion.

Despite considerable information about French and Vietnamese mechanized operations provided by the French Army and by U.S. observers of French operations in Indochina, the dominant conviction of U.S. force planners was that Vietnam was just another Pacific jungle. They saw it as a place where tanks and other armored vehicles could not function effectively and, if used at all, then only in support of dismounted infantry. Thus, U.S. infantry divisions were initially deployed to Vietnam minus their assigned tank and mechanized infantry battalions and armored cavalry squadrons. Once on the scene, deployed division commanders normally sent for the mechanized units they had left behind, and advised the commanders of units deploying after them to bring their mechanized units along. The largest single mechanized unit deployed was the 11th Armored Cavalry Regiment. As noted by Lewis Sorley in Chapter 10, it was not until a comprehensive examination of mounted operations by a Chief of Staff-chartered task force in 1966–67 that the battle utility of mechanized forces was set forth objectively. By that time, however, most important force structure decisions had been programmed, making substantial change very difficult. Nevertheless, that war's senior infantrymen clung to their antiarmor bias and their long-standing image of the individual rifleman as the principal weapon of combat to the bitter end.

New to battle in Vietnam was the helicopter. While frequently set aside in considering the mechanization process, helicopter-lifted units, especially so-called air cavalry units, must be included in considering mechanization at the tactical and operational levels of war. The airmobile concept was the brainchild of the 1960s Howze Board. Composed of a group of farseeing cavalrymen headed by Maj. Gen. Hamilton Howze, the board experimented with a concept called air cavalry. In it the helicopter was to replace ground vehicles as the means to move Soldiers and selected weapons quickly over the battlefield to bring force to bear at unexpected places and unanticipated times. A ready-made cavalry idea, it would no doubt have delighted General Howze's father, Robert L. Howze, who as a major led the 2d Squadron, 11th Cavalry, into Mexico with General Pershing's 1916 Punitive Expedition, in some of the most strenuous and demanding horse-mounted operations ever recorded. The helicopter was surely a more than adequate surrogate for Major Howze's exhausted horses! Full blown into a division-sized force, the air cavalry operational concept was both sophisticated and complex. Above all it demanded well-trained units and highly effective command and control.

Vietnam experience with one air cavalry division led to two conflicting observations. One says that helicopter lift has freed infantry from the tyranny of terrain. As a result, the rifleman can now be moved quickly and put down with supporting weapons in a new location in jolly quick time, rested and ready to fight. The counter view is that heliborne infantry suffers from serious limitations. Deploying onto unfamiliar ground and usually uncertain about the enemy, the airmobile force is less mobile, once landed, than its on-site opponent. The situation is aggravated by an individual replacement system that virtually guarantees lack of excellence in unit training. It is made even more difficult by a command-control lacuna from company level upward. Fearful of losing control either of assault forces or their support, most airmobile units prescribed that battalion-level and higher commanders remain in command posts where communications were in place. This practice frequently inhibited higher level knowledge of the situation at lower levels, slowing decision cycles and reducing leader presence forward in the battle area—both of which increase risks to mission success.

Planning for redeployment of U.S. forces from Vietnam began in late 1968. In April 1969 President Nixon issued National Security Study Memorandum 36, requiring, among other things, plans for the initial redeployment of twenty-five thousand of the nearly 549,000 U.S. troops in Vietnam and long-term plans for future redeployments. Limited mobilization, already noted, meant that most redeploying units would simply disappear from the force structure and their manpower from the end strength. Despite General Abrams's strong objections, redeployment of individuals rather than units was the solution chosen by Chief of Staff William Westmoreland. His decision, after all, had historic precedent. It was subsequently decided to rebalance the force in Vietnam by sending home long-serving Soldiers from all units, inactivating designated units, and transferring individuals with longer time remaining to other units. This practice increased personnel turbulence in the remaining units to unacceptable levels, adding to other factors already militating against unit effectiveness. Worse yet, large numbers of battle-weary Soldiers came home alone to a country where few, except their families, either knew or cared where they had been or what they had been doing for the last year or so, or gave any indication of being glad they had arrived home safely. That experience left a generation of young—and some not so young—Soldiers with scars that are yet unhealed.

Virtually hundreds of technical innovations were applied to existing equipment during Vietnam operations. Most were Soldier-invented and applied. Many were as simple as sending home for spring-loaded clothespins to use as triggers for Claymore mines. Others were more complex. None were very expensive. The Army created a Limited War Laboratory to aid in developing modern technology solutions for battlefield problems. The Vietnam Mechanized and Armor Combat Operations study group that met in 1966–67 found most laboratory contributions to be relatively ineffective, poorly designed, hard to use, fragile, or some combination of those factors. A case in point with mechanized units: mine/countermine operations, that is, equipment for the high-speed search, detection, location, identification, and clearing of antivehicular mines. It is a complex problem with no single technical solution. The search for technical solutions has been trickle funded for many years. As a result, the problem remains unsolved. But one example among many, this situation was further demonstration of the inability of the traditional research, development, and acquisition system to respond to demands for timely fielding of new and innovative technologies.

Battle damage was extensive. The Vietnam War ended in 1973—at least for the United States. Its wounds, however, fester to this day. The national treasure in young lives sacrificed by death, and by physical as well as psychological trauma, has scarred generations. The revolt of the young against all authority so characteristic of most of the important war years has changed our society irrevocably in ways yet to be understood.

Post-Vietnam military attention turned back to the nation's commitment to NATO Europe. We discovered that the Soviets had been very busy while we were preoccupied with Vietnam. They had revised operational concepts at the tactical and operational levels, increased their fielded force structure, and introduced new equipment featuring one or more generations of new technology. With tactical and operational level nuclear forces deployed, the Soviets embraced the notion that they could fight and win at the operational level of war with or without nuclear weapons. Their preferred solution: without.

As always in Europe, numbers spoke loudly—so much so that several distinguished SACEURs in turn opined that, while the conventional force might hold out for as long as ten days, when that threshold was reached there would likely be a compelling case for nuclear release. Depending on the resulting status of the battle at the operational level, a decision would have to be made to use or not to use thermonuclear weapons delivered by intercontinental ballistic missiles, bombers, or both.

While the Soviets had been building toward winning at the operational level, the disparity in theater force strength and the need for mobilization in many NATO countries, including the United States, inevitably tied the NATO operational level nuclear decision to an extremely high risk of an all-out thermonuclear weapons exchange. It was almost impossible to rationally project NATO strategy to any outcome except Armageddon.

As U.S. forces in Vietnam redeployed, military thinkers recognized the need for a new objective force for a new era. It was an era characterized by the expanded threat in Europe, a growing threat of conflict in the Third World (especially the Middle East), increasing worldwide economic interdependence, greater difficulty articulating political goals for the planners who design military activities to achieve them, and intrusive and abrasive media probing into all aspects of military operations.

One bright Pentagon morning in 1972, toward the end of Vietnam redeployment, force planners, of whom I was in charge, awakened to the reality that as force structure and end strength had declined, only that structure and strength specifically committed to NATO, either deployed or reinforcing, had been retained. There were, therefore, at that moment, no more than perhaps twelve divisions remaining in the structure, with a strength authorization of about 765,000. By comparison, the pre-Vietnam force structure had included sixteen divisions and nearly 986,000 Soldiers. Army Chief of Staff Abrams, General Westmoreland's successor, personally intervened with Secretary of Defense James Schlesinger and got the green light to retain sixteen divisions. However, additional personnel strength would have to be justified in follow-on budget negotiations with Congress. While the pre-Vietnam strength had admittedly included some residual holdings from the early 1960s partial mobilization in response to the Berlin Wall buildup, clearly the right answer was more than 765,000 but perhaps less than 986,000. Since increasing strength requires more money, only nominal strength increases were ever realized.

The U.S. Army was somewhere close to its nadir. Soldiers and units deployed to USAREUR saw themselves as minor speed bumps for Soviet forces en route to the Rhine River and beyond; they did not believe they could defend successfully, let alone win.

While there were many reasons for this, three stand out.

First force modernization—that is, the promulgation of new doctrine, fielding of new equipment, organizational changes, and improvements in training and education for officers, noncommissioned officers, and Soldiers—had been at a standstill for nearly ten years.

Second, absent significant mobilization for the Vietnam War, it was necessary to use the entire Army—including the Army in Europe—as the rotation base for Vietnam. Combined with the one-year tour length in Vietnam, this increased personnel turbulence in units to a level well above that at which reasonable unit effectiveness could be achieved and sustained. The militant youth revolt against authority that characterized the 1960s in the United States had infested the country's armed forces. Military jails were full. The drug culture rampant in the country pervaded all but the best military units. Largely because of these factors, among others, conscription had been shut down in 1972—more than a year ahead of the draft law's expiration—without any sure knowledge that the Army could recruit sufficient numbers to fill its force structure.

Finally, Soldiers and leaders returned from a war in which they had won all the battles, only to find the nation had lost the war. There was a crisis in confidence; Soldier confidence in leaders, leader confidence in themselves as well as in the nation's political leadership, and Soldier and leader confidence in their Army and its units.

There was a widespread consensus that the Army needed substantial rebuilding. That rebuilding began with General Abrams's appointment as Chief of Staff in the fall of 1972. By the summer of 1973, reorganization of the Army's command structure was underway. That reorganization included, among other changes, dividing CONARC into FORSCOM and TRADOC. The former was responsible for all Army forces stationed in the United States, including the U.S. Army Reserve; the latter established doctrine, developed equipment and organizational requirements, and provided training and education for all ranks.

Born in the summer of 1973, TRADOC was commanded by Gen. William DePuy. At the same time I was posted to command the Armor Center and School at Fort Knox, armed with instructions from both Generals Abrams and DePuy that I was to define what the first and succeeding battles of the next war might require of the Army.

As that work began, seven thousand miles away the Arabs attacked Israel in October 1973. Within a few weeks General Abrams dispatched me and Brig. Gen. Bob Baer—the program manager of what would later become the Abrams tank—to Israel. There, Bob and I walked battlefields with the IDF Commanders who had fought on them, seeking answers to many questions about the future U.S. Army. The questions we posed focused on documenting requirements for the Ml tank, especially the need for a larger caliber gun than the planned 105mm, and, most importantly, on critical operational lessons of the Yom Kippur War. Answers to those questions framed the beginning of what grew into, some nine years later, the doctrine called AirLand Battle, a concept of war at the tactical and operational levels that U.S. and coalition commanders employed in Operation Desert Storm.

The armored battlefields of the Yom Kippur War yielded striking lessons about what to expect in first and succeeding battles of the next war.

First, we learned that the U.S. military should expect modern battlefields to be dense with large numbers of weapons systems whose lethality at extended ranges would surpass previous experience by nearly an order of magnitude. Direct-fire battle space would be expanded several orders of magnitude over that experienced in World War II and Korea.

Second, because of numbers and weapons lethality, the direct-fire battle will be intense, resulting in enormous equipment losses in a relatively short time. Significantly, we noted, combined tank losses in the first six critical days of the Yom Kippur War exceeded the total U.S. tank inventory deployed to NATO Europe—including both tanks in units and in war reserves.

Third, the air battle will be characterized by large numbers of highly lethal aerial platforms—both fixed- and rotary-wing—and by large numbers of highly lethal air defense weapons.

Fourth, the density-intensity-lethality equation will prevent domination of the battle by any single weapons system; to win, it will be necessary to employ all battlefield systems in closely coordinated all-arms action.

Fifth, the intensity of battle will make command and control at the tactical and operational levels ever more difficult. Effective command-control will be further degraded by the presence of large numbers of radio-electronic combat systems aimed at inhibiting effective command-control.

Sixth, at both the tactical and operational levels the complexity of modern battle demands clear thinking. Thinking takes time, and in battle there is no time to think. Therefore, to the extent possible, likely battle circumstances must be thought through in advance to reduce the chance of surprise and to ensure prompt, timely, and relevant decisions.

Finally, regardless of which side outnumbers the other, regardless of who attacks whom, the outcome of battle at the tactical and operational levels will be decided by factors other than numbers and other than who attacks and who defends. In the end, the side that somehow, at some time, somewhere during the battle seizes the initiative and holds it to the end is the side that wins. More often than not, the outcome of battle defies the traditional calculus employed to predict such outcomes. It is strikingly evident that battles will continue to be won by the courage of Soldiers, the character of leaders, and the combat excellence of well-trained units—beginning with crews and ending with corps and armies.

For those of us who crafted new doctrine to reflect the new environment, one single statement became the goal: The U.S. military must decide how to fight outnumbered and win the first and succeeding battles of the next war at the tactical and operational levels—without wasting Soldiers' lives, and without having to resort to the use of nuclear weapons to offset the military's likely numerical disadvantages or for any other purpose.

The first try at new doctrine focused on the tactical level of war. Several important facts drove that decision.

First there was the advent of long-range ATGMs in the early 1970s and their appearance and initial success early in the Yom Kippur War. Lethal at far greater ranges than any other infantry weapons, ATGMs promised to provide forward-deployed forces a new dimension in defense: the ability to maneuver to seize the initiative.

Second were General DePuy's convictions about inadequate tactics, ineffective training and inept leadership in small units in World War II, and most importantly the terrible cost in Soldier lives lost unnecessarily as a result. Drawn from his traumatic experience as a young officer in the 90th Infantry Division, recorded eloquently in his post-retirement oral history, and described in some detail in Richard Swain's discussion of the development of AirLand Battle doctrine in Chapter 11, those convictions burned deep in his soldier soul. While commanding the 1st Infantry Division in Vietnam in 1966–67, he was notorious for relieving commanders, mostly at battalion level—a practice that was widely and bitterly criticized. In truth, however, his practice had serious purpose, best described in a statement he once made to me: "I'm just not willing to trust the lives of the soldiers to the command of officers in whom I don't have confidence."

Finally, in organizing TRADOC, General DePuy had assigned a disproportionate number of officers to write doctrine at the CGSC at Fort Leavenworth, where normally lay responsibility for writing operational level doctrine—the capstone ideas on which the writing of tactical level doctrine was based. The aim was to write the pacesetting doctrine at Leavenworth, while TRADOC's branch schools would write appropriate tactical doctrine for their respective branches. Early on it was quite clear that what was intended of Leavenworth would probably not be forthcoming, for reasons fairly objectively set out in Chapter 11. This created a situation in which doctrine, normally written at Leavenworth, was largely being put together elsewhere, primarily at Fort Knox, where work had begun early and in earnest to solve the Army's most important challenge: how to fight and win on the armored battlefields of NATO Europe. Time was of the essence. The DePuy watch at TRADOC could be expected to last four years; he wanted to finalize what had to be done within that window. In addition, General Abrams, while ill and before his untimely death in the fall of 1974, made imperative his approval of TRADOC's new direction.

Thus it was that, in just a bit more than two years, a new capstone doctrine was drafted. Called "Active Defense," it appeared in a 1976 revision of FM 100-5, *Operations*. Since it had been written by a handful of officers at Fort Knox, and in General DePuy's closed sessions with his center Commanders at Fort A.P. Hill, the new doctrine attracted many critics—both inside and outside the Army. At root, most criticism stemmed from the fact that the doctrine had not been written at Fort Leavenworth. Some disdainfully called it "DePuy Tactics."

Active Defense was published in mid-1976. Earlier that year, armed with draft copies of the new manual and of the implementing armor force manuals written on my watch at Fort Knox, I took command of USAREUR's V Corps. Since I had authored critical parts of the new doctrine and was not completely happy with what had been written, I took as a first task to somehow test the ideas on the ground.

Issued in draft to all levels of command, Active Defense became the basis for staff rides—terrain walks in which division, brigade, and battalion commanders met with me on-site to determine if the doctrine based on the Yom Kippur War's lessons was about right or all wrong. After more than six months of evaluating tactics, weapons, and organization using analytical models provided by BDM International, commanders at all levels had about convinced themselves that, with a little luck, the Soviet first echelon could be defeated well forward in the main battle area. It was quite clear to me at that juncture that, as I had feared earlier, the operational level problem had yet to be confronted, and it was the corps commander's responsibility to do so.

In May 1977 I returned to Israel's battlefields to revisit action at the operational level and then translate that experience to Europe's environment. This led to a concept for extending the battlefield in time (the campaign) and distance (the theater of operations). Most importantly, it resulted in requirements for long-range surveillance and target acquisition systems and long-range weapons systems with which to find and attack Soviet-style follow-on echelons. These systems would be used to disrupt and delay their advance into the main battle and to destroy as many as possible before they could reach assigned objectives deep in the main battle area. The primary goal was to involve the Army in the operational level of war business.

Two months later I succeeded Gen. Bill DePuy as TRADOC Commander, faced with the twin tasks of somehow defusing vocal criticism of Active Defense doctrine and expanding doctrine to integrate the tactical and operational levels of war. In addition, it would be necessary to build a broad consensus about the correctness of whatever ensued. Change is not possible without consensus. At least part of our problem with Active Defense was that it reflected the fact that the idea was developed quickly and in a fairly closed forum. Those who considered they had no voice in the matter tended to reject it out of hand. It was also obvious that somehow whatever was done had to be accomplished in the TRADOC schools, beginning with Leavenworth.

To defuse criticism and change the perception that new doctrine could only come from the big leather chair in the front office, I created in the TRADOC headquarters the position of Deputy Chief of Staff for Doctrine. There had been none before. I appointed Brig. Gen. Don Morelli, a very bright and persuasive officer, to that position. Morelli, assisted by a briefing team, did very little else for four years but expose the developing concept to staffs in the Congress and academia, even as the details were being written. Those who did not agree were invited to provide suggestions, with the assurance that their suggestions would, to the extent possible, be included or dealt with in the final product.

At Leavenworth, attrition had taken its toll of the reluctant phalanx of disappointed doctrine writers, and Lt. Gen. Bill Richardson, the new Commandant, gave the task of writing what became AirLand Battle to some very bright lieutenant colonels—Huba Wass de Czege, Don Holder, Richard Hart Sinnreich, and Richmond Henriques.

AirLand Battle doctrine, as first written in the 1982 edition of FM 100-5, followed closely the concepts set forth in "Extending the Battlefield," a doctrinal essay taken from my briefings and published in *Military Review* in March 1981. The concept derived from my observations as corps commander, from visits to Israel's battlegrounds, and from a study of deep attack targeting for nuclear weapons done by Dr. Joseph V. Braddock of BDM International for the Defense Nuclear Agency. From the latter came convictions that advanced weapons delivery and fusing accuracies had reached the point at which we could now attack with conventional munitions what once required nuclear weapons to achieve desired target effects. A search for enabling surveillance technology found synthetic aperture radar. Funds were found to accelerate development based on requirements which we could now write, for what became the JSTARS and ATACMS systems, designed to be the corps or joint task force commander's surveillance and target acquisition and weapons systems, respectively. Thirteen years after we first set down requirements for those systems they performed with stunning success in Operation Desert Storm.

AirLand Battle, in my view at least, was an operational level concept; it combined the best tactical lessons of the Yom Kippur War with operational-level schemes designed to defeat Soviet operational-level concepts: mass, momentum, and continuous land combat—with or without nuclear weapons. Beginning with the Bolshevik Revolution in 1917, Soviet military thought had focused on operational-level concepts, largely driven, some say, by the trauma of their defeat in World War I and subsequent civil war. In the West, as suggested earlier, while mechanization offered operational-level opportunities to all nations, only the Germans took advantage of the offer.

From AirLand Battle doctrine came requirements for equipment. The long development time characteristic of the normal materiel acquisition cycle almost guarantees that either threat or technology, perhaps both, will change significantly during development, changing requirements and further prolonging development time, thus aggravating the inefficiency of an already severely dysfunctional system. At the outset, TRADOC inherited several ongoing programs. Principal among them were the armored reconnaissance scout vehicle (ARSV), the infantry fighting vehicle (IFV), the UH-60 Black Hawk troop-carrying helicopter, and the AH-64 Apache attack helicopter. In addition, requirement documentation was in being for a main battle tank, the XMl, and was being developed for a division-level air defense system. Proponency for the ARSV, Apache, and the tank resided at Fort Knox.

Shortly after I arrived at Fort Knox in 1973, ARSV prototypes were delivered for testing. There were two candidates, one full tracked and one wheeled. One look was sufficient to suggest that both were wide of the requirement, by then some ten years old. Having just forced a decision to take the unsatisfactory M114 scout vehicle, and the equally deficient M551 Sheridan airborne assault/armored reconnaissance vehicle, out of the inventory, I was extremely reluctant to buy into another uncertain program. We tested the candidates at Fort Knox and recommended the program's termination. In retrospect it might have been better if I had kept the program alive and tried to "fix" one of the candidates to meet the changed requirements, for despite considerable testing to develop relevant requirements, the Army still does not have a satisfactory ground scout vehicle.

Not long thereafter the MICV prototype was delivered for testing—again ten years after the requirements documentation was written. Just on observation it was worse than the ARSV. However, having just terminated ARSV, we feared cancellation of another major program would eliminate TRADOC as well. So it was decided to "fix" the MICV by redesigning the power train, adding armor, mounting the TOW ATGM system, finding a suitable cannon and fire control system, and including firing ports for mounted infantry. To help hold down rising costs we added the ARSV acquisition objective numbers to the IFV numbers to help reduce unit cost. Hence the Army eventually fielded two versions of the Bradley fighting vehicle, one for infantry and one for cavalry scouts. However, neither model met the requirements.

Recognizing the Bradley's shortcomings for fighting the central battle alongside tanks, despite the serious upgrades just mentioned, the vehicle was inadequate for the task. Therefore a Heavy Infantry Fighting Vehicle Task Force was convened to draw up requirements for such a vehicle based on study of the Arab-Israeli wars, and IFV systems in other armies. We then considered revising the XMl tank design to provide space inside for an infantry fire team, a concept similar to that of the Merkava tank then being developed for the IDF. Design change of that magnitude would have severely delayed the XMl program, a risk we decided not to accept.

In 1974 an Advanced Scout Helicopter (ASH) Task Force was convened at Fort Knox to set forth requirements for a replacement for the Hughes OH-6 and Bell OH-58 scout helicopters used during the Vietnam War. The new scout aircraft would be a companion to the AH-64 Apache, then being developed. A highly successful ASH Task Force effort suffered rejection in some bitter bureaucratic infighting in the Pentagon. Nearly twenty-five years later, the Army scout helicopter requirement is being met by upgraded Bell OH-58s, with the Comanche intended as a future dual-purpose scout and attack helicopter.

Early in TRADOC's life the Air Defense Center at Fort Bliss, under Maj. Gen. C. J. LeVan, developed requirements for an air defense gun system based on observations of the Yom Kippur War. Twin Bofors 40mm cannons would be mounted on a refurbished M48 tank chassis, and an armored turret and F16 tracking radar added to make the Sergeant York Division Air Defense (DIVAD) gun system. The winning prototype candidate was defeated by last-minute changes to requirements—changes designed to make the cannon system less effective than required at extended ranges. Air defense missile enthusiasts who replaced the air defense gun enthusiasts were responsible for changing the criteria on which the DIVAD gun system was tested and subsequently failed.

Studying Yom Kippur War battles with IDF commanders led to a concept for an improved "over the hill" battle surveillance system, more recently styled as an unmanned aerial vehicle (UAV). Early requirements visualized a low-light television system mounted on a remotely controlled model airplane to demonstrate feasibility. The IDF did just that, flying the system in operations in southern Lebanon for several years and then employing two systems—Scout and Mastif—in the 1982 invasion of Lebanon, dubbed Operation Peace for Galilee. Meanwhile, in the United States, the artillery and intelligence communities started a fatal quarrel over a system called Aquila. In the end it was overloaded with sensors of all kinds, costs were considered exorbitant, and the vehicle displayed a penchant for crashing directly after launch, causing program termination. Only recently has the U.S. military developed a UAV capability—in a world where such machines have been fairly commonplace for nearly twenty years.

As Active Defense doctrine unfolded, General DePuy undertook a division-level test of doctrine and tentative organization at Fort Hood. Called the Division Restructuring Study (DRS), instrumentation was provided by an early version of the Multiple Integrated Laser Exchange System (MILES). As is normally the case with large-scale tests, the attempt to gather definitive data about too many things with inadequate instrumentation produced less than adequate results. Returning to command TRADOC in 1977, I found the study had exceeded time and budget targets while producing suspect data outcomes. The study was terminated and TRADOC center commanders met frequently with me over the next eighteen months to develop alternatives based on what had been learned in the DRS evaluation and in manual battle-lab exercises.

The resulting organization for heavy forces was called Division 86. Study and test outcomes strongly suggested a need for smaller combat units from platoon upward—with more leaders per soldiers led. However, the idea ran contrary to personnel management dictums since it increased the numbers of officers in the division. The question of numbers of tanks at platoon, company, and battalion was clouded by uncertain performance of the Ml tank power train in tests then being conducted at White Sands, New Mexico. Resulting compromises produced an organization that was inconsistent with the organizational concepts in mind at the outset. We therefore styled Division 86 as an interim organization with opportunities for further improvement.

Training for all ranks underwent considerable change consistent with developing doctrine. Revised soldier training began in 1974 with one-station unit training for initial entry Soldiers at Fort Knox. A noncommissioned officer education system (NCOES) was implemented over several years. Tied to job performance, it became the basis for assignment and promotion, ending long-standing inequities in unit-level NCO academies. Meanwhile, officer branch advanced courses were halved in length—largely a budget decision.

The Combined Arms and Services Staff School (CAS3) was established at Fort Leavenworth for all officers in the grade of captain in the 1970s. It was designed to make up for the foreshortened advanced courses and to accommodate the more than half the officer population that would not normally attend the year-long CGSC course at Leavenworth. The CGSC curriculum in turn was gradually changed to reflect the need for officer education at the operational level of war. A few select graduates of the year-long CGSC course were allowed to remain at Leavenworth for a second year in the School for Advanced Military Studies program, devoting themselves entirely to in-depth study of war at the operational level.

The resulting product, some seventeen years later, was a U.S. force that went to war as part of the coalition that executed Operation Desert Storm. During those seventeen years, almost everything about that force, especially the Army, had been remade from what existed in 1973 at the beginning of the post-Vietnam recovery.

It is well to remember that Desert Shield-Desert Storm was undertaken at a time when the United States was in the process of "downsizing" in the wake of the collapse of the long-standing Soviet threat in Europe. While the ultimate shape of U.S. post-Soviet national security arrangements was not at all clear, coalition operations to remove the Iraqi Army from Kuwait provided some useful calibrations.

Desert Shield-Desert Storm was a limited war. Its political goal was made clear at the outset and its accomplishment was contingent on joint operations by U.S. forces and coalition allies, and public support at home. When the announced political aim was accomplished, forces were redeployed. Despite some grumbling that more should have been done, the deployed military force did what was set forth at the beginning and then withdrew—for better or worse.

The military force fought outnumbered and won its first and succeeding battles, without invoking the specter of nuclear weapons, and virtually without casualties. The part of the force that brought the war to a successful termination was a corps-sized combined arms mechanized force employed with lightning-like speed and devastating lethality. The mechanized force was supported by deep surveillance and attack systems. Attack helicopter units were integrated into the tactical maneuver scheme, and fighter aircraft protected the battlefield from attack by enemy air forces and provided fire support—deep and close. Everything worked—battle tactics, Abrams tanks, Bradley fighting vehicles, Apache helicopters, the MLRS, JSTARS, and the ATACMS. New organizational schemes proved more effective than their predecessors. Soldier and leader training was more than adequate to the demand. In summary, the equipment, organization, and training designed to support AirLand Battle doctrine was an unqualified success—one might say in spite of and not because of all the compromises and shortfalls just recited.

There are hidden caveats, however. First, there was the mobilization of considerable numbers of small units or detachments whose numbers were nominal but whose skills were essential. Three National Guard mechanized brigades were called to active duty. Tested at the National Training Center at Fort Irwin, they were found not ready for combat.

The annoying and persistent truth about the Army's reserve components, especially National Guard combat and combat support units, is that no one knows what it takes to bring those units to a satisfactory level of readiness. Changing force requirements by the active Army have historically resulted in reorganization of National Guard units—to the end that they are forever unready. However, the reserve components have become so politicized that, despite an obvious need for dramatic systemic change, hardly any change is possible. In former years there was talk of "One Army." Conceptually it was a seamless structure including active, National Guard, and U.S. Army Reserve units—all ready, willing, and able to go to war quickly and effectively. However, concept and reality were poles apart. There was much rhetoric and little accomplishment. More recently "One Army" has given way to the "Total Army." Unfortunately, the conversation/reality gap is wider than ever. This is largely so because in the post-Soviet national security spectrum there are fewer deployed forces, hence increased requirements for ready forces and the means to deploy quickly to a more variegated selection of geography and contingency situations. Systemically, the Army Reserve Component system is a legacy of Secretary Calhoun's expansible Army concept and its eventual implementation (with conscripts instead of militia) and the mobilization system developed as a result of the reforms initiated by Secretary of War Elihu Root in the early twentieth century.

Second, unit performance, especially in Operation Desert Storm, strongly reflected several months of training in the region before the Hundred-Hour War began. Unit training, few if any individual replacements, plenty of ammunition, range area, and time, all enhanced unit cohesion and leadership. After some months, several unit commanders, satisfied with unit proficiency, observed that it was time for the war to begin. Once the Hundred-Hour War began, unexpectedly low casualty rates, especially in ground forces, spelled extremely low personnel turbulence rates and continued excellence in unit performance. Unit effectiveness is ever the victim of the individual replacement system, a fact commented on strongly and adversely after every war. The individual replacement system is a residual of the Army's Industrial Revolution heritage and of Secretary Calhoun's expansible Army idea. In the modern era it is clearly dysfunctional—an unacceptable inhibitor to unit effectiveness.

Third is logistics. Months of in-theater train-up before the Hundred-Hour War enabled the establishment of a robust logistics infrastructure. Logistics support for the U.S. VII Corps and its supporting units in the great end run around Iraqi forces in Kuwait was surely one of the most impressive military logistics accomplishments ever. The period of preparation concealed both the complexity and difficulty of what was done. Since it all worked so well, hardly anyone is now looking closely for buried problems. There are some, however. Problems ranged from the enormous transportation effort to support such an operation over tremendous distances to the vulnerability of large on-ground stores to enemy attack, especially ballistic missile attack. The U.S. Army logistics system is another significant part of the traditional mobilization system that, like the personnel replacement system, begs for change.

Fourth is the research, development, and acquisition process. None of the marvelous new weapons used in Desert Storm appeared overnight. The Abrams tank had its beginnings in the MBT70 program. Conceived as a combined American-British-West German venture, it broke up in the late 1960s because of increasing costs and national differences over weapons and other integral systems. A unilateral U.S. version, the XM803, followed, but was unilaterally terminated by Chief of Staff Westmoreland in the wake of congressional termination of the Cheyenne helicopter program.

A user task force at Fort Knox in 1970–72 drew up the requirements for what became the Abrams. Eventually included was the exciting new British Chobham armor technology. Requirements were rigorously reviewed and amended based on the IDF's experience in the 1973 Yom Kippur

War. Developed by competitive bid to industry, prototypes appeared in a relatively short five or six years, but the audit trail goes back to the late 1950s.

Requirements for what became the Bradley were first laid down in 1964 as a Mechanized Infantry Combat Vehicle (MICV). After substantive changes to armor, weapons, and power train, the Bradley finally appeared nearly twenty years later.

The acquisition system takes far too long, is far too encumbered by a complex maze of regulatory milestones, and is far too vulnerable to the funding whims of a Congress convinced that all weapons cost too much—as does the Defense Department establishment as a whole. These concerns—along with cost, competitive contracting, and ethical practices by defense contractors—head most congressional agendas. While the Goldwater-Nichols Defense Reform Act of 1986 sought to reform the system, the basic systemic flaws remain, despite the reform legislation.

The world has continued to change since Operation Desert Storm in 1991. Indeed, it changes even as we assemble this account of three-quarters of a century of mechanization in the U.S. armed forces. Nonetheless, some things do not change. Several of those have been cited above—for emphasis, more than once. There are others that have not been cited. However, I think it is necessary here to make two additional enduring observations.

The first is about technology. We like to think of ourselves as the most technically advanced nation on earth. Without debating that premise, permit me the following observation. An almost childlike faith in the marvels of modern technology is as much a part of U.S. history as is Manifest Destiny. Techno-utopianism has pervaded the last half of this century. This is true of society as a whole, but it is especially true of defense undertakings because of the size and visibility—and hence the vulnerability—of the defense budget. Techno-utopianism notwithstanding, we seem too frequently to deny out of hand some important achievement of modern technology that offers new capabilities and fresh opportunities. Again consider the tank, the centerpiece of modern mechanized warfare. The U.S. Army's Abrams tank has only one peer in world tank fleets, the Israeli-designed and produced Merkava. The Ml Abrams and its variants embody lessons about tanks and tank warfare learned by tank Soldiers beginning in 1918, when the Army borrowed French light tanks so that Sergeant Starry and his Tank Corps comrades in C Company, 329th Tank Battalion, could cross the trenches and make it through the wire ahead of the infantry. Although considerable zeal has been displayed by those working to improve tanks technically, even more zeal has been shown by those trying to make the tank technically obsolete. The arrival of tactical nuclear weapons in the middle decades of this century was widely heralded as the death knell of the tank. However, with better understanding of nuclear weapons effects, and upon sober reflection, tanks were deemed probably the very best place to be on the nuclear battlefield. The appearance of ATGMs in the 1973 Yom Kippur War—along with considerable misinformation about their contribution to tank losses in that war—was widely heralded as marking the end of the tank in battle. After collecting and analyzing the facts, it still took considerable time and effort to persuade some, but not all, techno-zealots that ATGMs accounted for only about 8 percent of the IDF's total tank losses. Beginning with my time at Fort Knox, after visiting the Yom Kippur battlefields and examining closely the history of tank-on-tank engagements in that war, I created a brief lecture called "Tanks Forever." In it I set forth a concise statement of fact regarding the allegations of the ATGM enthusiasts.

However compelling it may have been initially, it was subsequently necessary to revise and revive that briefing about every four years for the next fifteen years or more.

The second observation is about Soldiers. Modern battle is extremely intense and complex. The next modern battle will certainly be more intense than the last. It will witness increased range and lethality of a growing menu of weapons of all kinds, along with increased complexity of command and control amid a growing deluge of information. Added to this picture are ever more difficult strategic and operational level challenges in a yet dangerous world. As a result, Soldiers and their leaders will continue to constitute the deciding element. The systems people, the nuclear people, airpower advocates, maritime enthusiasts, and information technology wizards all too frequently misunderstand, ignore, or would forget that. Scientists will ever contend they have just discovered some new technology that makes courage, leadership, and Soldier and unit effectiveness obsolete. Let us hope that one day this may come to pass. History, however, instructs that it is unlikely to be so. The history of mechanization as recounted in these pages is a record of the search for the best doctrine, equipment design, force structure, organization, and training for soldiers and leaders. Yet battles will still be won by the courage of Soldiers, the character of leaders, and the excellence of unit training conducted before battle. Nothing in the superb performance of U.S. forces in Operation Desert Storm, indeed nothing in the half-century-long standoff with the Soviets, goes against that eternal truth.

Taken altogether, we have come full circle with ideas set forth in the introduction to this book. In preparing for battle we must ever recognize the challenge of taking available technology and making with it the best force that can possibly be fielded. Clearly, as history has demonstrated time and again, the better tank is the one with the better crew—almost regardless of the level of technical sophistication. On the other hand, technological advances continue to provide new challenges and new opportunities. This relentless technical march can neither be avoided nor set aside. So it is that those who prepare for war must ever be mindful of their obligation to capture and harness technical advances to the doctrinal imperatives of battle—especially at the tactical and operational levels of war. That reality is why we began this anthology. We were impelled by the notion that: "we are . . . not so much in search of dominating technology as we are in search of the intellectual power to understand the possibilities and limitations of burgeoning technology, and the moral courage to step out in new directions. While there is no lack of new technology, the intellectual power and moral courage to use it properly seem ever wanting." That search for wisdom, courage, and leadership must never end.

Operations Analysis for Armor

Armor Magazine
March–April 1970
Co-Authored with Arthur R. Woods

It was the scientist Archimedes who, at the request of the King of Syracuse, set to work finding a way to raise the Roman naval blockade of that city over two thousand years ago. Since that time, as no doubt before, the scientific hand has, more than once, been lent to the art of war. In more recent times, as early as Frederick the Great war games playing out on a map the course of battles and wars received considerable attention. Later some bright American military minds of the nineteenth century brought the science to this country. The Englishman, F.W. Lanchester, in 1914–1915 described mathematically relationships between attrition, firepower superiority, and victory in air war, resulting in the now familiar Lanchester equations. American Thomas Edison, of electric light fame, working for the Navy in World War I made analytical studies of more effective ways for ship convoys to evade submarine attack. Not until the Second World War and after, however, was there widespread application of the scientific method to military problems. Blackett's Circus, a mixed bag of scientists and military persons trying to improve United Kingdom air defenses in Hitler's War, really gave birth to modern military operations analysis, or operations research. And the Atlantic War against the German submarine fleet saw the scientific method applied in operations of the U.S. Navy's Tenth Fleet.

From these beginnings has grown today's profusion of agencies and groups dedicated to scientific evaluation of military problems. Business and industry have applied operations analysis to many activities—transportation, inventory control, merchandising, and servicing; and to the curricula of colleges, universities, and business and military schools. The recent past has seen the introduction of courses to train enough people in a wide variety of disciplines to meet the increasing demand for analytical talent in government and industry.

With respect to solving military problems, it is often said that new analytical methods of the scientific approach outdate more traditional systems of military decision making. There is extensive commentary about: the incompatibility of military thinking and the scientific method; the alleged military preoccupation with subjective and solution-oriented thinking; and the scientist's all too leisurely collecting, sifting, and sorting facts into shaky and inconclusive hypotheses.

The soldier (says the scientist) blunders with precipitate haste into premature decisions with inadequate data, blindly resisting change, even after better alternatives become apparent. The scientist (says the soldier) dawdles and ignores or obscures the commonsense obvious, delaying decision and avoiding responsibility for choice until decision's time is long past.

The military record of operations analysis is spotty, even though the black art had its modern genesis in solving military problems. Too often descriptions of operations analysis methods are in terms of rolls of the dice, turns of the playing cards, or revolutions of the roulette wheel—real, practical military applications being much harder to define and describe. So, while there is a [profusion] of literature describing tools and techniques in fundamental terms, or at best related to business applications where profit or loss statements suffice as measures of effectiveness, there is but limited writing about operations analysis couched in useful military contexts.

Making decisions about military operations is a military business. Of necessity, it seeks problem solution, not so much out of intellectual bankruptcy as from the ever present demand to get the job done—and on time. Military decision making is the unique task of the military man. It combines experience, judgment, and technical competence, acquired through years of schooling, service assignments, and combat duty. Since 1945 an exponential growth of science has made available new and more powerful methods of analysis. These analytical methods are the unique field of the scientist. They combine education, experience, and technical competence in a field of expertness into a set of skills that have been, and can be increasingly, useful to the soldier.

Each year the Army gets more technical and the soldier becomes more reliant on the civilian scientist to help decide what is needed for military tasks, to help design, test, and produce weapons and equipment, and to help find better ways to use military equipment. It is nearly impossible, if not undesirable, to try to train the military man to stand in for the scientist. It is far more practical to educate the soldier into the scientist's ways just enough to combine scientific method and soldierly judgment in more precise statements of requirements; in conducting realistic research, development, testing, and engineering; in analysis of operational methods; and in the inevitable dialogue with the scientist.

This is the story of an association between scientists and soldiers lasting almost two years. It is a good demonstration of the potential of scientific skills applied to operational methods. It will not be necessary to identify units and persons except to say that the authors were parties to the dialogue—one a scientist, the other a soldier; that the setting is a Seventh Army tank battalion; the locale the historic German invasion routes in the Hessian Corridor; the time 1962–1964. The scientists were a varied group of mathematicians, physicists, ballistics experts, electronic wizards, and psychologists, sometimes in platoon strength, who off and on during the period invaded the domain of the soldiers. The soldiers were the standard issue variety who have come to be furnished by the government for an increasing variety of purposes over the years.

Some of the data are hypothetical, but representative—for obvious reasons. Some experiments were not completed, and are so reported. Some ideas presented are embryonic and should be further investigated. All are related, along with completed work, to show the potential as well as the accomplishment. The story is told from the two viewpoints by a soldier and a scientist.

SOLDIER: It was a meeting engagement at point blank range. Blessed by the division commander, whose wisdom (as usual) became more apparent as time wore on, they came with charts, big words, a scientific air, and a proposal to experiment with our operations. There were to be cameras, simulators, recording devices, and much analysis of small unit tank operations. There was a least a platoon of them.

SCIENTIST: We went to Europe that spring to collect, for the first time, good tank duel hit-kill data to use in a computer-run war game analyzing small armor unit combat actions. We had an idea that, with movie cameras strapped to tank gun tubes, we could record gun lay, acquisition, identification, and fire times, gun trunnion cant, and accuracy of lay at time of firing. Thus we could better identify component elements of tank duels and, by further analysis, get more accurate performance data for the war game. We briefed the soldiers on our problem.

SOLDIER: We huddled for a quick estimate. It was a big project, and would eat up training time. But we were scheduled for platoon and company exercises, and some of us felt that we could use their gadgets as gunnery training devices at a time when limited range availability had made us apprehensive about maintaining gunnery proficiency. So we set to work.

SCIENTIST: After some preliminary work with movie cameras we finally settled on 35mm data recording cameras capable of several thousand frames a minute. The camera attached to the tank searchlight mount was activated partly by an assistant instructor (civilian analyst) riding the tank, and by tank commander-gunner firing controls. It took individual pictures of targets at acquisition, during ranging and gun lay, and at the time of firing. At the bottom of each picture was a data panel (Figure 1) on which twelve event marker lights indicated, when illuminated, a specific operation and time of performance. Seven events were marked manually by the assistant instructor, others automatically by commander-gunner controls. Range data were fed direct from the computer.

SOLDIER: The almost perpetual haze that shrouds the Hessian landscape, a year with a dark wet spring, and the inevitable graininess of 35mm negatives blown up to 8x10 for evaluation combined to make it hard, if not impossible, to spot a target in a picture, even though the crew had acquired, identified, and fired on the target. We used up our training stock of TNT for smoke and flash. And, since blank tank gun ammunition was not available, we turned to the scientific platoon for help.

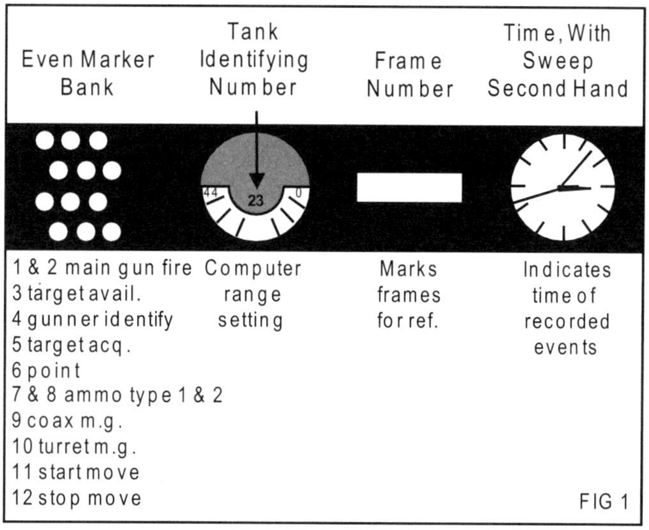

FIG 1

SCIENTIST: Using some 3-inch diameter blank steel stock in 8-inch lengths, we drilled for half the length a half-inch diameter hole and mounted the gadget on a large plank. These we fastened to the glacis plate of the tank, and into the tubes we stuffed some standard flash-bang simulators with detonators hooked to the tank firing circuit. The simulator generated enough flash and smoke to make target identification positive on film and incidentally, we were told, made crew training more realistic.

SOLDIER: Although good searchlights were among our assets, we wanted a way to use the cameras at night, and we needed to know more about night firing without illumination. The problem we put to the scientists was simply to give us a night capability equal to the camera potential in daylight, [then] we could give them their data.

SCIENTIST: This was more difficult. We wanted night firing data, and this was a good idea. After some false starts we hit on the idea of ultra-violet target illumination which would not be visible to the human eye and yet would burn through searchlight illumination to record on film. Ultra-violet light tubes were arranged on targets in "V" or "L"-shaped patterns. Gun lay was scored by indexing with reference to the light pattern on the resulting picture. We tested it and it worked.

SOLDIER: By this time several months had passed and we were ready to try our fledgling training device in field exercises. We convinced the scientific platoon that they needed this kind of data too, and we finally agreed on a series of tests.

- Individual tank in quasi-combat course similar to gunnery table VIII day and night.
- Tank platoons in a quasi-combat platoon course day and night.
- Tank platoon quasi-combat courses with free maneuver against random targets.
- An exercise to develop data about battlefield agility. We had observed that for some reason our gunners' lead applications on moving targets varied a great deal, and that our lead doctrine, or possibly our training, needed examination.

We also felt that if we could not improve film processing times to make reasonably prompt debriefing possible, much of the training value of the camera system would be lost. We were not insisting on the immediate debriefing which was our custom on table VIII, but some reasonable approximation had to be achieved. The scientific platoon had been developing film, printing a set of 8x10s, and scoring each frame with a scoring grid. We suggested that, if they could get a quick processing device, we could settle for projecting the negatives on a screen in the debriefing shack, scoring as we went along. With several projectors we could display a whole platoon situation at one time by projecting identically timed frames from all platoon cameras. In this way we could score platoon coverage of the target area, see how the platoon leader distributed his fires, and at the same time score each individual tank for gun lay and fire time.

SCIENTIST: Hundreds of firing sequences from the first three tests outlined above produced a mass of data which reduced to these statistics:

POSTURE	TOTAL FIRE TIME (SECONDS)
Offensive	12
Defensive	8
Overall Average	10

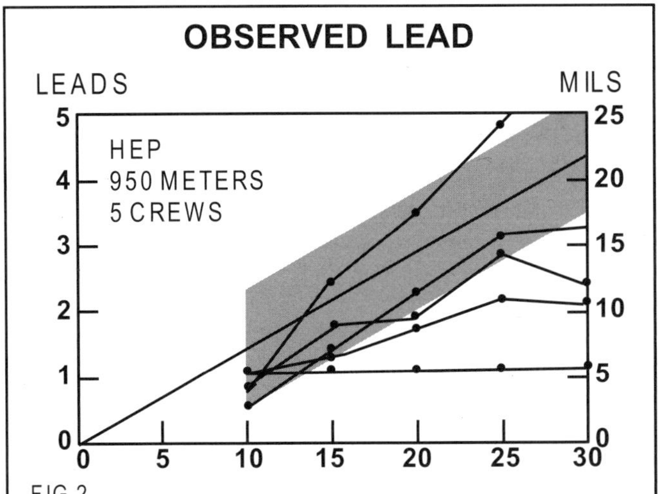

FIG 2

Gun lay data indicated a high probability of first round hit which, coupled with laboratory lethality data, worked out to a 70 percent probability of hit-kill on the first round. There seemed to be wide variance in results, depending on who fired the first round in a tank duel. It appeared from camera data that the tank firing the first round enjoyed about an 80 percent better chance of a kill than its adversary.

Graphical representation of parts of the agility experiment are shown in Figure 2. This portrays data concerning a platoon firing at a moving target. The shaded area represents acceptable lead variances for a lethal hit. Lines on the chart represent actual lead setting applied by gunners as target speed varied.

In night firing tests targets were flash-bangs to the front of the firing tanks, offset enough to require acquisition and gun lay. A lethality grid keyed to target range and type was used for scoring effects of the standard night firing sequences prescribed by gunnery doctrine.

The film processing problem we recognized ourselves from the beginning. Finally we found a suitcase-size commercial film processor to develop film at a rate of six feet a minute. No darkroom was required, and battery power made this an ideal field kit. With standard issue 35mm slide projectors rigged for film strips we could begin debriefing within ten minutes of an individual tank run and within twenty minutes of a platoon run.

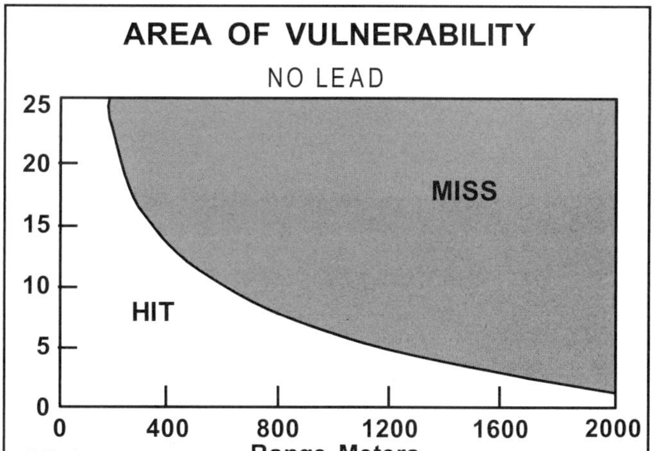

FIG 3

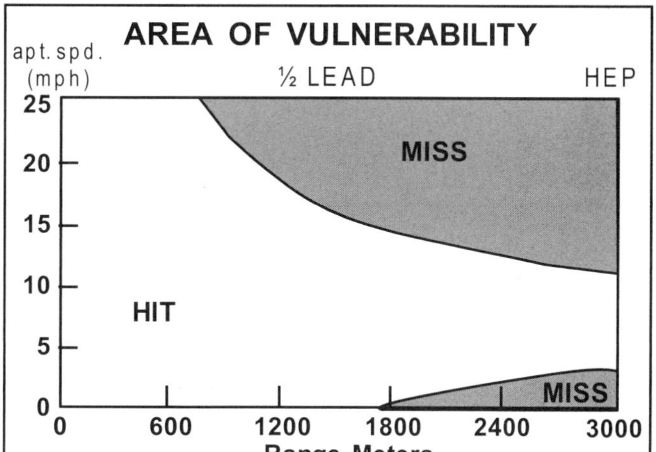

FIG 4

SOLDIER: At the same time the camera tests were in progress, part of the battalion was firing gunnery qualification tables. To verify data from the camera tests we made a complete record of live fire to match the nonlive data. Fire times for initial rounds were slower by more than a second than in camera-recorded sequences, averaging about 11.5 seconds. This is probably accounted for by crew reaction to live fire. Using a lethal target area grid on live fire targets, and the laboratory lethality data, we were able to verify the 70 percent first round hit-kill probabilities suggested by the camera tests. It was of course not practical to verify that the tank firing the first round in a duel enjoyed an 80 percent better chance of kill, although by this time some were willing to suggest that a crew of scientists should oppose a crew of tankers for a live test.

Results of the agility experiment caused detailed analysis and revision of our lead doctrine. We noted that at no reasonable range was it possible to track fast-moving targets accurately. This suggested that in the tank duel, if one is caught exposed, it may be the best tactic to simply turn flank to the enemy and run at top speed for a covered position from which to attack him. For, at apparent speeds over 20 miles per hour, it appeared that there is no gunsight system in the world that could track accurately enough to consistently bring fire into the lethal band. We constructed a series of charts (Figures 3 and 4) to analyze our problem and work out solutions.

The camera pictures also showed clearly a problem we had only suspected before—that of trunnion cant. The guns always seemed canted a little, but how much and with what effect? We suggested an experiment, first to find out what average cant is under field conditions, [then] a live fire test to determine the precise effects of cant on trajectory.

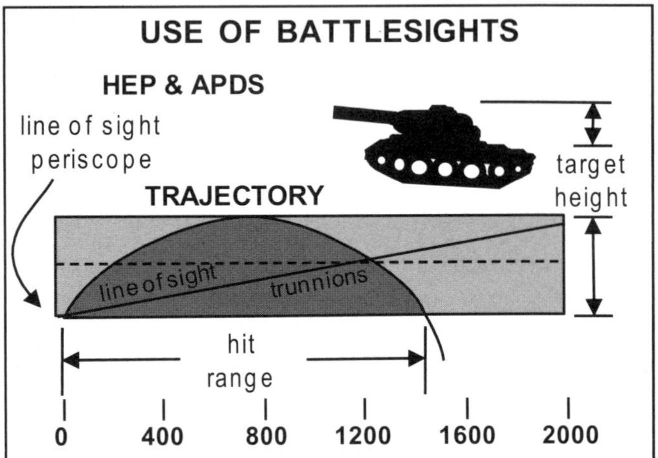

FIG 5

Scoring thousands of firing sequence pictures with the scientists showed us a wide variety of range and intervisibility factors. We knew now how much better it was to fire first, and were understandably anxious to reduce first round fire times. We also knew that tank commander ranging was the most time-consuming element in the acquire to fire sequence. These factors combined to cause reevaluation of our battlesight doctrine. Battlesight is a way of reducing initial round fire time, but the other factors made us wonder if we had the best SOP battlesight. Which setting was better, and how to find it? The problem is shown graphically in Figure 5.

SCIENTIST: To find average trunnion cant we bought several dozen builder's spirit levels, fastened them to commander's cupolas, [and] attached a small acetate-covered board and a grease pencil to each. The commander recorded his tilt at each firing position before moving off. Over several weeks of maneuvering and firing, average cant rounded out to about five degrees. Continued tests showed this figure varied with terrain as one would expect, but that the relationship between average slope and average cant was not linear. In other words, regardless of the terrain, tank commanders habitually sought fairly level firing positions. With live fire we measured the effects of trunnion cant from two to 20 degrees in either direction. Right cant throws the round up and to the right, left cant down and to the left. Rounds with lower muzzle velocities are of course more affected than high velocity rounds.

The battlesight problem was more difficult. By this time the soldiers had sufficient insight into the way we worked to really hold our feet to the fire. The question they posed in this problem is illustrative: "Give us the range finder setting at which we get grazing fire for the greatest possible distance, with the round rising no more than five feet off the ground, which is generally the hull down elevation of an enemy tank." We set up a linear program to analyze thousands of ordinates trajectories at hundreds of ranges for several rounds. But we had to get a computer to work it out. We instructed the computer to optimize range against a primary constraint of an ordinate no greater than five feet.

SOLDIER: Since they varied so widely from doctrine, we tested the computer results with live fire against a spaced array of targets from 100 to 2500 meters, with better results than the computer had predicted. Aiming at the bottom of mass, at a range finder setting of 1700, for example, the computer figured the round to fly no higher than five feet from 150 to 1700 meters. We found by test a much better performance, with the round flying no higher than five feet out to about 2000 meters. HEP, with a slower muzzle velocity, on the other hand, was not susceptible to battlesight and required ranging for each round because of its high looping trajectory. We revised our gunnery doctrine generally as follows:

In the chamber: APDS
Range finder indexed: 1700 meters APDS
Initial fire command:

COMMANDER	GUNNER	LOADER
"Gunner"		
"Tank"		
	"Identified"	*"Up"*
"Fire"	*"On the way"*	

Average fire time:
First round—5 seconds
Second round—11 seconds (total)

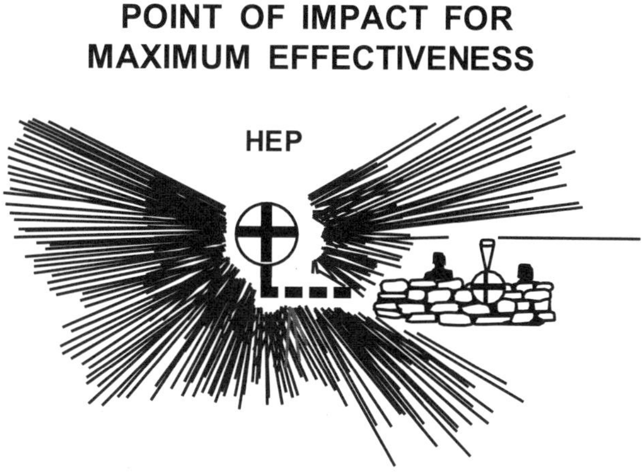

FIG 6

Cutting fire times in half had licked the better part of the problem of how to get off the first round, and thereby increased effectiveness by a factor estimated to be about 60 percent. Should the initial target require ammunition other than APDS, we decided to fire the round to get the advantage of firing first, [then] change ammunition and index or range for the second round.

Our evaluation of trunnion cant effects showed that for faster rounds they were generally within expected dispersion patterns for normal ranges, so we decided to ignore them. With slower ammunition the problem was complicated by the unique bursting characteristics of the HEP round. A base detonating fuze on the round we were using caused much of the lethal spray of the round to go to the rear and sides of the point of impact. Before adjusting for average cant we wanted to have an integrated picture of the total effect of cant and burst characteristics, especially on horizontal targets. So we asked the scientific platoon to instrument a test firing and give us the data.

SCIENTIST: To instrument the test of ammunition burst characteristics we fabricated targets shaped like personnel in dug-in antitank positions. Target cloth envelopes were sewed up in appropriate form and by insertion and inflation of a large balloon we created a target with area, volume, and resistance to fragmentation approximating those characteristics in the human body. Body count we got by counting deflated balloons—those which fragments had penetrated sufficiently to rupture the target cloth envelope and the balloon skin.

SOLDIER: Recognizing only the bursting characteristics, maximum effectiveness could be gained by aim off as indicated in Figure 6. The cant problem was analyzed by studying a series

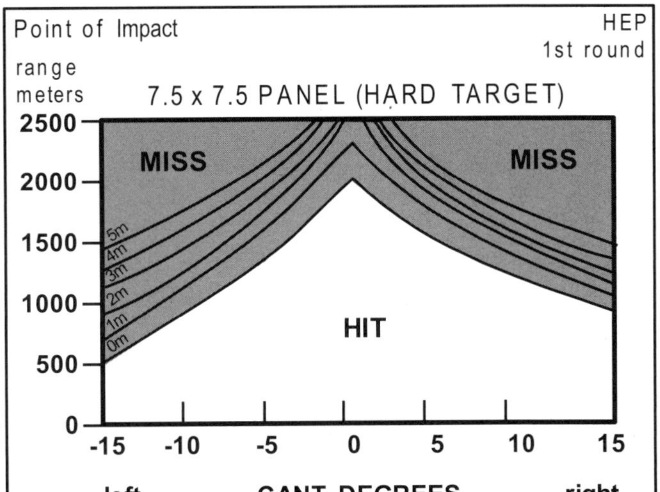

FIG 7

of charts like Figure 7. We were working on a reticle calibrated in oranges, apples, lemons, or similar geometric figures, and fixed to indicate where to aim for average cant condition when the project had to be shelved. More work needs to be done on this problem.

About this time the helicopter was being widely recognized as a machine of great potential—especially in Vietnam. Having no experience or data to evaluate, we asked if cameras could be used in helicopters and on tanks to record the events of tank versus helicopter duels. By this means we hoped to get a preliminary view of what problems we might encounter fighting an enemy equipped with aerial platforms of the helicopter variety. So we put this problem to the scientists.

SCIENTIST: We were fascinated by this idea. No research had been done, and little if any real data was available. We resurrected the movie cameras with which we had begun our tank duel experiment, mounting one on the barrel of the tank's cupola machinegun and another in a helicopter weapon system mount. With a hundred or so runs of helicopters against tanks in a variety of tactical situations we began to develop fairly consistent statistics and hit-kill probabilities. More analytical work needs to be done on this subject. Since we were near the end of the time available for field experiment, we were unable to explore this any further.

SOLDIER: Helicopters got some good shots at tanks, and tanks were quite effective against helicopters. This experiment convinced us that we had better potential against aerial platforms than we had been led to believe, and that it was probably better to fire than to hold fire against helicopter-like platforms. It certainly showed us the dangers of not looking up as well as all around, and it demonstrated the vulnerability of some of our formations to attack by aerial weapons platforms while moving, and from the rear 120 degree sector. We acknowledged these facts in revisions to the training program, and by adjusting our battle drill combat formations.

SCIENTIST: Our time was up—in fact we were overtime. We had stayed on to investigate the new avenues that seemed to open up all the time. We had stimulated more ideas than we could handle, more knotty little problems that needed investigating, and were more than ever convinced that good scientific analysis is fundamentally the product of good real world performance data. But our data bank was full, thanks to an imaginative bunch of soldiers who were willing to try anything, and who many times showed opportunities for investigation that our more stereotyped vision just overlooked.

SOLDIER: Detaching the scientific platoon was a reluctant process. We had come to appreciate their tremendous capability to analyze our problems. Summing up, we were convinced that the camera system had potential as a training device. It permitted precise measurement of performance without firing a shot, and within reasonable tolerances of actual performance, at a time when training resources—area, time, and ammunition—were all in critical supply.

At the same time we were equally convinced the 35mm data recording cameras were much too sophisticated. They were not the kind of equipment we could be expected to care for properly without additional skills, equipment, and personnel. They had capabilities far exceeding what we needed for training. In fact the old reliable movie cameras, with a little improvement, could serve our purposes very well. We had also investigated miniaturized television cameras, available commercially, for use as data recorders and for instantaneous assessment of performance, but this had to be set aside for the moment.

The scenarios of all tactical tests designed to gather data had to be drawn very carefully and with precision, to include measuring ranges, siting targets, positioning equipment, and rehearsing enemy scenarios beforehand. While initially we saw this as burdensome, it came through clearly before long that, the more meticulously training is planned, the better is the training. It should not be necessary to relearn this lesson, for it is much taught, but the idea of precision in planning, precision in measuring performance, and precision in analyzing effectiveness stuck with us. It permeated everything we did tactically from that day on, forming the basis for platoon and company test exercises, for equipment evaluation, and even for an analysis of performance of crew and small unit commanders. It caused us to reevaluate at one time or another almost everything we did. Better load distribution of our bulk fuel and ammunition loads was the product of a linear program set up for us by the scientific platoon, but which we were able to operate ourselves once they were gone. The prescribed load list of spare parts was the subject of a thorough analysis, though this project had to be set aside in the face of insurmountable administrative barriers impeding implementation of what we had found necessary to improve. A scheme to navigate cross-country at night using a radio station grid system was put aside, but promised great potential. The battalion training program underwent a thorough analysis in which we found such a lack of available human performance data that we had to collect data to provide the basis for analysis and design of a new program.

In short, the scientific platoon had been quite an asset. They taught us to apply the scientific method to military problems, and to temper it with a little military common sense—an invaluable combination. We learned we could make many analyses ourselves by just collecting good hard facts, and with them make some cupola estimates. The lesson was clear. We had a good synthesis. The scientists provided the technical knowledge and understanding of how to collect, relate, and interpret data. The soldiers contributed the sense of urgency, of purpose, of insistence on clear goals, the knowledge of how human beings and machines behave in battle, and a certain hopefully useful cynicism about the relationship between theory and practice in everything to do with war. We learned to tolerate one another's idiosyncrasies in the search for well reasoned, properly documented, soundly analyzed, and thoroughly practical results, results to improve our combat effectiveness.

Sheridan Critique I

Letter to Colonel Lane Carlson
***Soldiers* Magazine**
11 May 1972

I have read with great interest your two-page spread on the nonmetallic cartridge case for the 152mm gun-launcher in the M551 Sheridan on pages 44–45 of *Soldiers*, May 1972 issue. While you undoubtedly had this spread cleared by the appropriate authority, I have to take issue with it from a standpoint of the facts it presents and the credibility problem it raises.

Your article indicates that the caseless cartridge is relatively insensitive to jolting, jarring, dropping, and so on. This may be the case in whatever tests have been conducted, but is certainly not true in combat where fire and explosion hazards presented by this cartridge case make it extremely dangerous.

The fact is that the nonmetallic cartridge case usually breaks up when the vehicle hits a mine, and always shatters when the vehicle is hit by armor-defeating ordnance which penetrates into the area of the ready racks. In combat the round is really quite frangible. In addition, once it breaks up the nonmetallic cartridge case becomes, in effect, a collection of powder pellets on the turret floor. These explode and burn from the heat and/or blast of a mine detonation or from penetration of the vehicle by armor-defeating ordnance. Sheridan crews are extremely sensitive to the danger of fire and explosion, and in combat simply do not stay with the vehicle after it is hit. In my experience in Vietnam, while commanding the 11th Armored Cavalry Regiment, every Sheridan hit by an RPG exploded and burned because the ammunition broke up and detonated within 15 seconds, destroying the vehicle and seriously injuring or killing the crew. In almost every case of a mine hit one or more rounds in the ready rack fractured, and in several the detonation vented the hull and ignited the fractured cartridge cases.

In addition to combat environmental problems, a rigid "no smoking" rule has to be invoked in all environments—training and combat—with caseless ammunition. While it is a good idea to have no smoking rules around any kind of fuel-consuming equipment, soldiers will inevitably violate the rule. With the nonmetallic cartridge case smoking in the turret inevitably causes fire, even if the rounds are bagged. The nonmetallic cartridge case is therefore a long way from being soldier-proof.

So I consider your spread on the nonmetallic cartridge case to be misleading and nonfactual. Soldiers who have fought in the Sheridan will read it and it will become another case of lack of credibility in an official Army publication. Soldiers who have never seen a Sheridan will read it, talk to their buddies who have fought in a Sheridan, and the Army will lose credibility with them too. I understand the R&D community's commitment to the idea of the nonmetallic cartridge case, but from a practical standpoint it is beyond the state of the art at this time, and from the user's point of view it is a substandard piece of combat gear. Therefore the less we say about it the better. And, if it is necessary to say something, tell it like it is!

The Role of Cavalry

Blackhorse Association
Fort Knox, Kentucky
20 May 1972

. . .

The Vietnam experience created considerable confusion about the role of infantry in our Army. Many now believe that the role of infantry is as a reconnaissance force whose mission it is to find the enemy, then destroy him with external firepower. Experience proves that there is no substitute for a force that can and will close with and destroy the enemy. If infantry no longer has this mission, what force does?

Vietnam experience also proved armored cavalry to be much more than the reconnaissance and security force doctrine prescribed as its primary role. The cavalry role in Vietnam was as a combat maneuver force; it was the most cost-effective such force on the battlefield in terms of killing power per man exposed to hostile fire.

Extrapolating from Vietnam, it appears reasonable to conclude that, with the possible exception of a major war against the Soviets, armored cavalry is the most capable, and in the long run the most economical, force to meet requirements of the President's coalition strategy.

There are several reasons why this is so. Cavalry is a ready-built combined arms team, integrated at levels most likely to be needed in combat with almost any enemy except the Soviets. In terms of capability to fight in almost every environment, cavalry is a true economy force against all but a heavily armored enemy. The armored cavalry regiment, with a couple of battalions of airmobile infantry, can control an area of operations normally requiring a division for less than half the manpower, equipment, overhead, and dollars required by the division-sized force.

In terms of the problems that confront the Army today—the confusion of experience, the confusion of values—cavalry training, organization, and methods of operation are tailor-made systems for problem solution. There are several reasons why this is so. Cavalry is the best environment in which to train young officers and NCOs in the integrated employment of combined arms at all levels of command, lessons the young officers especially will desperately need later in their careers.

Cavalry is the greatest challenge to soldier and officer alike in terms of variety of experience, range of work to be done, and professional proficiency required in performance of a variety of missions and tasks. Cavalry is a ready-made environment for testing new doctrinal concepts relating to the mix of combat arms in the combined arms team, especially the ground-air mix, and for testing new equipment concepts for all combat arms. Cavalry is a ready-made capital intensive force where improved technology has made and can make a more effective force for the same or less manpower.

Cavalry units present the single greatest challenge to leadership of any Army organization in terms of variety of mission, organization, and equipment. There is no better leader training ground than a cavalry organization. However, cavalry today is largely organized and equipped to refight World War II. We need to look long and honestly at our organization, equipment, and doctrine to better configure ourselves to meet today's demands. The following hypotheses highlight several dimensions of the challenge.

If the role of cavalry is to be primarily that of a combat maneuver element, then there is a need to rethink the roles of its components, especially scouts and infantry. In all probability we should combine these two elements into ground reconnaissance teams who fight mounted for the most part, but who can also dismount to patrol, ambush, and so on.

The third dimension—air cavalry—needs to be tailored to the demands of the mission environment and to recognize the limits imposed by our aircraft maintenance policies. If there is a normal operational requirement for x gunships and y scouts daily, then there must be 1.6 of each assigned to the air cavalry fleet to allow for maintenance and backup. The mission of airmobile infantry in air cavalry should be reassessed. If there is a need to secure downed aircraft and to fight as airmobile infantry, aerial rifle elements might be organized quite differently than if they were to perform one or the other but not both missions.

The need for more infantry, and the role of the infantry in cavalry units, must be restudied. The basic decision that has to be made relates to the role of infantry in the Army as a whole, then that role has to be analyzed in the context of the cavalry mission. Is infantry only a recon force, or is it also to close with and destroy the enemy?

Cavalry units all have a vast communications capability, both AM and FM. Now that tactical FM is nearly as capable as AM, the whole command-control-communications setup could be revised, eliminating the AM capability and saving thereby an enormous amount of equipment, training, manpower, and money.

The cavalry vehicle family must be integrated. No longer can we afford to develop one type vehicle without carefully ensuring it is compatible with other members of the family in which it must fight. Before we develop a scout vehicle and an infantry vehicle separately we should resolve the role-of-infantry problem. There is a strong possibility that a single piece of equipment could serve both elements equally well, even if their roles remain different, but more especially if their functions are combined.

Cavalry needs tanks, but not medium tanks whose operating characteristics are so widely apart from other vehicles in the cavalry family. What kind of tank should this be? What weapons should it mount?

The vehicle recovery fleet in cavalry units is inadequate to the demands of recovery operations in area war. Cavalry units need more, and more effective, recovery vehicles. The cavalry supply vehicle fleet must be compatible with the rest of the fleet and integrated into the organization. The cavalry DS maintenance capability needs to be organic, but should be decentralized to eliminate one echelon and make DS ordnance units primarily supply activities. Maintenance standards for cavalry should be developed to optimize fleet OR rates. It is no longer appropriate simply to judgmentally set 85 percent as an acceptable OR rate, for this may not be the optimum rate at which the supply system operates.

After every war in our history, probably starting with the Civil War, we have reorganized and restructured to better fight the conflict just finished. and so it is today. However, the contemporary environment is significantly different. We are now faced with a change in national strategy of significant proportions, requiring reevaluation of our concepts of military organization, mobilization, and deployment, forcing us to find a whole new role for the Army. Historically our geographical isolation has given us sufficient time to think the problem through and develop solutions almost at our leisure. Distance will no longer buy us time. Our most powerful enemies are but seconds of missile flight away across our horizons.

More than 30 years ago a group of visionary and imaginative Army officers, drawing on their own experience and thinking, on their observations of British and German armored experiments in the 1930s, and on observations of early German successes with armor in World War II, devised basic United States doctrine, organization, and tactics for mounted combat. We have not added much of substance to their bold innovations in the ensuing 30 years. The time has come for us to move ahead with imagination, honesty, and fortitude—drawing on the lessons of mounted combat in World War II and Vietnam, and on the changing United States posture in today's world, to design forces and concepts of force employment that will meet today's and tomorrow's needs. My message is that doctrine, concepts, and organizational and tactical principles embodied in armored cavalry are central to the solution of many if not most of the problems with which we are confronted.

Sheridan Critique II

Letter to Major General Winant Sidle
Chief of Information
23 May 1972

I appreciate your 19 May response to my 11 May letter to *Soldiers* regarding the nonmetallic cartridge case for the 152mm gun-launcher. Also I am aware of the Army Materiel and R&D community claims that everything is now A-OK; they made the same claim about the rounds we were using in Vietnam. As best I can find out there has not been any change in that particular cartridge since the experiences which prompted me to raise a voice in protest. Also, since we no longer have Sheridans in combat, the RPG and mine hits which caused the difficulties I alluded to have gone away—as the developers allege, the thing works fairly well in training.

Since the materiel people as proponents of the round and the combat folks as users of the round will never agree, I am extremely reluctant to start an open dialogue that really is not capable of compromise. As a commander in combat I've simply lost too many good men to this substandard piece of gear to believe any claim the materiel people ever make about it. What I was trying to suggest was that the feud between user and developer is so bitter on this point that *Soldiers* should not be a party to any effort by the development community to whitewash their product. And I am not willing to be a party to letting them respond that what I complain about has been fixed—I don't believe it—nor can anyone who has seen a Sheridan hit by an RPG or mine and explode within seconds, killing or maiming the crew. My own very parochial and admittedly emotional view is that had I the authority I'd prosecute those who persuaded us the round was safe to use in the first place, and turn my deaf ear to the same group now claiming they have fixed up their original booboo.

Tanks Forever

Armor Magazine
July–August 1975

Everyone is talking about tanks.

Armor soldiers—users of tanks, examining modern battle, view the tank as a multipurpose weapon with a variety of essential combat capabilities. The October War confirms their views, and demands improvements in tank capabilities. Other observers—budget analysts, antimilitarists, skeptics—for a variety of reasons, view the tank as an anachronism, a system rendered useless by recent advances in numbers and effectiveness of long-range antitank systems. The October War, they say, suggests that tanks can now be replaced by large numbers of antitank guided missiles (ATGMs).

Are tanks necessary, or are they not?

In answering this question, two things must be said.

First, modern war is a contest of measures and countermeasures. For every modern weapon system, there is an effective countersystem. For aircraft, there are surface-to-air cannons or missiles; for tanks, there are other tanks and ATGMs; for artillery, there is counterbattery; for infantry, there is direct and indirect fire suppression by tanks and artillery.

It is quite like the children's game of "rock, scissors, and paper." Rock breaks scissors, which cut paper, which, in turn, covers rock. The goal in battle is to apply the tactic which best utilizes the capabilities of each battle system, while minimizing its vulnerability to countermeasures. As in the "rock, scissors, and paper" game, a mixed strategy enables a win. We do not refuse to play the game just because each tactic has an effective counter.

Armor soldiers have never viewed tanks as a self-contained battle system; tanks have always been a part—an essential part—of the combined arms team. We learned this lesson at Cambrai; it has been reinforced by every tank engagement since. No one denies that on today's battlefield, unsupported tank attacks face mass destruction from accurate and lethal ATGMs, as well as from other thanks.

Therefore, the question really is—are tanks a necessary part of the combined arms team?

Second, tanks were created in an attempt to restore mobility to battle, enabling the side using them to seize the initiative from the enemy. Tanks were the first element of the combined arms team to become other than foot or horse mobile. However, the essential lesson of the need for and value of mobility as a means to seize initiative was drawn from lessons history taught about the effectiveness of mobile cavalry, dragoons, horse- or elephant-mounted infantry in battle. Therefore, the question really is—are tanks necessary as a part of the mobile weapon combination to seize battle initiative, or can some other systems do the job?

How, in modern battle, would an army fare that did not use tanks? While the answer to this is a function of threat and environment, modern war games show that a force in which tanks are either not present, or present in insufficient numbers, simply cannot fight successfully against an enemy equipped with even a modest number of tanks. Light infantry units equipped with the latest ATGMs are only marginally effective against armor. It is necessary to balance the

---

combined arms team in order to have sufficient staying power, and enough mobile integrated firepower to wrest the initiative from the enemy. In summary, we don't fare well without tanks in the combined arms team. Tanks are necessary.

Survivability

Can the tank survive? Again, this depends on threat and area; but what concerns us all is the allegation that modern ATGMs have driven the tank from the battlefield. There is no question that when tanks are employed alone against a combined arms force in terrain such as that in Europe, or the Mideast, their survivability is greatly reduced.

In the early stages of the October War, when the Egyptians crossed the Suez, and the Israeli Defense Force (IDF) was trying to contain the crossing, a pure force of 50 IDF tanks lost 40 in a local counterattack against an Egyptian defense based on dug-in *Saggers* and *RPG-7's*.

When the IDF crossed the Suez, and attacked the well prepared defense, it was with a combined arms force, using time proven combined arms tactics.

In the breakthrough, the IDF lost 25 percent of its attacking tanks, destroying 30 percent of the defending Egyptian tanks. When the cross-Suez battlefield became fluid, the IDF, without ATGMs, destroyed 90 percent of the defending Egyptian tanks with no IDF losses, sweeping up the canal banks, destroying Egyptian ATGM positions where crews had been destroyed or driven from their positions by suppressive fires of artillery or by infantry.

How well can an individual tank survive a hit from another tank, compared to a hit by an ATGM?

We know that overall, our tanks have a higher probability of surviving a hit from a Soviet *Sagger* than from a kinetic energy round fired from a *T-62.*

But the fact remains that the most lethal antitank weapon on the battlefield is the high-velocity tank cannon, and within range, tanks defeat tanks much better than do ATGMs.

How well can an ATGM survive on the modern battlefield?

Studies tell us that a division subjected to a 45-minute artillery preparation can expect to lose 25 percent of its ATGM teams. The tank's armor protection makes it relatively invulnerable to artillery fire.

How effective are ATGMs?

Although antitank guided missiles are generally considered to have high hit probabilities at ranges from 500 to 3,000 meters, experience in the October War does not reflect a high hit probability. It is estimated that several thousand missiles were fired at IDF tanks, yet *at most* only a modest number of tanks destroyed were victims of missile hits.

It is often said that antitank guided missile systems are much cheaper than tanks.

On the surface, tanks are much more expensive. For example, suppose a tank costs $400,000, its kinetic energy (KE) round costs about $150, and the tank can fire about 400–500 rounds before the gun tube needs replacing. This gives a very rough cost of about $470,000 for the tank, or about $1,000 per KE round.

But suppose an ATGM costs $3,500. The hardware cost of 400–500 missiles, including one launcher system mounted on a vehicle[,] would be about $2 million or $4,000 per missile. Thus we could fire almost four KE rounds at a target for less than the cost of one guided missile. Therefore, the cost of the several thousand antitank missiles fired in the October War could have bought 40 to 60 tanks, a force which could be used on a variety of missions rather than a special antitank mission.

Furthermore, the additional tanks could fire enough kinetic energy rounds to have a high probability of hitting more than 5,000 enemy targets.

Properly employed, the tank not only can survive on the battlefield, it will dominate the battle. Without tanks, we don't accomplish much against an armor threat. The tank can survive better than an antitank missile system. Tanks are a relatively less costly way of destroying the enemy than are antitank guided missiles. As a bonus, because of their relative invulnerability to small arms and artillery, there will be fewer personnel casualties among tank soldiers than among antitank guided missile crews.

Secret of Winning

It is the nature of our democracy and its armed forces that U.S. Army units deployed or sent overseas at the beginning of a war can expect to find themselves outnumbered. For a number of reasons, it should be expected that future wars will be shorter and more violent than in the past, and their successful outcome will depend considerably on results of those first violent battles. Winning the first battle(s) is critical, and they will have to be won by U.S. Army forces fighting outnumbered.

Now, critics of the tank say that in modern armor battle, our probable adversaries outnumber us so grossly that we can't hope to match their numbers. While this is true, we believe it should be regarded as an advantage rather than an encumberance. This is so because tactical systems of our major potential adversaries have as a basic premise that the side with the biggest numbers must inevitably win. To support this view, they quote work of the distinguished British mathematician, F.W. Lanchester, whose early work in aircraft combat exchange ratios still provides the mathematical bias for many weapons systems exchange models—tank as well as aircraft. Unfortunately, more recently developed mathematical descriptions of combat exchange ratios arrive at a Lanchester-like outcome, even though by a different process. All these models predict, by whatever method, that the side that enters the fight outnumbered is foredoomed to defeat. While all this is academically interesting, and perhaps even logical, careful analysis of several hundred tank battles tells us that their outcomes defy predictions of any existing mathematical methods of combat.

In fact, it appears that the side which is outnumbered wins more frequently than not, and that probability of victory seems to hinge more on which side manages to use its mobility to best advantage to seize the initiative.

Therefore, we believe it most fortuitous that our potential adversaries have concluded that the ultimate truth is in Lanchester. For as in a thousand other battles, the IDF in the October War demonstrated again the bankruptcy of that idea.

The secret to winning is not in numbers. Mobility provides the means to mass in time and place arriving at a reasonably matched force ratio, say three, four, or six to one. Then by intelligent use of terrain and mobility maximizing one's own capabilities while at the same time minimizing one's own vulnerability, exchange ratios of six to one or better can be achieved, and indeed should be expected. Six to one exchange ratios demolish the neatness of Lanchester's squares, and are indeed quite in line with what really happens when masses of armored vehicles lock in mortal combat in situations in which the stakes are high.

And so tanks are essential, essential to the combined arms team, whose task it is to win the first battle(s) of the next war—win outnumbered, win by extracting from the enemy's overwhelming hordes exchange ratios of five or ten to one. It will not be easy. It can be done. It requires a certain cleverness, obstinacy, persistence; even more, it demands a thorough understanding of the dynamics of modern battle.

Improved Fighting Ability

It is becoming increasingly apparent that we could be much more productive were we to concentrate on how to improve the fighting ability of our mobile combined arms team, instead of spending the inordinate amount of time now dedicated to proving to antimilitary skeptics that we need tanks at all. Proceeding along this line of reasoning, *what needs to be said?*

Traditionally, we have begun speculations about what to do next with any armament system with an analysis of what systems our potential adversary will have in the field. This global mindreading is called "threat analysis." It tries to read the minds of a group of men who probably haven't made up their minds yet. And so the further away from today one goes, the less useful this process becomes.

What is most instructive is to begin a "whither tanks" study with a technical analysis—a systematic evaluation of state-of-the-art developments in a number of technologies. What is the purpose of this study? It is to sum up where we are, and where we might most profitably go by pursuing one or more technical approaches.

Let me be specific. In the field of gun-ammunition, it now appears that we have the technical capability to produce armor that can defeat chemical energy rounds which depend on the shaped charge for penetration, in diameters that can reasonably be used on a mobile weapons platform. True, with a 10- to 12-inch diameter cone, even advanced armor might be penetrated. But even the most voluble tank enthusiast would probably be reluctant to suggest a gun that large. So what this tells us is that our technical problem is now to optimize kinetic energy systems that can defeat modern armor. For if we have the armor technology, we must assume our major adversaries have it.

We also know that a kinetic energy system can be optimized using advanced penetrator design and materials technology, and that it can be done in calibers smaller than those now considered necessary.

Propellant technology analysis suggests that we can exceed burning rate limits imposed by today's powders, and by so doing increase penetrator velocities, and thereby penetration ability.

Therefore, technology analysis tells us it is both necessary and possible to build smaller, lighter, gun-mount combinations with much improved lethality.

A look at fire control technology suggests that we can provide our smaller, more lethal gun-mount combination with vastly improved fire control capabilities, rounding out the range-lethality equation. If we can increase the hit performance of tank cannon in the 1,500- to 4,000-meter range band, then the utility of ATGM systems will have been considerably degraded, and the old "rock, scissors, and paper" game has to be played again—with new rules.

Further, if we can mount such a system as I have just described on a more agile platform, the system itself could be more survivable, and therefore more lethal.

Again, technical analysis tells us that we have probably exhausted torsion spring technology, and that if we are to dramatically improve the way a vehicle meets the ground, some other technology has to be explored. We also know that technically, hydropneumatics, especially hydropneumatic energy storage systems, may offer a new agility dimension—hyperagility. For years we have insisted on higher horsepower-to-ton ratios as a means to greater agility. However, in World War I, my father's tank outfit marched to battle at speeds about the same as today's tank battalions, despite tenfold increases in horsepower per ton.

So we have to start asking the right questions—questions about agility and how it relates to survivability. Thus we must explore battlefield intervisibility segments—lengths, discontinuities, silhouette heights, acceleration rates in the low speed band, and other agility related parameters, in the end describing survivability in terms of ability to escape enemy fire control systems. Once this is done we can perhaps marry up our new, more lethal gun-mount system with a more agile, survivable platform—a tank for the year 2000.

While we are doing all this, we can reasonably expect others to be doing the same thing. So by the year 2000 we can expect to find ourselves, as we are today, with competing hardware systems which, despite some differences in sophistication, are relatively equal in battle.

Meanwhile, we can expect that for every *tankerbangerboomer* someone develops, there will soon appear an *antitankerbangerboomer*, and so "rock, scissors, and paper" is a game destined to continue. And about this phenomenon we must make one final observation.

The clear lesson of war is that in the end, the outcome of battle depends on the excellence of training, the quality of leadership, and the courage of soldiers. It is also quite clear that the side that thinks it will win, usually does.

Conversely, the side that thinks it may lose, or whose soldiers are not convinced that they can and will win, regardless of the odds, usually loses. We simply cannot permit ourselves to be seized with the defeatist malaise which underlies the antimilitarist dialogues now in vogue in our country.

For the U.S. Army must confront its foes in the first battles of the next war with soldiers whose state of training, whose confidence in themselves, and their leadership, whose confidence in the excellence of their equipment and tactics, and whose understanding of the dynamics of modern battle are such that they can fight successfully at odds of ten to one or more and win. Win through excellence in training, tactics, and weapons employment. Win because they are better led, and because they are convinced they can win the first battles, win outnumbered, win using the combined arms team built around tanks.

Tank Platoon Organization

Letter to Colonel Edward P. Davis
Fort Knox, Kentucky
10 May 1976

Tank platoon organization. I talked with General DePuy about this, but believe your observation to be correct—he is convinced we have to look at organization from platoon up. He told me that he's going to develop a paper to present to the Chief about how armored forces should be organized to fight in the 1980s. He is also convinced, I'm certain, that a platoon of either three or four tanks is better than a platoon of five. In this he is persuaded by the Israeli, supporting arguments provided by the Brits, Sovs, and Germans. He said to me he is convinced our lieutenants can't successfully command five tanks on the modern battlefield; that the company commander has to become the primary controller of maneuver-fire; and that the battalion commander must become much more intimate with the battle. He may be right. He certainly is right about getting company and battalion commanders more closely involved in the fight—but it doesn't necessarily require reorganization of the platoon to cause that to come about. There's more, but you get the idea.

For all those reasons, as well as the fact that I had sort of struck a bargain with him that if he'd support my position on the cavalry, we at Knox would tackle tank and mech organization next, I suspect he'll press on. Therefore the decision you all must make is whether you'll lead the way or follow along. Unless something has happened since 15 February, I suspect you are still strapped for people, but he'll do it at Monroe or Leavenworth if you demur. The serious problem will be one of judging relative effectiveness of tank organizations under one or the other schemes. No model exists sensitive enough to measure command-control of fires and maneuver at platoon level. So it will be a force structure exercise, and the cost-benefit equation will be largely driven by the larger dimensions of structuring considerations, whether or not any clear case can be made for five, four, or three tanks per platoon. Nonetheless my personal judgment is that Knox ought to lead the way—if you don't, he'll give it to someone else, and that's worse than the work you all take on by picking up the tab.

Tank Unit Organization

Letter to Lieutenant Colonel Dale K. Brudvig
Carlisle Barracks, Pennsylvania
18 May 1976

Your conclusions of course lead to the larger issue of how the combatant elements themselves should best be organized to fight, then how we should support them. The Germans are doing some interesting tests with combined arms battalions and are beginning to test battalions with various numbers of tanks in each company—a function of how many per platoon. Theirs is a political problem with the force structure, but at the same time the military guys are seriously trying to find out whether one number is better than another. We must get on with this ourselves one day, for it is probably true that lieutenants, captains, and lieutenant colonels can't control the fires and maneuvers of units as large as our companies and battalions on battlefields as dense as those of the Golan and the Sinai.

Tank Platoon Organization and Training

Letter to Arthur R. Woods
Fort Hood, Texas
16 August 1976

By means which I cannot disclose I have come into possession of a paper you assembled with regard to testing at TCATA incident to the TRADOC Division Force Structure Study—specifically the matter of three versus other numbers of tanks per platoon. The paper weighs just under a pound and a half. Following the "Starry Law of Inverse Bureaucratic Squares" which says that the quality of the work varies inversely as the square of its throw weight, throw weight being of course a function of the energy required to get the package into a suitable receptacle, I have some observations. . . .

I need not tell you that you must somehow measure effectiveness of command-control. Nor need I point out that this will be most difficult. However, the test must be designed so as to measure something relating to the hypothesis on which the force structure itself is based. That is that the battlefield is now so dense with high quality systems, especially where the main battle is being fought, that the latitude of action afforded lower-level commanders—platoon, company, and battalion, is considerably reduced from the latitude allowed before—and on the basis of which, know it or not, our present platoon, company, and battalion organization and command-control arrangements are based. Put more simply, the platoon leader will probably no longer have the time, nor will he be capable of controlling the fire and maneuver of a two-section platoon. His platoon must now fire and maneuver as a unit, directed by the company commander in concert with other like platoons. Now, having said that, there must be established a requirement against which to measure the relative performance of the two types of organization. And there must be made some judgment about how much is enough when it comes to the organization's ability to meet the requirement. In that context we here are using the target servicing concept as an analytical tool not only to describe the battle in each battalion battle position, but to describe how many of what kind of systems we must have in that position to service targets, how long they have to get it done, how the fight must be commanded and controlled, and many other factors.

Target servicing simply says that, given the way the enemy says he will fight, we can expect on a given piece of terrain a certain target mix of a variety of systems to arrive to be serviced—within range, at a certain rate. It is really a big queuing problem and is therefore amenable to partial solution by at least some familiar techniques. The question is therefore how the servicing system reacts to the arriving queue—and that's what you have to measure.

The present organization can be made much more responsive to the demands of target servicing if several things are done. For example, if you cut out the company net control station and require all company combat systems to operate on the same net, a significant improvement in command-control can be made. I have seen battalion-size units immobilized both in training and in combat because they let net control operators run the system. There was so much useless chatter in the net that commands could not be passed. Platoons should not use their platoon nets except under circumstances wherein they are necessary to command—night, bad weather, reduced visibility. Over the years we have developed the platoon net as a full-blown part of the command-control system, when in reality it was not intended for that in the first place. The battalion has the same problem with the NCS. So don't go out and measure a system that is operating inefficiently, although according to SOP, without first considering what could be done to improve it within existing capabilities, discipline, and so on. If you do you will lose at least part of whatever delta you may find for the smaller organization over the larger one. At the same time we could take the proposed organization and, by inserting net control stations, so constipate its command-control that any advantages to the organization itself would be completely absorbed by the gross inefficiencies of its command-control setup. . . .

There are also some interesting aspects of this problem relating to our ability to train units organized one way or the other. We are trying to build a tank platoon fire and maneuver live fire course at Graf. We are having a hell of a time for many reasons, not the least of which is that there simply isn't enough room nor is there enough time to do that little drill often enough to maintain anything like proficiency.

This is the second time in my career that we have tried to do this. In the early 1950s General Bruce Clarke, then commanding the Constabulary, tried to transplant the Bergen Höhne ranges to Graf. Abe's tank battalion S2 and S3, me, and another fellow tried to lay the thing out. We even bought a couple of expensive target-pulling devices. But when we got it all together it didn't work. There wasn't enough room, and we couldn't get enough other areas for practice and dry run. A couple of years ago Knox did a study that showed the only two places where the US Army could live fire and maneuver platoons and larger with any degree of realism with the new ammo and with XM1 are Bliss and Irwin. The expense of doing that for an army as large as ours is prohibitive. So there's an added argument for the new organization that says if we can't train the organization we have, then we should look carefully to see if we can find another organization that is as effective, perhaps better, and which we can afford to train on several bases instead of just a couple.

Tank Duel Analysis

Letter to Bryant Dunetz
Ballistic Research Laboratory
23 August 1976

Way back at the beginning of the year we swapped notes about your proposal for an operational test of the tank duel. I haven't forgotten that I promised to write you more once I got here and surveyed the situation. So this is sort of a report about what I've managed to get started so far, and some observations about the feasibility of an experiment conducted in this environment.

We've spent considerable time trying to model the Corps Battle. To do that we have used the target servicing concept. That is, how many targets of what kind and in what deployment appear within range of a set of defending systems and in what time sequencing; what systems are there to service them; and can the job be done before they close with the defending systems? The core of the servicing problem is the battalion battle position, composition of the battalion task force fighting in the position, and judgments about what mix of what systems need be deployed in the defense to service the enemy array.

We picked the battalion because that's the lowest echelon at which the commander has all the resources of battle at his disposal—air, helicopters, artillery, mortars, and so on. Obviously, however, the tank duel is a subset of that larger set, the battalion battle. Related to the small unit tank-antitank duel, then, are all the problems of intervisibility segment lengths, discontinuities therein, and attendant servicing opportunities based on suitable target exposures over given terrain and movement paths. . . .

TRASANA has done some recent work on this problem but more needs to be done. John [Weisz] was here recently and I proposed to him that we simply get a bunch of cameras, strap them on some tank gun tubes, rig them against a standard time base, put some sort of grid on so we can tell where each gun is aimed, and run at a platoon or company so equipped, something like a standard enemy target array. Then we would play the pictures back frame by frame and begin recording exposures, firing opportunities, times, targets serviced and not serviced, and all that. It could be a cheap experiment—cheaper than some, at least. It could be done in a reasonable time, done on our terrain, and our troops could get a lot out of it. . . .

Meanwhile I've been working with ARPA to try and break loose some DDR&E money to buy a system like one the Germans have built and are using in conjunction with their high mobility test program at Lichtenau. It is computer scored and operated, based on duel data provided by laser guns firing at laser sensing devices on the target vehicles. They have built one that runs at platoon level—any combination of systems totaling about 10 as I recall. If we could get hold of one of these without having to do the development ourselves, we could simply move into the experimentation phase without a lot of hardware building and computer programming. Unfortunately, the German system is not quite fully developed. However, in all probability by the time we got the money and were in a position to spend it, the system would be ready enough for our purposes.

So that's where we are here. Lots of talk and not much progress. The biggest problem over here is that commanders don't stay on station very long—at least in positions where they can begin, carry into execution, and see to the end something that takes a lot of time and money and a consistent view of what we are after. I had hoped to get moving much more rapidly, but there was much to be done that I hadn't anticipated, so had to start with one thing at a time. I guess I'm saying that I'm still vitally interested in what you propose, however our biggest difficulty will be instrumentation and for that we need lead time and money.

New Main Battle Tank

Letter to David L. Funk
Office of the Deputy Chief of Staff for Operations and Plans
1 September 1976

The XM1 decision is a bloody disaster. But I wasn't consulted, and had I been I'd probably be retired by now for sounding off. Just wait until they start asking the Congress for money for their "standardized" tank, and try to explain how it costs about half again as much as the original design-to-cost prototype!

Tank Development

Letter to Lieutenant Colonel Edward Halbert
Office of the Deputy Chief of Staff for Operations and Plans
2 September 1976

Many thanks for the note and papers you sent me regarding XM1. I am of course more aware of what the Germans are up to than what has happened in Washington since February. The Germans told me a year or more ago that they were certain their government would not let them have Leo unless they could sell all or part of it to someone—hopefully the US. The reasons were quite simple—first, it cost too much. The other big motivator was technical—they had not solved the laser rangefinder and night sight problems on Leo. They desperately wanted our technology in these two areas, since we were very far ahead of them. Therefore we became the logical target for their hard-sell operation. With characteristic Teutonic efficiency they went at it, and now they may have their way.

I still believe that we will never be able to explain satisfactorily to the Congress why we now have a tank that costs more than the two we built—substantially more, just in order to have something "standard." Thereupon I expect they will withhold the money and we will again be without a tank. Then the Germans and ourselves will both be in deep effluvia.

Anyway, for whatever it is worth, we did what Abe told us to do—we got our own stuff together so that if the program got screwed up it wasn't the Army's fault. I wonder how he would have handled this one?

Chief of Armor

Letter to Brigadier General David K. Doyle
Fort Knox, Kentucky
11 November 1976

Thanks for your note about establishing a Chief of Armor. As I believe I have told you several times, the longer I was at Knox the stronger became my belief that an answer to all the problems you cite would be establishment of an armored force with a commander at Knox. The pattern would have to be much the same as the charter Chaffee was given in 1940—but read that document sometime, then tell me if you really believe, in these days of overcentralization of everything, such a charter could ever be written.

I should also point out that even the Commander, TRADOC, was not consulted on many of the issues which you cite as having been taken in the absence of consultation with the commander at Knox. General DePuy and I have talked about this and not arrived at a solution either of us considered possible given the current climate in Washington. It would help, I suppose, to have a real Director of the Army Staff—a man with the determination to bring it all together, and with the conviction that the ultimate wisdom about nothing resides in the Army Staff. Short of that kind of acquiescence at the top, I doubt seriously that we will ever get any more attention in Washington than we have now.

Tank Fire Control Development

Letter to Dr. John D. Weisz
Aberdeen Proving Ground, Maryland
17 December 1976

This responds to your note of 16 November accompanying the Armor Team position paper on "A Human Factors Approach to the Development of a More Effective Tank Fire Control System."

Since the contents of that paper echo concerns expressed individually and jointly by Andy and me over the years, I'd be hard put to argue against any point made—at least in principle. What does concern me is the thrust of your paper in the context of where we are now in tank fire control design and development. Let me elaborate.

We have just taken a very important decision to go ahead with the version of XM1 which incorporates the very complex, but in my mind not very sophisticated (in your context), Hughes fire control system. I was not privy to the fire control analysis conducted by the SSEB, therefore have no knowledge of who rated what. Thus I feel free to philosophize a little on fire control on my own. In doing so I am not being critical of the XM1 decision—either tank was a winner in my book, and the SSEB evaluated the whole tank, not just fire control. However we must be candid—the Hughes fire control system represents the ultimate in complexity—it is the latest word in the long struggle to cope with the demanding parameters of that specious bit of technical dogma known as the BRL error budget. The fact that putting a man in the loop seems to upset the neatness of the error budget solution—my personal observation, matters not! The delightful mathematical cleanness of Hughes' solution to each of the nine or more critical elements of the error budget is quite appealing analytically. While the Delco system on the GM tank did not solve the error budget problem with comparable mathematical precision, it was not as appealing, despite the fact that with a man in the loop it worked as well or better than the other—again, my observation. So we have just made a decision that goes counter to all the arguments set forth in your paper. And so, with a very complex Hughes fire control system about to be fielded in M60A3, and an advanced version of that complexity aborning in XM1, where does that leave us?

It seems to me that to pursue the line of logic set forth in your paper is at this point counterproductive. This is so for two reasons. First, it gives the appearance of poor-mouthing the XM1 decision and, whatever views one may hold about system components, it was a good decision and we all need to get behind the program and press on with all due dispatch. Secondly, it swims upstream against the tide of a very major Army decision taken after an inordinately expensive and time-consuming series of analyses. I doubt, therefore, that requests to spend money on programs to further develop your thesis would be favorably received. It's sort of like taking a leak on the windward side of a vessel at sea—it all blows back in your face!

So what to do? It seems to me our biggest problem is our astounding lack of very basic data with which decent analyses may be made. As you know, in the course of the TSSG deliberations we found BRL wanting in several very critical areas of ballistics—many of which put any error budget analysis in less than a favorable light. In the M60A3 tests we found gross errors in firing tables—reflecting the fact that BRL does not draw up those tables based on full-range firings, but rather on the basis of extrapolations from subrange, subcharge firings. Now I know just enough about modem ballistics to know that, while that might have been OK for cannonballs, it won't cut it for APDS.

As a final example may I cite our attempts to define the benefits of target behavior in terms of gunner ability to acquire, track, and bring fire to bear in the STAGS test. The frustrations of our inability to get a clean analysis out of that test's mountains of data only prove to me that we need to do lots more of that kind of work, and soon.

It's always easy to tell someone else how to do his business, but were I you, it would seem to me that the proof of usefulness of HEL—and indeed its funding survivability, in the near term at least, lies in programs designed to strike at the core of our collective problem—lack of good basic data. Your problem is that you need a test vehicle—units willing to spend time and resources in a lot of rather grubby, unexciting work to give you what you need. We have a bunch of those in old V Corps and so long as I'm here, at least, we're ready to do something to help this effort which is rather critical to us both. So this comes back to our conversation when last you were here. And my suggestion is that we together embark on a very low-key but comprehensive program to find out answers we need to have in order to be not persuasive, but conclusive, in our arguments about fire control for any system—tanks included.

Tank Suspension Systems

Message to Major General John McEnery and M60 Program Manager
4 March 1977

1. Recently I became aware of some of the testing you all are doing on hydropneumatic suspension components. My information may not be complete, but in the tests described to me there was no mention of a survivability to mine damage analysis.

2. It is certainly true that hydropneumatics give a more stable platform for firing and riding across a wider range of speed and terrain conditions than do torsion bars. But the overriding fact that drove us to push hard for putting hydro on the fielded fleet was my conviction that the modern mine threat made it imperative that we get torsion bars out of tanks. Therefore, unless we are doing at least a parametric analysis, and hopefully more, of this very important matter, we are neither fully exploring our problem nor fully evaluating the alternatives. What say you?

Tank Suspension Mine Damage

Message to Major General John W. McEnery and M60 Program Manager
17 March 1977

1. Thanks for your 112100Z March response to mine on this subject. The principal reason for my raising the point was that none of the Vietnam mine damage data quantify two very important facts. The first has to do with the difficulty of extracting torsion bar remnants and installing a new bar when the old one has been broken or bent by mine damage to the hull. Secondly, my own experience in Vietnam was that tank hulls after two or three mine detonations became warped to the extent that torsion bar hull openings and anchors were out of alignment so badly that the tank had to be junked. Could we have but fixed the suspension, we could have continued to use the vehicle. All my experience was with the standard 23-pound Soviet/Chinese antitank mine. All my experience convinced me that we had to get torsion bars out of tanks—at least in the front stations, but preferably in all stations. In theory, if we could fabricate a bolt-on suspension, then by simply cleaning off the debris, redrilling, and shimming as necessary we could get a lot more mileage out of our vehicles.

2. Your message seems to suggest I'm concerned with torsion bar fragment damage inside the tank, and that by encasing the bars somehow we could prevent this. That is not my concern, and I know of no evidence to support a finding of considerable interior damage from torsion bar fragmentation.

3. Any externally mounted suspension system is bound to be more vulnerable to fragments and direct fire than are torsion bars. However, here we're using the wrong measures of effectiveness. To begin with, suspension components are pretty well below that center of mass hit distribution pattern derived from the October War. The probability of a direct fire hit is very low, of a frag hit equally low. Secondly, externally applied suspension components should add protection for the crew along the hull sides. Crew protection is a more important MOE than suspension protection. Third, if properly designed, ease of replacement of externally affixed suspension components is a big plus in their favor. Ease of replacement should be an MOE. These three factors must be accommodated in your vulnerability analysis.

4. Don't forget that Detroit and BRL are torsion bar proponents. They will make every effort to arrange their analyses to prove their preconceived notions. It was over their strong protests that I started the hydropneumatic experiments in the first place.

5. I've discussed this whole matter at length with Generals Tal and Peled. They too prefer externally mounted suspensions for the same reasons as do I.

Cavalry Element Tanks

Message to Multiple Addressees
29 April 1977

1. Recently I came into possession of [Major General] John McGiffert's memorandum establishing the Special Tank Task Force to conduct a comprehensive review of the Army Tank Program prior to FY79 budget development and review. One issue cited in the enclosures to John's memorandum has been the subject of considerable confusion since I have been associated with the tank fleet problem. That is what tank should we put in the cavalry.

2. We never addressed this issue head-on while I was at Knox. At the time the very idea that we wanted to retire the Sheridan fleet was unthinkable. We therefore tacitly assumed that the cavalry tank would be XMl; but nowhere was this confirmed in writing, and in various staff echelons there has not been a consensus, let alone a consistent view of the matter. At the risk of intruding into some sacred deliberations, let me state what I believe the cavalry tank requirement to be, to wit:

a. Divisional and regimental armored cavalry squadrons should be equipped with the XM1 tank.

b. The basis of issue should be 45 per divisional squadron, 62 per regimental squadron. These numbers are important to note; they represent an increase over cavalry TOE submitted by Knox after the Cavalry Scout Study. The reason for the change is that we have become convinced here that cavalry troop commanders must also be mounted in tanks, as should their FIST teams. Therefore there must be two additional tanks in the cavalry troop headquarters section. Three additional tanks in the squadron headquarters are for the commander, S3, and fire support coordinator. Our studies here show that none of these commanders will survive long mounted in M113s.

3. Until the ultimate requirement to equip cavalry with XM1 can be satisfied, cavalry should be equipped with the best tank we have in the field. This is so because of the urgent need for deployed cavalry forces to deal with first enemy attacks. Whatever advantage we can give cavalry, in almost any functional area—firepower, night vision, or whatever, will pay enormous dividends. Following that logic, we are working with USAREUR staff to take the first issue of M60A1 RISE Passive tanks, due for issue to USAREUR beginning this summer, and put them in the cavalry to replace the Sheridans. The BOI would be as cited in 2b above. For the moment we would beg the issue of TOW vehicles and other scout equipment, being willing to wait for the improved TOW vehicle. When ITV is available, we should then complete reorganization of cavalry with the new cavalry platoon organization. For the immediate time, however, the passive A1 will give the cavalry a significant improvement over Sheridan in armor protection, firepower, and night vision. As soon as M60A3 becomes available, it too should be issued to cavalry units first priority. We can anticipate considerable resistance to this part of the proposal. However, I'm convinced we have to suppress the resisters and press on!

Tank Gun Selection

Message to General Walter T. Kerwin Jr.
Vice Chief of Staff
16 January 1978

1. As you are aware, I am not scheduled to attend any further meetings on the tank gun selection issue. It is my understanding that you will be present. Therefore I believe it necessary to tell you of the two issues that concern me as the user. I have discussed what follows in past steering group sessions.

2. My basic concern is that we not make a premature final commitment to a 120mm system. It centers on two areas: adequacy of the recently completed test program and the XM1 fielding schedule. First the adequacy of the recently completed test program. These tests demonstrated 120 system potential. However, they did not provide a basis upon which we can commit ourselves to either of the foreign candidates. Testing of caseless ammunition and resulting system vulnerability is a critical area that requires further examination. I don't need to dwell on our caseless ammunition experience on the Sheridan with you. Related to this is the increase in system vulnerability that will occur if we maintain a high stowed load with a larger caliber round. This issue becomes one of particular concern when we recognize that the Soviets will probably field some form of pyrophoric round. A comprehensive test program to resolve these issues must be conducted before any final decision is made. Related to this is the problem of reduced stowed load in the face of what we considered a solid requirement for 60–65 rounds. We compromised at 55 rounds in our Tank Special Study Group analysis. Now we are talking of less than 50 rounds. Clearly we're well below the margin. Second, the fielding schedule. To date I have heard no one suggest that fielding the system be delayed to permit incorporation of a 120mm gun. However, this alternative will surely be suggested as our tank program is reviewed in future budgets. We should insist that any decision to select a foreign candidate contain a commitment to field 105mm XM1s as soon as the production schedule will permit. In my judgment it would be a serious mistake to allow any delay in fielding the XM1.

3. In summary, if a decision is made to adopt one of the foreign systems I am convinced that we must execute a complete test and evaluation program and our current production schedule must be maintained.

Light Armored Vehicle Losses in Battle

Message to Dr. Wilbur B. Payne
TRADOC Systems Analysis Activity
2 June 1978

Last week I played Battle at Newport, where tanks per platoon, platoons per company, and companies per battalion are under investigation via simulation. Incidentally that's going very well; they've a report in the mill; it's one of the best studies I've seen, especially for the resources we've invested. One matter of concern to me, however, is what appears to be an inordinately high rate of "mobility kills" on light armored vehicles. That is mobility kills that are not at the same time catastrophic kills. It doesn't matter so much with the M113, for it contributes little to the battle anyway, but with the BMP, BRDM, and other weapons-bearing platforms it could be skewing outcomes considerably. It is irritatingly reminiscent of the problem we encountered with the AMSAA model in the Tank Special Study Group work on XM1. You may remember Glenn Otis discovered using that model it took something like 20 rounds to kill an APC or equivalent vehicle moving across a given path, whereas an M60 tank taking the same path could be killed by something like six rounds. As I recall it, we finally determined the problem to be one of having assumed catastrophic kills on the light vehicles only when the attacking ordnance encountered suspension system or power train parts—a low proportion of the total vehicle profile. Vehicle profile itself was said to be less than that of the M60, and a reduction in hit probability resulting therefrom further reduced kill given a hit probabilities. Common sense, IDF experience, and a host of other evidence say that it is an infrequent hit on such a vehicle that does not result in a catastrophic kill of vehicle and/or crew inside—to the extent that the main weapon system can't be employed in any event. You may recall that we raised considerable fuss over that problem at the time of the TSSG; I was assured after a lot of bureaucratic obfuscation that the problem was being solved. However, the Newport Battle experience is so much like the problem we encountered in the TSSG that I have the uneasy feeling that the model has not in fact been fixed, and I've been the victim of another analytical pat on the head by the model makers whose low opinion of me is surpassed only by mine of them.

XM1 Tank Testing

Message to Major General Thomas P. Lynch
Fort Knox, Kentucky
10 June 1978

1. Friday the ninth I talked with both Don Babers and Bob Baer to express my concern about the repair parts situation and contractor support for the test vehicles at Bliss, not only with regard to the situation in which we found ourselves at week's end, but also with regard to the ITDT problems, tools, and Chrysler's poor management of their own team as well as their subcontractors. I am extremely displeased.

2. I want to make it clear that this is to be a tough, honest, objective test. If it is to be that, then both sides, user and developer, have some obligation which must be met. On our side we must ensure that the training is right, that we've done everything in our power to go more than halfway. We will not become hypercritical of the contractor. On the other hand the contractor has some overriding obligations. To date he's not meeting them in my judgment. While we must bend every effort to avoid hypercriticism, we should also be very tough, insistent, and demanding that he meet his obligations. If we aren't, he won't. Fielding this tank is critical to the Army. But I won't be party to fielding one that in any way compromises what we need. If there are design deficiencies, they must be corrected. Patchwork fixes won't do.

3. It is the responsibility of the TSM, Lynch and Starry to make sure all the above happens right. If we all have to take station at Bliss to do it, so be it. For the time that I'm travelling I want a report every two days on the test vehicle status. My office at Monroe can tell you what SSO stations can deliver. That report should come from the TSM to you and me. For the duration of the test difficulties it's apparent that Zuke [Colonel Frank L. Day] will have to spend more time at Bliss. So may you. And so may I. Meantime, let's make sure we're doing everything in our power to help where necessary and to raise hell when that's required.

Wheeled Reconnaissance Vehicles

Letter to General Bernard W. Rogers
Army Chief of Staff
19 January 1979

This responds to your question to Shy Meyer as to why the US Army does not use wheeled reconnaissance vehicles as do many foreign armies.

First, some background is probably appropriate as a perspective on how we got where we are. All our studies, including those of the Yom Kippur War, validate the need for the cavalry mission to be performed. And so we have a need for units that can fight while also acquiring and reporting information. The equipment these units need to accomplish their missions emphasizes speed, range, communications, and firepower; over the years it has included a combination of wheel and track vehicles. During World War II there were as many as 12 track and 4 wheel vehicle models in cavalry units. The wheel vehicles were both scout jeeps and armored cars. By the late 1950s we had no more armored cars and had introduced light tanks in cavalry platoons. This development was largely the work of General Bruce Clarke who, as Assistant Commandant at Knox in the late 1940s, introduced and got approved the cavalry platoon combined arms team idea—tanks, personnel carriers, scouts in jeeps, and mortars in jeeps or trucks. While he was antiarmored car, General Clarke believed we needed an armored vehicle for the scouts—but a tracked armored vehicle.

After considerable development work in the 1950s we introduced the M114 in the early 1960s. It proved totally unsatisfactory and, by 1964, we had stopped procurement and initiated action either to find or to develop another vehicle to fill the scout role. Meanwhile the M113A1, modified as the Armored Cavalry Assault Vehicle (ACAV), was the scout's primary vehicle in Vietnam. Meantime, a project manager's office was established in 1966; a requirement document [was] developed and approved in 1970, and two scout vehicle prototypes, one wheeled and one tracked, were built.

One of my first duties at Knox, in 1973, was to evaluate candidates produced to meet the scout vehicle requirement, the Lockheed wheeled vehicle and the Food Machinery Corporation tracked vehicle. General DePuy and I were responsible for terminating the Armored Reconnaissance Scout Vehicle (ARSV) program in November 1974, $39.8M having been spent to that time. Several reasons for this, but I'll cite only a couple. The wheeled candidate was engineered to withstand forces of about seven or eight times gravity; a more reasonable engineering figure would have been twelve or thirteen times gravity. It was a RAM nightmare. The tracked candidate looked very much like the WWII M5A1 light tank, but did not perform as well and was outgunned by the M5Al. These and other considerations led General Kerwin to terminate ARSV as a development program.

Meantime I had set to work two task forces, one to define the role of cavalry in modern war, the other to evaluate some fifteen vehicles which we saw as scout vehicle candidates. These included the two ARSV vehicles. While the work of this latter task force continued for nearly two years, it was clear by the fall of 1974 that the M113 was a better performer than either ARSV candidate, albeit it had not their armament. None of the candidates could show a significant mobility differential over the fleet they were supposed to support—in fact the two ARSV candidates were for the most part outperformed by the M60 and always by the M113. This realization led to two things. First, General DePuy and I agreed that we should adopt the

MICV as a scout vehicle and put all our infantry and cavalry eggs together in that basket. At the same time we decided to put TOW on the MICV, and the additional buy offered by including the cavalry as a user was most attractive. However, MICV was some way off; we needed a scout vehicle soon. This realization led to the proposal to put TOW on some M113s, cannon on others, and so have an interim scout family. The Improved TOW Vehicle was the result. Although we still need the cannon vehicle, we just couldn't gather the support we needed, and so it died aborning.

All that left us with the MICV—by this time an IFV, and its cavalry version the CFV, as the preferred family of vehicles for infantry squads and cavalry scout teams. Not that we consciously decided against wheels. There just weren't any good candidates among the fifteen vehicles we tested. . . .

The apparent resurgence of interest in wheeled combat vehicles is due to several factors. These factors address a range of scenarios which could be faced by the US Army; however, the mobility and survivability requirements of the NATO Central Battle continue to favor tracked reconnaissance vehicles.

The analysis [done by TARADCOM] addresses itself to the rear area security role against a predominantly light armor threat. It is assumed that, while the wheeled combat vehicle would have antitank weaponry, it could not have substantial armor protection and thus could not participate in tank duels. It would rather be employed from a defilade position, using hit-and-run tactics or standoff tactics, similar to those foreseen for the High-Mobility Weapons Carrier (HMWC).

With wheeled vehicles, weight is an especial problem. For examp1e, the Army Mobility Model for Thailand-type terrain shows a sharp drop in wheeled vehicle mobility just beyond 30,000 pounds of weight. This weight threshold is important because armor protection and firepower sufficient to cope with the armor threat means the vehicle will weigh this much and more. On the other hand, in the Fulda area during the wet season the wheeled vehicle could negotiate approximately 92% of the area compared to 96% for the XM1 and nearly 98% for the Infantry Fighting Vehicle. These data reflect one vehicle pass over virgin terrain with no steering induced. Once these criteria are altered, wheel performance decreases rapidly. Thus, given our requirement for a high-mobility vehicle in a worldwide environment, wheels lose out in all soft soil conditions.

Now vulnerability. Two comparable vehicles—wheel and track—are about equally vulnerable to small arms, HE shell fragments, and mines; and each has a limited limp-home capability, which tends to neutralize any vulnerability argument. There have been some promising advances in tire technology, however, which warrant our continued interest.

Acquisition costs of wheels versus tracks do not provide the discrimination one would imagine; wheel vehicles are not a cheap answer. It is only from the engine fly wheel to the ground that real cost differences emerge. With a weapon station accounting for about 50% of the cost, a cost differential of as much as 20% (from calculations) translates into a cost difference of not more than 10% in favor of the wheeled vehicle.

Significant differences in operational and support costs were hypothesized based upon M113A1 sample data collection and on field data and extrapolated field data for a wheeled vehicle. Excluding common weapon station costs, the track costs annually to maintain perhaps as much as three times the costs of the wheel.

Reliability, availability, and maintainability of chassis components of the M113 APC and V150, a wheeled Armored Car, were compared. M113 data are derived from a 15-year experience factor, and the V150 data are calculated from tests conducted at Yuma and Aberdeen Proving Ground. Although 50% more man-hours are required for the tracked vehicle, and operational readiness of the wheel is better (.92 versus .76), there is an insignificant difference in reliability of the two (.94 for tracks and .96 for wheels) in completing a 100-mile mission (52 miles primary roads, 42 miles secondary roads, 3 miles on trails, and 3 miles off roads). As the distance increases beyond 100 miles, the tracked vehicle reliability will be better.

Foreign systems have been considered as candidates for reconnaissance vehicles and continue to have our interest. Most recently evaluated was the German-made LUCHS, an eight-wheeled vehicle. LUCHS is old technology, especially compared with the Cavalry Fighting Vehicle. LUCHS does not provide space for our five-man scout crew; LUCHS armament (20mm) will not defeat the BMP; the vehicle does not have an antitank capability. The thermal sight on the CFV will see through smoke, haze, fog, and dust; night vision equipment on the LUCHS will not. The CFV is projected to cost $370K per copy in production; LUCHS is estimated at $650K per copy. Both vehicles can be fielded about 1981.

So in sum the Army has had and still has a requirement for a scout vehicle; we have, in the past, had a mixed bag of wheels and tracks to satisfy that requirement. Our materiel acquisition process dictates that we specify operational requirements and pretty much precludes [our] specifying details such as suspension type unless it can clearly be shown that one type is to be preferred over another. The last time a formal requirement was established for a reconnaissance vehicle, both wheeled and tracked candidates were tested. The wheeled candidate fell out early, and the tracked candidate followed soon thereafter. This admittedly was a small sample of wheels, but it would be difficult to spend the resources to evaluate a sufficient number of various wheeled prototypes to determine the wheels versus tracks issue forevermore.

The TARADCOM analyses show offsetting advantages and disadvantages in weight, size, vulnerability, mobility, acquisition costs, and performance that tend to make any decision a close one. When one considers such factors as life cycle costs and RAM, wheeled combat vehicles are cost effective in selected roles. For example, the High-Mobility Weapons Carrier (HMWC) would provide TOW squads in light divisions a quantum jump in mobility and survivability over the 1/4-ton and the mule. A track vehicle could not do as well.

Generally, countries now buying substantial numbers of wheeled combat vehicles—Singapore, Malaysia, Saudi Arabia, and others—are buying them for their special roles. The Brits and the French both harbor a long tradition featuring wheeled vehicles in recon units. To some extent the Germans share this bias. In some measure this bias reflects the notion that considerable survivability can be had by speed, agility, and mobility. Nearly six years of test and evaluation in this country have convinced us that notion is not true today, if indeed it ever was. Further, no recent vehicle development, wheel or track, has been able to demonstrate a degree of mobility, agility, and speed significantly better than the primary fighting vehicles of the supported fleet—XM1 and IFV.

There is no prejudice toward either wheels or tracks. We will continue to look at available technologies and select the most cost and operationally effective solution for our ground scout vehicles.

M60A2 Tank Assessment

Message to Lieutenant General E. C. Meyer
Deputy Chief of Staff for Operations
9 March 1979

1. Understand there is a decision memorandum within your shop which recommends that the Army:

a. Maintain status quo on the A2 redistribution,

b. Activate the second A2 CONUS battalion, and

c. Reevaluate the phase-out question again during 1QFY81 based upon the status of the XM774 105mm KE round and a reevaluation of the threat.

2. As you know, I am not a big fan of the A2. Based on what we knew of the threat at the time, Lou Wagner's study group in September 1977 recommended that five battalions be fielded in Europe instead of six as a means to resolve the rotational base imbalance problem. Dutch Kerwin made the decision in October 1977 to reduce USAREUR frontline assets to four A2 battalions in FY78, which has been completed.

3. Based on new assessment of the threat, we looked at the A2 and it is a marginal performer at best against T72. When the Soviets field T80 in the early 1980s, the situation gets worse. The Tank Forces Management Office picked up on our briefing to the VCSA on 8 December and on 14 December 1978 recommended that the M60A2 be the first priority (FY81–82) MBT asset to be removed from the Army inventory. It was also recommended that the second CONUS A2 battalion at Fort Hood not be fielded. The recently completed "Anti-Armor Strategy Executive Summary," which was forwarded to the Vice on 25 January 1979, reiterates our concern with the M60A2 Shillelagh missile and conventional round performance against the T72 threat. I know of no compelling logic to keep the A2 in our inventory. It is now a questionable target servicer, has always been a maintenance nightmare, and should be considered for removal as soon as possible.

4. My staff at CACDA has just completed a brief analysis of what to do with the A2 assets when removed from the inventory. While not an exhaustive study, it does show we can make other uses of the chassis. I will forward the results to your staff soon.

5. We have frequently discussed how bureaucracies tend to keep us from cutting away our marginal programs. My experience and intuition on A2 is now supported by threat and analysis. Do not believe we should activate another battalion this year. We should start to get the A2 out of our system with deliberate speed.

Mine Detection

Letter to Dr. Robert R. Fossum
Defense Advanced Research Projects Agency
18 June 1979

1. The purpose of this letter is to request DARPA assistance within the area of countermine warfare.

2. For the past 30 years very little progress has been made in the US countermine efforts. During this period there has been a continuum of analyses and studies of countermine programs without the fielding of a single operationally effective system. Hand-held mine detectors, line charges and hand-emplaced explosives form the nucleus of our current countermine capability. Although some countermine systems show promise, promises are all we seem to be getting from the countermine development community, especially when it comes to remote detection of minefields. With the advent of scatterable mine technology, our inability to detect and neutralize minefields presents a risk to our ability to move about the battlefield that we must counter.

3. As you know, Soviet doctrine makes extensive use of mines in both the offense and defense. The Soviets have fielded a complete family of minelaying equipment, from simple truck-mounted chutes to full-tracked armored and sophisticated helicopter-mounted systems. Their mines are effective and diversified and they have fielded them in quantity.

4. The nature of mines and their mode of employment combine to confront us with a problem that, in all probability, cannot be solved by a single mine-detecting device. The question then becomes—what is needed? Although we need help in both detection and neutralization of minefields, the greatest vacuum of ideas seems to be in the area of remote detection. It is difficult to understand that the US scientific community cannot devise a means to remotely detect a small box of metal on or slightly below the ground surface. Looking at the threat and what must be done to counter it, I see an urgent requirement for a high-speed remote mine-detecting mechanism. This remote mine-detecting system must apply a technology or technologies that can detect both metallic and nonmetallic mines, whether influence, pressure, or tilt-rod fuzed. We should aim to identify mined areas, then if time and resources permit and the mission dictates we should have a close-up detection capability that is able to identify individual mines. Remote detection must be provided over a wide area, at high speed, in real time and with reasonable reliance. If you care to participate in the other problem of mine neutralization we need assistance there also.

5. Considering the growing disparity between the Soviets' ability to mine and our inability to detect, we need a successful effort that will impact on the near to midterm in countermine development. This is a chance to truly enhance our battle fighting capability while simultaneously revitalizing an area grossly neglected over the years. I am certain that DARPA's effort will help fill the current vacuum. I am ready to offer any assistance or additional information you may need.

XM1 Tank-Mounted Barrier Equipment

Message to Major General Thomas Lynch
Fort Knox, Kentucky
16 July 1979

1. I have in hand your 111215Z July message: XM1 Tank-Mounted Dozer Blade.

2. It is a masterpiece which could have been written only in the bowels of a deeply rooted and completely unimaginative combat developments bureaucracy. In that regard, congratulations.

3. Now let's discuss what has to be done. First of all I hypothesize that, however it might finally be stated, one outcome of our operational concept regarding barrier warfare will be a need for the mounting of a dozer blade, a plow, and a roller on the tanks of each company. In the new organization that will call for the mounting of some kind of barrier-defeating device on one tank in each platoon in every tank company in the Army.

4. To explore that hypothesis it is first necessary to find out what happens when we mount those devices on XM1 tanks. Always before we have rejected mounting such mechanisms on the front of tanks, largely because they were cumbersome, hard to operate, and in fact reduced the mobility of the M60 tank. The IDF equipment about which I've been talking with you recently does inhibit mobility in the M60, but it seemed to me that it does so to a lesser extent than have US versions of similar gear in the past. If that is true, and we need to find out if it is, then we should press on with the test program I outlined some time ago to see exactly what those pieces of equipment will do for us under our operating conditions. At the same time there is an urgent need to find out what happens when we put 1500 horsepower behind the dozer, plow, and roller. It will for certain be a different kind of operation—the question is how different, and what does that difference do for our notions of whether or not we can and should mount such devices on tanks to the platoon level?

5. All that tells me that we must do several things: first, we have to flesh out the operational concept. I've tasked Jim Kelly with that, but you and Dave are the principal users, so the three of you better get locked together quickly. Second, we have to get some of that equipment and test it—on fixtures attached to the front on M60 tanks on the IDF mode, then on an XM1. The Armor-Engineer Board should do that. We should do that as quickly as possible, without wandering through the maze of the CD process, and without letting DARCOM get their hands on it any more than is necessary. What we need is to know if the stuff will work, and then can we go ahead with the operational concept we think we need to adopt, or must we seek some alternative.

6. I would like to talk with you about this on the 20th when I am at Knox. It is quite clear that, if we let your combat developers carry on with business as usual, nothing will ever happen with this whole thing.

Armor for Armored Vehicles

Message for Dr. Walter LaBerge
Deputy Under Secretary of Defense (Research and Eengineering)
2 January 1980

1. For some time now we've been reviewing the status of armor protection on armored vehicles in the armies of the world and armor technologies under exploration in several countries. All this is an effort to determine what developmental strategy we should next pursue in the constant to and fro between measures and countermeasures.

2. What comes quite clear is that the weight of our tanks will increase with the advent of special armor on XM1; it will increase again as we try to install the 120 gun on XM1. By that time we will have a tank in the 65-ton weight class—perhaps more. And we will have reached reasonable limits to which many of us believe we should allow the weight to grow. It is not that the engine developers couldn't power a 70-ton tank at 30 horsepower or better to the ton. Indeed they are standing by eagerly at the instant, hoping they can persuade someone to buy their 2,400-, then 3,000-horsepower engines. It is, however, that we have reached practical limits with regard to bridging, air and ground transportability. Every weight increase in armored vehicles, especially tanks, now will bring with it an enormous burden in bridge redesign, support vehicle redesign, and transport redesign. We should avoid these burdens if at all possible. All that seems to us to argue that the next logical step in tank design is to figure out how to get the same or better protection than we have now with no increase in weight, or better yet with a weight reduction.

3. It appears there are technologies in ceramics and glass that seem, at the moment at least, to offer the opportunity to explore lighter weight-same protection armor. It may be that this is the route the Soviets have elected to take, and we would merely be following suit. However that might be, I'd like to suggest that this is a fruitful area for ARPA to explore. Some monies are being spent at Lawrence Livermore Labs in this regard, but I would judge them not sufficient to bring the matter to resolution in anything like a reasonable time. I would welcome the opportunity to open a discussion with you on this business at your convenience.

M551 Evaluation

Message to Lieutenant General Glenn Otis
Deputy Chief of Staff for Operations
4 January 1980

1. We are responding to your request to assess the impact of reconstitution of the M551 training base. As I read the SecDef's memo he has left the door open for other solutions. I believe we have substantially better solutions to the RDF requirements than the M551 and urge that these be pursued as the Army recommended response to the SecDef memo. Before we address alternatives, let's resurrect why we eliminated the 551.

2. First of all the ammunition being proposed (standard HEAT) is only marginally effective versus the T72 and may be even less effective versus the T80. We stated this in our antiarmor strategy report as approved by the CSA. The PIPed TOW will shadow the effectiveness of the 152.

3. Removal of the M551 two years ago was based upon unacceptable OR rate, more vulnerable than MBT, limited basic load, slow missile rate of fire, excessive crew training costs, no firing sight for vehicle commander, no missile night sight, signature vehicle, etc., to name a few.

4. The Sheridan was phased out of Europe this past July. We are over half complete now in CONUS. The vehicles have been turned in "as is," so they are in miserable shape. The only vehicle PIPs being procured are those for the residual force and NTC. The training base is next to nonexistent, which was addressed more specifically in separate correspondence.

5. Before we proceed any further let's review the bidding for the RDF. The Army contribution consists of both heavy (4th Mech/194th Armor Brigade) and light (82d/101st/XVIII Airborne Corps) forces. If we are proposing equipment for these forces, then this factor is significant. They don't need unique vehicle systems in low densities. Our heavy forces are already faced with the introduction of XM1, IFV/CFV, ITV, GSRS, M901s, etc. Therefore, the RDF equipment must correlate with its capabilities, i.e., rapid response, mid-intensity, and sustained operations in a bare base undeveloped theater. None of this equals M551. In fact, the 82d is already destined to give up the only 551 battalion left in the force structure.

6. For the RDF there are more operationally and cost effective alternatives than the M551. For the heavy forces my vote goes to ITV. It's being fielded in USAREUR today. The production lines will be running hot for the next 3–4 years. Funds are in the budget to support it. It has been proven through a COEA, DT/OT I, DT/OT III, FOE, IPT, etc., to be cost and operationally effective, including survivability. More importantly, the RDF heavy units, 4th or 24th Mech and the 194th Armor Brigade, are programmed to receive ITV in FY80 and 81.

7. For the light forces we have the HMMWV. It's in R&D with an IOC of 4Q83. The HMMWV technology is here today. Given priority and support, particularly by Congress, the program could easily be structured to permit receipt by RDF units within 12–18 months. The 82d and 101st HMMWV TOW carriers are number one and two to receive the vehicle. Incidentally, it is important to note that the USMC intends to replace every ¼–1¼T wheeled vehicle in its entire fleet with HMMWV. The USMC is also a key member of the RDF. USAF is also buying into the program for base security, combat control teams, and forward air controllers.

8. The dramatic impact on reduced tactical and strategic deployability with HMMWV is readily apparent and deployability is the name of the game. Air sorties will be critical for the RDF.

9. ITV and HMMWV will cost us approximately $78–80K and $20K, respectively, a copy. The training and logistical bases for ITV are established. HMMWV offers attractive reductions in both training and logistics over the long run. ITV and HMMWV weigh 26,500 pounds and 7,200 pounds respectively versus the 38,000-pound 551. Our vigorous and funded TOW PIP program will further enhance both systems.

10. We are prepared to assist in the formulation of any alternative (materially, organizationally, or doctrinally) to enhance the RDF requirement.

11. Need I say more?

Alternative Engine Programs for the XM1

Message to General John R. Guthrie
Commanding General, Materiel Development and Readiness Command
7 January 1980

1. In light of our diminishing resources and the latest tests results, I feel compelled to express my feelings concerning the XM1 backup engine program.

2. The AGT-1500 turbine has experienced some growing pains; however, I do not feel that the problems have been abnormal and I am especially pleased with the latest test results. The reports of both the 4,000-mile three-tank test at Fort Knox and the 1,000-hour two-engine laboratory test show that tremendous improvements have been made through the identification of deficiencies and the timely development and application of fixes. As a result of these improvements we have now exceeded the MR 1 milestone in both the reliability requirement and durability threshold.

3. I am convinced that the Army cannot afford the manpower, time, or money to pursue an unnecessary alternate engine program. We must concentrate our resources and efforts on continued testing and improving of the XM1 with its turbine in order to field the best possible system and do it in a timely manner. Accordingly I urge that we band together and take a strong position against spending scarce resources on an alternate engine program.

Light Tanks

Message to General E. C. Meyer
Army Chief of Staff
7 March 1980

1. For some weeks we have noted some modest increase in traffic concerning light tanks. In your building it is taking on the imprint of a major new initiative. There is enthusiastic support from many quarters. Before all this gets too far along I am obliged to signal for you the most important inconsistencies in what we are saying and apparently trying to do.

2. First, in the ACTV program we have now the HIMAG test bed and the HSTVL experimental design vehicles. While it is well that such programs are under the same umbrella, it is also necessary that we remember what they were started for in the beginning.

a. HIMAG is an attempt to fill in the gaps in our knowledge about survivability as a function of mobility and agility derived from a host of previous experiments. It was and is a test bed. It is not a prototype. It may or may not tell us all we wanted to know about mobility/agility and survivability. [Terry] Covington would have his viewers believe that the outcomes of tests performed with that vehicle are beginning to show a great return for what he calls mobility. What he is measuring is speed made good; therefore what he shows is the advantage of using sprint to cover as a tactic to reduce exposure time to enemy fire. The advantage he claims is thus a function of the firer's inability to fire more or fewer rounds during a period of greater or lesser exposure. It does not, repeat not, deal with the fundamental outcomes of the test which show that the rate of change of apparent acceleration of the target, as viewed by the firer, is the important variable. Nothing we've learned in HIMAG tells us that we can achieve rates of change in apparent acceleration that will buy much more survivability than we have with XM1. But if you use rates of fire and exposure times, as Covington has done, then you can make a case for what you wanted to prove. Those are not unimportant variables. They are not, however, the only ones, and any analysis which treats with them only simply ignores other far more complex and equally important variables which in the end have a greater effect on the gunnery problem than speed made good and exposure times.

b. HSTVL began as an ARPA program to demonstrate a 75mm gun on some kind of light chassis. It was never said what was to be the role of this vehicle. In the mind of its original proponent it was to be a light tank, even though the Army said it didn't need such a vehicle at the time. The light tank bias has now pervaded the ACTV program. The Army is being accused of dragging its feet and so on—on what? On an experiment? Surely not on a development, because it was never intended to be such, except in the minds of the DARPA office which invented it and of the head of DARPA who very cleverly insulted the Army leadership into making HSTVL a CSA-sponsored program. There is therefore considerable confusion about what this whole program is about and what its component elements are supposed to be proving out. And there are far too many exaggerated claims for what has been found out in testing, and so [for] what is possible in further development.

3. Second: While it is never said exactly what threat the light tank is being designed to handle, Covington's briefing uses the T72 as a typical target. Several years ago we were forced to take on the 120mm gun for XM1. We finally justified that by saying it was to shoot T72s. When that debate was hot, General Abe sent me and Bob Baer on a trip around the world with

instructions to report back on what we should do. Based on all the evidence we could collect, we recommended that we believe the 105 could handle the then threat and with improved ammo handle the future threat. Abe then said he would go persuade Jim Schlesinger that we were making a mistake. Before he got that done, he fell ill and died; to the best of my knowledge that intended conversation never took place. So the Army had to twist itself around to rationalize its arguments to support an OSD program that most of us in the business didn't believe we needed in the first place. Now comes Covington. He says all we really need against that future target is a souped-up 90mm cannon. He has no evidence, just some paper estimates. He admits the 75mm won't do the job, probably beyond 1,000m or so, but that if we'll just make the gun a little bigger it will be okay. I don't know how we intend to rationalize this latest argument with the arguments we have just finished making that we couldn't cope with the threat unless we had a 120mm gun. That's just the most glaring kind of inconsistency; it is terribly transparent; it is in my judgment intellectually dishonest.

4. Third: In Covington's briefing he comes round to a vehicle of 22–24 tons and says that a missile system won't do the job, it has to have a gun on it. The chart he uses to prove that is again based on comparative rates of fire of a selected slow-firing missile system versus a fast-firing cannon. Again, if only parameters of exposure time and fire rate are used, one can reach such a conclusion. Again, however, there are other parameters equally important, and in the aggregate overriding. Several years ago we made a very tough decision to delay the MICV program in order to mount on the MICV the TOW missile system. We did that because we recognized too late that MICV armor could not survive in the threat environment of the tank, and with a short-range cannon the MICV would be at a serious disadvantage, even against enemy lightly armored vehicles. With the lighter vehicle then we needed a longer-range system so that survivability could be improved by increasing the standoff range. That was a very costly decision. We agonized long and hard over it. A lot of us have testified, argued, and defended ad nauseam. Now we are going to stand up and say that was all wrong; that a missile system is not needed for standoff on a vehicle that's more lightly armored than the IFV, and that all we really need is a gun, for which we've made other exaggerated claims cited above. Again, this is the most glaring kind of inconsistency; it too is terribly transparent; it too is in my judgment intellectually dishonest.

5. No vulnerability estimates in Covington's briefing. There's no way for a vehicle that light to stand up and fight even a T55 or T62. Ignoring that fact simply won't make it go away. The main reason we say we're buying XM1 is for its armor protection. But now we're saying we don't need that protection after all, the light vehicle will be okay. We just can't have it both ways, boys.

6. We are, in response to a DCSOPS tasking, once again trying to say whether or not the Army needs a light tank. Personally I've been down that primrose path so many times it sickens me to consider another trip. We will produce the study. I seriously doubt that much has changed from the last iteration, and we will once again conclude that, all things considered, we don't have such a requirement.

7. I know we are trying to capture some of the RDF money. About that I would say a couple of things—so far no one has convinced me that there will be any; I would not recommend capturing money only to use it for a system for which we haven't a real requirement. To date Covington's project is just a lot of very, very spongy predictions looking for a requirement.

8. If we want to capture a share of the RDF scenario I would suggest that we:

a. Acknowledge the Marine role as the amphibious part of that force with a mission of establishing a lodgment from which larger forces might operate.

b. Claim for the Army the airborne, airlifted part of the lodgment mission, working in conjunction with Marine forces.

c. Claim for the Army the role of reinforcing and operating from the lodgment area, either with heavy forces or with light forces, depending on the threat. If the threat permits the use of lighter forces, then we have a requirement for:

(1) Airmobile maneuver and firepower forces.

(2) Ground mobile firepower and light maneuver forces mounted in a family of light armored vehicles like the V150 family of armored cars. Something we can perhaps afford, and which could be bought instead of developed.

(3) Advanced EW systems targeted against enemy C3 of maneuver and firepower forces and mounted in V150 variants.

(4) Advanced antiarmor systems—fire and forget, supersonic, PGM warhead ATGM of one kind or another, and perhaps in some combination, and mounted in V150 variants.

9. In any event we just must remember the background of all these lines of development and avoid doing something dumb. It would also be refreshing to see the staffs take a somewhat more mature, reasoned approach to these matters in place of the near-hysteria that is reported to me at almost every ring of the phone.

Tank Design

Letter to First Sergeant George D. Chamberlain
Somerset, Kentucky
14 April 1980

Your recent letter, reference armored development, covered a subject that you know is always foremost in my mind. The reduction of the weight of our armored vehicles is now the most important direction for our subsequent vehicle development.

The protective skirts on the sides of the XM1 are designed to provide standoff protection to the often exposed upper portions of the tracks. Admittedly, they do add weight, but our experience and tests show the protection trade-off is worth it. The skirts are hinged in sections so they can be raised to work on the suspension system.

The threat we face, in the Soviet T-72 tank and its potential follow-ons, has dictated that larger guns are necessary. We were, as you know, outgunned in World War II, but overcame the difference with numbers and better mobility. Numerical advantage is a luxury we will not enjoy in the next conflict. We are looking at the growth potential of various calibers from 75mm to 120mm. They each have their advantages and disadvantages, but ultimately the threat will dictate the choice. There is, as I'm sure you know, an upper limit on just how much you can "soup-up" a small caliber gun.

The HIMAG and HSTV-L programs are test bed programs that promise some utility value in the future. Gun size is only an offshoot of these programs. Their main purpose is to find solutions to mobility problems.

Finally, I agree with your comments on crew size but the manpower situation, now and in the future, does not look favorably on a five-man crew. We have tested this concept in combat in two wars and most recently in peacetime. It is difficult to keep the crews filled because of personnel shortages. We will not cease our efforts to solve the problem, but the obstacles are difficult to overcome.

Kick in the Grill Doors

Armor Conference
Fort Knox, Kentucky
14 May 1980

Armor conferences are usually described by outsiders as examples of a mutual admiration society, a corporate closed mind, or an association for the preservation of obsolescence. Obviously insiders have a different view. However, as I observe Armor today from my vantage point, there is an air of complacency—one officer called it a ho-hum attitude—about us that is alarming and I believe out of character. So let me talk about that a little tonight, just to get your attention and keep your thoughts moving. If I make you angry, so much the better. Perhaps you'll accept the challenge and do something instead of resting on your laurels.

First, we need to be reminded that Armor was built on ideas formulated by mavericks. They were the professional visionaries and malcontents of the 1930s. They refused to be ho-hum. They strove to develop new doctrines for war and to become excellent in the application of those doctrines. They believed that a relatively small, highly trained force could defy the tribal wisdom about ways to win wars.

Actions on both sides of World War II battlefields testified to the wisdom of their judgments. What won in that war was not always calculated in terms of better weapons, more supplies, or more equipment.

Indeed, the history of war testifies that battle outcomes are most often determined by factors other than a disparity in numbers of people, numbers and types of weapons, or force ratios. All the counting you want to do cannot explain Cannae, Thermopylae, Alexander's Macedonian Army, Napoleon's Grand Army, Bastogne, or Inchon.

On the contrary, history tells us that the outcome of battles most frequently follows from the courage of soldiers, the quality of leadership, and the excellence of training. I've said that before. I didn't invent it. It's apparent in any study of the history of battle. Battle analysis tells us that well-trained soldiers in well-trained teams and crews in well-trained and led units win far more often than not.

What is it that wins? How do we define it? How do we get it? How do we know when we have it? Excellence comes from two things:

- Initiative of leaders in training.
- Willingness of leaders to take the initiative in operations.

Both of these are well-established fundamentals of mobile warfare. In training and operations, but especially in operations, initiative equates to mobility. The side whose leaders seize the initiative, the side that is more flexible and mobile, is the side that most often wins.

Flexibility and mobility of units in training and operations comes from the mental flexibility of leaders. It is the ability to create, to be innovative. It is the willingness to act rather than react. It is the ability to comprehend, to understand meaning and intent rather than seeking after and clinging to rote formulas.

Doctrine for Armor in its early years was never dogma. There were few books on Armor operations before World War II, but that didn't prevent operations from going forward. Leaders understood the concept and the meaning of flexible, mobile operations. They fought with new weapons and new means, but they understood the meaning of the task before them. Understanding does not come easily and, for leaders, it requires several things:

- First, there must be a sure knowledge that soldiers can perform their individual jobs to a high level of proficiency.
- Second, there must be an equally sure knowledge that crews, teams, and small groups can perform their collective jobs to a high level of excellence.
- Third, units must be trained in their operational lessons to a high level of proficiency.
- Fourth, leaders themselves must hold a sure confidence that they know how to put their units into motion in battle to seize the initiative.
- Finally, in all that, the leaders must develop and share a common set of battlefield values with their soldiers.

No one can count on personal charisma to provide these requisites. The supply system doesn't issue them; the personnel system can't find them. They must be developed in the essential system—the human system. They must be sought through hard work and concentration, by training, just the way they have always been found.

Most important to all these notions are four elemental qualities, the qualities that are part of that last notion, a shared set of common values. These are, in soldiers' terms, competence, commitment, candor, and courage.

First, professional competence. Professional military competence includes a willing acceptance of disciplined professional responsibility; it acknowledges willingness to sacrifice. In the micro sense, professional competence involves, among all the other details of a leader's job, developing the ability to train the requisite values into our young soldiers of today. For, without them, you can't have an effective military organization, large or small.

Commitment is a sense of obligation to something larger than yourself. The profession of arms is a commitment, an obligation. As rank increases, so grows the commitment to larger issues and purposes. But, closer to the bottom, the obligation is to the unit of which we are a part and to the men and women we lead. In either case, the commitment is to something larger than ourselves. There is no room for careerism or "What's in it for me?" or "Look out for old number one."

Characteristic of today's changing society is the way in which the language is used to diffuse the truth. It may be we don't tell the truth very much any more because it's most often unpleasant. It may be that it's just harder to discern truth because today's issues are too complex. In any case, the military profession must hold in high merit the virtue of candor, the willingness and ability to discern and tell the objective truth. In political-military deliberations candor, with regard to the capabilities and limitations of military force in pursuit of political objectives, is essential. Had we had more of it, perhaps the legacies of Korea, the Bay of Pigs, and Vietnam would not today be so burdensome. A willingness to tell the unvarnished truth is similarly an essential ingredient of leadership in units of a military force. In battle, it is always necessary to tell the truth. Someone's life usually depends on it.

Finally, there is courage—the courage necessary to tell the unpleasant truth, the courage to make a commitment to something larger than yourself, the courage to insist on that higher order of values essential to a successful military profession, the courage to understand and articulate convincingly the extent to which military force has utility in the pursuit of national objectives, the courage to insist that those objectives be defined and made clear so that some decent assessment might be made of the best way in which they might be sought, attained, and secured.

None of these values is more important than any other. More important, none can exist for very long without the others. They are truly interdependent. If you ask how these fit together, I would have to answer that I'm not sure yet after over 30 years in military service. But I can give you an example from which you may draw your own conclusions.

When I commanded the 11th Cavalry in Vietnam, it was routine for me to interview each new officer reporting for duty with the regiment. Normally, this was done at night, long after the day's fighting had died down and after my battle with the paperwork was done. Among other things, I asked each of them if he was afraid, for I believe that all of us are afraid in that world of battle, but only those who are willing to admit it are capable of coping with it. I had a lot of strange answers to that question. Most people simply hadn't faced up to it, for it is one of those very, very complex questions of life. Faced with the question, some backed off and refused to take command of their units. Others passed it off with a shrug of bravado. Most concluded with, "Well, now that you ask, I guess so."

One young lieutenant I remember especially. His response was, "Yes, Sir, I am afraid. But I've thought a lot about it. I've even asked God's help in finding an answer. God's answer wasn't too clear to me, but I believe this is what I have to do, and that I'm about as well prepared to take command of my platoon as the Army can make me, and I've decided that there are worse things than dying for your country."

He was a good platoon leader, one of the very best. A few months after our talk, he was killed as he led his platoon on a reconnaissance down a dry creek bed near Loc Ninh. But I found, as I looked into the battle, that he had been at the head of his platoon—leading. He had seen the ambush, alerted his men, and his platoon had survived because of his quickness. Not only did they survive but so angered were they over the death of their lieutenant that, without outside help, they proceeded to destroy the entire North Vietnamese ambushing unit. Then they marched out of the jungle, bearing on their shoulders the body of their fallen leader—a man who had the candor to admit fear, the courage to control it, the commitment to succeed, and the competence to do so, values he shared with his soldiers.

In my mind Armor once epitomized all that in our Army. But as a group, a group of soldiers with the same basic notions, it seems to me we've lost that perspective. No longer are we sought out as the leading purveyor of new ideas. We've become bureaucratic, soft. We lack innovation. We don't suffer well those who do take the initiative. We have a ho-hum attitude.

So I leave you with a challenge. For you older officers, those who remember from whence Armor sprang, it's up to you to think back and see where it went off track. Only you can create the climate that fosters the battlefield values that win and the leaders who will do so. Armor needs a shot in the arm in spirit, and if that doesn't work, maybe we need *a good solid kick in the grill doors*.

To you younger soldiers, learn your history well, for we must recapture our spirit. Only you can do that. Once we were very cocky because we had some good ideas. The ideas are still good; what we need is some renewed dedication to our fundamental values and renewed determination to be willing to act, to seize the initiative—in training, in operations, in battle, in victory.

Tank Main Gun Service Ammunition

Message to General John R. Guthrie
Commanding General, Materiel Development and Readiness Command
12 June 1980

1. The Army is again faced with a disappointing turn of events in tank ammunition development with the recent in-bore projectile breakups of XM774 APFSDS-T. This represents a continuation of past tank ammunition development/production problems.

2. This problem first surfaced in the early 1970s with the discovery of in-bore breakup of M392 APDS-T. The first APFSDS-T round (M735) continued the trend so that today there are more M735 rounds being held in CONUS ammo plants as a result of lot failure than there are deployed in USAREUR. The latest XM774 failure could significantly delay its fielding date. XM833, the third-generation round, is still some years away (1983–1984). However, its problems may be beginning with the recent BRL acknowledgment that it may not achieve its predicted muzzle velocity.

3. Increasing armor protection on the T72/80/80FO Soviet tanks will be met, to some degree, by upgunning XM1E1 to 120mm, but that $2–5 billion program will do little toward extending the service of thousands of M60/XM1 tanks remaining in the inventory.

4. The Army cannot provide a viable antiarmor force in the near term unless effective ammo can be provided for the M68 105mm tank gun. The 105mm tank fleet represents a significant investment, not only to US forces but to our allies. It must remain a potent asset throughout the remainder of the 1980s and well into the 1990s. If the XM774 program slips one year, the round will be fielded in the same timeframe as the T80 tank. Against the frontal 60 degree arc it will be marginal to ineffective (defeats turret at 1.3 kilometers but will not defeat the glacis). The Army must take action now to improve our current record of multiple-generation R&D ammunition programs, each of which is obsolete before it is fielded.

Mine-Clearing Issues

Letter to Major General W. Russell Todd
TRADOC Combined Arms Test Activity
19 June 1980

I received your letter on the Track Width Mine Roller and fully agree with you that we need to field this piece of equipment immediately. Ultimately need a total system that will give us an effective mine-clearing capability.

Our countermine efforts, or lack thereof, have been a great concern of mine for many years. Since World War II very little progress has been made in the US countermine program. There has been a continuum of analyses and studies of countermine programs without the fielding of a single operationally effective system. Our current countermine capability is limited to hand-held mine detectors and hand-emplaced explosives.

Meanwhile, the Soviets have improved their mine warfare capability by fielding a complete family of minelaying equipment, from simple truck-mounted chutes to full-tracked armor-mounted and helicopter-mounted systems. Their mines are every bit as good as ours and they have fielded them in quantity. Simultaneously they have made great strides in their countermine program. For detection, they have a series of detectors better than ours. For clearing, they have plows and rollers mounted on tanks to platoon level. Their doctrine integrates countermine and barrier warfare into their notions of how they'll fight the combined arms team.

Operationally, the mine threat is twofold: fortified positions, strongpoints or belts; and scatter mines delivered at random by artillery, rockets, or aircraft.

On our side, we suffer from the "lacks." Lack of an operational concept, lack of priority, lack of money, and lack of interest.

All this tells me that we must do several things. We need to flesh out the operational concepts. . . . We need to field the equipment that will comprise the system I've alluded to. The nature of mines themselves and their mode of employment combine to confront us with a problem that can in all probability not be solved by a single device, either for detecting or clearing.

Looking at the threat and what must be done to counter it, and looking then to our own use of mines, I see two basic requirements: a high speed mine-detecting mechanism, and a rapid mine-clearing capability.

Mine-detecting systems must first of all be able to detect mined areas. Then, if time and resources permit and the mission dictates, we should also be able to identify individual mines. Detection equipment must operate over a wide area, at high speed, in real time and with reasonable reliability.

As to clearing, it is also obvious that no one device will do the job. This is because of the multiple fuzing of the mines themselves, and the threat's way of integrating mines with other obstacles, such as antitank ditches and berms. Therefore the clearing capability must be a total system. Ideally it should include the ability to roll, plow, blow and/or scrape away mines, whether in a minefield or randomly sown; the ability to cross a ditch; and finally the ability to fill the ditch and thus completely defeat the whole barrier system. The equipment to do all this needs to be deployed to company level in armor and mechanized units.

Conceptually, such a system would include rollers, plows and line charges (man-portable and vehicle-launched) for clearing single lanes; vehicular-mounted bangalore systems for clearing wider lanes; then dozers and tank-propelled bridges for overcoming ditches and berms. One outcome of our operational concept will be the need for the mounting of dozer blades, plows and rollers on some tanks of each tank company. This new organization would call for the mounting of some type of barrier-defeating device on one tank in each platoon of the tank company.

Ultimately, many obstacle breaching problems could be solved by an armored counterobstacle vehicle. It would be equipped to do many things. One of the most important would be to fill the ditches and reduce the berms using a backhoe.

All of the above will be part of an integrated system employed in pursuit of a well-thought-out operational concept.

I could not agree with you more that a reluctance to accept a piece of countermine equipment that does not meet, to the letter, every desired operational characteristic, is in part why we have no countermine equipment. Therefore I intend to see the roller funded, procured and fielded soonest, while simultaneously tasking my doctrine and combat development guys to finish the concept and field the system.

Cavalry Vehicles

Letter to Major General George S. Patton
Washington, DC
29 August 1980

This responds to your . . . letters regarding cavalry vehicles, with Hap's comments included.

Your observations concerning the CFV reflect identified shortcomings which we are working with the FVS Program Manager to correct. . . . We know the vehicle's faults and we're working at making it the best system possible prior to production. We are, however, strapped for time and money; if we don' t field it as soon as possible we're likely to have even tougher longer range problems. We can work out the inconveniences later; but, for the interim, we'll have a more than capable combat vehicle in the hands of the troops who need it now.

With regard to the doctrine problems: I believe that in modern war the traditional cavalry role needs to be performed; also I'm certain that mission requires a somewhat different mix of sensors, and air and ground vehicles, than has been the case in the past. All this gets all too frequently confused in our minds as we contemplate the opening phase of a war in Europe in which the cavalry as a covering force will have a tough, intense fight on its hands from the outset. Recognition of that fact has, of course, caused us for many years to make antitank regiments of our Europe-based ACRs. And indeed they won't survive for long unless we do just that. I would also say that, as the Middle East becomes more and more a theater where heavy forces will fight, cavalry needs heavier, longer range weapons. The CFV is a compromise, and it may well be I did the wrong thing by insisting on its adoption. But at the time it was that or *nothing*, and I couldn't let it go at nothing. Tactically we'll just have to train ourselves to use it correctly, and that includes using it with the tanks. We'll have eventually to separate the tanks and scouts into two platoons—the nine-vehicle platoon that grew out of the Cav Scout Study is too big. We knew that at the time, but were operating against a constant manpower base and so couldn't afford the extra officers. We'll have to add sensors in some configuration, and we'll have to figure out the right mix of air cav to put with the ground elements. What we have now is not enuf, but I'm not prepared to say how much "enough" might turn out to be.

Do we need a wheeled vehicle? Yes, I believe we do. What should it be? Well, for the moment the CAD Gauge Systems would do. In the longer term we probably need something else. Can the Materiel Acquisition Systems provide it? I doubt it—certainly not if we let DARCOM run the development in traditional fashion!

The V150 is a candidate vehicle for the RDF and light infantry division. Gerald Bartlett's letter reinforces data we have been provided by the contractor. While V150 has significant advantages over tracked alternatives, it compares unfavorably with some other wheeled candidates. Its unit cost, tactical transport ability, strategic deployability, weight, silhouette and other characteristics are cases in point. In short, we'll be a long time sorting this one out. But we'll keep after it.

We'll keep on this one—we have to have that vehicle.

Tank Track

Message to Major General Duard D. Ball
XM1 Program Manager
21 November 1980

I am well aware of the durability problems with our current MBT track. Additionally it requires excessive maintenance and weighs a lot, which in turn impacts upon power train and suspension RAM-D. We all recognize the desirability of interoperable track, but it is obvious by the actions of the FRG that Diehl 570F with its high vibration characteristics is not the way to go. The Gerlack G5O does have reduced vibrations, which at least is a step in the right direction. I am aware of our developmental efforts that have produced the XT152. While I am interested in how well it performs during tests, I am concerned, as I know you are, about its weight. Any track that adds 1.6 tons to the vehicle weight must be looked at carefully. In addition to our own in-house work we must continue to explore international equipment and technology, and if necessary intensify joint development programs.

Winter Track Pads

Message to Multiple Addressees
12 February 1981

Following is text of message I sent John Sherman [Crow] on receipt of his. It sums up my perspective on the issue.

Quote

. . .

2. Thirty years ago last fall a team of "experts" from the Tank Development Group in Detroit visited Europe to inquire of all the field "experts" there what we should do about tank track for winter war.

3. Tank and cavalry units in Europe in those days were loaded with "experts." Almost all NCOs and officers present had fought in armored units in Europe during the war. Since I was not one of those "experts," I listened carefully to all that transpired so that I might become wiser about the whole matter by listening to all that experience.

4. The constabulary CG and the commander of the battalion to which I was assigned at the time convened a three-day meeting to decide on the matter. All the "experts"—sergeants, officers, and others—came. Our battalion commander even invited a number of his German friends against whose Panzer units he had fought in the Great War just finished. All us nonexperts watched and listened.

5. The steel track and the rubber track guys quickly separated into two camps about equally divided. The grouser guys, the special track pad group, and the inverted center guide guys formed splinter groups. They attached themselves at random to the two main groups, depending largely on personal or unit loyalties more than any other reason.

6. After three days of vigorous discussion and sometimes violent debate which extended well into the beer halls following the daily meetings, it was concluded that it was not possible to reach agreement on the matter. However, feelings ran so strong that everyone had to be accommodated. So a decision was made to buy two sets of tracks for every tank—one steel, one rubber. Rubber was to be used in peacetime and during maneuvers along civilian roads and so on. Steel was to be used in wartime and in maneuvers in training areas where road damage from steel tracks was not so much a problem as along highways and byways off the reservations. And for some time after that we changed tracks from rubber to steel when we went to training areas—a terribly time-consuming and frustrating process!

7. As time wore on most of us came to the conclusion that, under a wide variety of conditions, the differences in track performance as between steel and rubber weren't all that great. Certainly they are not great enough to warrant the cost of issuing two sets of tracks for every tank, and the trouble and time to change back and forth all the time. To the best of my knowledge nobody of date before or since would conclusively argue against that judgment. Considerable testing has been done—all of it about as conclusive as that three-day meeting in 1949.

8. The term "aggressive track" is a phrase invented by a track salesman. The only "aggressive" track I've ever seen was one that came off a tank (when the front idler came off for some reason) and rolled down a hill, then through a gasthaus at the bottom, heaving up several cases

of beer and scaring hell out of the owner's daughter and her boyfriend, who were having a class in anatomy in Braille in the room where the beer cases were stored.

9. You need not be reminded that the track salesmen have a very, very poor track record.

10. While I deplore as much as do you the unfortunate death of one of your soldiers due to an accident on icy roads, may I remind you that the best way to cope with these matters is to reverse a few of the center guides, if time permits, and to conduct a very good training program for tank commanders and drivers. Whatever the road conditions, training can prevent accidents that would otherwise surely occur, especially under adverse road conditions. There is no development program for "winter track," and I'm not going to start one. We don't even know how to make "summer" tracks yet!

Tank Electrical Power

Letter to James E. Sayers
Somers, Connecticut
17 April 1981

I think I can best respond to your letter by first telling you what we have already done and will be doing to help prevent electrical power problems, and then let you know what a tanker can do should he find, for some reason, that he no longer has electrical power.

As you can imagine, the Ml does have a large electrical power demand. We have already done several things to ensure that adequate power is available for the electrical components. To begin with, new technology has made present-day electrical circuits much more efficient than earlier circuits. Second, the batteries that we use today have a longer life and are more reliable than earlier batteries. In addition, we have installed sensors that will tell the driver when the batteries require servicing. To assist in servicing, we have moved the batteries from under the turret floor to a more accessible position just below the back deck.

However, under high power demands, even with these improvements, we must periodically start the main engine of the Ml to recharge the batteries. One alternative that we think has possibilities and that we are now testing is an auxiliary power unit. Essentially it is a separate generator that is powered by a small turbine engine and will probably be located in the left rear sponson box. It can provide enough electrical power to operate the tank even while sitting in a defensive position with the main engine off, thus preventing the batteries from running down.

The next question is what can we do if for some reason we find ourselves in a tank with dead batteries. We could start the engine of that tank using the electrical power from another vehicle simply by connecting them using a slave cable. In an emergency, even with dead batteries and no way of starting the tank, all machineguns and the main gun can still be fired.

Your last question was concerning naming a tank after me. Although I would consider it an honor, I am certainly not in any hurry since in most cases the tanks are named after a deceased person.

Impact of the M1 Tank I

Message to Multiple Addressees
15 May 1981

1. Watching the M1 test this week confirmed in my mind the need for a close look at what we say and teach about tactics, command and control, and logistics support for the combined arms team when M1 tanks are a part thereof.

2. There are several reasons for this. Primary among them are:

a. M1's mobility differential over companion vehicles of the combined arms team is so substantial that other vehicles simply cannot keep up. The situation is analogous to that we faced in the early 1960s when we fielded the M60 tank as a companion to the M59 APC. It was not until we got the M113 a few years later that we had a team in which there were vehicles with common mobility characteristics. In the interim we had to modify our tactics. It made no sense to slow the M60 down to the march rates and cross-country mobility of M59s, thereby depriving ourselves of the mobility advantage afforded by M60. At the same time, the need to operate in the all-arms mode made it necessary to carefully calculate how to get the mech infantry around on the battlefield so that they could be employed at a time and place where they could be useful members of the combined arms team. That circumstance is upon us again—this time in spades, for the mobility of M1 is so much greater than that of the rest of the fleet that the problem is at least an order of magnitude greater than that we faced 20 years ago.

b. The M1's night and daylight fog-smoke-dust capability for both fire and movement is so much greater than that of other combat vehicles, with the limited exception of M60A3, that there is no way that companion vehicles of the combined arms fleet can keep up with M1 in night movement, night combat action, or daylight movement and combat in conditions of extreme obscuration.

c. The problems cited above for the combat vehicle fleet are even more acute if we consider command and control for combined arms operations. For M1 tank units there is no choice; company and battalion commanders must ride in and operate from their tanks. Since that's where they should be anyway, despite some opinion to the contrary, the problem is how mech, artillery, brigade, and other command and control and support elements keep up with the movement of and physically follow along in a battle featuring the M1.

d. Logistics support, for all the reasons cited above, must be much more aggressively handled and directed, with a lot more anticipatory planning than has ever been the case before.

3. Several years ago we addressed ourselves to this problem at Knox when first we had the second-generation night vision equipment on the M60A3. Then we set about to try to solve the problem by adding at least night driving devices to other combat and command and control vehicles. I am afraid that idea died aborning for budgetary reasons. However the thermal sight for movement at night, as well as in dust, smoke, fog, and obscuration during the daytime, gives M1 so much more of a mobility and firepower advantage than did the A3 fire control and driving viewer that it's time we looked once again at what we should do with regard to these three essential elements of combined arms operations.

4. At your earliest opportunity I'd like for the three of you to put your heads together and tell me what you think we should do about this. As a point of departure use the work we did at Knox with the A3; describe what's been done, if anything, since then to improve the night and obscuration movement capabilities of the rest of the fleet; then lay out what we should do about the M1 problem doctrinally, and in terms of equipment fixes, to try and improve the ability of the rest of the vehicle fleet to keep up with M1.

5.Your consideration should include how to use the AAH and the M1 together; operating together will surely give us a tremendously powerful and mobile striking capability that we've not had before. I'll reserve judgment on whether or not the M2 and M3 can keep up with M1; my immediate judgment is that they cannot, although they will have the night and obscured vision capability.

6. We should also remember that last time this circumstance confronted us the Department of the Army did nothing about it; that is, the school system did not doctrinally investigate or provide to us in the field any suggestions about the scope of the problem or how to cope with it. We had to figure the problem out in the field for ourselves; some did it better than others; none did it very well; nor was there any consistency to the solutions. Everyone marched off to his own drum, making combined arms actions very difficult unless one worked with the same unit all the time. While that is a course of action for us to consider in this instance, it probably is not a relevant one—we are on the verge of a substantial new combat capability here, one we simply must learn to exploit if we are to win; it's that problem I'd like you to tackle and report on to me NLT mid to late June.

Impact of the M1 Tank II

Message to Multiple Addressees
15 May 1981

1. Earlier I sent a message to some of you expressing my concern over the problems of combined arms employment—tactics, command and control, logistics—brought on by the advent of the M1 tank's considerable mobility differential over the rest of the fleet, and the marked advantage the M1 enjoys for night movement and fighting and for daylight fighting in obscured conditions—all afforded by the thermal sight.

2. As soon as we have in hand the significant results of the M1 test at Hood, especially those relating to the three functions cited above, we must move quickly to infuse the important test outcomes related to those matters into our combined arms instruction. Staff planning in divisions and brigades armed with or supported by the varieties of equipments we are about to field will become a much more complex operation than ever before. Officers of all ranks, but especially those who hold staff jobs at battalion, brigade, division, and corps, will have to know in some detail about types of ammo loads for artillery munitions, tank gun ammunition, and a variety of subsystem support items for subsystems unique to combat and artillery vehicles by type—almost by side number. My judgment is that we should try to get the most important relevant test outcomes from M1 testing integrated into our combined arms instruction starting this fall. That means some tentative planning factors, and a detailed accounting of the factors with which we want to begin integrating these considerations into the instruction.

Implications of M1 Tank Capabilities

Letter to Lieutenant General Sir Robin Carnegie
15 May 1981

After leaving you I went down to Fort Hood to see the M1 tank trials underway there. Going very, very well with a full battalion under test. It is however quite obvious that what we've suspected for some time has come to pass. That is that the mobility differential of this machine is so much apart from the companion vehicles in the combined arms fleet that we shall have to look again at our tactics, command and control and logistics support. M1 just outruns everything else on the field.

I followed them through a night road march in which they made about 80 kilometers in an hour and a half with no problem at all. All the infantry carriers, command and control and support vehicles were hours behind, struggling along, and the tank battalion had swung into the attack! It is quite like the time 20 years ago when we first fielded the M60 tank and had the M59 armored personnel carrier to go along with it—a vehicle with scarcely half the capabilities. We had a devil of a time until we got the M113, which could keep up at least most of the time.

Additionally the M1 thermal sight capability at night and in conditions of obscuration in daylight really gives it a marked advantage over anything else on the field. Somehow we must cope with that fact as well. If indeed we succeed in fielding another generation of fighting vehicles, it is difficult indeed to imagine what new capabilities that might portend; we'll have the devil's own time simply adapting to bring everything else we own up to M1 capability levels.

M1 Tank Assessment

Message to Major General John B. Blount
Fort Monroe, Virginia
4 June 1981

Have . . . your 021400Z on M1 follow-on durability testing. Okay. Let's do what we can to help. We have to have that tank—it is just super!

Tanks Forever and Ever

American Defense Preparedness Association
Fort Knox, Kentucky
23 September 1981

Some years ago, you were kind enough to invite me to speak at a gathering similar to this one. The meeting was held at Fort Belvoir. The 1973 Yom Kippur War was then a recent event, so recent in fact that there had not been time for sober reflection and objective analysis, and in the wave of misinformation and mistaken notions that followed that war, there was loud clamor in this country that the antitank guided missile had at last done the tank in. The day that dawned at Cambrai had finally come round to a twilight.

While the world around us basked in that technocrat's euphoria, we talked about whether tanks are or are not necessary, whether they have or have not been done in by technology.

There may be no one here present, besides Jimmie Leach and me, who was present at that earlier meeting. However, even should there be, the time appears propitious to lay out those earlier arguments, which I called "Tanks Forever," in an updated version. Several things prompt that decision.

First, there is resurgence of the idea that technology is about to make the tank obsolete. While it would be easy to dismiss that as yet another example of our infatuation with technology, it might be unwise to do that just yet.

Second, what was said at that earlier meeting about the tank as the central actor in the combined arms team is still true—perhaps more so today than then. The history of the last 65 years tells us that, despite considerable devastating evidence, the concept of combined arms operations with the tank as the central system is one that, although we have paid it great lip service, has enjoyed, at best, a retarded growth in our forces. There are many reasons for this. Some will be dealt with in a moment.

Third, in that earlier talk I tried to set forth some challenges with which we all must cope in order to take advantage of technology and to challenge technical people to think seriously about what they should be doing to provide us with a vastly improved battle capability. Today I'd like to repeat, but broaden, those challenges.

Fourth, I elected to revisit our earlier arguments because of strategic mobility difficulties now besetting us that inhibit our ability to deploy heavy forces quickly enough to respond to what is foreseen as our most likely contentious problem. Because weight and volume inhibit moving mechanized equipment over long distances, we have elected to deploy light forces that can be moved more quickly with limited lift. Acknowledging the lift problem, which inevitably we must solve, deploying light forces may be exactly the wrong thing to do, even though it may be the quickest done.

Now, let's talk about all that. First, a perspective on the tank-antitank debate. As with all elements of warfare, tank-antitank battle is a contest of measures and countermeasures. For every system there is a countersystem. If none exists, rest assured one will soon be forthcoming.

The relatively free flow of information in today's technology world virtually ensures countermeasures will appear very soon after the new measure itself appears. If the materiel acquisition cycle of one side turns more quickly than that of another, this is surely to be the case.

Thus our constitutionally lethargic acquisition system, coupled with rapid technical information flow, practically guarantees a high expectation of seeing the other side's countermeasures before we have fielded our new measure.

All this is quite like the children's game of rock, scissors, paper. Rock breaks scissors, scissors cut paper, paper covers rock. As it is in the game, the goal in battle is to apply the tactic that best utilizes the capabilities of each battle system while minimizing its vulnerability to countermeasures. As in the game, a mixed strategy is required to win. We don't stand off, refusing to play the game, just because an effective countermeasure is foreseen for every measure we may take.

About technology we must recognize several things. First, as a nation we have a historic obsession with gadgetry. We are and have long been convinced that technology offers a solution to most of the problems that beset us. So it is that for years we've embraced the notion that it was okay for the other guy to outnumber us; our superior technology would compensate for the disparity in numbers. Not true. Probably never has been. Certainly not true today. The history of battle clearly shows, to the technocrats' continuing dismay, that it is neither quality nor quantity that necessarily ensures a win. There is great strength in numbers, some advantage in quality, but the greatest advantage is in the training of soldiers, leaders, and the units in which they fight. To so many of you, I've said that so many times before that I shan't embellish it further.

Second, to the scientist writing in the August issue of *Scientific American* about precision guided weapons and the demise of the tank, I have to recite a set of numbers reflecting the costs of tank-antitank duels using his new, exotic, terribly sophisticated and technically advanced munitions. Suppose a tank does cost a million and a half in today's dollars. Further suppose its kinetic energy round costs $500 and that the gun can fire 400 or 500 rounds before the tube must be replaced. This gives us a rough cost of $3,800 per KE round.

Suppose the antitank guided missile costs $8,000. The cost of 400 or 500 such missiles, with a single launcher for every 10 missiles, would be nearly $5 million or $11,000 per missile. Thus we could fire about three kinetic energy rounds at a target for the cost of one guided missile. Using the missile-ammunition expenditure rates from the Yom Kippur War, the several thousand missiles fired could buy 40 to 50 tanks, a force to be used on a variety of missions rather than one specialized antitank mission.

It is interesting that, no matter how costs increase, the relationship between those numbers stays about the same. This happens for a whole lot of explainable reasons—explainable, but all too frequently ignored in the spate of promises of what technology offers.

Now, let's consider the tank as the central actor on the combined arms stage. Armor soldiers have never viewed the tank as a self-contained battle system. The tank has always been a part—an essential part—of the combined arms team. One of the striking lessons the Yom Kippur War taught us was that the diversity and lethality of systems we can expect to see on that busy, busy modern battlefield is such that no single system, or even one or two systems, can possibly cope with the total spectrum of threat. So on the ground, in the air over the ground, in the electronic spectrum of the battle, indeed in all dimensions, combined arms are the order of the day. Tanks can't go it alone. It is the air-land team—tanks, infantry, artillery, close air support, armed helicopters—that wins. That team must fight—must fire, must maneuver in close, concerted, coordinated action—to win.

Some further words about tanks and the combined arms team are necessary. Tanks were introduced into battle to restore mobility to the battlefield, enabling the side using them to wrest the initiative from the enemy. Tanks were the first element of the combined arms team to become other than foot or horse mobile. They reflected the essential lessons history taught concerning the need for and effectiveness of mobile cavalry, dragoons, horse, or elephant-mounted infantry in battle. Modern tanks possess impressive firepower, so much so that critics now contend we have become an attrition-minded Army solely bent on destroying numbers of the enemy. But it is precisely that firepower that makes it possible to utilize the superb mobility we've provided our modern tank and its fellow combined arms team members. It's the balance of firepower and mobility in the system itself, and in the team, that is the secret to its battleworthiness. With regard to attrition, let me simply say that when there are a lot of enemy out there it is usually necessary to kill quite a few just to get their attention. Then some more must be killed in order to convince them we're serious. That magnificent reporter of war, S.L.A. Marshall, wrote in his classic little book *Men Against Fire*:

> . . . armies from civilized states are so strongly influenced by civilian thinking that in their desire to refrain from circulating any ideas which may be shocking to civilian sensibilities they sometimes slight their own first principles. That is one reason why the subject of fire is not given its just due. We are reluctant to admit that war is essentially the business of killing, though that is the simplest truth in the book. Indeed it is so simple that the thinkers on war have passed it up in favor of half truths which contain a greater element of mystery.

Words like deter, dissuade, fend off, delay are not the action words of warfare. Words like attack, defend, seize, hold, kill, win are the operative words of war.

Now, more technology. Aside from the lack of historical evidence that technology can relieve everything from the general's hemorrhoids to the national debt, other things need be said about the application of technology to battle. First, if ever we are to make technology work for us, and if we are to stop being slaves to it, we must proceed from the conviction that operational concepts drive the development and acquisition process. A concepts-based acquisition system is essential if ever we hope to produce what we truly need at anything like a reasonable cost. We simply cannot afford to heed the siren sound of hucksters peddling wares for which there's no operational concept. If we do that, we're back to the old Ordnance Corps mentality: "We'll design and build it. You grubby soldiers leave us alone. When we've got it done, we'll give it to you. It shouldn't take long to figure out how to use it."

Second, new technology should be applied to developing weapons for the purpose of rendering ineffective costly investments by the enemy. Some at least of our potential adversaries do this very well against us; we should be ashamed that our marvelous technical capability hasn't given us a similarly impressive track record.

Third, new weapons technologies should not simply seek to match or counter something the enemy has developed. They should rather seek to confront the enemy with systems that open new dimensions of combat. It cannot be the purpose of technology simply to enable us to meet the enemy on quantitatively or qualitatively equal or near-equal terms.

Finally, new technology should seek to make battle outcome less, not more, calculable. Instead of restoring some balance to a neat Lanchester Square or firepower score equation, technology should seek to introduce new imponderables into the calculus of battle—new and difficult imponderables, ones that will cause the enemy untold anguish and expense in an effort to cope.

Now a few words about today's need to deploy forces to areas other than Central Europe in support of our vital national interests. Traditionally we have considered the contingency mission one in which we could and should employ only light forces. While this is so in part because, until recently, that's what most contingencies called for, it is also so because our limited capability to deploy heavy forces further reflects our traditional military bias toward insertion forces—airborne, airmobile, marines. While each of these forces has its place in the structural arsenal of our military complement, that place is very heavily affected by enemy, terrain, weapons, and the need for mobility once inserted. Despite the ease with which they can be transported, it is probably counterproductive to set down light insertion forces in an environment in which they are likely to encounter substantial armored forces and in which they cannot be expected to survive. Today's contingency operations must take account of the substantial growth of heavy forces in areas of the world currently in the news.

With nearly 3,000 tanks each in Iraq, Syria, and Israel, with nearly 2,000 in Egypt and Libya, and with other armored vehicles, artillery, and high-performance combat aircraft in numbers proportional to the tank fleets, the deployment of light infantry to operate in that highly mechanized environment, however rapidly it may get there, must be viewed as a risk operation of the highest order.

Recognition of this problem has led to proposals for fielding a lightweight armor-defeating vehicle. Some development has taken place, some of it here at Fort Knox. While we do indeed have a requirement for a lighter weight armor-defeating vehicle, three things must be made quite clear. First, given the state of current armor technology, no known armor formula can give us the protection we need in such a vehicle at an affordable weight. Second, the myth persists that light weight brings with it a certain agility that can be used to offset the lack of armor in the armor envelope. This is a self-fulfilling prophecy and, of the shibboleths of modern times, the hardest to slay. Yet a considerable body of experimentation and other evidence gathered over the last eight or nine years shows little if any advantage in survivability afforded by improved agility. Speed made good in a dash to cover—yes; agility as some sort of horizontal gavotte across the battlefield—no. Finally, if a lightweight tank defeater is what's needed, then that signals a missile system as the primary weapon. Weight, standoff, and many other advantages accrue. Contrariwise, our most recent experimentation has been with gun systems provided by technology and suitably huckstered by the sybaritic whispers of the technocrats.

A host of solutions abound, offering alternatives to deploying at high risk those light forces. One, we have made extensive use of pre-positioned equipment in Europe. Pre-positioning elsewhere—afloat or onshore, practice deployments, and a host of other schemes are available to get heavy forces to where heavy forces are needed, to get the tank-based combined arms team onto the battlefield where its firepower, mobility, and survivability are sufficient to face the challenge.

One would hope that the administration's drive to strengthen our defense establishment will in due course redress the considerable imbalance in our strategic mobility capability. Until that time comes, however, we must seek alternative but perhaps less-effective ways to provide heavy forces in situations where their presence is essential to survival of the force and accomplishment of whatever mission it might be assigned.

So, as was the case last time we met to discuss the matter, we come to the end with conclusions much similar to those we drew before:

- Technology is not about to make the tank obsolete. The measures-countermeasures game is just in full sway. We shouldn't lose our sense of perspective in the excitement of the game.
- The tank is the central system in the combined arms battle. It cannot fight that battle alone, nor can other systems fight that battle successfully without the tank. The lethal firepower of the tank and its supporting systems is a vital factor in the firepower-mobility equation. That equation is key to the success of the combined arms team. Without its firepower, the tank's mobility is impressive but not effective; without its mobility, the tank's firepower is impressive but effective locally and for but a brief time.
- Technology must be harnessed to provide systems whose general characteristics are spelled out by a carefully structured operational concept of how the battle is to be fought. Technology should be harnessed to the tasks of identifying and developing the means to render ineffective heavy enemy investments in specific systems or capabilities. New weapons technologies should not just seek to match the enemy qualitatively or quantitatively or both. Rather they should seek to challenge the enemy in new, different, and demanding ways. Technology must make the outcome of battle less, not more, predictable.
- Finally, until our determination to improve our defense posture can take effect in terms of new or additional capabilities, we must seek alternative ways of providing heavy forces where heavy forces are needed. Just because we find ourselves unable to get there quickly with heavy forces, we cannot ignore the need for them. Certainly we cannot do so at the risk of deploying light insertion forces that are clearly inadequate to the task that they've been assigned.

So, once again, it's tanks forever. One would hope that in six or seven years it will not be necessary to make this speech again. But it probably will be. Institutional memory is short-lived, and it's most difficult to proceed on a steady course. And that's why I elected to challenge you once again with tanks forever—forever and ever.

Antitank Missiles

Letter to Alton B. Parker
Tampa, Florida
7 December 1981

With regard to the antitank debate . . . let me cite some facts which may put the matter in some better perspective. There were about six thousand antitank guided missiles fired at tanks during the war. Most were fired by the Egyptians against the Israeli Defense Force tanks in the Sinai. Losses of tanks to those six thousand missiles amounted to about ten percent of the total tank losses experienced by the Israeli. The cost of those missiles, launchers, crews, and peripheral equipment would have bought for the Egyptians nearly sixty tank battalions—formations with much greater mobility, killing power and survivability than the antitank weapons, crews and vehicles.

The truth of modern battle is that no single system will dominate. It will continue to take all systems operated in concert to do the job. Combined arms operations are still the order of the day. For every new measure the scientists think up, someone soon fields a countermeasure—it's like the children's game of Rock, Scissors, Paper. You may not be old enough to recall it; some of us are! The rock is covered by the paper, which is cut by the scissors, which, in turn, break on the rock and the cycle begins again. So it is with weapons systems. Nonetheless, we always find some zealot predicting that his particular gadget will conquer all. Hasn't happened yet; not likely to do so in the foreseeable future.

Evolution of Doctrine: The Armored Force Example

Army War College Committee
Carlisle Barracks, Pennsylvania
10 June 1982

Don Marshall assigned me the imposing task of presenting the "theme address" tonight. He described the theme as "how to make relevant our theoretical concerns and intellectual activity in helping solve the real problems of battle commanders and of military and national planning." Some initial difficulty understanding just what you're trying to do led me to a search for some definitions. What is meant by "general theory of combat"? A "philosophy of war"?

Let's take theory to mean "systematically organized knowledge applicable in a relatively wide variety of circumstances." Philosophy, let's say, is "any system of motivating concepts or principles." Now, we search for theory and philosophy about war. Despite some difference of opinion, we can probably say war is both an art and a science.

If art is "a system of principles and methods employed in the performance of a set of activities" and science is any "activity that appears to require study and knowledge," then we can argue that war is an activity requiring study and knowledge (science) and that our study involves a system of principles and methods employed in the performance of military activities (art). Further, we would like our study to result in a system of organized knowledge applicable in a wide variety of circumstances (theory) and a system of motivating concepts and principles (philosophy). So, with that heading, how do we proceed? Well, the object of our study should be that total military system embracing military strategy, operational art, and tactics.

This is particularly so if we consider that *military strategy* is the art and science of employing the armed forces of a nation to secure military objectives commensurate with national political policies or strategy by application of military force or the threat of such force. *Operational art* is the marshaling of fighting systems—weapons systems, organizational systems, tactical systems, supporting systems of all kinds—in large units. In practice, it is the planning and conduct of campaigns with those systems. *Tactics* is the set of specific techniques used by smaller units to fight and win battles and engagements. Key in the definitions of theory, philosophy, and art was the word "system," in this instance the military system, which begins with strategy, proceeds through operational art, and continues into tactics.

With those definitions in mind, may I rephrase the theme question? Isn't what we're trying to describe how to bring the results of our intellectual activity and theoretical concerns to bear on the process of change in our military system? If we have theoretical concerns that spur intellectual activity about that system, it is quite likely so because we perceive a need for change. So the basic question, then, is how to effect change in that system of strategy, operational art, and tactics by which the political aims of the nation are translated into action by the nation's military forces acting as one of the several instruments of national power. So may I just argue along those lines? The history of change in military systems of the world is instructive. A few examples are illustrative.

In this century the history of the development of concepts for mobile all-arms warfare involving the tank is most instructive. This is particularly so in the British Army, where the idea had its genesis and was the subject of much early development and experimentation. The story of Swinton's tank and the strategy, operational art, and tactics it represented is a familiar one. I'll not recount it here except to note several aspects of that struggle relevant to our question of how to effect change. First, it will be recalled that a succession of single-minded tank and mobility enthusiasts persisted in developing the concept of all-arms mobile warfare built around the tank striking force, and they did so in the face of persistent opposition by most of their peers and frequent adverse personnel actions by their less-imaginative superiors.

Second, most of the reformers were "loners." For the most part they were argumentative, assertive, and hardly ever in agreement with one another.

Third, despite support from the doughty Winston Churchill, they chose or were forced to work around an organizational system that, like all such systems, abhorred change. In frustration, many went public with their arguments, as did Fuller and, in many cases by doing so incurred enmity among their superiors sufficient to bring on their early retirement from active ranks or relegation to some inconsequential posting.

Fourth, although trials were held to demonstrate the new concepts, those to benefit most from the trials were the Germans, who spawned the blitzkrieg concept based largely on their study of tank actions in World War I and of the British trials on the Salisbury Plain in 1934.

And so it was finally that war came to Europe in 1939 to find the British Army with an imperfectly perceived operational concept and inadequate tactics, organization, equipment, and training to implement a military strategy and operational art they themselves had invented. Kenneth Macksey's new book *The Tank Pioneers* gives a first-rate accounting of this whole train of events. I commend it to your study.

In the American Army the concept of mobile war fared even less well than in the British Army. The pioneers were fewer in number, and the institution proved considerably more resistant to change than the British Army. A succession of Army Chiefs of Staff rejected the idea out of hand, that imperious cavalier Douglas MacArthur testifying before the Congress to the effect that one should not buy too many tanks, for they were terribly expensive and so soon obsolete. Strongest among the opponents was that bastion of mobile thinking the United States Cavalry, whose last chief, John K. Herr, was the most stridently outspoken opponent of the idea of all-arms warfare built around the tank. There were really only two heroes of this drama in our army—Generals Adna Chaffee and Daniel Van Voorhis. Without General Chaffee, the American Army quite likely would have had no tanks at all in 1940, and without Van Voorhis there would not have been the foggiest notion of what operational concept framed the world of mobile warfare. As Nicholas Katzenbach concludes in his fascinating little paper, "The Horse Cavalry in the Twentieth Century," the Army of the most mechanized nation on earth came to the threshold of World War II firmly wedded to strategy, operational art, and tactics deeply rooted in the 19th century.

On the other hand, the Germans seem to have had, in what Trevor Dupuy calls their "genius for war," a much more impressive willingness and ability to adapt to change. Tim Lupfer's fine Leavenworth Paper Number 4 describes well the German Army's ability to change operational concepts and tactical schemes in a matter of months in World War I. Guderian, reading reports of the armored force trials on the Salisbury Plain, demonstrated the concept with a small force for Hitler at Kummersdorf in 1934. With the Führer's *nihil obstat* on the concept, Guderian then in 18 short months produced an all-arms panzer division. It operated within a fairly well-spelled-out concept that included a strategy for mobile warfare; a general campaign plan of how the larger forces would fight; and the organization, tactics, and at least a preliminary array of the type equipment needed to bring the concept from theory into reality. In his marvelous book *The German Army 1933–45*, Albert Seaton describes the German Army's ability to adapt to change in those turbulent years. Ken Macksey describes well how the German tank pioneers seized on and matured the preliminary British work on all-arms warfare built around the tank.

How did they do it? How were the Germans different from the British? The Americans? Trevor Dupuy can speak more eloquently about this than can I, but there stand out in my mind several facts that frame the answer and that outline a set of requirements I would set forth as requisite to the ability to effect change.

First, the Germans had a General Staff element whose primary function was to examine the need for change and, when change was decided upon, to draw up the necessary changes to strategy, operational art, and tactics to make it happen. True, this capability became diffused as Hitler fragmented his Army Command into *OKW* and *OKH*. Indeed, some of the bitter antagonisms that arose between those two organizations in World War II survived until recently between the Heeresamt and the *Fu Ha* (*Fuhrungshaptquartier*) [Headquarters of the German Armed Forces] staff in the Bundeswehr. Nonetheless, for the critical developmental years, there existed an institutionalized framework for examining the need for changing the system of strategy, operational art, and tactics; describing the changes needed; and producing change through the office of the *Inspekteur*.

Second, the German mavericks were all products of the enormously demanding and rigorous officer selection and training program characteristic of the German system to this day. Mavericks they may have been, but all had been taught to think logically about tough problems, taught in the same way in the same schools. Compelling logic to one was, therefore, equally compelling to all, making a consensus much more easily arrived at. And consensus by some means there must be or change simply cannot be effected.

Third, the principal instigators of reform remained for years in positions related to implementation of the changes they espoused. Follow Guderian through the solution of blitzkrieg, for example. Change was further facilitated because the senior leadership, to include most importantly Hitler himself, was quick to seize on the strategic advantages Germany could gain over its potential foes by changing the basic ingredients of its military system.

Finally, trials had been conducted—by the Germans in Russia, by the British on the Salisbury Plain, and by the Germans and Russians in the Spanish Civil War, and the closely observed lessons fed back into the system of strategy, operational art, and tactics.

So, recounting then, here we have a set of generalized requirements for effecting change:

- There must be, in the institution, the mechanism to identify the need for change, draw up parameters for change, and describe clearly what is to be done and how that differs from what has been done before.
- The educational background of the principal staff and command personalities responsible for change must be sufficiently rigorous, demanding, and relevant to bring to the solution of problems a common cultural bias.
- There must be a spokesman for change. The spokesman can be a person—one of the mavericks, or an institution such as a staff college or staff agency.
- Whomever or whatever it may be, the spokesman must build a consensus that will give the new ideas and the need to adopt them a wider audience, converts, and believers.

- There must be continuity among the architects of change so that consistency of effort is brought to bear on the process.
- Someone at or near the top of the institution must be willing to hear out arguments for change, agree to the need, embrace the operational concepts, and become at least a supporter if not a champion of the cause for change.
- Changes proposed must be subjected to trials, their relevance convincingly demonstrated to a wide audience by experiment and experience, and necessary modifications made as a result of trial outcomes.

It seems to me this is a requisite framework necessary to bring to bear our intellectual activity and theoretical concerns to effect change in the system—strategy, operational art, and tactics that are the operative elements of a theory of combat, a philosophy of war.

Now may I describe for you an example of one attempt to do what I've been trying to describe. In the Army reorganization of 1973, we addressed ourselves, in part at least, to the institutional side of the problem we're examining. In those years the Army needed many changes. Some were purely managerial, reflecting our apprehension of a lot of structure and too little manpower. More important, however, we realized we needed to change our concepts of warfighting—the strategy; the operational art of larger units that perforce would be fewer in number; tactics; and the organizations, equipment, and training to bring us out of the Vietnam trauma and moving ahead—briskly, we hoped—into the last quarter of this century.

We found ourselves confronted by principal antagonists who were almost always sure to outnumber us and a growing militarization of conflict in the Third World. The Soviets, impelled by their obsession with numbers, were obviously in possession of a mature operational concept embracing mass, momentum, and continuous land combat. Convinced by the realities of our then and impending resource constraints that we could not afford a like concept, we set about to look for ways to win even though fighting outnumbered. While Russell Weigley would argue that was more of a radical departure from our antecedents than some might agree, it was nevertheless a crucial first thing to be decided on. Doctor Bob Helmbold suggested that it was possible; the Yom Kippur War provided a marvelous and most fortuitous field trial; the lessons drawn from our study led us to conclusions about the requisite military system—strategy, operational art, tactics, organization, equipment, and training.

The end result of our early intellectual activity and theoretical concerns was set forth in what became the 1976 edition of FM 100-5. Its primary emphasis, at least as viewed by its critics, was on an operational concept we called the active defense.

However well we may have done that early work, it met with considerable criticism from within the Army and without. Some reflected institutional resistance to the notion of change; some reflected intellectual and theoretical concern. However, the experience demonstrated that we had done all too little consensus building and that we had in hand a concept perhaps not quite as mature as it should have been. The results of that realization were severalfold.

First, we set about revising the principles of war as used by the United States Army. An early criticism of the 1976 edition of FM 100-5 was that it did not recount our principles of war. While there were good reasons for this omission at the time, the decision cost us more in adverse reaction than it was worth. So principles of war were set forth in a new book, FM 100-1, called *The Army*. After two iterations, it came to embody a set of principles of war that spell out the components of a military strategy compatible with our needs today.

While that development was in train, our operational concepts evolved through a succession of changes familiar to many of you by the titles: the Corps Battle, the Central Battle, the Integrated Battle, the Extended Battle, and finally the AirLand Battle.

One lesson of that experience was that we had imperfectly designed the institutional framework to accomplish change. In 1973, TRADOC absorbed the Combat Developments Command. There were several good reasons for that amalgamation, some related to resources, others related to perceived shortcomings with the output of that command. In any event, while strong on equipment development and organizational matters, the new Combat Developments Directorate of the TRADOC staff was weak on conceptual work. So the bulk of the concept work reflected in FM 100-5 was done by a handful of people, none of whom was assigned to the Combat Developments Staff at TRADOC Headquarters itself or in the schools. Realization of this omission in our original concept of how TRADOC was to do its business caused us to create a principal doctrinal development staff element in TRADOC, a Deputy Chief of Staff responsible for identifying the need for change and for describing the conceptual framework of the change itself. Without that orderly process at the beginning and one agency directly responsible for it, the need for change would always be ill-defined and the conceptual direction of change cloudy at best.

Going back now to the beginning, we laid our work out for our Chief of Staff, General Abrams. He made many amendments but supported the concepts we were trying to explain to ourselves and to the Army. After General Abrams' untimely death in 1974, General Fred Weyand confirmed General Abrams' support, and that support from the top has continued with both General Rogers and General Meyer. Then we set about designing tactics, organizations, equipment needs, and training systems to support the new concept. This resulted, among other things, in the Division Restructuring Study and the field trials of its organizations and their tactics at Fort Hood. Because the concept was not yet mature and because, in the trials, we attempted to measure performance differentials at the margin with an instrumentation system and a test scheme not adequate to that degree of precision, the trial outcomes were too ambiguous to attract widespread acceptance. At this point it was apparent that we had to begin anew. Particularly would consensus building be necessary as organizational development proceeded. So, for two and a half years, we gathered school commandants and representatives of the Army Staff, major commands, supporting organizations, and other services, and at Leavenworth we hammered out what we now know as Division 86.

One final consideration—in the process of bringing about change, a first need is for a conceptual notion of what must be done to fight successfully in the battle environments of today and tomorrow. That conceptual thinking can only result from close, detailed, and reflective study of a wide spectrum of technology, threat, history, world setting, and trends. That kind of thinking can only be done by imaginative people who have trained themselves or been trained to think logically about tough problems. That kind of intellectual development is one of the most important functions of our Army school system, especially at the staff college level. It is perhaps here that we have not yet fully equipped ourselves with the requisite elements to achieve change. For, of them all, in the German system, the intellectual prowess and staff brilliance—the cultural commonality that was brought to bear on the process was the most impressive, if not the most effective, catalyst in making change possible, and quickly. As you know, we have begun work at Leavenworth in both the long course and the course now styled as "CAS3" to do just that. But some years will be required before it is possible to judge just how well we have done.

So, while much remains to be done, we have in place today most ingredients history would suggest as necessary to effect change. And we are in the throes of changes produced by that system, changes designed to move us into the last two decades of this century.

The need to change will ever be with us. The fact that we may have analyzed the process, framed its essential parameters, and made some considerable progress toward arming ourselves with systemic mechanisms to permit change to take place in no way ensures either that change will occur or that it will be an easy, orderly process. And so the intellectual search, the exchange of ideas, the conceptual maturation must continue and be ever in motion.

We would be well served in the future if that process could include more sound thinkers in uniform and fewer academic and amateur military strategic gadflies. We would be better served as the process matures if we could somehow focus the intellectual prowess of the operations analysis community on our fundamental rather than our peripheral needs. We would be much better served in the long run if we could learn how to change our institutions from within instead of creating the circumstances in which change is forced upon us by civilian Secretaries of War, Defense, or whatever. We would be much better served in the end if we could develop and refine in our institution the cultural commonality of intellectual endeavor and the ability to think logically about tough problems necessary to develop new ideas, mature them as quickly as possible, and chart relevant action programs to effect change in an efficient, orderly way.

In short, we need institutional leadership as well as individual leadership, for without a requisite combination of both, history instructs us that the need for change is difficult to define; that what is to be done—the goals of change—is virtually impossible to circumscribe; and that the whole process takes so long that not much ever happens. In today's and tomorrow's worlds, we simply cannot afford the luxury of that kind of inefficiency.

Combat Vehicles

American Defense Preparedness Association Conference
Fort Knox, Kentucky
18 September 1986

You have gathered here to discuss the subject of combat vehicles for a balanced force. By that title I presume the principal focus is to be on what kind of armored vehicles to provide light forces and how, at the same time, to provide the heavy forces the equipment it so desperately needs to cope with our most contentious threats—the Warsaw Pact in Europe, the North Korean People's Army in Northeast Asia, and any one or combination of several well-equipped forces in the Middle East. While I am not wise enough to solve that difficult problem for you, perhaps a perspective on where we are and where we have been in the world of fighting vehicles might be useful.

In the early 1930s General Douglas MacArthur, then Chief of Staff of the Army, testified in congressional budget hearings that we should not buy too many tanks and other armored vehicles in peacetime, for they became so quickly obsolete. The Congress heeded his warning and, in the end, provided no funding for armored vehicle development. As the story goes, it was only through the budgetary legerdemain of General Adna Chaffee that the US Army had any tanks at all at the onset of World War II.

In the wake of World War II and the Korean War, while modest resources were provided to develop and procure combat vehicle fleets, there never seemed to be enough to satisfy operational requirements for the numbers of such vehicles necessary to meet our national commitments. Our strategic largesse, it seemed, ever exceeded our willingness to loosen our purse strings sufficiently to buy what was needed to make our strategy viable. Faced with that dilemma, we embraced the notion that we would somehow overcome numbers on the other side by technology on our own side. Hence the rationalization that we really need not match our foe's quantity, for the quality of our guns, aircraft, armored fighting vehicles, ships, et al.—quality provided by technology—would more than suffice. And so it came to pass that theater nuclear weapons, a technology advantage at the time, came to represent the apogee of technology's force-leveling capability.

It is not necessary to cite to this audience what has happened to dispel the neatness of that myth. Not only has the threat quantitative advantage continued to grow, but capabilities have grown apace as well. Once nuclear parity, or near parity, was achieved, the Soviets turned the resource hose loose on their conventional capabilities to the end that we remain outnumbered, outranged, and outgunned in new and impressive dimensions from armor to guns to missiles to helicopters to electronic warfare systems. And so it is that, in theaters wherein we face our most contentious threats, the force balance is ever more precarious and the technology advantage, if ever we had it, has become one of history's most magnificent self-sustaining myths.

In 1974 the Chief of Armor spoke before an ADPA conference at Fort Belvoir on armored fighting vehicles. Pointing to the lessons of the Yom Kippur War, not yet a year past, he cited the then-popular notion that antitank guided missiles—a new technology used in significant numbers in the October War, had made the tank obsolete. That speech, often called "Tanks Forever," reaffirmed several fundamentals drawn from careful evaluation of October War lessons:

- The need for the tank as the pivotal actor on the combined arms stage, its staying power and mobile integrated firepower necessary to the successful performance of the armored fighting vehicle family.
- The relative ineffectiveness, especially the cost inefficiency, of ATGM against armored fighting vehicles, especially tanks, and especially when compared to the lethality of kinetic energy projectiles launched from armored fighting vehicles.
- The fact that the history of battle teaches that within reasonable limits—say one on six, six on one—the outnumbered side is not necessarily foredoomed to defeat, that the side that seizes and holds the initiative wins more often than not in the end.
- It was, however, noted that numbers do count; that, despite some spectacular success by vastly outnumbered forces, as a general rule numbers beyond the 10-to-1 range became very, very difficult to contend with. Hence the notion that some combination of technology, tactics, and the operational art of war could be, and indeed needed to be, invoked to successfully cope with the growing quantitative and qualitative threat.

In that latter context, "Tanks Forever" suggested some technical avenues whose successful pursuit could provide an advantage that, if incorporated into a properly employed armored fighting vehicle fleet, could spell the difference. Those technical avenues were:

- Chemical energy munitions designed to take into account the increased survivability afforded by modern composite armors.
- Kinetic energy systems designed to provide vastly increased velocities necessary to defeat modern chemical warhead resistant armors.
- Hypervelocity ammunition of all types made possible by emerging propellant technology.
- Fire control systems designed to extend hit/kill accuracies in the close battlespace regime and to make the night/weather equation equal to the clear day fire control solution.
- Agility capabilities designed to provide vastly improved dash-to-cover capabilities to increase the survivability delta.

It should not be necessary to recite to this audience that we are this day struggling to make real the potential suggested in that 1974 recitation.

Time meanwhile, as it ever will, marched on. In the ensuing six or seven years, the Soviets increased their quantitative advantage in all systems and, at the same time, introduced new capabilities. This was a period in which four new tank systems were fielded on their side, one on ours; two new infantry fighting vehicles appeared in their fleet, none in ours; very capable attack helicopters obviously designed for close air support appeared in three or four versions on their side, while we struggled mightily to upgrade our aging Huey fleet but fielded no new system; at least four new artillery systems appeared on their side, none on ours.

In sum, the strong signal of those years from 1974 to about 1981 was, and is, that the Soviets are outmodernizing us at a rate of about four to one. Driven by the relentless march of their Five-Year Planning System and the growing functional rigidity of their national economic system, their arms factories spew out numbers we simply cannot match. Additionally, each serial development brings the fielding of new technology, either on a new product line or applied to an existing fleet. Those were the years when their conventional buildup decisions of the late 1960s began to show a momentum that is both dramatic and alarming.

During the same several years, we began to see the Soviet quantity/quality modernization rate reflected in their arms exports to the Third World. Heading an impressive list of other exports, let me just cite tanks to illustrate the point. Over two decades ending in the early 1980s, the Soviets exported to seven countries—four in the Middle East, one in south Asia, and two in this hemisphere—about 11,000 tanks. That's more tanks than we have produced in all serials and mods of the M60 fleet. Other armored vehicles, artillery, and aircraft follow the same pattern, to include most recently armed helicopters.

It is customary to dismiss this phenomenon with: "That's okay; it's all old, obsolete stuff." Not true. The facts are that, as soon as the Soviets themselves begin issue of a new item of equipment to their own forces, they apparently are willing to release the next-oldest item in the export fleets. So it is that we find T-72 tanks replacing Syrian T-62 tank losses in the 1982 War for Peace in Lebanon, a move made possible by the issue of the T-80 to Soviet forces just a year or so earlier. So the equipment in the hands of client states tends to be reasonably new, and the same quantity/quality growth equation noted earlier with regard to the Soviets themselves is now reflected in the arsenal of many, if not most, of their arms customers.

All that development from 1974 onward caused the Commander in Chief, Readiness Command, to come before an ADPA conference at Fort Knox in the fall of 1981 with a speech titled "Tanks Forever and Ever."

Noting the qualitative and quantitative changes of the previous seven years, CINCRED further observed two important and conflicting circumstances:

- In the world into which he was most likely to be required to deploy forces in his role as CINCRED/Director, Joint Deployment Agency, there was a strong likelihood that US forces would encounter threat forces fairly well armed with an increasingly capable array of armored fighting vehicles, artillery, and even armed helicopters.
- The historic bias of the United States, and the US Army in particular, toward light forces. The presumption that, in most contingency areas, light forces would suffice, and the notion that, with limited lift, light forces were the most capable that could be gotten anywhere in any numbers very quickly.

The sybaritic appeal of that latter logic, and its utter rejection of world realities, need not be commented on further—at least until some light force runs afoul of even limited guns and armor in some faraway place in some ill-conceived caper against a vastly underrated threat.

Also noted were several acceptable alternatives that could put heavier forces onto those distant fields in about the same time as light forces—in fewer numbers, but with far more combat capability, especially against the increasingly well-armed threat. Pre-positioning options, mobile maritime deployments designed to keep heavier forces at least partially deployed and therefore closer to most likely trouble areas, were among the available courses of action.

"Tanks Forever and Ever" included a further technology challenge from that laid down in 1974:

- It cannot be the purpose of technology simply to enable us to meet the enemy on equal qualitative disparities. History testifies to the bankruptcy of those traditional notions.
- Rather it must be the role of technology to provide weapons systems that render ineffective costly investments by our foes, not simply to try to match something the other

fellow has just fielded. The latter course foredooms us to be reactive; our very survival depends more and more on our ability to, on the contrary, be proactive.

- With new weapons we should seek new dimensions of combat, following well-thought-out operational concepts with carefully designed organizations and with precisely tailored support concepts.
- In sum, technology should seek to make battle outcome less, not more, calculable. Instead of restoring some balance to a neat firepower score equation, we should introduce new imponderables into the traditional calculus of battle.

Five years have passed since that September morning in 1981. We are only now beginning to think seriously of developments that would in the long term provide some relief from the dilemma into which we have gotten ourselves and meet at least some of the challenges laid down on that long-ago day. Meantime the threat has continued its relentless quantitative and qualitative advances and the export fleets have grown in size, capability, and diversity in an increasing number of Third World military forces.

So what is to be done now? Last year the Defense Science Board convened a Summer Study task force to assess the state of the armor/antiarmor competition and lay out for the Secretary of Defense what might be done to redress whatever imbalances might be identified in its evaluations. Without reciting the details of that report, let me just note a couple of its principal conclusions, then speculate in that context on some relevant responses to the question of what is to be done now.

- The armor/antiarmor balance today favors the Soviet Union and has done so for 20 years, perhaps more.
- This imbalance has always existed on the quantitative side; now it exists on the qualitative side as well.
- In addition, the Soviets have exported to a substantial and growing number of client states an impressive array of modern military gear that in the aggregate puts at considerable risk the light insertion force strategy embraced by the United States.
- At its root, this problem is a combination of our unwillingness to accept reasonable intelligence projections of what is under development in the Soviet Union and the rate at which they are modernizing; our growing ineptitude in management of our RDA and budget processes; and our unwillingness to acknowledge the urgency of the new risks to our already risky strategy.
- Our most urgent need is to develop and put in place the management schemes to redress the imbalance by first catching up technically, then by a combination of well-managed programmatics and technology to establish and sustain a modernization rate at least equal to, if not better than, that of the other side.

Of the action programs growing out of the Summer Study, none perhaps are quite so exciting or promising as the set of DARPA armor/antiarmor programs that, in concert with industry and the DOE laboratories, is just now starting up, and the Army's program to define requirements for an integrated fleet of combat vehicles to meet the challenges set forth above.

Without prejudging or predicting the outcome of any of that work, may I just suggest some notions that seem to me, as one of the principal investigators in the armor/antiarmor evaluation, to be reasonable and relevant ways of responding to the "what is to be done now" questions.

First, an operational concept. In full scope, the constellation of ideas laid down in AirLand Battle is relevant:

- An active defense in depth, well forward in the battle area.
- A simultaneous attack into the interstices of the follow-on echelons to delay, disrupt, and destroy the momentum, velocity, and mass of the follow-on force attack.
- All should be designed to enable us to fight outnumbered and win at the operational level of war with conventional levels of weaponry.

New technology offers the opportunity to expand the close battlespace—the FLOT battlespace, to five kilometers or beyond with hypervelocity, armor-defeating weapons. Moving at five kilometers per second or greater velocities, hypervelocity weapons—surface-to-surface, surface-to-air, air-to-surface, and air-to-air—offer significant new opportunities to quickly moderate force ratios in the FLOT battle, enabling the outnumbered force to seize the initiative and maneuver to bring the battle to a quick and decisive conclusion. The challenge to tacticians is to harness this opportunity to their singular advantage. I have only sketched in and smoothed out the sandbox. You can do the rest.

Now equipment. The so-called "balanced force" fleet is, as ever, built around tanks. Constructed of advanced composite ceramic laminate armors giving protection about 2½ times that of the M1A1, interior volume of no more than 350 cubic feet, mounting a hypervelocity, nonturreted overhead launch system—an electromagnetic gun or hypervelocity kinetic energy missile launcher, delivering about 2½ to 3 times the muzzle energy of M1A1's 120mm cannon, with a crew of two or three men, powered at the 35 horsepower-to-ton ratio with an advanced-design power train. The vehicle would weigh in at about 30 tons. The day/night fire control system would operate against NOE and low-level helicopter targets as well as surface targets.

The companion vehicle is an infantry assault vehicle mounting a hypervelocity missile system with an advanced design CE warhead missile and a companion assault fire system—a cannon system of about 60mm with a wide variety of warheads for use against light armor and personnel. It has tank-equivalent armor protection, a fire team of four, plus driver and commander/gunner. At about 450 cubic feet interior volume, it weighs about 25 tons, has an identical or derivative power train, and is also antihelicopter capable with day/night fire control.

A reconnaissance version of the first two vehicles, at 200 cubic feet interior volume and protection at the tank level weighs about 20 tons, is powered at the 45 horsepower-to-ton ratio, holds a cavalry crew of three, and mounts the 60mm cannon system of the infantry assault vehicle. An advanced FLIR/PNVS-type combined navigation fire control system provides limited antihelicopter capability and surface-to-surface day/night fire control.

There are mortar and artillery firepower systems on the same basic components and also command control vehicles and forward supply vehicles. VTRs and bridge layers round out the fleet. The heavy assault force will consist of a tank, infantry assault vehicle, recce vehicle, artillery and mortar vehicles, armored recovery vehicle, engineering counterobstacle vehicle, and armored resupply vehicles, all of which have the same degree of armor protection and that utilize common chassis components.

Organization

The fighting vehicles operate in platoons of three. Each platoon is commanded by an officer. His wingmen are highly trained NCOs at the E7/E8 grade level. Three platoons are the company. The company command and second-in-command vehicles make 11-vehicle companies. Battalions of 3 companies include 3 command vehicles, for a total of 36 fighting systems.

Battalions are employed pure. There is no cross-reinforcement of companies from battalion to battalion. They are one axis of advance battalions. Brigades are fixed and numbered members of the regimental systems.

Brigades have assigned DS artillery battalions, heavy mortar battalions—120mm, air defense batteries, recovery and maintenance companies, bridge layer companies, recce troops. Heavy divisions have three fighting vehicle brigades and an aviation brigade consisting of an air cavalry squadron, two attack squadrons, a C&C squadron, and lift squadron.

The Army at about 780,000 men—all nonsmokers! Smaller than the US Navy in manpower, it has 12 heavy divisions and 6 light infantry brigades on the model of the Vietnam-era separate light infantry brigades. For insertion operations, light infantry brigades and armor brigades are deployed under a Joint Task Force command and control headquarters provided by the Readiness Command.

Training the force is accomplished by training up spare battalions of the regiments (brigades) in the training centers: the total armor force at Knox—tank, mech, recce; the total light infantry force at Benning—light, airborne, airmobile, ranger; artillery, including mortars, at Sill; aviation, including lift aviation, at Rucker; medium- and high-altitude air defense is handed off to the US Air Force and FAAD operations and training integrated into the maneuver forces. Fort Bliss becomes the home of Armor and the site of a second National Training Range extending across the whole million acres or so of Fort Bliss and the White Sands Missile Range combined. Combat engineers are part of the maneuver system and trained at maneuver force schools. Belvoir trains vertical and horizontal construction and public works engineers. In sum, we train like we intend to fight.

Support

This modern force requires a considerably different logistics support system. The present system is too large, too layered, too cumbersome, too nonresponsive to support modern mobile warfare. The demand-supported logistics system is replaced by a prognostics-driven log support system. It is a total quality process system driven at the beginning by a power train lease system in which warrantied power packs are automatically delivered, based on an hourly usage warranty system augmented by an expert systems prognosis capability that checks at periodic maintenance for warranty standards and automatically changes out the about-to-fail component before failure. It is a proactive, not a reactive, system. One version is called Rent-a-Pack. It is vastly cheaper than the existing system in which ownership of everything, coupled with a reactive maintenance concept, virtually ensures layer upon layer of logistics organization, much of which is unnecessary and all of which is so expensive it represents a luxury we can no longer afford.

With Rent-a-Pack as a successful pilot, the concept is extended to other major components of large systems and in less complex systems to the major components themselves.

With all that—operational concept, equipment, organization, training and support, lean and austere, tailored to the resource dilemma, designed as a total quality system from people and machines to the processes that deploy them to the field in effective fighting organizations and sustain them there in highly mobile combat operations—we can provide our country with an effective Army within the resources it is apparently willing to expend.

Next year, for the first time in my lifetime, US Navy end strength will exceed that of the Army—the result of the work of an aggressive, charismatic, and very clever Secretary and a series of very, very capable Chiefs of Naval Operations. This year, for the first time in history, the US Air Force R&D budget equals the Army procurement budget, and next year it will exceed Army procurement dollars—the handiwork of several competent Secretaries of the Air Force and their very capable Chiefs of Staff. The Army meantime has reached a new low—its share of the Defense budget the lowest since the terrible lean Eisenhower years, which in turn were the lowest since the leaner yet MacArthur years cited at the beginning. With the highest quality junior enlisted force we've probably ever had, with an increasingly competent officer corps, it does seem that we owe ourselves and the country a better Army. And that, at the root of things, is what must be done.

The Threat and Armor Development

General Dynamics Corporation
Mackinac Island, Michigan
23 September 1987

You asked that I talk about the threat and armor development. Certainly you have heard from several other speakers about the same subject. So what new might I bring to this debate?

Since your concern is with tank design and development, your primary interests are with technical aspects of the armor-antiarmor competition. While I am prepared to, and will, talk about this problem from a technical vantage point, it is important to consider technical problems from a baseline broad enough to put the technical situation in a somewhat different perspective. For technology is only useful if it contributes to the successful pursuit of a concept for the conduct of operations. While it is true that, from time to time, technology will offer a surprise operational opportunity, more often than not it is most useful for operational concepts to lead the way—indicating to technology what new capabilities would help most in making possible new and improved warfighting operational capabilities. Not everyone, especially not every scientific type, will agree with that statement. But I would argue that we have not the technical resources—time, people, or money—to spend simply hobby shopping around the laboratory in the vague hope that what gets done will somehow be found to have utility in some yet to be defined operational scheme. So taking the intellectual initiative is one very important responsibility of those who must establish operational and technical requirements.

Regarding the threat and armor development, how, first, might we view the threat? In this case the threat likely confronts us with a three-pronged dilemma—operation, tactical, and technical. A fourth, a strategic prong, may also be present, but we'll save that for another day.

For this argument, let's consider Soviet operational, tactical, and technical concepts as the baseline threat. This is neither to say the Soviet threat is the only problem we face nor to contend that confrontation with Pact forces in NATO Europe is the most likely eventuality. It is simply to acknowledge that the Soviets and the confrontation with the West across Alliance boundaries represent our most contentious problem. It is also to acknowledge the presence in other theaters of significant military capabilities built around Soviet strategic, operational, and tactical concepts and employing Soviet or Soviet-designed equipment.

What is the dilemma—operational, tactical, technical? Operationally we face a threat whose fundamental biases are:

- That numerical advantage is essential, if not key to winning, especially at the tactical level but in large measure at the operational level as well. This applies to forces at conventional and nuclear levels and to all types of forces necessary to fight in modern battle.
- That victory in battle goes to the side that, in the course of battle, seizes and holds the initiative, and that the offense is the form of maneuver best suited to the application of superior numbers to the problem of seizing the initiative. Further, surprise is an essential complement to numbers and other capabilities in achieving the initiative by offensive action.
- That technology must be harnessed to the task of providing equipment that will permit the rapid assemblage of numerically superior forces and their commitment in offensive battle in such a way as to achieve surprise and the maximum synchronization-correlation of forces.

These basic notions have not changed much over several decades. The means of accomplishing them have changed dramatically several times and are still changing as technology either raises up some perceived challenge on our side or permits some improved operational capability on theirs.

Although we are a nation whose fundamental military strategy, at least from the Civil War through 1945, was simply to overwhelm the enemy with numbers—especially in equipment, but in manpower as well if that be necessary—we have struggled mightily since 1945 to avoid the costs of providing sufficient numbers of forces or equipment to meet the requirements of our strategy. We have, instead, invoked the gods of technology to offset the growing numerical advantage of our principal foe and to try to defuse his offense-surprise-correlation of forces in order to deny him accomplishment of his principal objectives.

When we enjoyed nuclear monopoly, this may have been an appropriate notion. It may have been useful when we enjoyed nuclear superiority. In a time of nuclear parity, however, it loses much relevance. This is the genesis of our operational-tactical-level dilemma—how to fight and win outnumbered at the operational and tactical levels of war without having to invoke the immediate threat of the use of nuclear weapons.

The operational-tactical-technical merry-go-round begins. In the decades of the 1950s and 1960s, we fielded nuclear weapons in large numbers. Ten thousand of them in Europe were originally said to be required to make up for the disparity in numbers. Operationally and tactically the Soviets responded by making their forces more mobile so as to be able to operate successfully on a battlefield that they, for a time, even saw as all nuclear. Finally, by echeloning their forces so as to bring fresh formations into the battle in order to sustain numerical advantage, they added the final operational-tactical building block in their essential battle equation—mass, momentum, continuous land combat.

Technically at a temporary disadvantage in the nuclear field, they struggled mightily to add substantive tactical and operational nuclear capabilities to their arsenal. As with other capabilities, numerical superiority was paramount—a condition they describe as "parity." By the end of the 1960s, the evidence is that they had come to believe they were "almost there" in the nuclear world, and nuclear firepower became simply another element in the all-important correlation of forces.

At the same time, and much lower on the tactical scale of matters, alarmed by our fielding of the M60 series tank in the early 1960s, they sought to regain lost technical ground by improving their antiarmor capability in two important ways. First, they developed and fielded long-rod kinetic energy penetrators for tank guns. Second, they developed and fielded large numbers of antitank guided missiles (ATGMs), the most famous of which was the Sagger, later of Yom Kippur War fame. The Yom Kippur War gave lie to the notion that the modern ATGMs sounded the tank's tocsin of doom, this despite considerable uninformed blather in this country over the advent of the TOW in the early 1970s. It was, however, true that the long-range accuracy of the ATGM, with all its guidance and vulnerability problems, meant some considerable added advantage for a defending force hunkered down with terrain advantage. In addition, the ATGM raised a new technical challenge to armor developers to provide the means to defeat the shaped-charged warhead, especially when mounted on the long-range ATGM. This was the genesis of a tactical-technical-level dilemma that continues to this day.

The doctrine of active defense, as set forth in the 1976 edition of FM 100-5, was a tactical-level doctrine that sought to say how to fight and win though outnumbered at the tactical level of warfare, and with conventional means, largely by taking advantage of the large numbers of ATGMs that had been, or were being, issued to infantry units of the US Army. Active defense simply acknowledged nuclear parity for whatever it was at the time, recognized the need to significantly raise the nuclear threshold, and sought once again with improved long-range antitank means deployed in depth to attrit mass, disrupt momentum, and cope with successive application of numbers in continuous land combat. The framers of active defense essentially did not address themselves directly to the question of how the concept fit into the operational level of war.

If the ATGM signaled a technology-facilitated tactical revolution of sorts, it was short-lived indeed. For, no sooner had the tacticians begun to reap returns from technology's invention than the technology genie rose again. Based on US lab experiments of the 1960s, the Soviets developed, and began to field in the mid-1970s on the T-64B model tank, a ceramic laminate armor that could defeat the best shaped-charge warheads fielded at the time—those on TOW, Milan, and HOT. Precursor by about five years to fielding of similar armors in the West, this capability spelled the beginning of the second tactical-technical revolution of the 1970s. Successive models of Soviet tanks, the T-72 and T-80, feature improved ceramic laminate armors, as do successive models of M1 and other main battle tanks in the West.

On the operational level the Soviets responded to active defense with a maneuver group concept based on their 1944–1945 doctrine. Featuring autonomous, independent, highly mobile maneuver forces—an army for a front, perhaps a division for an army, the Operational Maneuver Group (OMG) was to strike deep in order to disrupt ATGM defenses, command-control, reserves, and the rear. In Europe especially, it was quite clearly designed to get through the dense ATGM belts quickly and disrupt the integrity of the defense from the rear.

Meantime US operational-level thinkers, striving to fully develop active defense into an operational-level battle concept, produced (circa 1982) a fully developed operational concept styled AirLand Battle. Recognizing the need to fight outnumbered both tactically and operationally, win, and do so below the nuclear threshold, AirLand Battle concentrated on simultaneous attack of assault and follow-on echelons and early identification and attack of OMG. Its technical means were sensors to be embedded in systems like JSTARS and JTACMS and the PGM system—derivatives of Assault Breaker and Tank Breaker designed to disrupt, delay, and, to whatever extent possible, destroy follow-on echelons. It was and is a full-blown operational concept for warfare at the operational and tactical levels of war. Technology—fielded technology—has not yet caught up with the operational concept, a phenomenon typical of the lethargy and nonresponsiveness of materiel acquisition systems in the West, especially in the US.

Now let's focus closely on the tactical-technology level of matters, reviewing that history so that we have a more certain view of where we have been, are, and what our challenges for the future might be. Considering Blue Bullets and Red Armor for the moment, we come out like this. On relative scales it will be seen that chemical energy warheads historically have been better penetrators than kinetic energy warheads by a factor of at least two to one. That situation reversed itself in the mid-to-late 1970s when, with the advent of ceramic laminate armor, and later of active/reactive armors, chemical energy penetration became extremely difficult for fielded, or about to be fielded, systems. Today it is true that ceramic laminates and active/reactive boxes have made obsolete every fielded antitank guided missile system in the world, as well as PGMs that seek to attack more lightly armored sectors of armored vehicles—top and deck armor. Specifically, for chemical energy warheads, Soviet armors overmatched TOW, even when TOW was first fielded, and continue to do so today. On the kinetic energy side, we see that we stuck with the same stubby penetrator for the M68 cannon on the M60 tank despite the fact that Soviet armor achieved overmatch early on and got better over time.

Two phenomena also appear from a Blue Bullet-Red Armor comparison. First, we have tried to catch up, only to find that developments on the other side have moved ahead apace to the end that, once our catch-up programs are fielded, we are still behind, for threat developments have again outreached us. Second, we find that, despite several attempts to do so, we have never been able to establish a modernization rate equal to, or better than, that of the threat. In sum, we are behind, have been for some time, and are getting further behind—on balance at a rate of about one year for every four years of elapsed time.

Considering Red Bullets and Blue Armor, we again find the inversion in chemical and kinetic energy as the penetration mechanism of choice. It also appears that, in fielded systems, we have been overmatched for a long time and that the situation is not improving. In sum, we are behind, have been for some time, have tried to catch up only to find that developments on the other side have put threat ahead as we finished our catch-up and have not been able to establish a modernization rate that will match or better that of the threat. The box score is Reds 4, Blues 1, and growing in Red's favor. What to do?

A recent article in *Defense News* set forth a view of the acquisition process that I'll share with you on paper but won't detail here. The short of it is that we've a gross management dilemma. It's at the top. The Goldwater-Nichols reforms may or may not solve the problem, but don't count on it. Meantime much needs to be done, for we're behind and getting further behind at an alarming rate.

So, back at the operational-tactical-technical levels, what needs to be done next? It appears that the next tactical revolution lies in the regime of expanding close battlespace. There are at least two good reasons for this. One is that the advent of CE-defeating armors shortened the close battlespace to the effective range of KE weapons. So how to restore that long-range high hit/kill probability? The other reason is that technology can now provide velocities that promise improved penetration at range and thus the opportunity to effectively expand close battlespace.

From many trials on instrumented ranges, we have considerable testimony to the value of opening fire at long ranges. In terms of the differential in loss exchange ratios, there appears to be an advantage of about four to one for a defender opening fire at 3,000 meters as opposed to an opening range of 1,000 meters. This says that, if you're outnumbered at the outset and believe you have to win outnumbered without flinging a bunch of nukes around, you'd better open fire as far out as possible. Don't let him close on you, for your four to one advantage at three kilometers quickly erodes to odds on at one kilometer. Analysis says that it is not only important to restore the tactical close battle ranges to those provided by ATGMs, but that it is essential to go even further, opening fire at four, five, six kilometers or beyond—whatever fire control systems can acquire and identify. Why not do that?

One problem—a significant one—with ATGMs was their relatively low velocity. With line-of-sight systems, given normal intervisibility segments, the target had gone to ground before the weapon could fly the intervening distance. So now we have the full range of the ATGM problems—ineffective warheads and times of flight that will not normally find targets still exposed when the missile arrives because of long flight time versus normal intervisibility segment lengths and average speed made good in a dash to cover across those segments.

If now we could increase the velocity of the round flying to the target, we could increase the probability that the target would be available in the intervisibility segment to be hit when the round arrives. Suppose we could move the bullet at three, four, five kilometers per second. Instead of 20 to 30 seconds flight time for an ATGM or 5 to 6 seconds for a tank gun round, we could close the gap in 1 second or 2, no more. Now, we have loss exchange ratios on the order of 10 or 12 to 1 in a defender's favor.

That's attractive, useful, exciting, worth exploring. So the tactical value of expanded close battle space is clear. The need for velocity in order to expand the close battlespace is evident. Can technology provide the velocity? Without plumbing the technical depths or exposing the classified data, it appears that the answer to the question is "yes." It's only a question of how much velocity is really needed. Velocities in the three-, four-, and five-kilometer-per-second regime are easily achievable. More is possible: it's a question of at what cost. Today's electrothermal and electromagnetic systems, although in embryo, are clearly capable of providing this capability.

A second and allied factor relates to the value of velocity at the target. Is $\frac{1}{2}MV^2$ valid for KE penetrators at all velocities, or is there phenomenology that takes over beyond some threshold velocity that limits the effectiveness of more velocity? Some empirical evidence says more is better; some theoretical evidence says it may not be so. There is considerable need for resolution of the issue in order that we may know which way to go next—technically, tactically, and operationally.

Finally a few words about what this all means. First off, it is but another cycle in the continual armor versus bullet game that has been played out on thousands of battlefields over thousands of battles. On the other hand, it represents a problem of considerable urgency: for we are behind the threat, have been for some time, and are getting further behind at an alarming rate. It is not that we haven't the technology, but that we haven't the will. It's not our technical ability that's in question. It's our decisionmaking ability. We have demonstrated a marvelous capability for indecision. Now we have the opportunity to expose the obverse side of that coin. If it, too, spells indecision, then we are in deeper trouble than I care to contemplate.

So, considering armor/antiarmor, I hope this provides you food for thought as you probe those important technical horizons that hold the key to tactical and, in part, operational success in battle.

Armor/Antiarmor in the Future of Land Combat

Center for Strategic and International Studies
Washington, DC
6 October 1988

To speculate in regard to the future, it is always useful to first sum up the current state of affairs and to say how the current situation came to be. Armed with that knowledge, it is then possible to project long-term trends through the data point that is "today" into some future but less clearly defined framework. So it is with this effort to develop relevant views about armor/antiarmor warfare in the future of land combat.

First, about the current state of affairs—where we are and how we came to be in our present circumstance. As a front-rank world power, the United States has committed and deployed military forces to ensure the political and economic goals of security and stability in NATO Europe and the Republic of Korea. In other so-called secondary theaters, the US also contributes stability and security in many ways but, in most cases, in ways short of on-the-scene military forces of significant size—forces whose instant involvement in air-land combat is virtually inevitable in the event of hostile action against those with whom we all allied and on whose landscape our forces are deployed. The ongoing naval deployments in the Persian Gulf committed to secure passage of oil-bearing ships through that body of water are, of course, an exception. In NATO Europe, the viability of the Alliance today is primarily a reflection of the strength and commitment of the partners providing the largest share, the US and the Federal Republic of Germany (FRG). This is not to belittle at all either the resolve or commitment of other member nations or the apparent willingness of France to join and participate militarily if needed, but simply to acknowledge from whence comes, at this juncture, the bulk of the forces and who pays, therefore, the largest share of the cost of the undertaking.

With more than one-third of its land forces stationed overseas, and roughly another third committed to reinforce deployed elements, the most contentious challenge facing US air-land forces at this writing is the Warsaw Pact/Soviet threat to Central Europe. That circumstance represents no change from what has been or what is likely to be going forward, notwithstanding changing attitudes and political and economic circumstances in this country and in Europe (particularly in the FRG), ongoing Soviet initiatives in regard to nuclear arms, and continuing efforts at arms control/reductions.

In Central Europe, Alliance forces, which include the deployed US air-land force, are considerably outnumbered by Warsaw Pact forces, at whose core is the imposing strength of the Group of Soviet Forces, Germany (GSFG). A situation extant from NATO's beginning, this is also a circumstance quite certain to continue going forward, absent some accommodation toward a less uncomfortable imbalance through the arms reduction medium.

Compounding the quantitative imbalance today is a qualitative balance favoring the other side. Where once the West clearly enjoyed a qualitative edge in fielded equipment, that no longer is the case. Of all mission areas in which quantitative and qualitative imbalances combine, redounding to the significant advantage of one side or the other, none reflects more clearly the extent of the problem than the combination of forces that represent the combined arms air-land force in close combat—the Central Battle. Here the combination of armor/antiarmor, indirect fire, air defense, mine/countermine, and electronic warfare capabilities clearly favors the other side. We are behind, have been for some time, and are getting further behind at an alarming rate.

While the present situation in this regard reflects several decades of growing imbalance, it is at root a reflection of a deeper malaise. In the US in particular, it derives from a materiel acquisition system that is and has been beset with a host of persistent and long-lived problems that, in the aggregate, are highly dysfunctional to the making of orderly, timely decisions in regard to modernizing our forces with new technology in fielded systems. Today, the Soviets are outmodernizing us at a rate of about four to one in the application of new technology to fielded systems. Each year, for example, the Soviets modernize a force the size of the total US heavy force; every two years, a Soviet force the size of the total NATO heavy force undergoes modernization. While, in part, this situation is a matter of scale—the Soviets having a much larger force to modernize—more important, it reflects a commitment to modernize their entire force over time at a rate we simply have not been willing to match, even with a smaller force. We are now, therefore, confronted with the cumulative results of at least 20 years of determined effort in the Soviet Five-Year Plans to achieve a conventional force balance that is clearly in their favor.

In regard to the nuclear backdrop against which conventional forces are viewed, the advantage is also largely with the other side. With tactical- and theater-level nuclear capabilities equal to or better than our own, and in that state for several years, the strength of the nuclear deterrent lies not so much in the threat that one side or the other might lose as a result of nuclear war, but that both sides would lose—badly damaged to a level that, when viewed soberly at the outset, would bring prudent men to at least pause before making the fatal decision to press on with a nuclear exchange—almost regardless of what dire circumstances might have been wrought as a result of conventional warfare.

In the Third World (the "secondary theaters," as they are sometimes called), one sees today an increasing trend toward militarization of conflict and increasingly with modern weapon systems. Most dramatic in this regard is, of course, the conflict situation and potential for conflict in the Middle East. But it must also be noted that, in nearly 50 nations of the so-called Third World, there are sufficient numbers of fairly modern weapons systems deployed to make conflict in those areas something more than a desultory exchange between ill-equipped antagonists, albeit yet something quite less than war in Korea, the core of the Middle East, or in Central Europe. Nonetheless, as one views trends in arms distribution into this Third World, it is quite clear that the larger arms producers have over time been providing their surrogates, clients, and allies with substantial amounts of modern weaponry. Contemplating strategies—economic, political, and military—for these regions, one cannot fail to take this into account.

Now, armor/antiarmor—where we are and how we have come to this point. It is perhaps best to examine the situation from a broader perspective, then to probe more closely into detail in order to adequately illuminate the broader thesis.

First, it is important to acknowledge that what has been set forth above applies not only to the armor/antiarmor equation but to the host of combined arms battlefield systems as well. Such also is the case with what follows. The armor/antiarmor competition is front and center here simply because that is the focus of this paper.

Historically, we in the US have been wedded to the notion that, to win in battles and so in war, it was necessary simply to outnumber the other side. Therefore the growing conventional strength of Soviet and Soviet surrogate or client state ground forces, especially those arrayed against NATO Europe's Central Region, has been and is a most vexing problem. In the beginning, in Europe, it was General Dwight Eisenhower who, as Supreme Allied Commander, Europe (SACEUR), set forth the requirement for 96 divisions and 9,000 tactical fighter aircraft to defend NATO. Rejected by member countries unable or unwilling to provide the staggering resources necessary, President Eisenhower finally settled for 25 divisions—12 of them to be West German—1,400 fighter aircraft, and 15,000 theater and tactical nuclear weapons. Of the latter, about 7,000 were deployed and some 4,000 remain. And so it was that technology, in the form of tactical- and theater-level nuclear weapons, was to be a surrogate for numbers.

In the beginning, and for some time after, the United States controlled an overwhelming strategic nuclear capability; NATO, under the US nuclear umbrella, was likely to prevail in a nuclear war. Certainly this was the case in a first-strike situation; it was true for an even longer time for a strike-second situation. This was so, of course, because the United States had the capability to destroy Soviet strategic forces and/or reduce a Soviet counterstrike to manageable proportions. Since the mid-1960s, Soviet nuclear capabilities have grown dramatically in all dimensions. Now, and for the foreseeable future, the United States no longer has the capability to reduce a Soviet counterstrike to reasonably tolerable levels. It was early perception of this developing reality that led the Kennedy Administration in the United States to abandon massive retaliation and embrace the flexible response doctrine in its stead.

For a time the Soviets themselves apparently believed that war between the superpowers would perforce be all nuclear—it would begin and end that way. Then, as time wore on, the awesome totality of all-nuclear conflict at tactical, operational, and strategic levels apparently overwhelmed their ability to think logically about the matter. At that point their attention turned once again to operational concepts and weapons systems for the conventional battle.

Since about 1966, when they had concluded that they had achieved nuclear "parity," the Soviets have embraced the notion that they could and should try to win at theater level, avoiding the use of nuclear weapons if at all possible. At first they filled gaps in their conventional capabilities with nuclear weapons until they could develop the numerical superiority and technical means necessary to win quickly at the theater level of war without invoking the nuclear specter in the process. Their conventional weapons developments in the Five-Year Plans have for 25 years been clearly focused on that goal. Perhaps that focus is more clear in hindsight than it was in the threat projections—some say yea, others not so. However, they are now completing the most recent conventional development that will make their nonnuclear theater-level winning capability a reality, at least as they view it.

Over the span of the relevant Five-Year Plans, we have seen the development and deployment of impressive capabilities in artillery, air defense, radio-electronic combat, combat helicopters, ballistic missiles, fixed-wing airpower, and force structure changes such as Operational Maneuver Groups (OMGs) and special action forces (Spetznaz). The missing link has been in the armor/antiarmor equation. For, with deployment and proliferation of antitank guided missiles in the West in the early 1970s, Soviet ability to break through NATO's antitank defenses without risking substantial loss and possible defeat was uncertain at best, doubtful at worst. So they set about to solve the technical riddle of the shaped-charge warheads used on all the world's ATGMs—TOW, HOT, Milan, even their own Sagger. Not only were those warheads roughly twice as effective for armor penetration as counterpart high-velocity kinetic energy penetrators from tank guns, but the long-range accuracy of ATGM systems nearly doubled effective gun engagement ranges and, in so doing, quadrupled the close-in battlespace that could be dominated by well-laid-down defenses built around integrated combinations of artillery, tank guns, and ATGMs.

By the mid-1970s the Soviets had begun to field tanks in the T-64B series featuring their initial response to the problem of how to defend against the ATGM shaped-charge warheads. For the T-64B and subsequent models—T-72, T-80, and now what the West calls FST-1 (Future Soviet Tank 1)—contained lighter weight glass composite or ceramic laminate armors that were more than a match for the early model ATGM warheads. While tank fleets with this advanced-level protection were growing apace—with production running 3,000 to 4,000 units per year, technology provided the means for even better protection in the form of what is popularly called reactive armor. Reactive armor in the form of appliqué boxes—fixtures about 10 by 12 by 4 inches in size containing specially designed explosive arrays that detonate on impact, destroying the integrity of the incoming ATGM shaped-charge jet—was first fielded by the Israeli Defense Force (IDF) in the late 1970s and early 1980s. It went to war in the 1982 Israeli operation Peace in Galilee mounted on IDF M60s and other more lightly protected armored vehicles. Some of these boxes fell into Syrian and so into Soviet hands. Soon improved models were available in Israel, and within the last two years, similar improved boxes have begun to appear on tank fleets in the Group of Soviet Forces, Germany. In fact, by some estimates the Soviets have equipped more than half the Soviet tank fleet in GSFG—T-64Bs, T-72s, and T-80s—with reactive boxes and can likely complete the fielding in the next couple of years.

The unhappy fact is that tanks with reactive armor on top of built-in glass composite or ceramic laminate armor will quite likely defeat the warheads of all deployed ATGMs in the world today—TOW, HOT, Milan, even Sagger. While the technology involved is not all that exotic, reactive armor in this form is a very attractive solution to what, for the Soviets since the early 1970s, has been a most vexing problem: how to quickly get through the deepening belts of antitank guided missile systems deployed forward in NATO without suffering inordinate losses in Soviet tanks. Here, then, is a relatively uncomplicated, fairly inexpensive solution, and one that can be achieved without building a new or rebuilding an older tank fleet. Further, it is a fairly rapid way to complete development of their conventional capability to break through quickly, strike deep to disrupt the integrity of NATO's defenses, and do so quickly enough to preempt a nuclear decision in the West. So now, not only are they in possession of all the capabilities long considered essential for quick victory in Europe without risking a nuclear exchange, but this all comes to pass at a most opportune time. It allows them to offer to the West the opportunity to apparently defuse the theater-level nuclear confrontation by taking out the theater-level nuclear systems that the Soviets have concluded they don't need any more but that we, given the conventional force imbalance and the technology situation in regard to armor, need even more perhaps than we did before. Thus we have a technology advance at the tactical level of war that has important implications for theater- and national-level strategies and capabilities as well. This is the unique feature of the armor/antiarmor imbalance, a feature that sets it apart from other similar quantitative and qualitative imbalances in the Central Battle.

The Soviets obviously believe that they now have the means to execute the operational concept they have embraced for so long, to achieve quick, decisive victory at the theater level of war and to do so with such thoroughness and dispatch that the nuclear decision in the West is yet in abeyance even as they consolidate their theater-level gains.

Were it not for the strategic situation, the INF Treaty agreement, and the growing clamor not only for additional nuclear arms reductions but for conventional arms drawdowns as well, it might be possible to look on this as just another of those technological flip-flops that occur now and then, one that in due course can and will be overtaken, perhaps reversed. In this case there's much more to it than that. For, from the Soviet perspective, eliminating the INF in Europe, coupled with their vastly improved antiarmor capability, significantly improves their capability to execute a long and consistently held operational concept. While, from our standpoint, we are without a theater-level nuclear deterrent that can range into the Soviet homeland, the close battlespace in our forward defensive areas has been shrunk to its pre-1970 levels, and we are apparently without a carefully thought-out strategy defining the linkages, if any, between the conventional imbalances at theater level and our remaining nuclear capabilities.

How did this situation come to pass? A 1985 Defense Science Board Task Force examining the armor/antiarmor competition between the Soviets and the US concluded that we are behind the Soviets and getting further behind for three principal reasons: the way in which we deal with the threat, our R&D-to-fielding rate, and our program budget system.

Looking first at the threat, the task force found that, on a relative scale, a good deal is known about what technologies are in research and development on the Soviet side. Scientific journals and other open literature collectively provide a fairly substantial body of data from which we can determine, by inference at least, what is in R&D at any given time. Moving along in time, the volume of information in regard to a given technology declines or even disappears. There is a tendency to say, "They've given up on that." While that may be the case, it is equally possible that they may have moved the technology into what, in the US, is called full-scale engineering development, a process in their system about which we know far less than we do about the R&D process. Time passes—as much as 8 to 10 years. Then, all too frequently, we see from overhead sensors a new weapon system in test or, in some cases, even being fielded, a system that fields that "disappeared" technology. Because of the general profile of the curve describing this phenomenon, the task force chose to call this decline of information during full-scale engineering the "Bathtub of Ignorance." We react—try to catch up. Historically catch-up has taken, conservatively, at least five years. Hence, depending on whether or not we are pursuing the same technology, and depending on how far along we may be with our development, we could be as much as 15 years behind in applying new technology to fielded systems. Historically we more often than not find ourselves about 8 to 10 years behind.

This is not an indictment of our intelligence system. For, despite the presence of relatively less information about what is in the "Bathtub" in the Soviet engineering development process, we do have sufficient information on which to base fairly reliable estimates. It is, rather, our inability or unwillingness to make timely and relevant decisions about the matter that is at fault.

Our system tends, for the most part, to react positively when confronted by hard evidence—usually photographs of fielded equipment—but negatively to intelligence community projections of what is likely in the Bathtub at a given time. Confronted by a fielded system, or an about-to-be-fielded system, a reactive strategy such as ours foredooms us to a "get behind and stay there" situation.

As a sidelight to this phenomenon, there is the question of the volume of what is in engineering development at any given time. At the time of the 1985 study, major systems in the Bathtub outnumbered those in full-scale engineering development in our own system by a factor of 8 or 10. Related to the threat estimate situation just described is the history of our ability, or inability, to field modern technology in a timely way. Out of some 30 technology developments examined, the task force found at least a dozen that had been funded at entry levels of development, 6.1–6.2, for 12 to 15 years. This situation illustrates the fact that, while we may be ahead of the Soviets technically, more and more the advantage is only on the laboratory bench. In fielded systems, we display a notorious penchant for "just a little more time and just a few more dollars and we can have so much more capability." Years and millions (or billions) later, we find ourselves with but a marginally better capability than we would have had had we fielded the technology sooner. Like intelligence and the threat, this is not an indictment of our technology. What it does highlight is our inability or unwillingness to make timely decisions in the acquisition process.

In armor/antiarmor systems, the situation is best illustrated using a comparison of Blue Bullets versus Red Armor and Red Bullets versus Blue Armor.

Considering Blue Bullets and Red Armor, on relative scales chemical energy shaped-charge warheads—warheads on antitank guided missiles, for example—have historically been better penetrators than kinetic energy, high-velocity, gun-fired penetrators by a factor of at least two to one. That situation reversed itself in the mid-to-late 1970s when, with the advent of composite, and later ceramic laminate, armor, and still later reactive armor, chemical energy penetration became extremely difficult. Today the combination of composite or ceramic armor and reactive appliqués has likely made obsolete every fielded system, especially every fielded antitank guided missile system that uses a shaped-charge warhead as a defeat mechanism.

Two additional phenomena appear from a comparison of Blue Bullets and Red Armor. First, we have tried periodically to catch up, only to find that developments on the other side have moved ahead apace to the end that, once our catch-up programs are fielded, we are still behind. Second, despite several attempts to do so, we have never been able to establish a modernization rate equal to, or better than, that of the threat.

Considering Red Bullets and Blue Armor, we again find that, in fielded systems, we have been overmatched for a long time and that the situation is not improving. From time to time we have tried to catch up only to find that developments on the other side have put threat ahead as we finished our "catch-up." And, here again, we have not been able to establish a modernization rate to match or better that of the threat.

Historically, therefore, we find ourselves almost always in a catch-up mode. By the time we catch up, momentum on the other side has put the threat ahead of us once more, so we hardly ever catch up. Worse yet, we are unable to achieve a modernization rate that can match or better that on the other side. The end result is that the Soviets are outmodernizing us at a rate of roughly about four to one.

Put another way, in general terms the Soviets apply new technology in upgrades to fielded systems about every 5 years and field new technology in a new system about every 10 years. By contrast, we field new technology in new systems about every 20 years.

The task force further singled out the US program budgeting system as an inhibitor to providing the US with an acquisition system competitive with that of the threat. The system, it was pointed out, does not demand and seldom permits "block modification" programming for major weapon systems—"block mods" designed to upgrade existing systems in a timely way so that product improvement of fielded systems is a constantly ongoing process. Nor does the system demand timely fielding of technologies that will overmatch the threat at time of fielding, at a systems/technology fielding rate that will ensure a competitive modernization rate—a rate at least equal to the expected rate of development in threat systems.

Summing up, we are where we are and have gotten to where we are due to a materiel acquisition system that has provided neither a point advantage, a competitive edge on our behalf, nor a rate of modernization competitive with that of the other side. Hence we should not be surprised to find ourselves in arrears and getting further that way.

While all the foregoing has been in train, we have over the past two or three decades begun to see the Soviet quantity/quality modernization rate reflected in their arms exports to the Third World. Heading an impressive list of other exports are tanks. To illustrate the point, over two decades ending in the early 1980s, the Soviets exported to seven countries (four in the Middle East, one in south Asia, two in the Western Hemisphere) about 11,000 tanks. Other armored vehicles, artillery, and aircraft follow the same pattern, to include, most recently, armed helicopters.

It is customary to dismiss this phenomenon with, "That's okay; it's all old, obsolete stuff." Not true; the evidence is that, as soon as the Soviets themselves begin issue of a new item of equipment to their own forces, they apparently are willing to release the next-oldest item in the export fleets. So it is that we find T-72 tanks replacing Syrian T-62 tank losses in the 1982 operation for Peace in Galilee, a move made possible by the issue of T-80s to Soviet forces just a year or so earlier. So the equipment in the hands of client states tends to be reasonably new and the same quantity/quality growth equation noted earlier with regard to the Soviets themselves is now reflected in the arsenals of many, if not most, of their arms customers.

This is a circumstance that calls attention to two important circumstances. On the one hand, in the world into which we are most likely to be required to deploy forces, the so-called Third World, there is a strong and growing likelihood that US forces will encounter threat forces fairly well armed with an increasingly capable array of armored fighting vehicles, artillery, and armed helicopters.

On the other hand is the traditional presumption that, in most contingency areas, light forces will suffice, a presumption largely based on the difficult reality that, with limited lift, light forces are the most capability that can be gotten anywhere in any numbers very quickly. If so-called light forces are to be employed in Third World endeavors, the evidence is that it is increasingly necessary that they be equipped with systems that will enable them to survive and to prevail in pursuit of their Third World missions.

However the future, with all its uncertainties, may unfold, the fact is that, given current world inventories of fighting vehicles designed for combined arms battle—tanks, infantry fighting vehicles, artillery, helicopters, others—it is probably safe to say that the conventional force balance will be measured for some time to come in times of systems at the core of the combined arms team: armor/antiarmor, air defense, mine/countermine, and indirect fire. That being the case, it is well to ask what might be the nature of battles to be fought by these formations.

To answer this question, one turns to lessons of the most recent battles of which we have extensive knowledge, the 1973 Yom Kippur War and the 1982 operation for Peace in Galilee.

Summing up those lessons we find:

- The staggering density of the battlefield at critical points.
- The presence on both sides of large numbers of modern weapons systems.
- The presence in the air over the battle of large numbers of sophisticated air and air defense weapon systems.
- The increased criticality of command and control against the increased difficulty of command-control communications due to the presence of large numbers of sophisticated electronic warfare means.
- The inability of any single weapon system to prevail, reaffirming the essentiality of all-arms combat.
- The outcome of battle reflecting, more often than not, factors other than numbers.
- In addition, it is starkly obvious that large-scale destruction in a short time is a highly likely outcome of early battles in modern war. Especially is this the case when surprise is an operative factor at the outset.

So the combined arms team of armor/antiarmor, air defense, artillery, mine/countermine remains the pivotal force on the conventional warfare stage. At its core is the survivable mobile integrated firepower of the armored fighting vehicle family, especially the tank. This circumstance is certainly to be the case in Central Europe and Korea where large forces in high states of readiness face one another across international boundaries. Further, it is likely to be the case in the Middle East and in other areas of the Third World to which the world's largest arms producers have provided significant amounts of fairly modern military equipment.

Operationally, as well as tactically, conventional force warfare—armor/antiarmor battles of the future—will be set in the context of competing operational concepts. And so it will be necessary first to contend with Soviet or Soviet-style operational concepts—mass, momentum, and continuous land combat—by dealing with two important matters at one time. First is the fight against the assault echelons. Second is the attack against follow-on echelons in order to disrupt, delay, and destroy sufficient of their velocity and mass to prevent them from building momentum. If friendly forward echelons can fight successfully, and if threat follow-on echelons and OMGs can be prevented from loading up on the front-end battle, then the friendly commander fighting the forward battle has half a chance to seize the initiative, maneuver to the attack, and win. At the front part of the battle, then, there is a need for an active defense that essentially deals with enemy assault echelons. As the active defense begins against first-echelon forces, a deeper campaign to delay, disrupt, and destroy follow-on forces and OMGs is brought to bear. Reconnaissance, surveillance, and target acquisition means are directed so as to detect, follow, and target follow-on forces attempting to close on the forward battle.

In Central Europe, as an example, allied corps commanders should, as a general rule, be cognizant of, and following, enemy forces that are within about 72 hours of being able to intrude into the corps-level battle. On Central European terrain, that time equates to distances of as much as 150 kilometers. Time is, however, the important feature, for of prime concern is the time of closure, the time within which the enemy is able to intrude himself into the corps battle, requiring the marshaling of forces and fires or posing a threat to corps operations.

The missions at that depth are to delay, disrupt, and destroy. Attacks are made against chokepoints, command control nodes, logistical support, routes of movement, means of deployment, and troop concentrations. It is essentially an air and long-range missile battle. As follow-on echelons close to within about 24 hours of the forward battle, perhaps as much as 75 to 100 kilometers away, depending on the terrain, the divisional commanders begin to follow their progress and to attack as appropriate. Here the mission is to disrupt, delay, and destroy as possible. Real-time or near-real-time target acquisition is most desirable. The attacking force is beginning to have fewer maneuver alternatives—the battle is an AirLand Battle with close air support, battlefield air interdiction, and ground maneuver forces, as appropriate, being applied against the follow-on forces.

Finally, follow-on forces closing to within 12 hours of the forward battle are the responsibility of forward-deployed brigade commanders. The mission is to destroy and disrupt those forces, defeat the echelon in contact, and go in against follow-on echelons, as appropriate, to destroy or render the force ineffective before the next follow-on echelon can appear. It is without question a land forces battle with a full spectrum of all-arms capabilities brought to bear to defeat the enemy as quickly as possible. It is the central and critical duel of the AirLand Battle.

That is the essence of AirLand Battle. Its most important ideas are:

- Deep attack and the close-in battle are inseparable.
- Seizing and holding the initiative through maneuver of forces and fires is essential to success.
- The objective of the battle is to win, not just to avert defeat.
- Successfully conducted, the AirLand Battle can, with conventional means alone, considerably postpone the time at which the defender must consider the first use of nuclear weapons, thus raising dramatically the nuclear threshold.

To successfully execute that operational concept, many developments are required. Among them all, three stand out as demanding first order attention. They are:

- The command control means to synchronize deep battle, close battle, and rear battle. Here it is quite likely that there is no serious dearth of information. Indeed, we may be suffering from information glut. What is needed is the disciplined flow of information from all sources to command-control nodes where the critical decision must be made; the making of timely decisions; the dissemination of decision information; and decision execution in timely fashion. "Timely" means the ability to turn the decisionmaking cycle "inside" that of the threat.
- The integrated deep battle sensor and weapon system capabilities to delay, disrupt, and destroy the advance of follow-on echelons into the close battle at the forward line of troops (FLOT), providing the ability to moderate force ratios at the FLOT in such a way that the commander fighting the close battle is provided the opportunity to maneuver, seize the initiative, and win the central and critical battle.
- Expansion of the close battlespace, the FLOT battlespace, to about five kilometers or beyond with hypervelocity, armor-defeating direct-fire weapons, and to 50 kilometers or more using indirect-fire rocket/artillery systems. Moving at three to five kilometers per second or greater velocities, hypervelocity weapons—surface-to-surface, surface-to-air, air-to-surface, and air-to-air—offer significant new opportunities to quickly moderate force ratios in the FLOT battle, enabling the outnumbered force to seize the initiative and maneuver to bring the battle to a quick and decisive conclusion. Here, electromagnetic launch technology offers new and exciting possibilities. We have only to take advantage of that in a timely way.

In each of the critical elements set forth above, technology offers solutions. It is highly unlikely that we will ever find the means to hold a quantitative advantage over the threat. We must, therefore, seek solutions in superior operational concepts enabled by our technology. In that regard, the challenge to our technology is severalfold. Consider the following notions:

- It cannot be the goal of technology simply to enable us to meet the enemy on equal qualitative terms or to make up quantitative disparities. History testifies to the bankruptcy of those traditional ideas.
- Rather it must be the role of technology to provide weapons systems that render ineffective costly investments by our foes, not simply to try to match something the other fellow has just fielded. The latter course foredooms us to be reactive. Our very survival depends more and more on our ability to, on the contrary, be proactive.
- With new weapons we should seek new dimensions of combat, following well-thought-out operational concepts with carefully designed organizations and with precisely tailored support concepts.
- Technology should seek to make battle outcome less, not more, calculable. Instead of restoring some balance to a neat firepower score equation, we should introduce new imponderables into the traditional calculus of battle.

Summary

- Today the conventional force balance, at the heart of which is the armor/antiarmor balance, favors the Soviet Union, has done so for some years, and is growing in their favor at a significant and alarming rate.
- This imbalance has always existed on the quantitative side. Now it exists on the qualitative side as well.
- In addition, the Soviets have exported to a substantial and growing number of client states an impressive array of modern military gear that, in the aggregate, puts at considerable risk the light force strategy embraced by the United States for Third World operations.
- The rate of modernization of the conventional forces of the contending sides favors the Soviets by a factor of about four to one.
- At root, this problem is a combination of our unwillingness to accept reasonable intelligence projections of what is under development in the Soviet Union and so the rate at which they are modernizing; our inability to move new technology from laboratory bench to fielded systems in a timely way; and our inability to make our program budget system provide a competitive modernization rate in fielded systems.

Our most urgent need is to develop and put in place the management schemes to redress the conventional force imbalance by achieving a competitive position, then by a combination of well-managed programmatics and technology to establish and sustain a modernization rate at least equal to, if not better than, that of the other side. Absent that, we are foredoomed to be ever further behind our antagonists and confronted with a conventional force imbalance that not only favors the other side but allows our foes to seek and obtain political goals based on the significant and imposing capabilities of conventional forces.

Light Combat Vehicles

Tank-Automotive and Armaments Command
Conference on Light Combat Vehicles
Naval Postgraduate School
Monterey, California
7 March 1989

It is a considerable challenge to talk about light combat vehicles to this or any audience. There are two reasons for this. First, recent advances in armor for survivability of combat vehicles have only made the protected vehicles heavier and heavier. Even the most recent of these advances, that of so-called reactive armor, has, while adding considerable protection, demanded a concomitant weight penalty. Perhaps that will not always be the case, but it is for the moment. So the idea that one can have a light vehicle that can also fight, especially fight on the modern armor-dominated battlefield, could be written off as self-contradictory.

Having pondered that problem, I sought out the keeper of the US Army's requirements for light combat vehicles. There I found mixed circumstances. First were those who said the Army does have a requirement for such equipment, but who were not quite certain what in total that requirement might be. Then there were those to whom I was referred by the first group who, when pressed, admitted that the Army really has not gotten around to setting forth explicitly the totality of its light combat vehicle requirements. That led logically to another important question: If we believe all research and development must support some real or perceived user need, and the user has not expressed a need, then why are we here and why are you in the labs spending money working on this?

Now it may be unwise of me to pursue that line of reasoning further, since I would consider it presumptive of me to try to draw up a set of even speculative needs for light combat vehicles. May I just for the next few moments try to sketch out what seem to me to be the most important challenges facing us in the continuing contest between survivability and firepower, and do that in the context of so-called light combat vehicles, being not more specific than to use the term light—meaning something less than heavy, something less than the 55- to 65-ton weights that we now enjoy with our heavier vehicles.

Let's first consider survivability. To do that, we have to consider the threat. It seems to me that the threat to modern armored vehicles comes in three varieties. There is, first, the wide range of direct-fire armor-penetrating devices that use kinetic energy as a kill mechanism or that use chemical energy warheads for defeat. Second, there is the growing threat from artillery, especially in NATO. Artillery includes weapons of longer range, improved lethality, richer force structure, and a variety of warheads, possibly to include sensor-fused smart munitions. Finally there is the class of weapons that seek to destroy or disable not by attacking the armor itself but by attacking the crew or less-well-armored portions of the vehicle. Some call this soft kill.

Direct Fire. The history of the firepower survivability contest between armor and direct-fire antiarmor systems is best described in the outcomes of the Defense Science Board Summer Study of 1985. An unclassified version of these outcomes was published in the Senate Armed Services Committee hearing reports for March 1988 in which Generals Don Pihl, Dutch

Shoffner, and I testified concerning the problem and what the Army is doing about it. Many if not most of you are familiar with that work, but let me just review it briefly.

For years, chemical energy warheads have provided about twice the relative armor penetration as have kinetic energy warheads. From a survivability standpoint, the score was evened with the advent of, first, composite armors, later ceramic laminate armors, and most recently with reactive or explosive armor. At the moment, with reactive armor as an appliqué to composite or laminate armors, it is quite likely that all of the world's fielded chemical energy antiarmor warheads, especially those on antitank guided missiles, can be defeated. It also appears that such armors can be made to be, to at least some degree, effective against kinetic energy penetrators, although not to that same degree that they are effective against chemical energy warheads. In the worldwide armor/antiarmor balance, the Soviets have fielded far more of the reactive boxes than have we in the West, and they have a much larger fleet of tanks with laminate or composite armors built into the basic armor envelope. They have not yet, at least as far as we know, exported reactive armor to their client states. Little comfort here, however, for the reactive boxes are not difficult to fabricate and the technology is not exotic or difficult to master. To their key client states, they have already exported tanks with composite armors. Given the pattern of their support for their clients, it is quite likely that those exports will continue, and sooner rather than later we shall see more and more modern fleets in key client states featuring advanced armors in the basic armor envelope. All, of course, could well have reactive appliqués added should they elect to do so.

On the obverse side of that coin, we have the contest of penetrators. With chemical energy warheads at the moment at least rendered fairly ineffective against composite and reactive armor, we find on the other hand a family of increasingly capable long-rod penetrators. Starting with long rods in the 115mm gun on the T-62 tank and developing from there, the Soviets have fielded, one after another in quick succession, a family of long rods that represent a formidable threat to any modern armor. In this country we moved early to the use of depleted uranium for penetrator material. Shunning this for reasons not quite clear at the moment, the Soviets have continued to develop other materials, notably tungsten, for advanced penetrators. These penetrators, coupled with the Soviet willingness to use more of the potential energy from powder guns in velocity at the muzzle, makes the threat from Soviet-built tanks firing kinetic energy long-rod penetrators much more formidable than ever before. The increase in gun calibers alone, as they have moved from T-62 to T-64B and T-72 to T-80/FST-1 and -2, is manifestation of this increasing capability.

So the fundamental questions are: What must we do to armor in order to improve survivability? And what must we do to penetration mechanisms to increase penetrability?

On the survivability side, it seems to me that the challenge lies in improved materials, in better understanding of the physics of penetration in order to improve armor design, but more than all that, it seems to me that now is the time for a new approach to the engineering of armored vehicle design. So we should regard today's reactive armor boxes as but an interim step toward the time when we build armor modules containing special, reactive, and other advanced design armors for modular emplacement on basic envelopes of our armor vehicle fleet. Bolted, welded, or fastened on by other techniques, these modules can be changed out as the threat changes without the need to rebuild the entire vehicle. With the penetrator/threat changing rapidly, it seems to make better sense to provide ourselves the ability to change armor as technology and the threat provide the opportunity and the need to change. In advanced armors to insert in these modular arrays, we must look to including not only reactive armors but active armor as well, to include electromagnetic armor designs—designs that defeat by energy reaction within the armor as opposed to active explosive mechanisms that project from the armor to defeat incoming rounds before contact.

On the penetration side of the equation, it seems to me that there is a need to move away from powder guns or other chemical explosive propellants and move in the near term to electrothermal launch. Achieving velocities of 2.5 to 3 kilometers per second with electrothermal launch systems is quite likely possible in the near term—five years or so. This would give us in existing guns of various calibers the ability to penetrate appropriate threat armors on the variety of combat vehicles that can be arrayed against us. In calibers from 25mm to 120mm or larger, electrothermal guns would provide a quantum improvement in our antiarmor, antiair, antihelicopter capabilities, and in air-to-surface and air-to-air weapons as well. Over the horizon of the century, there is, of course, the possibility of moving to electromagnetic launch as a follow-on to electrothermal in the near term. My judgment would be that, with pulse power technology moving as rapidly as it is, we have a near-term opportunity to avoid the need to regun everything and to move to electrothermal in existing tubes in a fairly expeditious manner.

Earlier I suggested that the second most important threat to light combat vehicles, and I should add to noncombat vehicles as well, is the threat of artillery. Again the most impressive threat in this category is that of the Soviets in NATO Europe, but it increasingly applies to a wider and more comprehensive menu of client states as well. In the series of Soviet military power handbooks released by the Defense Department over the past several years, one can read the evidence of increased range, caliber, numbers, and force structure on the other side. The Soviet revolution in artillery, starting with their late 1960s decision to move to war-winning conventional capabilities, is every bit as dramatic, and perhaps even more so, than their advances in armor-antiarmor. In the 1970s, for example, they self-propelled much of their artillery and made most of it nuclear capable. More recently we find them increasing caliber in direct support systems and changing force structure to provide more tubes at direct support levels than ever before. By contrast, with the advent of tactical nuclear weapons, the West took the view that fewer and fewer artillery pieces were needed, and so money was ostensibly saved by taking out artillery structure.

Now we find ourselves lean to the point of alarm in counterfire and counterbattery systems in the face of severalfold improvements in range, throw weight, and lethality on the other side. Some argue that the Soviets now have sufficient artillery to support their famous breakthrough attacks without the need to resort to tactical airpower. Indeed they apparently have the capability to range into and cover most of a defending division's forward area with artillery, which is the equivalent of what 10 years ago required artillery and tactical airpower combined. This means that many activities and deployments normally considered fairly invulnerable on our side because of range and lack of adequate throw weight are now within lethal range of forward-deployed artillery and ballistic missile systems. Activities such as command posts, forward area helicopter rearm and refuel points, battalion and brigade trains, division support areas—almost everything forward of the division rear boundary, is now candidate for armor protection of some kind. Similarly, rear area defense by combatant troops, military police, or whatever requires these forces be mounted in armored vehicles.

There is, as some of you have heard me say many times, a need for an entire family of armored cars and other armored fighting command and control and transport vehicles in rear areas for functions as diverse as fighting against special forces—Spetznaz, medical evacuation, and rearming and refueling combat vehicles from helicopters to tanks. While these vehicles would not need be as heavily armored or armed as those further forward in the battle zone, the use of armors and armor-defeating systems—laminate and modular armors and electrothermal launch systems—is every bit as important, if not more so, as it might be for vehicles further forward in the battle zone.

Finally there appears to be an increasing requirement for protection against, as well as for, an offensive capability for what is commonly called soft kill—defeat of the system by attacking the crew or parts of the vehicle or system vulnerable to blast overpressures of, say, more than two to three pounds per square inch. While soft kill, according to some definitions, includes chemical and biological systems, those can be set aside for another discussion. That's not to say they are not an important consideration; it is just to acknowledge the complexity of the matter and the fact that it would require another forum on another day to adequately discuss the problem substantively.

So let's consider the overpressure problem. What is the threat? Apparently the threat is from an increasingly capable set of tactical ballistic missile systems that use fuel-air explosive warheads to produce overpressures in the target area roughly equivalent to those of smaller yield tactical nuclear weapons. In the Soviet literature of the past few years, there has been increasing commentary on what they call "nuclear-like effects" achieved by nonnuclear means. So a salvo or two from a tactical ballistic missile battery could likely produce in the target area overpressures sufficient to disable personnel and, just as in a nuclear blast, by translation damage alone injure additional personnel and achieve extensive damage to soft material targets.

So I would argue that the combined user-development community needs to look seriously at how best to protect against such lethal mechanisms. Overpressured vehicles of all types could provide adequate protection against chemical and biological attack and, if properly designed, could likely at least help protect against damage and casualty-producing effects or overpressure-producing systems targeted against people and soft targets.

Countering this threat—both armor-defeating and soft kill—is not an easy task. Fortuitously a partial solution at least appears in the form of electrothermal launch systems for artillery, the same technology cited as potential for significant near-term improvements in direct-fire systems. For, with electrothermal launch from existing tubes, range can be at least doubled, likely more. Here again there is across the threshold of the new century the possibility of even more substantive improvements with electromagnetic launch. For the moment, however, electrothermal launch for counterfire and counterbattery artillery is more than feasible in the near term, say five years. While such an improvement will in no way solve the artillery force structure problem, it does provide the technical means, if only the services have the foresight to take advantage of it.

Unfortunately for its participants, battle seems to be getting more dangerous by leaps and bounds. Technology appears also to be advancing by leaps and bounds. Unhappily, we have a notoriously poor track record at being able to field our so-called advanced technologies in a timely way. The end result of several decades of this trend is that the Soviets seem to be able to put their so-called old technologies in fielded systems much more rapidly than we can. So, if we have a technical edge at all, it is more and more on the lab bench, and we seem historically unable to pry it off the bench, out of the hands of the researchers into the hands of the engineers and builders of new systems. This set of circumstances has led to development of a gap in modernization rates between the two sides and indeed between ourselves and the primary clients of the Soviets. So it is that the Soviets are outmodernizing us at the rate of about four to one—every four years that pass we wind up about a year further behind. While this conclusion is drawn from our look in 1985 at armor-antiarmor systems, it is equally true of artillery, mine-countermine warfare, air defense systems, and many others as well.

What I have tried to suggest today is that, in the areas of survivability, direct-fire, and indirect-fire systems, we have the opportunity today to field new technologies in new systems, and in a fairly timely way so that, four or five years hence, we don't find ourselves another year or so further behind. Catching up all the time is no fun at all; it's much more exciting to be out in front, holding the initiative and so the opportunity to call the shots. My sincere hope is that you'll find ways to do that in the course of your deliberations here.

Armor on the Modern Battlefield: The Way Ahead

Officers of the Royal Armored Corps
Bovington Camp, United Kingdom
15 November 1989

The task assigned me is to take a look ahead and to describe as best I can what might be the future of armor on the modern battlefield. So that we might have a reasonable chance of success in a reasonable amount of time, let us restrict this discussion to matters at the tactical and operational levels of warfare. That is not to ignore the realities of the commingling of operational and strategic matters in some theaters, as well as in the operations of some threat forces—especially the Soviets. But it does help us focus with more precision on some of the more relevant matters likely to confront us as practitioners of the art of war.

We must begin with battle fighting concepts, in this case AirLand Battle in its many variants. Basically, the concept is framed by three principal notions. They are:

- An active defense, well forward in the operational zone, seeking to capitalize on the defender's terrain and position advantages, and on the technical capabilities of long-range antiarmor guided missiles. And recognizing that the latter part of this advantage, the guided missiles, may have been diminished considerably by the advent of reactive armor, for the moment in appliqués, but quite likely in the near future in modular armors as well.
- A simultaneous attack against the follow-on echelons of threat force, seeking to delay, disrupt, and destroy their ability to join the fight at the forward line of troops and so prevent a force buildup sufficient to overwhelm the defenses with numbers. Its basic purpose is to moderate the force ratios at the FLOT so that the commanders defending forward have a reasonable chance of success. They can yet maneuver forces and fires in an attempt to gain the tactical initiative.
- The above of course designed primarily to enable the defending force to fight and win outnumbered, and to do so well below the nuclear threshold.

How might we have to modify that concept in light of ongoing speculations about conventional force reductions in Europe? It seems to me that, however the CFE talks affect the overall balance of forces in NATO Europe, two imperatives remain for whatever forces NATO may be allowed under the terms of the agreement. These imperatives are:

- The need to cope with a Soviet invasion capability. Given distances and forces, it is likely to remain certain that the Soviets will retain a force generation capability at the operational, or in their terms the strategic operational, level that exceeds that of the West, considering mobilization and force deployment problems across the Alliance.
- The need for measures to ensure against a surprise attack from the East. Surprise being an ever important part of the Soviet battle lexicon and its advantages in modern warfare having been demonstrated vividly in the Arab-Israeli wars, it surely remains an essential consideration.

Do these considerations require a change in operational concepts on our own part? Possibly. Are there technologies that support evolutionary operational concepts in a manner that would help us cope with the invasion/surprise equation? Probably.

What might those conceptual changes be? And what might the technologies be that would enable new operational concepts to be inscribed in our battle books?

First, we must note what technology has done to us operationally since we embraced the notion that we could deepen the forward battle zone, the area behind the FEBA or the FLOT, by using long-range and very accurate fires of antitank guided weapons. Recall that, in the time before the long-range guided missiles, the direct-fire battle, in Europe at least, was a battle determined by the ranges of tank guns and the relationships between those ranges and the intervisibility conditions on the terrain we were defending. While those parameters vary somewhat from sector to sector and perhaps even from weapon system to weapon system, it is on the average fairly safe to say that the forward battle was one in which most engagements could be expected to take place at ranges on the order of 2,000 meters or less.

The advent of the long-range antitank guided missile virtually doubled that range. In so doing, it expanded the battlespace covered by such systems by a factor of four. It was this expansion of the close battlespace, the battlespace at the FLOT, that caused us to adopt what we in the United States called the active defense, a concept adopted in some form by both British and Bundeswehr forces as a result of collaboration in the staff talks between our respective countries.

The Soviet response to the long-range antitank guided missile was, of course, the reactive armor boxes that now characterize the pictures of the latest Soviet tanks. These boxes have made obsolete every fielded antitank guided missile system in the world today, so for the direct-fire battle, we are just about back to the starting point—where we were before the ATGM appeared in considerable numbers in the early years of the 1970s.

So the first order of business is to seek ways in which we might regain the extension of the close battlespace originally provided by the long-range ATGM. At the moment, it does appear that the technology to improve chemical energy warheads so that the ATGM might regain its utility in the close battle is just not at hand. Turning, then, to kinetic energy systems, there appears the opportunity to develop hypervelocity missiles using kinetic energy penetrators to afford both the improved first-round hit probabilities at range and the penetration capabilities at range that come from improved velocities—say on the order of 2 to 2½ kilometers per second.

The most impressive technical opportunity lies, of course, in electric guns—electromagnetic launch and electrothermal or electrothermal-boosted launch. For here, in electrothermal boost, we have the opportunity to achieve velocities of 2 to 2½ kilometers per second with existing gun calibers. This sort of performance is available today and could likely be applied to a fielded system in the next five to seven years, were we to put our minds and our money to the task. Electromagnetic launch of penetrators moving at 5 to 10 kilometers per second is likely somewhat further out in time, say beyond the turn of the century.

However, with electrothermal boost in guns of existing caliber, we could restore the range advantage and hit probability at the range that we had with the long-range ATGM, but which went away with the advent of reactive armor boxes. Not only could the range and area of coverage advantages be regained, but additional range and accuracy, say to five or six kilometers, with electrothermal-boost systems provide an even further advantage operationally.

What is that advantage? If we examine carefully Soviet-style operational concepts for operations at the tactical level, we find that they describe their assault echelon operations in terms of regiments operating in an area about 10 kilometers square, with two battalions on line and one in the follow-on role. Deployed in 10 kilometers of depth, this means that the follow-on battalions of those first-echelon assault regiments are unencumbered by a direct-fire threat from a force defending with powder tank guns, in present calibers, alone. But if we could extend the direct-fire threat to the assault echelons to, say, five to seven kilometers, then the force defending can cope not only with the assaulting battalions but with their follow-on battalions as well. In this fashion it is possible for a blue battalion to defeat a red regiment, blue brigade to defeat an assault division, and so on. If we are to fight and win outnumbered at the conventional level of battle, this is precisely the capability we must build.

In addition to considerably enhanced capabilities for direct-fire tank guns, electrothermal boost, as well as electromagnetic launch, offers enormous potential for artillery and air defense systems as well.

Consider for a moment tube artillery—155mm—existing tubes. Electrothermal-boost launch from those tubes could double the range. This would enable a defending force to bring counterbattery and counterfire onto the vastly increased array of indirect-fire systems that the Soviets in the last several years have begun to deploy forward in the battle area.

Consider air defense gun systems. Systems like Phalanx, for example, could feature ranges of accurate fire on the order of seven kilometers or better, thus providing the force defending forward with a close-in air defense capability against Havoc, Hokum, and older systems that the forward-deployed force does not have today.

The artillery capability with electrothermal-boost launch could quite likely be made available in the five- to seven-year horizon. Air defense systems using EM launch might be available in perhaps 8 to 10 years. Thus, we could well provide ourselves in a reasonable time horizon a technical capability in direct- and indirect-fire systems for the close battle that not only would expand the close battlespace to its former dimensions, but also could provide a considerably enhanced capability as well.

So much for the tactical and operational considerations. What about organization? At the tactical level, it seems to me that all this calls for more smaller tactical entities. Can a tank platoon leader or commander cope with battlespace that is at least four times larger in area? Likely not. So I would suggest that tank platoons consist of two—certainly no more than three— tanks. The platoon commander and his wingman would be the basic fighting team. There would be three such platoons in a company—probably eight tanks. The battalion would have perhaps two such companies, no more than 20 tanks per battalion. The brigade would be assigned perhaps two such battalions, with the capability to control more should the situation demand.

Direct support artillery would consist of perhaps two nine-gun batteries per battalion. These battalions would be equipped with 155mm howitzers firing electrothermal-boost rounds, about 60 rounds aboard, auto loader, NBC protection for the crew of four, digital interneted fire control or a quick halt, fire at about nine rounds per minute. All this can be done with some modest modifications to existing 155mm SP systems.

Bringing infantry to battle under armor is yet a problem. This would be done with an infantry assault team—panzergrenadier infantry, mounted in a vehicle protected to the same level as the tanks it accompanies and mounting an electrothermally boosted cannon of about 60mm in caliber for close work against light armor and personnel. Again, because of improved firepower and the need for an improved leader-to-led ratio, there would be more smaller panzergrenadier battalions, each of which would be as capable as two or more of today's battalions.

So we would have a much "lighter" force that would be every bit as capable as today's "heavy" force. In the end, however, it would depend, as has every military force since the beginning, on the courage of the soldiers, the quality of the leaders, and the excellence of the training they are provided before battle begins. Command-control will remain the central problem in battles of extended ranges and expanded areas. A thorough grounding in the doctrine for battle with these new capabilities must be provided at all levels. All ranks, especially the leaders, must have prepared themselves thoroughly for battle before it begins.

Given all that, with a little luck, the armored forces of our respective armies will, I am certain, once again prevail in battle.

3. Army Aviation

Page

Air Cavalry in Europe 146

Scout Helicopter Development 147

Aviator Careers and Training 148

Army Aviation Issues 149

Army Aviation Requirements 150

Air Cavalry in Europe

Letter to Brigadier General Paul S. Williams Jr.
Fort Knox, Kentucky
5 April 1976

The aviation thing was about a fait accompli when I got here, as we feared it would be. Whilst all the changes of command were in progress, the Army Air Corps was whittling away at it a piece at a time and finally had their way. I did manage to salvage keeping a few more scouts in the organization than had been proposed originally. This way I believe they can task organize so it will work. It is a serious mistake to hang those scouts out there by themselves. But, with nothing but TOW birds in the organization, what else is there to do? General DePuy is due here this week and one thing I want to hit him with is some proposed tactics for organizations configured like ours will be, with no gun birds. The divisional antitank units don't worry me so much—it will be a long time before we get the birds and a lot can happen along the way. My immediate concern is what to do about the cavalry. We should communicate about this and I will propose to General DePuy that we get together—Knox and V Corps, and try to produce some usable tactics to fill the lacuna.

Scout Helicopter Development

Letter to Philip C. Norwine
Bell Helicopter Textron
17 December 1976

You correctly observe that we need a scout helicopter—desperately in my judgment. The death of the ASH was a severe blow. The USAREUR Aviation Study which effectively wiped out air cavalry was likewise a disaster from which we'll be a long time recovering. And so what you saw was our almost complete ineptitude in this very critical area.

Aviator Careers and Training

Message to Major General James Smith
Fort Rucker, Alabama
11 October 1978

1. The DCSPER reports the staff is down to two alternatives in considering aviator officer careers. If I understand correctly, the alternatives are these:

a. First the officer is commissioned in a branch, attends basic and advanced branch courses; between these he has aviation assignments; after advanced he declares a second specialty and rotates between aviation and that specialty in the ensuing career pattern. The specialty can be armor, infantry, comptroller.

b. Alternatively, the officer is commissioned in aviation, goes only to aviation schooling, at which branch immaterial basic and advanced courses are administered. He follows only aviation assignments until after advanced, when he acquires a second specialty which can be any of the specialty fields—but not the combat arms as before.

2. In neither case would we have an aviator trying to maintain skills in three specialties—only in two.

3. The only substantive difference in the two alternatives under consideration is whether there is a branch affiliation, and branch schooling, or simply aviation with one of the normal specialties tacked on.

4. I may not have all that precisely right, but it's close. Also am aware that once there were four alternatives. The chief has ruled out two, so I'm talking about the two that remain.

5. From a TRADOC standpoint it seems to me this causes us to reopen the whole issue of who's in charge in aviation matters. Either alternative in effect recreates the Army Air Corps. This being so, a case is to be made for moving all aviation systems proponency and all aviation training to Rucker and enlarging Rucker's role as the hub of Army aviation. For reasons we all understand only too well, both General DePuy and I have avoided doing anything about this, although it's an issue that continues to present itself for resolution. Now it's about to be upon us once more. The DCSPER plans to go to the Chief no later than March 1979 for a decision concerning which of the two career management schemes to adopt. I have promised the Chief to accompany the DCSPER submit with an evaluation of what to do about Army aviation as a whole given the new complexion of the aviators' careers.

6. By 16 October please give me proposals as to how we might best address ourselves to this problem. We really have two alternatives—one, a status quo with minimum accommodating mods; two, put it all at Rucker—materiel and doctrinal proponency and training proponency. We'll need to set those up with cost estimates and so on. What to do and how to get on with it are things I'd like to have from you first off.

Army Aviation Issues

Message to Multiple Addressees
22 February 1979

1. This responds to the Chief's 152236Z February this subject [Army Aviator Career Pattern].

2. TRADOC concurs with proposals for training and use of aviators set forth in the Chief's message.

3. Have read George Blanchard's response and share his concern with regard to the importance of carefully identifying additional LTC and COL aviator positions to provide a viable career pattern. In this regard, you should be aware that one of the initiatives that we are looking into with our Division 86 effort is the establishment of a fighting aviation brigade (commanded by an aviator O-6) for each division with two or three squadron (battalion)-size units (commanded by aviator LTCs). Such an approach would take full advantage of the new capability afforded by attack helicopters and at the same time provide meaningful aviator LTC and COL positions. In addition, for the first time, it would make Army aviation force structure comparable with the aviation organizations of our sister services.

4. Do not, however, agree with George's conclusion that we should move toward an aviation branch. For all the reasons discussed when we met on this subject, it simply is not in the Army's best interests.

Army Aviation Requirements

Message to General George Blanchard
Commander in Chief, US Army, Europe
7 May 1979

1. Your interest and continued support in modernizing our aviation systems has been key and I fully endorse your assessment of the critical importance of the AH-64, ASH, CH-47D, and UH-60 programs to USAREUR and the Army. I think my comments as echoed by the VCSA in closing the aviation program review at Fort Rucker this past December reflect the TRADOC's and Army Staff's attention these programs will receive.

2. The structure of Army aviation as part of our Division/Corps 86 effort is beginning to take form. Focusing on the European heavy divisions and corps, we are settling in on organic air cavalry attack brigades (ACABs) to complement the ground antiarmor forces. And, to ensure that we can get Stateside assets to Europe timely, self-deployment is being vigorously pursued. In this regard, a flight of four CH-47s is being planned to depart CONUS 1 August for USAREUR, land in Heidelberg, and immediately perform a tactical mission. This demonstration will be also timely since the US/German international staff talks are scheduled in September.

3. The Army salutes your unyielding support for not only aviation but everything.

4. The Battlefield

Page

Armor Conference Keynote Address 152
Junior Officer Tactical Awareness 158
Influence of Terrain 159
V Corps Situation 160
Battlefield Smoke (Thurman) 162
Command Post Concepts 163
Combat in Built-Up Areas 165
Realistic Battlefield Environmental Conditions 167
Battlefield Smoke (Meyer) 169
Continuous Land Combat 170
Corps Battlefield Command 172
The Battlefield (Valz) 174
The Battlefield (Burdeshaw) 175
Artillery Rates of Fire 176
Battlefield Development Plan (Meyer) 177
Battlefield Development Plan (Blanchard) 179
Battlefield Development Plan (Kroesen) 180
Barrier Warfare 181
Battlefield Interdiction 184
Battlefield Function Terminology 191
Integrated Operations 193
Battlefield Development Plan (British Army Convention) 194
The Integrated Battlefield 199
Extending the Battlefield 201
Future Battlefields 213
Use of Phase Lines 217
Thermal Imaging Technology 218
TRADOC's Analysis of the Yom Kippur War 220

Armor Conference Keynote Address

Armor Magazine
November-December 1975

The first part of our meeting is styled as a keynote address. If we have a keynote message it concerns how we believe the Army should get itself ready to fight the first battles of the next war. Therefore I will describe for you what we are doing, and the logic behind what we are doing, because it pervades almost everything else you will hear.

For a number of reasons we believe that wars of the future will probably be shorter and more violent than in the past. So the important task of the US Army is to win the first battles of the next war, because those may be the only battles, as demonstrated in the Middle East in October of 1973. In order to decide what we must do to win the first battles of the next war, we must first describe the modern battlefield, and how the Army should expect to fight on it.

The single most impressive fact about the modern battlefield, drawn from observations of the 1973 Middle East War and analyses of what might happen in Europe, is that we are faced with masses of weapons.

We don't necessarily believe all wars are going to look like the October War, but some numbers from that war outline the dimensions of the problem. (See Figure 1.) It's important to point out that in the 12 violent days of that war, half of all combat vehicles on both sides were destroyed.

We were fairly well along in an analysis of the modern battlefield in October 1973, and what took place that month on the Golan Heights and in the Sinai served as a dramatic demonstration of what we thought we already knew about modern battle. Now, fighting outnumbered is a big problem. There are about a hundred thousand tanks in the world; as nearly as we can figure, about sixty thousand of them are of Soviet design or manufacture. So we are not only outnumbered at the outset, but there is no way for our factories to turn out enough equipment to catch up, let alone get ahead. In addition to the problem of numbers, modern weapons are of vastly increased lethality and effectiveness.

That's not all bad. Figure 2 shows four battles in October. The first column is the posture of the Israeli Defense Force in each battle—one night attack and three defenses. The second column shows the total number of tanks involved on both sides. The third column shows the odds at the beginning of the fight, Israeli versus Arab. In other words, the Israelis were outnumbered by 2 to 1 up to 6 to 1. The last column shows the loss ratios. Note that the Arabs, even though they outnumbered the Israelis at the beginning of one battle 2 to 1, lost at the rate of 6 to 1. In another battle the Arabs were totally wiped out. In still another they lost at the rate of 50 to 1. In almost every battle of that war, exchange ratios were very much like those shown. The point of this is that there is no pattern. Historically, we have described tank battles using mathematical formulas derived from the Lanchestrian equations traditionally used to describe the mathematics of aerial combat. Lanchester's laws, as most of you know, say that the side that is outnumbered at the beginning is foredoomed to defeat. Based on our preliminary examination of 15 to 18 battles in the October War, we analyzed about a thousand tank battles, and can find nothing in the history of mass tank warfare to validate the prediction that the side outnumbered at the beginning seems to have a better chance of winning than does the other side.

Reprinted with permission from *Armor* Magazine.

We think that is very encouraging. This is so because our potential adversaries have decided that the way to win is to just outnumber the other fellow and try to overwhelm him with masses. Since we know we will have to fight outnumbered, we need to develop some convictions about the fact that it can be done, and then we must figure out how to do it. So what we would like to describe is how to win outnumbered.

	Tanks	**APCs**	**Arty Tubes**	**AD Btry Deployed**
ARAB	4000	3000	3000	150
ISRAELI	2000	4500*	800	10-15

Note: All figures approximate

*Includes halftracks

Figure 1. Combat Vehicles Employed in the October War

IDF	**Total Tanks**	**Odds ADF:A**	**Losses IDF:A**
NITE ATK	870	1:2	1:6
DEFENSE	180	1:1	∙:TOTAL
DEFENSE	700	1:6	1:6
DEFENSE	100	1:2	1:50+

Figure 2. Four Battles in October

On the modern battlefield there are really three wars. The first is the General's war, and the General's first problem is to find the enemy. That is, get forces on the battlefield to find the enemy early enough and far enough away from his own forces so that he can do something about it. What does he do about it? He moves. He moves sufficient units to generate enough force in time and space so that, if he is attacking, at the point of attack he outnumbers the enemy about 6 to 1; or, if defending, he creates a force to ensure that he is not outnumbered by the enemy attack by more than about 3 to 1. If his units use tactics I will describe for you in a moment, the general should expect them to extract an exchange ratio on the order of 4, or 5, or 6 to 1 from the enemy. If that can be done, we think he has a good chance of winning the battle. That is the General's war. It is a war of movement. It is a war of mental agility. It is a war of command and control.

For example, a division commander defending a front of about 60 to 80 kilometers against a Soviet-type breakthrough attack—two motorized rifle divisions on a front of 7 to 11 kilometers, has to bring about 8 of his 11 battalions together in sufficient forces so that he is not outnumbered by more than 3 to 1.

Meanwhile what happens out on the flanks, in those other two or three battalions of that division, now covering the remaining 50 to 60 kilometers of divisional frontage? No doubt that is going to be a very sporting war out there. But one element that will help tremendously is air cavalry and attack helicopters. As was the case with finding the breakthrough attack, the only way to find the enemy on the flanks far enough away to move units against a secondary attack is to use air cavalry—backed up by attack helicopter units—to hold, or block, while moving forces in there on the ground to do something about it.

The second war is the one we can call the Colonel's war. The Colonel has to take forces the General moves into the battle area and feed them into the fight in such a way that they get where they are supposed to be, and get there in time to be part of a defensive, or an attack scheme, and win the battle, given the odds that the General has been able to generate.

The critical part of the battle, which we call the Soldier's war, is a war of lethal weapons and masses of enemy, a war of the combined arms. Now by improving arms and equipment we think we might improve the Soldier's fighting capability by about 10 percent. However, we believe most strongly that we can improve our fighting capability by 100 percent through training. Later today we will demonstrate for you some training devices which we believe will help us train soldiers better to do their job in their war.

The Soldier's war involves lethal weapons; understanding the lethality of those weapons is important. Today's M60A1 is about 10 times as effective as its World War II counterpart, the M4. In World War II, tankers had a 50-50 chance of hitting with the first round at 700 meters. That distance increases to about 2,100 meters with the XM1. And, of course, today's antitank guided missiles have a good first round hit probability to ranges of about 3,000 meters.

If we sweep those extended ranges over the area the tank commander can see, we find that today's tank commander controls an area 10 times the size of his World War II buddy. Putting it another way, in World War II one could count on a hit out of about 20 shots. Today we can hit with one of every two shots. And with XM1 that will increase to two hits out of three shots.

The air over the battlefield is also filled with lethal munitions. The range and effectiveness of modern air defense cannon and missile systems make the air over the forward battlefield a difficult place in which to operate. In order to fly there, one must avoid or suppress these weapons.

The lesson in all of this is that tanks can't go it alone. It takes a combined arms team. Generals, colonels, and soldiers have to understand the lethality and masses of weapons they face, and share a common outlook on what must be done about terrain, enemy, weapons, movement, and suppression in order win when fighting outnumbered.

Movement on the Battlefield

First, and most important, how do we move around on a battlefield like that? Any scheme of movement must recognize the lethality of modern weapons; that tells us that we must use the terrain—get down in the terrain, seeking protection from enemy long-range observation and fires. Movement must be on covered routes from one covered or concealed position to another; moving elements must always be overwatched by some other element in position to bring fire to bear on weapons which fire on the moving element. And so moving on the battlefield is governed by terrain and the enemy.

When enemy contact is not likely, we move by traveling. Picture a tank platoon moving along. One section leads. One follows. They may move fairly well spread out, or close up into a column-like formation for a defile, then spread out again. Both sections move. If the platoon leader thinks he is going to meet the enemy, he begins to move in traveling overwatch. Here the overwatching section always moves from one position to another in such a way that it can cover the other section if that section comes under fire.

If the platoon leader really expects to meet the enemy, but isn't exactly sure where he is, movement is by bounds. This is the slowest way, but the surest way. The platoon always finds the enemy with the least force possible, so that in that first 15 to 20 minutes of battle, when most losses occur, a lot of people are not killed unnecessarily because they just bounded out on the forward slope of a hill.

Once contact is gained, it is a simple matter to move directly from bounding overwatch to fire and maneuver. The only difference between bounding overwatch and fire and maneuver is that overwatching and moving elements open fire. The transition from movement to contact to hasty attack to deliberate attack is therefore very smooth. There is no mental gear shifting.

Now the same thing is true with mechanized infantry—the dragoons. The mechanized infantry rifle squad consists of a fighting vehicle and a fire team. Normally, of course, a squad has 11 men, but, as all of you know, those 11 men will hardly ever be present for duty on any given day. There will never be 11 men in that track, and we must recognize that. Therefore the driver and one or more gunners man the fighting vehicle, and the rest of the squad dismounts under the control of the squad leader and forms a fire team. Normally there is only one fire team. In the fire and maneuver, or bounding overwatch, at the rifle squad level the vehicle and its fire team work together. In the mechanized infantry rifle platoon, there are four fighting vehicles and four such rifle teams. The fire teams carry with them the antitank guided missiles and automatic weapons assigned to the squad. They take the most important weapons with them when they dismount, depending on how many soldiers there are to carry the weapons. Normally they are mounted. They may travel. They may move in traveling overwatch when they think enemy contact is likely; or, when they expect to meet the enemy, they move by bounds. And when they meet the enemy they go into fire and maneuver and assault the objective. They dismount when forced to do so.

We use the same tactics in cavalry platoons, and the same principle applies to helicopters. Part of an attack helicopter unit may move traveling along a covered route through a little draw covered by another element. Attack helicopter units also use traveling overwatch, bounding overwatch, and fire and maneuver. The point is this—the same tactics apply to every small combat unit on the battlefield. We can't find any exceptions, and the rule applies above ground as well as on the ground.

Aircraft must fly nap of the earth; the only way for them to survive is to operate in the ground battle environment. They are just like another ground vehicle, but they happen to be 10 or 20 feet off the ground moving at 20 or 30 knots. They can't fly low level, and they probably can't fly contour in the forward area. So they must be down in the trees. Now flying down there is difficult. Not everybody is skillful enough to fly down there. Not everybody is willing to fly down there. But it has to be done.

Major General C. J. LeVan at Fort Bliss has been very supportive of our efforts to get air defense weapons forward in our battle array. As most of you know, threat armies use air defense weapons such as ZSU-23-4s on the basis of one for every six tanks. And so there are a lot of air defense weapons out there. We don't have that many forward air defense cannon, but our air defenders are very enthusiastic about using what we have, and using them well up in the forward area.

Summary

Now to summarize what I have just said about movement to contact and attack. We move to contact, to find the enemy with minimum force; use covered routes; use terrain; key movements to terrain and the possibility of enemy contact—not likely, likely, or expect to meet him; then move into fire and maneuver for hasty or deliberate attack.

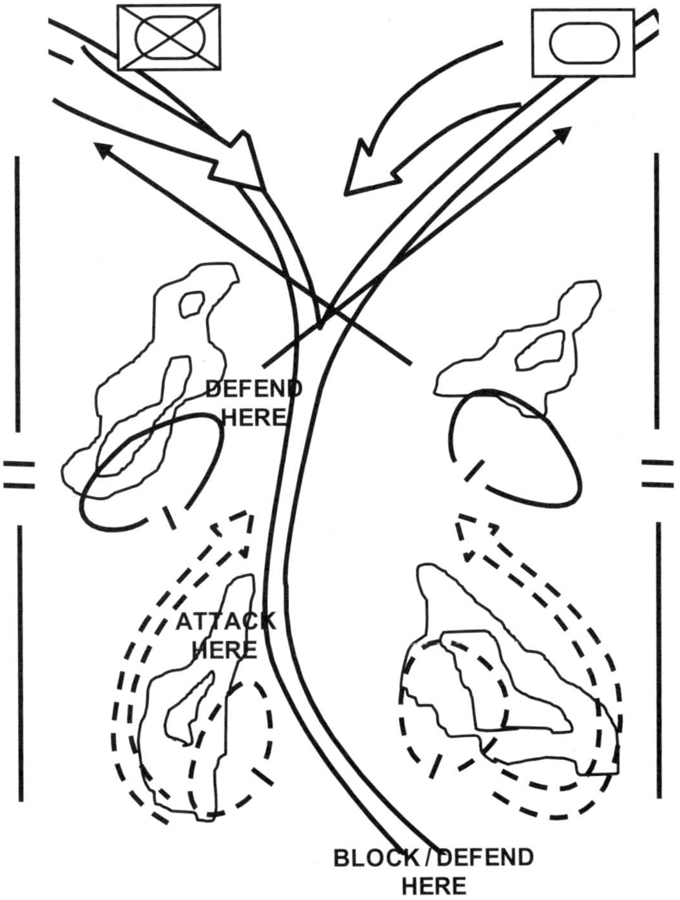

The purpose of all this is to break up the enemy's defensive system. If the General has managed to get forces together so that the odds are about 6 to 1 in his favor; the Colonel has fed battalions into the fight in an orderly fashion; the soldiers are down in the terrain, using their weapons to maximum advantage and minimizing their own vulnerabilities, enabling them to break up the defensive system; and the helicopters come in, go through the break, get in the rear, hit reserves, command and logistics installations—when all of this happens, we are convinced we can win outnumbered.

Now how do we defend in modern battle? Defense is a matter of understanding enemy weapons and tactics. Briefly, the important thing to know is how the enemy plans to attack by echelons. In using terrain he must seize on every advantage that normally accrues to the defender. In fact, the only advantage the defender doesn't have is the initiative. He regains the initiative by attacking at some point in the defensive fight. But when he attacks he must do it in a little different way than it has been done before. As the enemy comes into the defensive area, the commander must have his units in position, probably behind a hill firing across the forward slopes of the next hills so as to minimize their vulnerability and make maximum use of their long-range weapons capabilities. The defending commander should plan to counterattack, but only when he has a good chance of success. For when he attacks, he gives up many of the defender's natural advantages. He doesn't want to do that unless he can gain some advantage by it. He must plan to use smoke, suppression, and the counterattack to kill as many enemy as possible in sufficient time to get back into the next defensive setup before the second echelon comes along. He must know that 5 to 7 kilometers behind the echelon he is now in contact with there's another; he must tidy up what he is trying to do in order to get back into position in time to meet the next echelon.

The Soldier's war, defense or attack, depends on teamwork and training. We pay a lot of lip service to training, but one of our serious problems today is that we all live in a hostile training environment. We have many programs, most generated by agencies that aren't concerned with how to fight, all pressing in on training time. Not many contribute much to combat readiness. All eat into commanders' training time.

That sets the stage for what we are trying to do. The next speaker will tell you a little about what we are trying to do with cavalry and cavalry vehicles.

Junior Officer Tactical Awareness

Letter to Major General Willard Latham
Fort Benning, Georgia
18 May 1976

Finally, in questioning them [second lieutenants out of the Infantry Officers' Basic Course who are assigned to mechanized battalions] about tactics and watching them on field exercises, they seem to have taken on a lot of what we are teaching. Techniques of movement, what to do in the attack, and a vague perception of the panzer grenadier ideas seem to be taking some hold. What they are weakest in, however, is what to do in a battle position in the defense in the main battle area. There's a gap in what they do in the training area and what they do when you tell them to put it into practice in an actual defense battle position in a general defense plan. They all want to dismount, get the .50s off and dig them in, then send the tracks away somewhere so they won't clutter up the battlefield, and then get on with the leg infantry war. Much of this is a matter of education, and here again their captains and lieutenant colonels aren't too aware of what's in 71-1 and 100-5, so they too need educating. But I'd just ask you to look again at the part of your instruction that deals with mechanized infantry to make sure it tracks with what's in the manuals. You have a very difficult problem, and I recognize that. You have to teach airborne, airmobile, and leg as well as mech. But the fact remains that I suspect the bulk of your basic students will soon find themselves in that vast part of the infantry army which is mechanized. The tactics and concepts of employment are quite different, and they don't perceive this too well.

Influence of Terrain

Memorandum for V Corps G2

27 July 1976

1. I've been studying enemy literature about how he attacks in a march to contact and how in a breakthrough attack. It strikes me that at almost every echelon—corps, division, or battalion, real differences between the two attack scenarios in terms of numbers of targets that appear, distance laterally and in depth in which they can be expected, and the timing of their appearance are really not all that great. At battalion level it may be that, if we use our technique of fitting his formation to the terrain, whatever real differences there may be are in fact absorbed by the terrain. Not so in the desert, but so in V Corps. At corps level the ability of all those avenues of approach to accommodate his formations in the aggregate also limits total numbers that can be deployed and therefore, at that level also, terrain absorbs the deltas between the two styles of attack.

2. Using our standard threat, see if you can figure out some way of describing for me what the deltas are—first disregarding terrain, then applying terrain, perhaps that on the avenues of approach into the covering force area. Width, length, time, rates of arrival, and systems density are the key elements. Don't try to be elaborate—all I want is some rough calculations to tell me if my speculations in 1 above are about right.

V Corps Situation

Letter to Colonel Erwin Brigham
McLean, Virginia
8 June 1977

Thanks for sending along that battalion commander's analysis of what we ought to do in NATO. I think I know who wrote that paper; in the context of the time during which he held his command, I can't quarrel too much with some of what he says. In the year or more since he left command we have done several things. First, we have completely revised the GDP to more realistically reflect the tactics we believe will have to be used to win the first battles. Whatever reserves there were, and there were precious few, have been allocated to the corps so that we may build some depth in our defenses. Inadequate though we consider that at the moment, short of having more forces it's all we can do. In V Corps all our tank ammo is now loaded up on the tanks, and we have an active program going to load everything else up. Unfortunately it isn't moving as fast as we impatient fellows would like it to, but it is moving.

We've good personnel strengths—but I know the battalion commander's problem is the same evermore. I was a battalion commander here during the Berlin crisis in the early 1960s. For two years our division strength was about 112–115 percent of authorized, yet no battalion in the division ever had more than about 95 percent of its assigned strength and the duty strength was correspondingly lower. What it would take to get 100 percent manning on the line I do not know—all I can say is that in our system it appears to be well nigh impossible. But there are some things a battalion commander can do to live with that problem and come out pretty well in the end. As a matter of principle, I would agree that he probably shouldn't have to do them. As a realist I recognize that he always will have to live with them, and one measure of the successful battalion commander is that he has figured out how to do that well. The fact that this fellow bitches so much about it reflects his inability as a battalion commander to cope with it. The lessons of how to cope got lost over here in the Vietnam years, so I've had to go back and teach people about battle rosters and all the things you and I grew up with. Reinventing the wheel!

I do not agree with two of his assertions. First is that the side that's outnumbered is foredoomed to defeat. The history of tank warfare does not so indicate. But there are clearly some things the defender must do and others he should do to capitalize on the advantages he should enjoy as defender. The problem here in Europe was that, until I got here and 100-5 arrived with me, there was a general malaise—a defeatist attitude reflected by that fellow's statement. They had somehow concluded that the situation was so bad that they couldn't cope with it, and so had thrown in the towel—the same attitude he displayed towards his present for duty strength. My eldest son, Mike, served here as an artillery battery exec for three years from 1971–74 and he reported to me that the whole defeat mechanism permeated the atmosphere from the top down. It was appalling. I came here 16 months ago determined to change that. I am myself convinced from a lot of study and analysis that we can win outnumbered; now the problem is to share that conviction and develop the same conviction in my commanders. That I believe we have done.

Secondly, I don't agree with him that the Germans can raise the forces he describes as being necessary and within their capability. Here he is simply echoing the current in-vogue sentiment of the House Appropriations Committee, which continues to insist that we shouldn't pay anything until everyone else pays more.

Al Haig is leaning hard on these other countries to pick up their marbles. I dare say he won't have much luck with any but the Germans. So it is to the Germans and the Americans that falls the lion's share of the defense of Western Europe.

We should work on that theme. There are many things that can be done to improve our lot, and save money besides, before we need address ourselves to the politically very difficult task of increasing the German force structure. Force increases must be joint endeavors between ourselves and the Germans. Unfortunately, we haven't looked at the problem quite that way before.

So a great deal has been done to solve some of our young friend's problem. Much remains to be done. All of what needs done will take time, patience, and tough people working the problem. Meantime we have to have battalion commanders who believe they can win outnumbered, have figured out how to do it, and who can live with the ever-present problems of too little time and too few people.

Battlefield Smoke

Message to Lieutenant General John R. Thurman
Commander, Combined Arms Center
19 September 1977

Wilbur [Payne] has provided me with a copy of TRASANA's 21 July letter report on smoke tactics simulation. Believe CACDA has a copy. By any measure its findings are alarming, alarming enough to cause me to accelerate a program I have been forming up mentally for some time. I've become convinced that we must attack the smoke problem head on—threat estimate, friendly countermeasures, friendly capability, and performance deltas resulting from analysis, actions necessary to shore up deltas in near term/far term, equipment, tactical, organization and training actions and developments we must take in response to perceived needs. Would like to talk with you about this when next we meet. Meantime suggest careful reading of TRASANA's "Smoke Tactics Simulation." Glenn [Otis] may have in progress; let's review that. Review PM smoke and other factors germane to the issue.

Command Post Concepts

Letter to Lieutenant General David E. Ott
Commander, VII Corps
c. 5 October 1977

This letter forwards for your review our overall concept of command posts and describes in some detail the various elements of the tactical command post (TAC CP), their functional relationship, and how the TAC CP is to operate. All of this represents the results of our analyses and field tests—all part of our overall effort to improve the capability to command and control the corps battle. We now solicit your assistance in the preparation of a concept plan to be submitted to USAREUR.

Briefly, our concept of command posts entails manning and equipping four types of command posts—an emergency operations center (EOC), main command post (Main CP), rear command post (Rear CP), and a TAC CP. We believe the essence of a command post's ability to control and coordinate operations depends on three essential elements. First, information must be organized, current, complete, and accurate. Second, decisions must be made on the basis of this available information and in coordination with the staff and, whenever possible, the subordinate commanders. Third, communication of these decisions must be organized, current, complete, and accurate.

The EOC provides the focal point for control and coordination of activities within the corps during crises and initially during hostilities until either the corps Main CP or the TAC CP is fully operational and control can be transferred. Located in the corps' peacetime headquarters, the EOC provides the means for controlling elements of the corps during the transition to battle. The round-the-clock nucleus consists of the G3 Operations Center, with its Emergency Action and Alert Message capability, the G2's All-Source Intelligence Center, and the Staff Message Center. On order the EOC will be manned by a complete corps battle staff.

The Main CP is initially established in a hardened facility. Communications are established with higher, lower, and adjacent commands; control is transferred and EOC activities cease. We estimate that this will take about 30 hours. Operating from this facility the DCG and key members of the staff will have access to specific information dealing with all aspects of the corps operation. The initial emphasis is getting the main battle area forces in position. Once this is accomplished, the primary functions of the Main CP are intelligence, operations, and fire support. The key point is that the Main CP exists to support the TAC CP.

The Rear CP is a more permanent facility than is the Main CP due to its administrative and logistical functions. It is located well to the rear, preferably near an airfield, and collocated with the Support Command CP. Our estimate is 30 hours for it to be fully operational.

Our TAC CP is a small, mobile element capable of 24-hour operation from wooded or village locations. The nucleus consists of G3, G2, and FSE/ALO elements mounted in three M577 CP carriers and two M113s configured as commander's vehicles; a communications support platoon from the corps signal battalion; and an administrative, logistical, and security element. FM, UHF, and HF communications capabilities are provided. The physical size of the TAC CP is no larger than that of the CP of a maneuver battalion, while its electronic signature approximates that of a brigade CP. A jump capability has been built in, and the corps commander can quickly move out with one or both M113s, in a small command group, without disrupting the operation of the TAC CP. Our goal is to have the TAC CP on the move within the first two hours so that it can be fully operational in its forward location within six hours after the initial alert.

We have been developing all this since March 1976. The initial organization was fielded during REFORGER 1976, but it did not operate as part of the corps command and control structure. The REFORGER experience convinced us that the corps battle cannot be fought from the Main CP and we believe the evidence is sufficiently compelling for us to field an armor-protected TAC CP with sufficient equipment and personnel to track the battle and issue timely orders. Our concept has now been finalized as a result of our experiences during WINTEX 77 and FTX CERTAIN FIGHTER. In fact the TAC CP is essentially the same organization you saw during CERTAIN FIGHTER.

Now we are at the stage where we can brief your staff and would like to do so before the end of June. At the same time we would deliver a copy of the plan for your review with a further request that you provide us any comments or recommendations by 1 August 1977. Recognizing that the concept of command and control may vary between our corps, the only way V Corps will be able to get authorization for personnel and equipment changes to the current MTOE will be with the VII Corps support. And while you may not want to employ these added resources in exactly the same fashion as do we, they will afford an added capability with which to command and control the battle.

Combat in Built-Up Areas

Letter to Robert A. Moore
Deputy Under Secretary of Defense (Tactical Warfare Programs)
28 November 1977

This responds to your letter of 4 November concerning training facilities, materiel requirements, and simulations for military operations in built-up areas. There are several things you should know about what we are doing with regard to these problems.

First, we have under development with the Germans in our series of staff talks with them a concept paper describing operational concepts for operations in built-up areas. Earlier this month we met again in the fourth set of talks in the series and reached agreement on the contents of this paper. We plan to have the concept paper ready for our respective Chiefs of Staff to sign early next year.

Second, General Reichenberger, the Vice Chief of Staff, Heeresamt—my counterpart in the staff talks, and I have agreed that before we undertake any new weapons developments and create materiel requirements therefor, we need a better database concerning what the weapons we have on hand can do against structures in a built-up area. It may surprise you to learn that we don't know much about this. You might remember I had a hard time several years ago getting DARCOM to fire some gun ammo against a concrete building to try and determine whether or not we needed to keep the HEP round for fighting in cities. So the next step in the German-American proceedings is to test current weapons and communications equipment to see how it performs in cities. After that we will be able to conclude more precisely what we need to do by way of materiel developments.

On the training—our forces in USAREUR are using several facilities for training in combat in cities. The Berlin Brigade has one good facility. The Germans have one at Hammelburg which US troops have been using regularly. Here in the States we are trying to get a facility built at Fort Bragg. From time to time it suffers budget cuts from one or more of the many agencies who like to dabble in military budgets. It's currently under fire from OMB.

As for simulations of battle in built-up areas, we haven't any worth the title. I'll have someone contact George Schecter as you suggest and perhaps we can use some of his work.

I owe you a word on my own perspective about this matter. No question that we don't have a very well-developed idea yet of what we want to do in built-up areas. Nor do we know what our present weapons and commo gear will do, and so we don't know what best to pursue technically in order to redress whatever imbalances there may be. And finally we don't have a very well-developed simulation method for analyzing war in a built-up area. So where does that leave us?

There has been a lot of talk about this problem recently. For a time it was one of the big things to talk about—since overtaken by the RSI euphoria. From the perspective of the ground commander, especially in the early days of a war in Europe, neither side wants to fight in the cities. The Sovs will bypass them, leaving behind only sufficient force to hold in place whatever forces we are snookered into leaving in the cities. They would like nothing better than to have us stash away divisions in the cities, then they could march right by them, leaving them to rot. I'm talking about cities, now—not about urban sprawl—the two are quite different, and must be treated differently tactically. One trouble is that, in the developing dialogue on the matter, no one ever made quite clear enough the distinction between the two. Urban sprawl has to be treated like terrain—for that's just what it is. Even in the North German Plain this is the case. The city fighting I'm talking about is that which would take place in the old parts of the older and larger cities and towns—Nürnberg, Frankfurt, Giessen, Kassel, Würzburg, Mannheim, and so on. All these are places we'd like to stay out of, and all are places both sides would like to avoid having to attack because it's a tough and very expensive nut to crack.

How much of the time will we have to fight in such places? My guess is not very much. Which does not diminish the need to know a lot more than we do about the problem—it just says that, given all our other problems, this is one that we will pursue, but with some lesser urgency than we might perhaps tackle tank-antitank, command-control and counter command-control, smoke, night, and others, and the supporting simulations and materiel developments for those. We haven't enough resources of any kind—especially time and analytical talent, to do anything else.

So we are doing something, and we are doing it with the Germans. This always takes a little longer than it might otherwise, and so there is a tendency to allege that we're dragging our feet. If there's a better way, then send me the man who understands the problem and has a practical solution. I'll be only too happy to listen to him. Sooner or late we have to tackle this one with the Germans—our area of most serious concern is their country, so I'm convinced we're doing the right thing. Just how we'll do joint testing I'm not sure, but Reichenberger and I are to talk about it again soon in a few months and see what can be done. Meantime we are trying to figure out how best to go about it.

Realistic Battlefield Environmental Conditions

Message to General Walter T. Kerwin Jr.
Vice Chief of Staff
29 November 1977

. . .

2. We have been working on this problem [of the use of realistic battlefield environmental conditions throughout the Army] for several months. The first thing we find is that not much has been done in the past—lots of hand wringing, but not much analysis. The database is meager at best. There are many areas of disagreement about effects. The calculus of degraded visibility due to smoke, fog, haze, and dust does not exist. We even lack the most fundamental thing of all, an operational concept with regard to the use of visibility reducers over which we have some control, like smoke.

3. And so we have been trying to get hold of the problem. About a year ago we began, with night vision labs, to develop a detection model which contains attenuation factors measured in the European atmosphere. The model is now being used by CACDA in support of NATO RSG Panel 9 in determining the effects of smoke, fog, and night on NATO force effectiveness. CACDA, AMSAA, and NVL are working together on this. It is important to us for two reasons:

a. It is our first attempt to look at the problems brought on by limited visibility and its effect on TOW, Dragon, and tank systems.

b. It serves as a guide for future studies in that the NATO study group specified the atmospheric scenario to complement the tactical scenario under study. This work needs to be expanded and pursued; it is only a beginning.

4. What must be done is to get AMSAA and BRL working at high priority to specify the effects on enemy and friendly weapons and command-control systems of weather—fog, rain, snow, cloud cover, smoke, dust, haze, and various EW conditions, all in varying degrees and amounts. Using your message as authorization, I will soon send to Jack Guthrie [General John R. Guthrie] a proposal for a program to do what I've outlined above.

5. In TRADOC itself we are tackling the simulation problem—Roy Thurman [Lieutenant General John R. Thurman] and Wilbur [Dr. Wilbur Payne] are doing that. But I've a word of caution. I've told Roy to start with one condition and learn to do that well. Smoke may be the best place to start, since we have some control over that. When we are satisfied we've a model that correctly represents the condition, then we can match force on force in simulations and compare the outcomes to "clean" battles. What we learn from this is what we must have before we try to model an entire spectrum of environmental conditions. This does not suggest delaying any or all of the modeling things that must be done. It does, however, recognize that we have limited resources and that there may, and possibly will, be a high payoff in designing a pilot to serve as a guide for the rest.

6. While I appreciate the need to apply environmental degradations to weapon systems developments, my comments above about the condition of the database apply to this situation as well. It will therefore not be possible to make some universal application of degrading criteria as a matter of course in every systems development. We should try as best we can

for the time being, but we also must recognize the realities imposed by what we do and don't know. In addition, the overriding consideration in any system performance analysis must be the conditions under which the system operates best. Thus the base case for tank optics is the clear day in which the thing was designed to operate in the first place. Obviously if it's a night sight, night performance is the base case. We can't obfuscate already difficult decisions about systems performance by inserting entirely new subsets of performance criteria without clearly establishing what the base conditions are to be. Once the base condition comparisons are done, then excursions can be made into what happens under various conditions of degradation. Hopefully we will someday be able to describe a set of bounding parameters—limits which will help specify the variance in performance that needs investigation.

7. While I share the impatience of Shy's [Lieutenant General Edward C. Meyer's] staff who persuaded you to send the reference, I have spent considerable time examining the matter and believe it best to proceed as I've outlined above—in a kind of semiorderly way. It serves little purpose to mount a broad front attack on this problem. We know what the problem is. What we need is a coherent kind of program aimed at eating the elephant one bite at a time.

Battlefield Smoke

Letter to Lieutenant General Edward C. Meyer
Deputy Chief of Staff for Operations
29 November 1977

I know you are anxious to hear what we have to say about smoke, dust, and obscuration on the battlefield. I know this is so because it has been some time since you wrote me, and while we were trying to decide how best we might tackle this very difficult problem your impatient staff persuaded the Vice Chief to send out a message DTG 222225Z enjoining us all to get on with this work forthwith.

Which we will do. However our research has uncovered a wide disparity of opinion in the Army Staff about the matter. Your impatient henchmen, authors of 222225Z, obviously were unaware of this diverse opinion or, in their haste to goad us all on to greater endeavors, elected to ignore the rift at their level.

Accordingly it is requested as a matter of some urgency that your staff take the necessary action to resolve the dispute as between the Chief, Chemical Warfare Service, who, on the one hand, suggests it is a good idea to smoke up the battlefield, and Chief, Infantry, who, on the other, suggests he is not so certain. The attached unresolved correspondence in the matter is explanatory.

May I also suggest that the authors of 222225Z would be interested in this work—and indeed were they to set their hands to it could be kept more gainfully employed, thus dissuading them from antics of the sort characterized by their last missive to the field.

Continuous Land Combat

Letter to Ambassador Robert W. Komer
The Rand Corporation
7 December 1977

This is to acknowledge receipt of your note of 17 November and Emanski's Tech Report on *Continuous Land Combat.*

We are thinking and doing NATO—response to your note, indeed our detractors take us to task for overdoing the emphasis on war in Europe in the doctrinal products already produced during TRADOC's relatively brief existence, and those still ongoing in the system.

With regard specifically to the Emanski Report, I second Ken Cooper's assurance that we're aware of and doing something about the problem. There are, however, several things one must understand about this. The first and most significant is that continuous combat—the capability to do it operationally—is largely a force structure problem. The Soviet operational concept of continuous combat is supported by their force structure and to a lesser but growing degree by the performance capabilities of their equipment. No question that in areas where economy of force is applied, or even in areas of secondary or holding attack, they may not have sufficient forces to support their operational concept. But it is quite clear that their echelonment of forces, especially in the breakthrough area, and the force structure that supports that echelonment are both indeed capable of supporting continuous combat. We do not have, nor would I judge us willing to afford politically or economically, to try and solve the problem for ourselves using their solution—more forces.

The second point that needs to be made is that technology cannot provide the edge necessary to redress the imbalance—to permit us to cope with the Soviet application of the doctrine of continuous land combat. While it is true that we probably enjoy some potential margin of technical excellence—in the ability to fight at night and in bad visibility—a generation, perhaps, over what the Sovs have in the field—we are not yet fully equipped, and by the time we are it is quite likely that they will be too. And, even were both forces fully equipped, it's still a force structure problem. The super tank with super sighting gear for operations in night, fog, and smoke can't be fully utilized unless we can provide additional crews, or additional units, for round-the-clock, round-the-week operations. It is interesting to note that, while our US Air Force structure has always called for crew ratios of two or more crews per airframe, our Army aviation structure has always been based on but one pilot in each pilot's seat, and on the ground we have never considered the problem of more than one crew per tank, infantry fighting vehicle, or artillery piece. Beyond that structural problem is that of insufficient units in the structure to permit unit replacement, and the pretty certain unlikelihood that we could ever improve our force structure that much.

Now in those two considerations is the problem with Emanski's Report. While I do not quarrel with his thesis that the Sovs are organized for continuous combat and we aren't, I do quarrel with his blatant assumption that we're not aware of the problem, not doing anything about it, and his contention that improved technology can or will redress the imbalance. If there's no way for us to cope structurally or technically with the Soviet continuous land combat strategy, then we must be prepared to do so with superior tactics, better training, and an operational concept that seeks to break up the Soviet mass, destroy the momentum of the Soviet attack, and prevent the quick reinforcement by succeeding echelons. That done, we will have effectively destroyed their ability to conduct continuous land combat. In operational concept this means quick destruction of the first-echelon armies, the necessary recon, target acquisition, and surveillance to find the second-echelon armies, and the means to break up their mass, destroy their momentum, and prevent them from joining the fight before the first-echelon armies can be destroyed.

Obviously there are some considerable gaps in our now ability to do all this, but we are hard at work shoring up those gaps—in USAREUR more ammo deployed farther forward, uploading ammo on combat vehicles, improving our deployment capability, and other related programs all are contributors. During my time in V Corps the operational scheme of our General Defense Plan was revised to reflect the concepts set forth above. Just now we're hard at work between TAC and TRADOC to resolve the surveillance and target acquisition problem in a joint MENS for those systems.

In the end, then, if we can agree about the operational concept, the tactics that go with it, the technical requirements that flow from it, and the improvements that need be made in our force posture to accommodate it, we will have gone a long way toward developing a capability to cope with the continuous land combat doctrine without trying to match the Soviets in force structure, and without deluding ourselves that technology can redress the imbalance.

In light of all that, the Emanski Report is quite shallow in perceptions of the operational arena and concepts operating therein, and fails to suggest realistic, practical, common-sense solution sets that hold promise of anything more than a minor technical improvement here and there.

Corps Battlefield Command

Letter to Major Reginald H. Turner
Fort Leavenworth, Kansas
16 January 1978

This responds to your letter of 9 January asking my views on where the corps commander positions himself during the defense. First, I believe that, unless exceptional circumstances dictate, the corps commander should operate from a tactical command post. This activity consists of four or five operations tracks and about four commo vans of various types. The corps commander himself, his G2 and G3, or their execs run this operation. Generally it will be set up in a cellar, cave, warehouse, or whatever can be found with good thick walls and some overhead protection. Commo from both tracks and vans will be remoted. The vans in particular must be sited some distance away—1–3 kilometers from the TAC. The goal is to reduce the signature—IR, photo, physical, and electronic—to the level of a brigade, hopefully closer to a battalion CP or less.

Where should the TAC be located? In the covering force battle it should be located near the TAC CP of the covering force commander. In V Corps that would have put it somewhere near the TAC CP of the 11th Cavalry. The two headquarters should be close enough so that they can talk by hard wire.

In the so-called handoff, the corps TAC should stay with the covering force commander's TAC until a major part of the covering force has been chopped to the division(s) in the MBA. For example, should a main attack develop along the Fulda axis, then we planned to selectively chop covering force units in the Fulda Bowl to the 8th Mech. As that transfer occurred the corps commander would relocate himself and his TAC near the TAC of the 8th Mech.

In the main battle area fight the corps TAC is sited near the TAC CP of the division in whose area the enemy appears to be making his main thrust. If there is an attack on each of the two major axes of advance, then the corps TAC should be sited somewhere about midway between the two so that the corps commander can communicate with both division commanders and move quickly to the axis which seems to require his immediate personal presence.

Rationale? Study of thc history of battles wherein the fight was fast and intense, and my own experience in exercises and in battle, tell me that the single most difficult problem the commander must solve is that of finding out what's going on. The higher the level of command the more difficult that is. This is so for several reasons. Most important among them is that information is reported sequentially. Battle info comes to the corps commander from platoons through companies, battalions, brigades, divisions. Sensor info comes through nodes in lateral commo nets. Info from higher headquarters also passes sequentially in many instances. Each node through which the info passes slows its passage and garbles its content. Internally message centers in all headquarters further slow and garble traffic flow and info content. Some efficiencies can be achieved here, but the necessary disciplines are difficult to impose, require constant enforcement, and the whole system tends to revert to its slothful ways unless purged frequently. Externally we have made our commo systems slower, more cumbersome, and less reliable than ever before. In the FM nets we have interposed secure devices which reduce range and reliability dramatically. The RTT nets are no more reliable than ever before—in a track a

RATT rig doesn't work very well. And they are slowed by the cumbersomeness of the message priority and processing systems.

In Europe they suffer the added burden of the 25 WPM tape punch rate and the 66 WPM transmission rates of the Europe RTT equipment. The point to point and multichannel equipment takes too long to set up, tear down, move, tune up, and so on. It is very fragile. When it works it's OK, but it is too unreliable as a system to be depended on to do the job in a really intense fight.

Historically the successful commanders are those who have figured out how to overcome the burdens of sequentiality, the unreliability of commo systems, and the tyranny of the message center. Therefore I conclude that, if the corps commander wants to influence the battle, he had better be where the battle is.

The Battlefield

Letter to Colonel Donald J. Valz
National War College
1 March 1978

Your appraisal of the infantry dilemma is quite correct. It is most interesting to note that every army in the world is struggling with this problem—witness the Soviets with first the great antitank debate following the appearance of large numbers of antitank guided missiles, followed close on by their debate about the role of the BMP, the latter still in progress. Since I was party to the deliberations which led to incorporation of the TOW on the IFV, I suppose I have no one to blame but myself and General DePuy. However, in our corporate defense I have to recall that in the climate of the times then we were virtually told that if we didn't put the TOW on the damned thing we couldn't have it at all. So we did it—never quite, I think, realizing how acute was to be the ensuing trauma. We weren't and still aren't able to articulate very clearly what it or the people in it are really supposed to do in battle. Interestingly enough, in this administration we are being told that if we don't take the TOW off the damned thing we can't have it at all because it is too expensive. How things change! But we really need a separate antitank weapons carrier—the ITV will be such, but isn't very survivable. We may take TOWs off part of the IFV fleet, and that might not be all bad. Not sure yet how that will turn out.

Your artillery thesis is part of my Golden Tube hypothesis—FASCAMs, COPPERHEADs, ICMs, Nucs, who's firing Smoke and Illum, not to mention HE? I am trying desperately to set the mortar situation aright and help unstress the artillery. TACFIRE and the Battery Computer System will go a long way toward helping the artillery out of its dilemma, despite some obvious shortcomings in the hardware reflecting shortsightedness on the part of those who drew up the operational requirements. However all that might be, I'm certain we could find someone with sufficient numbers and guile in their use to prove whatever we decide unnumerically is right.

The Battlefield

Letter to Brigadier General W. B. Burdeshaw
1st Cavalry Division
1 March 1978

The problem I keep running into is that we have raised up a generation or two of people who want to think in the broadest general terms about subjects that I believe require the most detailed and meticulous kind of thought well in advance. Not that the predictions ever come true—indeed most often they do not. But to the extent that one has thought through the situation in a series of "what might happen," and done it well in advance—to that extent he is much better prepared for whatever happens in the end. Time and time again that has paid off—in training and in battle. Therefore I just have to believe it. Maybe all those other guys are so much smarter than I am that they don't have to do that, but I seriously doubt it. It's part of the political scientist syndrome with which we let ourselves get seized several years ago.

Artillery Rates of Fire

Letter to Colonel Earl W. Sharp
Fort Hood, Texas
11 October 1978

Thanks for your . . . letter about rounds per tube per day. Good data. What we're really confronted by is a need to fire more than 300 rounds per tube per day for perhaps three days. Then perhaps 50–100 rounds per tube per day for a couple of days. Then, as the second-echelon army arrives to be serviced, back to 300+ for a couple of days. If we believe we can't do that, then we need more tubes. How many? What do we trade off? Could we do a test to see if howitzer crews can fire 300 rounds per tube per day, moving four or five times a day? If you can sketch out a test scheme with some "about" dollar figures, maybe we'd better do it—50–85 rounds a day won't get it! If we can say 50–85 is a limit with some confidence, then maybe we don't need a test—we need only to look at the artillery structure and weapon systems. The GSRS will help—so do we have the right numbers of those, etc.?

Battlefield Development Plan

Letter to Lieutenant General E. C. Meyer
Deputy Chief of Staff for Operations and Plans
15 December 1978

This responds to your . . . letter regarding the BDP [Battlefield Development Plan]. Some observations about my intent as we move ahead with BDP II.

While I started BDP with [the] idea that it would provide a way of setting out our combat developments strategy, I now believe it has broader applicability, and that we should stretch it so that it sets forth our training strategy and strategy for sustaining the Army—logistically, administratively, and perhaps even "mobilizationally," although I have reservations about being able to do the latter with a vehicle like the BDP. In the end, then, three or four years from now, it would provide a single source of reference with regard to how the Army intends to do its primary business as we move ahead in time. It would take the place of all those weighty tomes once used by the combat developments community to look at the Army of the future. If we can keep it short, pithy, and to the point, it could be very useful. We should feed it with studies like Division 86, COEAs for weapons systems, and training studies like ARTS. The BDP would become then the cause for a study or evaluation to be done, and the vehicle in which we use study and/or evaluation outcomes to further refine the BDP strategy.

With regard to your specific concerns, may I offer the following:

We will never be able to filter out school and center parochialism, nor should we. My intent would be for us to use that parochialism to bring to the fore the essential issues requiring analysis—especially trade-off analysis. Because it is a document produced at Monroe, vice CAC or somewhere else, we should be able to keep such parochialism as there may be in context, under control, and at work for us. In this wise I would intend that we could surface for the Army Staff and the Chief the real tough issues facing the Army in a number of areas. Clearly these issues could become, and indeed are at the moment, the product of the mind set which create(s)(d) each new edition of BDP. And that's not all bad, either. It seems to me someone who is apart from the Army Staff mill should be tossing into that mill the tough grist that must be ground if we are to face up to our challenges directly and honestly. BDP tries to do that.

We have not now, nor is it likely that we will have, in the near future at least, the ability to establish analytically the best mix of weapons systems. Ideally, I suppose that's a useful goal. Practically it's an impossibility. We haven't a calculus that's powerful enough to cope with that problem, especially for the air/land battle. Nor are we likely to have such a calculus very soon. What we should strive for, therefore, is analysis that's good enough to inform our professional judgment, and develop the willingness to make that kind of judgment as well as to challenge the extant tribal wisdom—about almost anything.

The Air Force contribution to the air/land battle is indeed being taken into account. You're aware already of what TAC and TRADOC have done. Bill Creech and I are ready to move on now into the tougher problems. . . . My own view of this is that we can't just consider it an Air Force "contribution" to the air/land battle. Indeed, we've reached the point at which we can't fight the air/land battle without the Air Force. Therefore it must be a team effort. The TAC/TRADOC teamwork is inhibited a good bit by the tendency of Pentagon staffs to try and make of everything we do a roles and missions fight. We really ought to be big enough boys to cut that out. The problem we're addressing ourselves to is far too important for our country for us to waste our time polevaulting around over mouse dung.

Finally, the MACOMs will participate in the BDP process. To this end I've sent personally to each major commander a copy of BDP I soliciting his advice and counsel thereon. We worked with FORSCOM and USAREUR in producing BDP I, so it's not a new idea to them. This will get us started. I'm not at all certain that in the long term it's the best way to achieve MACOM participation in BDP. I really believe we need to have periodic strategy sessions with the major commanders—at least semiannually. These would be smaller, shorter, pithier meetings than the current annual fall commanders' conference meetings at which we would deal only with the substantive issues. If for some reason they cannot be conducted centrally with the Chief and the principal Army Staff present, then I've got to figure out how to do that myself—taking the plan around and talking it out with each person or group in turn. For the BDP isn't a TRADOC product. It's for the Army. The Army needs this plan, a strategy for its future, one decided upon and agreed to by its leaders of today. Somehow we've got to seek the necessary consensus, even if I have to act as the executive agent.

. . .

Long answer to some short questions. But then it's not easy to solve tough problems, and frequently even less easy to define them.

Battlefield Development Plan

Letter to General George S. Blanchard
Commander in Chief, US Army, Europe and Seventh Army
12 January 1979

I very much appreciate the fact that you took the time to comment personally on the first edition of the Battlefield Development Plan. You quite properly fingered most of its present shortcomings, which we've set about to correct in the second edition, on which work has already begun. I was seized with the dilemma of printing it now, incomplete as it is, or waiting another eight months to a year for a more complete version. Finally I decided it best to get the dialogue started as soon as possible, and for that reason I issued it, knowing the while that there were lacunae which needed filling.

I would like very much to make that an Army document—not just something that old TRADOC puts together annually. As you know, when CDC [was] absorbed into TRADOC in 1973 we promptly junked all the long-range objective planning documents on which that command spent so much time. Bill DePuy decided for good reason that what the Army needed to do was to concentrate intensively on the next few years. I agreed with that judgment, and still do. We both recognized at the time that sooner or later something of a longer-term range projection would be required. I believe that time has come—hence the BDP. By next edition it should have in it an objective force for ten years hence, training, interoperability, NATO ops, and a flavor somewhat different from the materiel orientation characteristic of this first edition.

Battlefield Development Plan

Letter to General Frederick J. Kroesen
Army Vice Chief of Staff
20 April 1979

While I started the BDP with the idea it would provide a way of setting out our combat development strategy, it has developed broader applicability. In addition to setting the course for developments efforts, it can also set forth a training strategy and a strategy for sustaining the Army. If it is to be useful as an Army strategy for the future, however, it must be decided upon and agreed to by its leaders today.

Barrier Warfare

American Defense Preparedness Association
International Symposium on Barrier Warfare
Adelphi, Maryland
5 June 1979

I have the uneasy feeling I've been here before. Eight and one-half years ago, as a colonel, six months out of combat command in Vietnam, I was asked to speak on "Countermining in Southeast Asia." There was much interest in the subject—great things were promised, technology was on the march, everyone had a solution, interest was keen. A report was drawn up by that conference. It was followed by other reports, but at longer and longer intervals, until interest subsided. So it was that, as Americans were landing on the moon with all the technical success that feat implied, no one could figure out how to find 23 pounds of explosive in a tin can buried six inches beneath the surface of a road, trail, or the ground.

For a number of reasons, US wars, internal and external, have been largely wars of maneuver. No matter how hard engineers have tried to enmesh us in close-combat siege warfare, US soldiers have resisted. In Civil War campaigns around Vicksburg, Petersburg, and Richmond, siege warfare was used. The world's largest land mine—a real mine—was exploded at Petersburg in an attempt to create the opportunity for mobility. The attack failed; the hole succeeded as a tourist attraction.

Following the Civil War, maneuvering and fighting on the vast plains of the West further contributed to whatever it is in our national makeup that considers obstacles, particularly those that indiscriminately kill or maim—like mines, booby traps, and barbed wire—unfair.

Coming late into World War I, the US Army found itself enmeshed in the sophisticated interwoven man-made obstacles of that war, buttressed by massive firepower—machineguns and artillery. No war of maneuver, just blood and dying on all those barriers. Something had to be done!

It was—the tank was born. Tanks were to restore mobility to a battlefield dominated by obstacles—mines, barbed wire, trenches, abatis, ditches, machineguns, and artillery. Then, countermeasures to the tank began to improve—particularly mines. And then there came counter-countermeasures—mine rollers, mine drags, and portable bridges to span tank ditches. After the war there was a resurgence in barrier warfare. Now we see permanent barriers, huge field fortifications stretching for miles, such as the Maginot Line in France and the West Wall in Germany. In their turn, they too would suffer countermeasures, principally mobility, and become permanent monuments to futility.

However, mobility could be affected by mines, and so on both sides in World War II there were mines and countermeasures—mine rollers, drags, and portable bridges reminiscent of World War I. There were also some new things—the flail, line charge, Bangalore torpedo, bigger bridges, and some crude detectors. However, the basic solution to the most dangerous obstacle in World War II—mines—was that some soldier had to get out, find, and disarm the mines manually, a task not conducive to gentle dispositions and long retirements.

When World War II ended, we noted the futility of permanent barriers and ignored the difficult problem of temporary barriers—mines. In Korea, against an enemy who used barrier warfare on a large scale, we were back to the combat soldier who had to do it all by hand.

The Vietnam War introduced a new concept in barrier warfare, random mining on a massive scale. Gone now were the neat, geometric minefield patterns of previous wars. Off roads and on roads, mines were laid anywhere, in any combination, in open fields and dense jungles. Unfortunately, our meager resources consisted of metallic and nonmetallic mine detectors—hand-held and operated for detection, and for clearing we were still limited to probing with hands, feet, and vehicles.

Statistics on the effectiveness of mines used by the enemy in Vietnam are striking: From November 1968 to May 1969, 73 percent of all tank losses and 77 percent of all armored personnel carrier losses were caused by mines. In approximately 12 months of operation along and across the Cambodian border, the 11th Armored Cavalry Regiment encountered over 1,000 antitank mines. Rarely were deliberate minefields encountered, and never in a recognizable pattern. Frantic requests from the field for effective countermine equipment did not bear fruit, and so, to the very end, combat losses to land mines continued to plague armored units in Vietnam.

Since Vietnam, barrier warfare has resumed its low profile with budget cuts and lowered priorities. In the field are some inadequate detectors, barbed wire, some standard land mines, and we're developing artillery-delivered scatterable mines, SLUFAE, and a warehouse full of contingency plans.

Now, I've been asked to speculate on the potential for barrier systems. First, what is the threat? The threat has the means to use barrier warfare on a large scale. The Soviets have *fielded—fielded*, not proposed, a complete family of mine-laying equipment, from simple truck-mounted chutes to full-tracked armored and sophisticated helicopter-mounted systems. Their mines are every bit as good as ours, and they have fielded them in quantity. Their mines are less sensitive to pressure. More and more, they are employing tilt-rods and influence fuzes that are difficult to actuate by pressure countermeasures.

For detection, they have a series of detectors better than ours. For clearing, they have plows and dozers mounted on tanks to platoon level. Their doctrine treats barrier warfare as integrated into their notions of how they'll fight the combined arms. Combat soldiers receive cross training in engineer work.

Operationally, the threat is twofold:

- Fortified positions, strongpoints, or belts.
- Scatter mines delivered at random by artillery, rockets, or aircraft.

Technically, there is an additional problem in that the Soviets' latest mines are less pressure sensitive than earlier models. Most recently, they have increased their reliance on tilt-rod mines, even going so far as to modify existing mines for this type fusing. They've done this to counter all the pressure clearing devices we're developing, to include SLUFAE, which isn't even fielded yet. In addition, influence mines—magnetic mines delivered by artillery or air, essentially defeat all our currently fielded mine-detecting and clearing systems.

On our side, we suffer from the "lacks"—lack of an operational concept, lack of priority, lack of money, lack of interest, lack of ideas.

So what to do about all that? Let's first try out a conceptual framework. Barrier warfare is an integral part of combined arms warfare. This is so because barrier warfare is a cheap combat multiplier that enhances the effectiveness of other parts of the team—tanks, infantry, artillery.

Targets destroyed, delayed, disorganized, or disrupted by barrier warfare are targets that either do not need to be engaged and defeated by direct-fire weapons or present themselves at a rate that can be handled by those weapons.

Modern barrier warfare is the ideal system for attacking enemy second echelons. Whether in the offense or defense, second echelons can be cheaply deprived of their essential capabilities—mass, momentum, and mobility, by the intelligent use of barrier warfare such as artillery-delivered scatter mines and DPICM.

Finally, the nature of the mines themselves and their mode of employment combine to confront us with a problem that can, in all probability, not be solved by a single device, either for detecting or clearing. A system of some sort is indicated. Looking at the threat and what must be done to counter it, and looking then to our own use of mines, we see two basic requirements. First is a high-speed mine-detecting mechanism; second is a rapid mine-clearing capability. Mine-detecting systems must apply a technology or technologies that can detect influence, pressure, tilt-rod, or nonmetallic mines. We should aim to identify mined areas; then, if time and resources permit and the mission dictates, we should be able to identify individual mines. Detection must be provided over a wide area, at high speed, in real time, and with reasonable reliance. Helicopters or RPVs are candidate platforms for such a system.

As to clearing, it is also obvious that no one device will do the job. This is so both because of the mines themselves and the threat's way of integrating mines with other obstacles such as antitank ditches and berms. Therefore the clearing capability must be a system. Ideally it should include the ability to roll, plow, blow, and/or scrape away the mines, whether in a minefield or randomly sowed; the ability to cross a ditch; and finally the ability to fill the ditch and so complete defeat of the whole barrier system. Against the Soviets, and in the face of their growing mine-dispensing capability, the equipment to do all this probably needs to be deployed to company level.

Conceptually, a system would include line charges, man-portable and vehicle-launched, for clearing single lanes, vehicular portable Bangalore systems for clearing wide lanes, then tank-mounted rollers, plows, dozers, and tank-propelled bridges, all part of an integrated system employed in pursuit of a well-thought-out operational concept. Ultimately many obstacle breaching problems could be solved by an armored vehicle equipped to do many things, the most important of which would be to fill the ditches and reduce the berms using a backhoc. A big armored backhoe on an M88 chassis would be an example.

Now that is a full plate, and I'm not sure that we will succeed with all of it. But, considering the trends and the threat we must face if we keep going the way we are, we have no real choice. We have a chance to truly enhance the combat potential of our battle-fighting capability. We owe it to ourselves and to all those soldiers who have IOUs on the promises we made. We can't afford to have those IOUs presented again when bullets are in the air, as we did in Vietnam. It's too late then.

Battlefield Interdiction

Electronic Industries Association
Arlington, Virginia
13 November 1979

Let me start by saying that I'm not a Johnny-come-lately to the problem of battlefield interdiction. As the US Army Armor Center commander in the mid-1970s, I participated in the tactical evolution that started in TRADOC and resulted in Field Manual 100-5, *Operations*, in 1976. As a corps commander in Europe, I wrestled with the problem with a genuine sense of urgency. In the Training and Doctrine Command, we've made the enemy second echelon and its interdiction our top doctrinal priority.

With all that concentrated study, you'd think we had all the answers. The first general observation I can offer is that we don't, and I haven't met anyone inside or outside the military—private or general or instant field marshal—who does. There are a lot of solutions floating around, most untested, and some with good ideas. The problem is that many of them are based on parochial viewpoints—buy this system or that—or require resources beyond the realm of reality—X amount more divisions in Europe will solve the problem.

So the second general observation I can offer is that there is no instant cure, no all-seeing, all-knowing system that will do everything and be all things to all men. If that's what we're searching for, then we might as well quit groping around now and get on to something we can handle. Every system or scheme, tactical or technical, has a countermeasure of some sort that degrades its effectiveness and still leaves something undone. Some of the schemes go beyond the possible into the magical. As a soldier, I'm not prepared to trust in magic done by mirrors and string.

Having said that, we can't just quit and ignore the second echelon because it's too hard. The fact that it's real leads to another general observation. Battlefield interdiction of the enemy second echelon, whether offense or defense, requires the use of all the arms and services, ground and air, as a part of the combined arms team. The Army can't do it alone, the Air Force can't do it alone, and in fact we just barely might do it together. So it's not a roles and missions discussion; we need each other, and much more of each other's specific systems than we have now. The problem of battlefield interdiction is a part of the AirLand Battle, and that is the only context in which it can be logically discussed.

Now, with these three observations in mind, let's review some enemy operational concepts as a background for US Army operations. Then, with our concepts in hand, we'll touch on the requirements for air-land interdiction. One more point is important: I'll discuss these concepts primarily from an enemy attack-US defense point of view. That's not because the US Army is defense-minded, or that I am, but because of the nature of things in NATO and many other places in the world, we believe that defense will be our first mission.

We've looked at Soviet and Soviet-related strategies; that is, the military strategies of those countries that have drawn their tactics primarily from the Soviet model and postulated their concept. I've tried to characterize it in the phrase "maneuver-based force disruption." In the formula that the Soviets use to describe battle, they rely heavily on maneuver. It is true that it is supported by vast amounts of firepower by our standards. But its primary purpose is to disrupt the enemy force in such a way that he can't get his defenses set or his attack set in time to carry it off effectively. By attacking, maneuvering, and supporting by heavy firepower, they disrupt whatever it is their enemy is trying to do before he can get it done.

Now, in order to do that, they espouse three fundamental operational concepts. One is called mass, the second is momentum, and the third is continuous land combat. In their terms, mass means bringing forces together at the point where you want to do something. I'll talk about the ratios that they think are appropriate in a moment. That mass generates enough velocity to have the mass times velocity equation equal momentum, which they describe in terms of making 40 to 50 kilometers a day in the attack. Continuous combat in their terms is the ability to stack up echelons of forces behind the attacking echelon in such a way that the opponent is confronted with a continuously generated mass of force moving at a relatively constant rate of speed.

Here are some ratios that they have traditionally said were necessary to do that: in a movement to contact, they like to generate force ratios of about three to one. If they want to break through, they say they need about six to one. If we are going to fight against an enemy who is on the attack, we must look at the entire battle. We consider that this battle consists of two primary elements—not two separate battles, but two elements of the same battle. One has to do with the fighting of the forces in the immediate assault echelons. The other is what we do to those follow-on forces that are coming to join the battle in order to prevent them from overloading the people who are trying to fight the assault echelons and so making their problem virtually insurmountable. The basic strategy that we think we should use to do that, I've described here. Remember I said the enemy used maneuver-based force disruption. What we have said here is that we're going to have to kill a lot of people in any event, mainly because we're outnumbered so badly. So, the strategy that we have adopted is what we have called maneuver and firepower-based force destruction and force disruption.

In other words, we are going to have to destroy some of them with firepower, probably enough of them to sort of get their attention. Then we're going to disrupt the forces, primarily the second echelons coming along behind, so that they do not join the first-echelon battle and so become a problem that is insurmountable in terms of the weapons systems that are available to take care of the first-echelon problem.

That concept has four operative notions: see deep; move fast in order to strike quickly; finish the fight rapidly before the second echelon closes; use terrain in all of this to your own advantage because, particularly as the defender, you have significant advantages in your ability to use terrain. But, whether you are attacking or defending, the clever use of terrain is very, very important.

Now, in order to talk about these operative notions, we've got to understand the second-echelon problem. A little earlier I mentioned that the enemy had some fundamental operational concepts. Mass and momentum were two of these. I noted an equation that viewed momentum as a product of mass, sheer numbers, times velocity, rate, and direction. If we intend to interdict the second echelon, then we must do something to change that formula.

In concept, then, mass can be altered by *degrading* the cohesion and effectiveness of the force as a whole by focusing on attacking unique elements of the formation—air defense systems, combat service support, command and control, fire support, and so on. We can, if we choose, also affect the mass by *destroying* the formation in general—primarily combat systems like tanks, BMPs and so on.

On the other hand, velocity can be altered though *disruption* by slowing or causing the formation to change direction. Velocity may also be altered by *delaying* or *stopping* the formation for a period or successive periods of time. This reduction of the velocity component can be accomplished through the execution of counterattacks, mining/cratering high-speed routes of approach and critical chokepoints, interdicting river-crossing sites and road-bound elements in march column, and significantly reducing the efficiency of his C3.

The dilemma in all this is to determine which target should be attacked, when, where, and with what intensity. Let's look at the enemy target array, his battle formations, and see if we can determine some answers. Consider an enemy army, the first-echelon army. These formations are described doctrinally by the Soviets in their literature in terms of how far behind one echelon another echelon is. The distances are not nearly as important to us as are the times. It turns out that, on Central European terrain, the second-echelon regiments of those first-echelon assault divisions can join the first-echelon battle in somewhere between 5 and 15 hours. The second-echelon divisions of the first-echelon army can join the first-echelon fight in something like 15 to 30 hours. Therefore, if we are going to prevent these echelons from joining the first-echelon fight, we have this amount of time to do something about them before they become a first-echelon problem.

The division commander who is trying to prevent these regiments from becoming a part of this battle has somewhere between 5 and 15 hours to get the forces and the firepower marshaled together somewhere in some way to do whatever it is he decides he wants to do about those regiments. That means he has to find them. He has to move quickly, strike quickly, and then finish the fight up before the second-echelon divisions join the battle. So it is the dimension of time here that's very important, more so than is distance, because on different terrain, obviously, the times vary even though the distances might be the same.

The first echelon is mostly combat and combat support elements, but the follow-on and second echelon have a greater proportion of combat service support. When you analyze that formation for potential tactical payoff, you must consider:

- The contribution of the elements to the overall momentum of the second echelon.
- The vulnerability of the elements (hard or soft) to friendly weapon systems.
- The location on the battlefield where the elements make their greatest contribution to the cohesion or integrity of the second echelon as it closes the FEBA.

What this tells you is that the contribution of combat systems to momentum is not fully realized until those systems are committed and their weapons are brought to bear on the defender. As a result, combat systems do not become the dominant cohesive element of the formation until they have closed the direct-fire battle. We also know that investment costs for weapons designed to acquire and engage hard targets at short ranges are substantially cheaper than those designed for acquisition and attack of hard targets at extended ranges. These factors tend to support an argument that emphasizes engagement of combat systems immediately prior to and during commitment to the direct-fire battle.

Moreover, it means that combat support and combat service support systems contribute to the velocity and cohesion of the battle at more extended ranges. At the same time, these systems offer softer targets that are easier to attack and harder to replace, both in terms of equipment and trained personnel. The loss of these softer systems will generally contribute much more to the disruption and delay of the second echelon that we are seeking—the alteration of the time schedules. While it may sound very attractive to defeat massive armor formations at a distance of 100 kilometers from commitment, the cost of doing that with high confidence may be substantially greater than the cost of delaying that force for 12, 24, or 36 hours and then degrading it by 25 to 35 percent prior to its commitment.

So a solution is that, in attacking the first echelon, we must, by direct fire, reduce both mass and velocity. Those targets that pose the next most dangerous threat to the direct-fire battle, the follow-on battalions of the attacking regiments, should also be engaged in a manner that will produce maximum alteration of both mass and velocity. However, engagement of subsequent echelons calls for more emphasis on the alteration of velocity and is not as time-sensitive as the attack of the follow-on battalions. Combat support and combat service support elements are the ones that contribute the most to velocity in the second echelon. They must be our primary targets.

Let me reinforce what I've just said about the vulnerability of the combat service support to attack. Of course we are after his combat systems when we can get them, but an enemy division requires a sizable amount of ammunition and fuel just to keep a tank division operational. Large volumes of ammunition must be stockpiled for preparation fires. This presents a fine target. If this ammunition can be destroyed or its movement forward interdicted, the attack will be delayed or stopped.

Considering the trucks required to transport the ammunition and fuel he needs to sustain the attack, the movement of a force of this size requires some rather detailed planning. His doctrine indicates that he intends to keep his command and control structure silent, but we believe that if we can get at these targets we are going to cause some delay—delay that will substantially interfere with his timetables, which is going to result in changes and cause his plans to begin to unravel. This will also cause him to surface his command and control for attack, exploitation, or both. The effect, then, is to begin the separation of the combat systems from combat service support and degrading of his command and control. Enemy tactics require deployment of field artillery and command and control systems three to four hours prior to movement of the maneuver forces. If these systems are destroyed or degraded, then the attack will be delayed or halted.

What happens if we don't attack the second echelon? As long as the subsequent echelons don't close, we can handle the first echelon. But if the second echelons are allowed to close unhindered and unaltered, our direct-fire target servicing ability is overwhelmed. Conversely, if we delay, disrupt, or destroy elements of the second echelon, we not only make the direct-fire target servicing problem manageable in terms of total targets—*mass*, we also control their presentation rate and location—*velocity*.

Now, I said we must see deep, move fast, strike quickly, and finish rapidly. Specifically, what does see deep mean? The question we must ask ourselves is how deep can we see and, having seen that deep, how far can we shoot to do something in order to disrupt the force, the second echelon force? A surveillance line reflects the maximum possible range of the side-looking airborne radar (SLAR) on the Mohawk aircraft. Now we don't have many of those. They are subject to weather. They are subject to terrain shadowing and so on. The truth of the matter is that the bulk of our surveillance assets can only see about as far as the front end of those second-echelon regiments back there. So we're not very good at seeing deep.

Well, you can say that's bad, but it isn't all that bad because we can't shoot that far anyway. Consider how far we can shoot with the nonnuclear Lance. Obviously, like the SLAR on the Mohawk, we don't have many of those, and the truth of the matter is that most of the cannon artillery in the divisions and the corps can't fire much farther than about where that surveillance line was. With the advent of the general support rocket system in the next few years, we will have the ability to fire into those regiments back there, just as we will be able to see with the systems like the SOTAS—the standoff target acquisition system. We will have a capability to interdict the second echelons, the second-echelon regiments and the second-echelon divisions.

We haven't neglected command and control of this battle, for it is tied directly to the problem of seeing deep. Doctrinally, we say that the enemy's first-echelon divisions, the regiments in front in the assault divisions as well as the follow-on regiments, are the property or responsibility of the defending division. If you were attacking, those same echelons are your responsibility. The brigade commander fights the first-echelon regiments. The division commander fights the first-echelon divisions. The corps commander fights the first-echelon army, because the corps commander's responsibility is to find and disrupt the advance of those second-echelon divisions before they become a part of the first-echelon problem. At the same time, the corps commander is very, very interested in where the second-echelon army of the front is deploying. He has to tie himself at corps level into the national target acquisition systems and other surveillance means to get information of where that army is and what it is doing. But his primary responsibility in battle fighting has to do with echelons farther back. Brigade-regiment, division-division, corps-army is the formula we are using.

You remember I said that we had to move fast to strike quickly. How fast do we have to move? If we use that 5 to 15 hours and 15 to 30 hours we mentioned a moment ago, the division commander's primary problem is force generation. It would be necessary against a main attack for him to concentrate something like six to eight battalions of his division in something like four to six hours if he is to cope with the advent of those second-echelon regiments into the first-echelon fight. So, if the division commander is fighting the enemy division as a whole, these are the time parameters and the numbers that he has to generate out there in order to cope with an enemy main attack.

Now all of this isn't going to be easy. Some of it needs equipment improvements before it can be done. Some parts will need changes in tactics, but most of all it depends on a quick, accurate flow of information and a sure, secure means of command and control. Unfortunately, this information search is also very complicated, because our technology achievements make it so. We have a multitude of information systems, just on the Army side of the house, that are proliferating in an almost exponential manner. The other services have systems that interface with or provide information to the Army. The commander's problem then becomes one of finding the information he needs quickly.

Today there is more information. Sometimes it moves faster, and it's somewhat more secure, but humans who must use it still have problems. Message centers, headquarters, and staffs are still constructed according to the shopworn, hierarchal, sequential principles of long ago. So much information is coming in that the system is in a state of overload, and we still can't find out what we need to know. Communications systems designed by communicators, not users, still don't bring us what we need to know. We, the users, unable to wean ourselves from the sweet sound of our own voices, cling to outmoded voice nets when digital data transmission technology promises so much more.

If today's problem is significant, beyond 1985 it is even more so. For, lurking just over the information technology horizon, are truly remarkable and necessary opportunities for improvement. Today, we have the potential to replace 1940–1950 voice transmission technology with digital data flow networks coupled with distribution techniques that send information to many users in one quick burst. There are multiple frequency switching procedures for both low- and high-speed transmissions that can make enemy electronic jamming and direction finding impossible. There is a potential to both send and receive from communications satellites directly to frontline units, providing commanders access to worldwide information. The list goes on.

But, unless we prioritize our requirements for information and stress those we must have, all our knowledge will be lost in the background static. Information, funneled through sure and accurate discriminating processes using discriminators provided by the commander, must get to the commander when it happens—in real time, not hours later after it's been administratively fumbled over by a myriad of well-meaning but essentially clumsy hands.

Ultimately, too, this system must be tied together in some kind of battle management system that allows the commander to get the essential things done, *maybe even in spite of the system.* If it all works right, then we won't be forever on the defensive. We'll be able to attack, for attack is a vital part of all defensive operations. Ultimately, the really favorable outcome of combat derives from the results of offensive operations. The best defense is still a good offense. Interdiction of the second echelon will provide us the opportunity to attack to disrupt the enemy forces, destroy his command and control, and disrupt his support.

What are the requirements to do all this? First and foremost, finding the second-echelon regiments, divisions, and armies is critical to commanders. We need systems that provide timeliness of response and responsiveness of coverage and that are guaranteed beyond doubt against significant degradation.

Second, we need systems that will attack to slow the momentum of the second-echelon forces. This means systems that will degrade, destroy, delay, or disrupt the vulnerable elements in the second echelon in accordance with the agreed upon air-land tactical plan. These systems must be both ground and air systems and must be complementary, *not competitive. No one service or arm can do the job alone*. It means an integrated, interdependent development, deployment, and employment of our scarce resources.

Third, we need agreement that the killing of hard targets—tanks—in the second echelon is not as conducive to the disruption of the momentum of the second echelon as is the killing of other qualitative targets. Tank killing in the second echelon is the most popular game in town, but not always the most productive. Ground commanders are glad for all the help they can get against enemy tanks, but you can only get them one at a time and usually with very expensive delivery systems. How much more effective would it be to have entire companies of enemy tanks stopping because they're out of fuel or ammunition or contact with their commanders?

This can happen if we destroy the combat support and combat service support in the second echelons, the trucks, supply areas, command and control centers, and artillery. I can promise you that, if these latter systems are eliminated, the ground troops in the main battle area can handle the tanks that do survive.

Fourth, we need command and control systems that are responsive and secure and can survive in the hostile electromagnetic battlefield of tomorrow. These systems must be keyed to the needs of the user, not the communicators. They must be simple and easy to maintain. Above all, they must be able to handle the tremendous amount of information that will be exchanged on the next battlefield.

One of the hard facts that we must face in NATO today is that we do not have great depth to our battlefield. West Germany is a narrow country and much of the really good defensible terrain is close to the eastern border. Defending forward is not a slogan, it is a necessity, but we need to add depth to the battlefield. What better way to add that depth than doing so on the enemy side behind *his* FEBA? The interdiction of the second echelons will give us the depth and allow us to conduct a maneuver battle. Conversely, it will seriously degrade the enemy's capability to do so.

There is a recent book by General Sir John Hackett titled *The Third World War, August, 1985* that describes a possible next war. There is a quote in the chapter titled "Reflections" that says, "The irony for the West . . . lay in the inability of military men to find the questions to which the electronics men already had the answers." I think this quote is about right. I've given you some questions; you'll get more here in the next few days. We desperately need some answers, for time is short and the risk is great.

Battlefield Function Terminology

Message to Multiple Addressees
10 December 1979

1. There have been some questions regarding the terminology to be used during the Army 86 studies as well as in the Battlefield Development Plan (BDP). The purpose of this message is to lay out the terminology to be used so that we can forge ahead.

2. The following set of terms and definitions which have been agreed upon by DCSCD and DCSDOC will be adopted:

a. Target Servicing. The task of neutralizing and/or destroying threat forces within line of sight which are capable of firing their primary weapon system on friendly forces. Targets include tanks, combat vehicles, ATGM, and dismounted infantry. Subtasks included are maneuver, target acquisition (by means integral to the target servicing system), battle control, target processing, target attack, and target attack assessment. Implied is the requirement to secure and hold terrain when necessary in order to service targets. Target servicing may also include the employment of supporting weapons such as mortars, field artillery, tactical aircraft, and electronic jammers as they contribute to the direct fire battle.

b. Counterfire. The task of suppressing, neutralizing, or destroying, by means of friendly indirect fire and electronic measures, threat indirect fire systems which are capable of firing their primary weapons on friendly forces. Targets include enemy mortar, cannon, missile and rocket systems, air defense, associated C3, target acquisition, and support systems. Subtasks included are maneuver, target acquisition (by means integral to the counterfire system), battle control, target processing, target attack, and target attack assessment.

c. Interdiction. The task of disrupting, neutralizing, and/or destroying threat forces beyond line of sight not capable of firing their primary weapon systems on friendly forces and other threat forces not directly participating in the direct fire battle. Targets include first-echelon units not directly participating in a direct fire battle, second-echelon regiments and other second-echelon units. Subtasks included are maneuver, receiving target information, battle control, target processing, target attack, and target attack assessment.

d. Air Defense. The task of destroying, nullifying, or reducing the effectiveness of enemy air assets, including fixed wing, helicopter, and missiles. Subtasks included are maneuver, target acquisition (by means integral to the air defense system), target processing, target attack, and target attack assessment.

e. Mobility, Countermobility, and Survivability. The task of altering the battlefield—terrain and atmosphere—to enhance survivability, to impede enemy movement, and to enhance friendly movement. Subtasks included are position fortification, decontamination, emplacement of barriers and obstacles, and overcoming natural and manmade obstacles.

f. Battle Support. The task of providing to committed forces those critical supplies and services essential to the successful conduct of combat operations. Subtasks included are resupply of ammo and POL, medical services, graves registration, battlefield recovery and repair, and control of these activities.

g. Reconstitution. The task of timely regeneration of the force in terms of people, organization, command structure, and materiel, during and in preparation for battle and the sustainment of the force through the provision of necessary administrative and logistic services. Subtasks included are resupply of all classes of supply, evacuation, recovery/maintenance, health services, personnel management, sustaining services, and those extraordinary measures taken to quickly restore a depleted unit to an acceptable level of combat effectiveness by critical item and critical personnel replacement.

h. C3. The task of timely command decision-making by analyzing information, assessing the situation, ensuring accurate information distribution, and directing and controlling the force during combat operations. Subtasks included are monitoring the enemy and friendly situation, planning and replanning, estimating, deciding, and providing for operations security.

i. ISTA (Intelligence, Surveillance, and Target Acquisition). The task of gathering and providing timely information regarding the disposition and intent of threat forces to the command decision-making process and directly to specific users. Integral to the ISTA task is the function of managing intelligence assets and conducting counter-C3 operations. Subtasks included are target development, situation development, weather and terrain analysis, and dissemination of timely information.

j. Force Movement. The task of preparing for and executing the rapid movement of troops and supplies about the battlefield to concentrate combat power at critical times and places. Subtasks included are coordination, planning, movement, and control.

3. Inherent to the change to ISTA is a modification of the intelligence task force composition to include USAARMC, USAFAS, USAICS, and CACDA, with the leadership of the ISTA task force centered at CACDA. It should be emphasized that jamming is not limited to one specific task. Jamming must be considered a weapon to be used in several tasks, i.e., target servicing, counterfire, interdiction, air defense, and ISTA. CACDA, in conjunction with the other task forces, will determine the appropriate location for management and direction of jamming operations in the Army 86 process. The mobility, countermobility, and survivability task force will be co-chaired by the commander of the Engineer Center and the deputy commander for training at Fort McClellan, whose responsibilities include the newly established NBC School. The force movement function will not be studied through the creation of a separate task force. It will be examined and reported on by LOGC through the use of their existing task forces. Accordingly, LOGC is responsible for the examination of the battle support, reconstitution, and force movement functions.

4. The above definitions and task force redesignations apply to all Army 86 studies.

Integrated Operations

Message to Multiple Addressees
9 July 1980

1. I am increasingly concerned regarding the reluctance and, in some cases, complete failure to accept and incorporate the idea of integrated operations on the modern battlefield.

2. We have been talking a long time about integrated operations. The direction was approved by the VCSA at the nuclear SPR and shortly you will have a formally written integrated battlefield concept to comment upon. We have a formidable task ahead in getting all this together; yet everywhere I go, I see and hear evidence that we simply aren't thinking through or attempting to work the problem. We persist in presenting our products in a conventional framework. We can't accept reasons that it is too difficult or that there isn't sufficient information available or the concepts haven't been fleshed out. It is in the attempt to work the integrated battlefield that these problems are resolved, i.e., the concepts get fleshed out and data becomes available!

3. It is extremely important that all doctrinal and training literature, studies, analyses, and requirements documents address appropriate aspects of integrated operations. What I prefer to see is us able to drop the word integrated and automatically understand the battlefield as including nuclear and chemical. If there is to be a study or analytical excursion, the excursion should be the exception, i.e., conventional. I recognize there are differences in approaching contingency operations. If you have ideas on how we can get on with it, I'm open. In the meantime we must put substance to our pronouncements.

Battlefield Development Plan

British Army Convention
Washington, DC
24 September 1980

For the next few minutes, I will trace the evolution of what we call the Battlefield Development Plan, or BDP, and how it contributes to the development of current Army 86 concepts and studies. I'll not dwell on the details or organizations. Rather, I'll try to give you a broader perspective of these efforts.

To begin with, I should mention that the BDP stemmed from an initiative that started in 1977. It appeared then to be a good time to step back from the efforts TRADOC had been undertaking in the systems management user/developer interface business and ask, "What's the direction? Where are we going? What's the environment? What is it we really need to be doing if we weren't quite so busy and could do the kind of goal setting that higher headquarters are supposed to do? Are we missing anything?" That's what BDP I was designed to do. In fact, the process isn't as much a plan as it is an assessment of the current environment, where we're going, and whether we're going to get there.

The BDP, first published in November 1978, is designed to be used as a road map for the future. It sets forth priorities and issues that require the Army's attention. The BDP is based on an assessment of selected Army near-term force readiness and midrange force modernization programs. It lists requirements necessary for program improvement. An assessment of US and Soviet combat readiness, force modernization, personnel, weapon systems, force mixes, technology, training, and production capabilities is also included in the BDP. Effects of technology on the Army of the 1980s are described, as are problems of training, personnel acquisition, and spiraling costs.

The BDP assesses battlefield activities and capabilities in terms of 10 battlefield functions. These functions establish a terminology to permit assessments of events occurring at the forward line of troops such as the collision of battalion and brigade forces in battle. Equally important, they allow for assessing activities that contribute to and sustain the continuation of battle. They are:

- Target servicing: The acquisition, engagement, and neutralization or destruction of enemy weapon systems.
- Counterfire: The attack of enemy indirect-fire systems and their control and support elements.
- Interdiction: Disrupting, neutralizing, and/or destroying threat forces, beyond line of sight, that are not capable of firing their primary weapon systems on friendly forces and other threat forces not directly participating in the direct-fire battle.
- Air defense: Designed to nullify or reduce the effectiveness of an attack by enemy airborne weapons systems.
- Mobility, countermobility, and survivability: Altering the battlefield—terrain and atmosphere—to enhance survivability, to impede enemy movement, and to enhance friendly movement.
- Battle support: Providing to committed forces those critical supplies and services necessary to the successful conduct of combat operations.

- Reconstitution: Timely regeneration of the force in terms of people, organization, command structure, and materiel, during and in preparation for battle, and the sustainment of the force through the provision of necessary administrative and logistical services.
- Command, control, and communications: The task of timely command decisionmaking by analyzing information, assessing the situation, ensuring accurate information distribution, and directing and controlling the force during combat operations.
- Intelligence, surveillance, and target acquisition: Gathering and providing timely information regarding the disposition and capability of threat forces to the command decisionmaking process and directly to specific users.
- Force movement: Preparing for and execution of the rapid movement of troops and supplies about the battlefield to concentrate combat power at critical times and places.

These 10 battlefield tasks are used as a logical framework in which to assess battlefield activities and functions. From this analysis, we derive indices for describing deficiencies in capabilities, US versus Soviet.

We use the present capability of the US corps as the baseline, compared to the projected 1980 Soviet threat. From all that, we get a "current deficiency index" for each of the 10 tasks. That's the gap between current capability and required 1986 capability. Some portions of each "current deficiency" will be resolved by programmed improvements in doctrine, force structure, systems modernization, and training. The gap remaining after successful implementation of programmed requirements is referred to as the "remaining deficiency index." In other words, that tells us where we've got to apply more effort in order to close that gap. Now, let's see how the BDP methodology was applied to our Division 86 studies.

First, some background. As with any study effort, we had to start with the environment. As most of you know, in recent years the Soviet Union has significantly increased its warfighting capability. Our quantitative inferiority has been evident for some time. An aggressive Soviet R&D program has now reduced the qualitative edge that we once enjoyed. The Soviets are equal to or ahead of us in the quality of most fielded ground combat systems.

Almost every potential threat mechanized army uses Soviet organizations, equipment, and operational concepts based on mass, momentum, and continuous land combat. Mass means numbers and concentration of forces. Momentum means sustaining advances of 40 to 50 kilometers a day. Continuous combat is echelonment of forces; when one unit is consumed in the battle, the next one is committed without a pause. Modern vision equipment on both sides enables the fight to continue around the clock, in smoke, and in bad weather.

Operationally, an enemy attack might find a whole tank division, more than 300 tanks, employed on a front as narrow as five kilometers, deployed in first and second echelons. Even moving to contact a motorized rifle division might be about 15 to 25 kilometers wide, divided in echelons—reconnaissance, advance guard, first and second echelon. The frontage is wider than the breakthrough, but speed is equally important.

Operational mass, a problem in itself, combines with a larger problem—our historical propensity for being outnumbered, losing the early battles, then mobilizing to outnumber our enemies and so win the war. A look at today's world strongly suggests that logic to be bankrupt.

Although we continue to recognize that NATO Europe is our primary commitment, it is clear to TRADOC that the increase in Soviet international activity places increasing demands on US ground forces beyond Central Europe. With this in mind, and in response to the Chief of Staff's challenge that our Army must meet threats to our vital interests outside of Europe without compromising the decisive theater in Central Europe, BDP II accommodates the broader view of this environment.

We are planning to meet the numerically superior and increasingly sophisticated threat of the next decade through new operational concepts and advanced materiel systems. Between now and 1986 almost all of the Army's major systems will be replaced, and new systems with improved capabilities will also be introduced. In order for new systems to be of value, organizations that can best employ them using the operational concepts developed are necessary. New systems also require new logistics support concepts and training strategies.

Here are some of the new materiel systems that will be entering our inventory:

- Infantry: XM2 (IFV), SAW, ITV, thermal sights.
- Armor: XM1, XM3 (CFV), M60A3, ITV.
- Aviation: Cobra/TOW, AAH with Hellfire, Blackhawk.
- Field artillery: Improved range 155mm (M109A2/XM-198), 8-inch (M110A1).
- Munitions: Scatterable mines, ICM-AT. TPQ-37, TPQ-36, CLGP with GLLD, TACFIRE and battery computer.
- Engineer: UET, GEMSS.
- Signal/C&C: SINCGARS, TRI-TAC.
- RSTA/EW: RPV, SOTAS, REMBASS, MAGIIC, TACJAM, ASAS.
- Air defense: Stinger, Roland, DIVAD, Patriot.
- USAF systems: A-10, PGM, EW systems.

Just managing the change from "old to new" systems on a one-for-one basis would be a major task. However, when we consider that doctrine, training, force structure, and manpower requirements and utilization will each be significantly affected by these systems, the problem becomes more complicated. Add to this budget pressures and limits on the size of the Army and it is evident that we are dealing with a complicated series of interrelated problems that cut across the business of TRADOC and the Army.

The magnitude of equipment modernization and the short time in which it will occur require a framework that will provide for an orderly transition from today's Army to one fully capable of operating on the battlefield of the middle 1980s. TRADOC has been engaged in designing that framework for some time. We call this effort Division 86. Division 86 has evolved from several events of the past few years—the Division Restructuring Study, Division Restructuring Evaluation, and Battlefield Development Plan.

The Division Restructuring Study was initiated in May 1976 to determine the optimum size, mix, and organization of the US Army heavy division in the FY 1980–1985 timeframe. A "clear alternative" heavy division organization and concepts for its employment, test, and analysis were developed; this was called the restructured or T-series division.

Proposals generated by the study group included integrate combined arms at battalion; leader-to-led ratios; develop balanced, weapons-oriented units; increase fire support; improved mobility-countermobility; move administration to battalion; weapons system-oriented logistics; and staff realignment for operations/intelligence and personnel/logistics. One brigade of the 1st Cavalry Division was restructured for test purposes.

A Division Restructuring Evaluation was initiated in February 1977 to evaluate the recommended organization. Restructured tank and mechanized infantry battalions, field artillery batteries and a field artillery battalion, an ammunition transfer point, a modified tank battalion, and one brigade were evaluated. Surveys related primarily to garrison activities and maneuver battalion gaming simulations were conducted. The results told us that both current and restructured divisions had features that should be considered for the heavy division of 1986.

Out of these earlier studies and reevaluations, we determined that the product of our Division 86 efforts should develop and disseminate operational concepts that capitalize on Soviet weakness, recognize battle environment, and modernize doctrine. Also, our efforts should build a balanced team that could facilitate near-term improvements, exploit new systems and facilitate management control and execution of the division's Central Battle and Force Generation tasks, and develop personnel redundancy for key tasks. Other requirements included the ability to exploit technology, deal with the division component of corps and above structuring efforts, provide the basis for resource decisions, and constitute a new Army planning base.

Considering that a problem-solving process was needed, there seemed to be two alternatives:

- Continue with the current approach, working year by year, system by system, and in successive budgets.
- Look to the future by selecting a point and planning backwards.

Because new operational concepts, systems, organization, and training methods are phased in over time, we have chosen to look to the future and work backwards. This process will provide yearly resource requirements needed for our budget programs.

The year 1986 was selected because the majority of our new systems will have been fielded by then. Having determined that we should "look out and plan back," the next decision was to select the organization to be examined. Since the "division" is the increment that drives our structure planning and level of command that will operate the majority of new systems, the choice became apparent.

The next step was to describe the interaction of weapons, organizations, tactics, and the operational concept necessary to defeat the enemy of 1986. What operational concepts can we employ? Being realistic about NATO Europe and a policy prohibiting first attack, we would begin a war there defending—just by the circumstances.

We consider it essential to use terrain to beat the enemy, exploiting his propensity to mass without regard to the ground. As you know, the defender loses the initiative to the attacker. But we also believe that, to win, we need to regain the initiative. Even if for short periods, it is necessary to attack—even while defending.

And so we decided that, first, it is necessary to see deep to find the following echelon, then move fast to concentrate forces, strike quickly to attack before the enemy can break our defense, and finish the fight quickly before the second echelon closes. All this was to be done using the defender's natural strength, terrain, to multiply the strength of the defense. Using this operational concept, then, we analyzed capabilities and deficiencies within those battlefield tasks I talked about earlier and were able to determine what the Division 86 organization structures should be.

As I've indicated, we've been at this for some time now, and we've about got the heavy division finalized. Its characteristics are these:

- Comparable in size to C-series divisions.
- Structured with redundancy and robustness.
- Improved command-control due to battle management concept, integration of shooters and jammers, and smaller, less complex fighting companies and platoons.
- Consolidated aviation assets.
- Increased fire support and air defense.
- Redressed tooth-to-tail imbalance.

Division 86 first addressed the organization of our heavy divisions, the armored and mechanized divisions. We have since started efforts to examine our corps and light division organizations under the umbrella of the Army 86 studies.

We have been talking about the threat, about organizations and concepts, about new weapons systems and equipment, and about battlefield functions. It is also necessary to look at the human dimension in battle, and the BDP/Division 86 approach provides for that—the man-machine interface.

Let me sum up by saying that in the end, as it has been throughout history, it is the individual soldier who will be the critical battlefield ingredient. Only if we have soldiers who are trained to fight, and trained to think, can we move our oncoming technology to victory. The Battlefield Development Plan and Division 86 are steps in this direction.

The Integrated Battlefield

Message to Staff Officers
25 November 1980

. . .

2. Concept treats primarily war in areas where there are large numbers of fairly well-equipped forces who use Soviet operational concepts/tactics. Therefore primary thrust of concept is towards the NATO-Warsaw Pact contention in central Europe; war in ME; in NE Asia.

3. Concept emphasizes the fact that, once political authorities commit military forces in pursuit of political aims, military forces must win something—else there will be no basis from which the political authorities can bargain to win politically. Therefore the purpose of military forces must not be simply to avert defeat, but rather to win.

4. In peacetime the purpose of military forces, especially in the context of operations in the critical areas, is to reduce to a minimum whatever incentives enemy leadership might perceive as favoring seeking military solutions to political problems. In NATO, in the Middle East, in Korea the defensive strategy must extend beyond simply denying victory to the other side; it must instead postulate definable, recognizable (although perhaps limited) victory for the defender. Enemy leaders must be made to understand clearly that, if they choose to move militarily, no longer is there a status quo ante bellum—something to be restored; rather the situation they themselves have created is one to be resolved on new terms.

5. As the strategic nuclear balance evens (teeters), so grows the enemy perception of his own freedom of action at theater levels—conventional and nuclear. Theater forces are not a bridge to strategic nuclear war; they are weapons to be considered in the context of a warfighting capability.

6. This dictates that NATO strategy must from the outset be designed to cope with the Soviet conventional-nuclear-chemical-combined arms-integrated battlefield threat. The growing threat of nuclear capabilities elsewhere suggests this strategy to be appropriate in other critical areas as well.

7. Warsaw Pact/Soviet-style strategy embraces two fundamental concepts:

a. In the first, mass, momentum, and continuous combat are the operative tactics. Breakthrough (somewhere) is sought as the initiator of collapse of the defender's system of defense.

b. In the alternative, surprise is substituted for mass in the daring thrust tactic involving, in NATO, up to thirty or more BMP regiments in independent attacks which, without warning, seek to deny to defending forces, especially in NATO, an opportunity to get set forward.

8. Both tactics are essentially maneuver-based schemes whose purpose is to disrupt the operational tactics of the defender, albeit by different methods.

9. NATO strategy (and defensive strategies in other key areas as well) must be designed to preserve the territory/resources/facilities of the defended area for the defender. In none of the critical areas of the world—those to which US forces are likely to be committed, is there sufficient maneuver room to accommodate a traditional defense-in-depth strategy. The defense must therefore begin well forward and proceed aggressively from there to destroy enemy assault

echelons and, at the same time, to slow/disrupt/break up/disperse/destroy follow-on echelons in order to seize quickly the initiative and go on the offense.

10. The operative tactics by which US forces seek to implement the operational concept set forth above must provide for quick resolution of the battle under circumstances that will allow political authorities to deal with their adversaries from a position of strength. This is so because the enemy generally enjoys a short-term advantage in ability to mobilize additional forces quickly. Therefore one purpose of the battle concept must be to preempt the possibility of prolonged military operations.

11. Further, the operative tactics seek simultaneously to:

a. Deny enemy access to the objectives he seeks.

b. Prevent enemy forces from loading up the assault force fight with reinforcing assault echelons and thus achieving by continuous combat what might be denied them by a stiff forward defense.

c. Find the opportunity to seize the initiative—to attack to destroy the integrity of the enemy operational scheme, forcing him to break off the attack or risk resounding defeat.

12. Because of the enemy advantage in numbers, attack of follow-on echelons must always begin when those echelons are relatively deep in enemy territory. If an outnumbered defender waits until his numerically superior foe has penetrated the defended territory to mount a counterattack, it is always too late to bring effective forces and fires to bear to defeat the incursion. Particularly is this the case if theater nuclear weapons are considered necessary to defeat the penetration.

13. Therefore, on an integrated battlefield, systems designed to defeat enemy assault elements, to disrupt follow-on forces, and to seize the initiative by attack must be able to deliver conventional and/or nuclear fires throughout the spectrum of the battle—throughout the depth of the battlefield.

14. Key to a credible warfighting capability on the integrated battlefield are:

a. Sensor/surveillance systems to prevent surprise attack in peacetime and provide necessary targcting/surveillance information in wartime.

b. Delivery systems—dual capable, with sufficient range, accuracy, and lethality to hold enemy follow-on echelons at risk in peacetime and successfully attack them in wartime.

c. Command-control sufficient to integrate all-source intelligence in near real time in peacetime and in wartime and to turn that intelligence and targeting information and information for maneuver force employments in near real time as well.

15. The operative tactics to support an operational concept of an integrated defense well forward are:

a. See deep and begin early to disrupt, delay, destroy follow-on/reinforcing echelons.

b. Move fast against the assault echelons.

c. Strike assault echelons quickly so as to prevent them from achieving their objectives.

d. Finish the opening fight against assault and follow-on echelons rapidly so as to go on the attack and finish the battle assault army(ies) before follow-on armies can join the battle.

Extending the Battlefield

Military Review
March 1981

The extended battlefield concept primarily deals with war in areas of the world where there are large numbers of relatively modern, well-equipped forces who use Soviet-style operational concepts and tactics. Quite naturally, therefore, the threat against which the concept is designed is typified by the Warsaw Pact in Central Europe, the larger aggregations of mechanized forces in the Middle East, or the threat from the north in Korea.

The concept emphasizes the all too frequently ignored or misunderstood lesson of history that, once political authorities commit military forces in pursuit of political aims, military forces must win something, or else there will be no basis from which political authorities can bargain to win politically. Therefore, the purpose of military operations cannot be simply to avert defeat, but, rather, must be to win.

This article does not propose new and radical ways to fight the battle to win. Rather, it describes an extension of the battle and the battlefield which is possible to accomplish now and which, if applied, will reinforce the prospects for winning.

The extended battlefield is not a new concept. It is a more descriptive term for indicating the full potential we must realize from our acquisition, targeting and weapons systems. The battlefield and the battle are extended in three ways: First, the battlefield is extended in depth, with engagement of enemy units not yet in contact to disrupt the enemy timetable, complicate command and control and frustrate his plans, thus weakening his grasp on the initiative. Second, the battle is extended forward in time to the point that current actions such as attack of follow-on echelons, logistical preparation and maneuver plans are interrelated to maximize the likelihood of winning the close-in battle as times goes on.

And, lastly, the range of assets figuring in the battle is extended toward more emphasis on higher level Army and sister service acquisition means and attack resources.

What emerges is a perception of the battlefield in which the goal of collapsing the enemy's ability to fight drives us to unified employment of a wide range of systems and organizations on a battlefield which, for corps and divisions, is much deeper than that foreseen by current doctrine. The word "doctrine" is used advisedly. It must be acknowledged at the outset that there is probably little set forth in this article which is not already being done, and done well, in some operational units. The purpose of this article is less to suggest innovation than it is to pull together many good ideas for making extended attack an integral feature of our combat capability—in all units.

In essence, our message can be distilled in four primary notions:

- First, deep attack is not a luxury; it is an absolute necessity to winning.
- Second, deep attack, particularly in an environment of scarce acquisition and strike assets, must be tightly coordinated over time with the decisive close-in battle. Without this coordination, many expensive and scarce resources may be wasted on apparently attractive targets whose destruction actually has little payoff in the close-in battle. The other side of this

---

coin is that maneuver and logistical planning and execution must anticipate by many hours the vulnerabilities that deep attack helps create. It is all one battle.

- Third, it is important to consider now the number of systems entering the force in the near and middle-term future. These are not just weapons of greater lethality and greater range, but automated systems and communication systems for more responsive command-control, as well as sensor systems to find, identify and target the enemy and to assess the effectiveness of deep attack.
- Finally, the concept is designed to be the unifying idea which pulls all these emerging capabilities together so that, together, they can allow us to realize their full combined potential for winning.

The extended battlefield is not a futuristic dream to remain on the shelf until all new systems are fielded. With minor adjustments, corps and divisions can and must begin to learn and practice fighting the extended battle now—during 1981. The payoffs in readiness for combat will be enormous, and implementing the concept today means that we are building the receptacle into which every new system can be plugged immediately, minimizing the buildup time to full capability.

To ensure that the extended battlefield concept is understood in the full context of the integrated conventional-nuclear-chemical battlefield, this article will first review, in a broad sense, major aspects of the concept. Then it will describe how, by attacking assaulting and follow-on echelons simultaneously, the prospects for winning increase dramatically.

The Concept

In peacetime, the purpose of military forces, especially in the context of operations in areas critical to US interests, is to reduce to a minimum whatever incentives the enemy's leadership might perceive as favorable to seeking military solutions to political problems. In NATO, in the Middle East, and in Korea, our defensive strategy must extend beyond simply denying victory to the other side. It must, instead, postulate a definable, recognizable (although perhaps limited) victory for the defender. Enemy leaders must be made to understand clearly that, if they choose to move militarily, no longer will there be a status quo ante-bellum—something to be restored. Rather, the situation they themselves will have created is one which will be resolved on new terms.

As the strategic nuclear balance teeters, so grows the enemy's perception of his own freedom of action at theater levels—conventional and nuclear. Theater forces should not be considered solely as a bridge to strategic nuclear war. They are weapons which must be considered in the context of a war-fighting capability. These considerations dictate that NATO strategy must, from the outset, be designed to cope with the Soviet conventional-nuclear-chemical-combined arms-integrated battlefield threat. The growing threat of nuclear capabilities elsewhere suggests this strategy to be appropriate in other critical areas as well.

The Warsaw Pact/Soviet-style strategy embraces two fundamental concepts:

- In the first, mass, momentum and continuous combat are the operative tactics. Breakthrough (somewhere) is sought as the initiator of collapse in the defender's system of defense.
- In the alternative, surprise is substituted for mass in the daring thrust tactic. In NATO, this could involve a number of *BMP* regiments in independent attacks which, without warning,

would seek to deny to defending forces the opportunity to get set forward. Both tactics are essentially maneuver-based schemes whose purpose is to disrupt the operational tactics of the defender, albeit by different methods.

The need for deep attack emerges from the nature of our potential enemies—their doctrine and their numerically superior forces. Whether our enemy is stylistically echeloned is not really critical. What is important is that superiority in numbers permits him to keep a significant portion of his force out of the fight, with freedom to commit it either to overwhelm or to bypass the friendly force. The existence of these follow-on echelons gives the enemy a strong grip on the initiative, which we must wrest from him and then retain in order to win.

NATO strategy (and defensive strategies in other key areas of the world as well) must be designed to preserve the territory, resources and facilities of the defended area for the defender. In none of the critical areas of the world, those to which US forces are likely to be committed, is there sufficient maneuver room to accommodate a traditional defense-in-depth strategy. The defense must, therefore, begin well forward and proceed aggressively from there to destroy enemy assault echelons and at the same time slow, disrupt, break up, disperse or destroy follow-on echelons in order to quickly seize the initiative and go on the offense.

The operative tactics by which US forces seek to implement the operational concept set forth above must provide for quick resolution of the battle under circumstances that will allow political authorities to negotiate with their adversaries from a position of strength. This is so because the enemy generally enjoys a short-term advantage in ability to mobilize additional forces quickly. Clearly, then, one purpose of the battle concept must be to pre-empt the possibility of prolonged military operations. Further, these operative tactics should seek simultaneously to:

- Deny enemy access to the objectives he seeks.
- Prevent enemy forces from loading up the assault force fight with reinforcing assault echelons and thus achieving by continuous combat what might be denied them by a stiff forward defense.
- Find the opportunity to seize the initiative—to attack to destroy the integrity of the enemy operational scheme, forcing him to break off the attack or risk resounding defeat.

Because of the enemy's advantage in numbers, attack of follow-on echelons must always begin when those echelons are relatively deep in enemy territory. If an outnumbered defender waits until his numerically superior foe has penetrated the defender's territory to mount a counterattack, it is always too late to bring effective forces and fires to bear to defeat the incursion. This would especially be the case if theater nuclear weapons were considered necessary to defeat the penetration.

Therefore, on an integrated battlefield, systems designed to defeat enemy assault elements, to disrupt follow-on forces, and to seize the initiative by attack must be able to deliver conventional and/or nuclear fires throughout the spectrum of the battle—throughout the depth of the battlefield.

Key to a credible war-fighting capability on an integrated battlefield are:

- Sensor/surveillance systems to prevent surprise attack in peacetime and provide necessary targeting/surveillance information in wartime.
- Delivery systems—dual capable, with sufficient range, accuracy and lethality to hold enemy follow-on echelons at risk in peacetime and to attack them successfully in wartime.

- Command-control sufficient to integrate all-source intelligence in near real time in peacetime and in wartime and to provide that intelligence and targeting information to maneuver force employments in near real time as well.

The operative tactics which support such an operational concept of an integrated defense well forward are:

- See deep and begin early to disrupt, delay, or destroy follow-on/reinforcing echelons.
- Move fast against the assault echelons.
- Strike assault echelons quickly so as to prevent them from achieving their objectives.
- Finish the opening fight against assault and follow-on echelons rapidly so as to go on the attack and finish the battle against the assault armies before follow-on armies can join the battle.

Areas of Interest and Influence

In the execution of such a set of operative tactics, there must be a division of responsibilities among commanders. Just as the means with which commanders see and fight the battlefield vary, so should their primary areas of interest.

Each level of command has a dual responsibility. Each must attack one of the enemy's echelons and must see, or determine the intentions of, a follow-on echelon. Doctrinally, we say that the enemy's first-echelon divisions, the regiments in front of the assault divisions, as well as the follow-on regiments, are the responsibility of the defending division.

In an attack, those same echelons would also be the division commander's responsibility. The brigade commander fights first-echelon assault regiments. The division commander fights the first-echelon assault divisions. The corps commander fights first-echelon armies. It is the corps commander's responsibility to find and disrupt the advance of second-echelon divisions of first-echelon armies before they become a part of the first-echelon problem.

At the same time, the corps commander is very interested in where the second-echelon army of the *front* is deploying. At corps level, he must tie into national target acquisition systems and other surveillance means to get information concerning where that army is and what it is doing. His primary responsibility in battle fighting has to do with the follow-on echelons.

Attacking the Follow-on Echelons

For such a division in areas of interest and influence to be effective in wartime, it must be frequently practiced during peacetime. It is critical for us to realize that, as the enemy achieves the echelonment so necessary for his success, he inherently creates vulnerabilities—targets. These same vulnerabilities provide us with the opportunity to put threat second-echelon forces at great risk. But only through repetitive exercise can we capitalize on his vulnerabilities.

What we must do is practice acquiring and targeting Warsaw Pact units now—during peacetime—so we will be prepared to attack them if need be. In addition, we can do careful intelligence preparation of the battlefield and thus be prepared to attack high-value targets. Such targets include fixed bridges and mobile sites that will cause threat follow-on echelons to bunch up and present themselves as attractive targets. Additionally, attacking other high-value targets such as combat service support facilities, which must exist to support rolling forces, or selected command posts, will also generate delay. Attacks directed in this manner will provide friendly forces time to finish the battle at the forward line of troops (FLOT).

There is a problem inherent in fighting against echelonment tactics. If the battle is fought with no directed interdiction, enemy follow-on echelons have a "free ride" until they enter the close-in battle. This occurs when follow-on echelons are ignored and allowed to stack up behind assaulting forces at the FLOT until a breakthrough is achieved. The enemy retains flexibility, initiative and momentum to apply his mass at a point and time of his choice. Deep attacks seek to deprive him of this freedom. There are three primary tools for a deep attack:

- Interdiction—air, artillery, special operating forces.
- Offensive electronic warfare.
- Deception.

In practical current terms, interdiction—principally battlefield air interdiction—is the primary tool of deep attack. At present, the range of jammers precludes effective use against follow-on echelons. However, jamming can be used in the close-in battle as a nonlethal substitute for fires and battlefield air interdiction sorties, which can then be freed for deep attacks.

We would like deep attack to destroy enemy forces before they enter the close-in battle, but, in today's terms, and in all probability tomorrow's as well, expense and scarcity of assets will limit the practically achievable effects to delay and disruption. Delay and disruption, however, must be aimed at more ambitious goals than just fractional attrition or harassment.

The real goal of the deep attack is to create opportunities for friendly action—attack, counterattack or reconstitution of the defense—on favorable ground well forward in the battle area. This can be done by avoiding piecemeal employment of acquisition means and attack resources. These resources must be concentrated on critical targets which have the most payoff in upsetting enemy plans and on creating situations wherein the friendly force can seize the initiative and win.

It is important to stress here that the deep attack is not just a tool of the defense. It is, if anything, even more critical in the offense. It is essential to winning because it creates opportunities to seize and retain the initiative. It is equally important that corps and division commanders fight this deep battle at the same time and in close coordination with the close-in battles. It is true that these commanders already have their hands full with the close-in battle, but the compelling reason for active corps and division commander involvement is that the number of targets we would like to attack and can acquire far exceeds available attack assets.

It is also essential, then, that attack means not be applied indiscriminately. Limited strike and acquisition means must be applied in a planned, well-organized and conducted scheme to support the plan for winning. Piecemealing long-range target acquisition and attack resources is a luxury that cannot be allowed.

The commander's choice of when to use deep attack means must be taken in such a way that it will create a window for offensive action some hours in the future. That choice must be based on a single unified scheme of maneuver and a plan of fires for the whole of the extended battle. The expected window for decisive action must be created in an area where previous plans have assured the availability of sufficient logistical support and fire support as well as maneuver forces.

This demand for careful coordination of present and future action throughout the depth of the battlefield dictates that the plan stem from the concept of a single commander. Separation of the close-in and follow-on battles invites the risk that windows will not be generated or that, if generated, units will be ill-prepared to identify and exploit them.

What emerges from this requirement for unity of command across the near and far components of the fight is a view of an extended battlefield, with well-defined depth and width, in which the commander is fighting not several separate battles, but one well-integrated battle with several parts highly interrelated over time. The depth of this battlefield beyond the FLOT is really a function of the commander's planning horizon expressed in hours.

The following scenario describes an integrated battle situation in which it would be greatly to the commander's advantage to fight assault and follow-on echelons simultaneously. From the outset it is acknowledged that, in this scenario, it would be advantageous to use tactical nuclear and chemical weapons at an early stage and in enemy territory. It is also fully realized, however, that authorization to do this may not be granted in timely fashion. And, that being the case, the battle will have to be fought with so-called conventional systems. Even though this somewhat reduces defensive combat power, the concept described here maximizes the remaining conventional power.

The corps commander's concerns in the deep battle are those enemy forces that are within 72 hours of the close-in battle. The corps commander needs to have a well-laid-out, flexible plan for 72 hours into the future in order to fight both close-in and extended battles, gain the initiative, win the fight and do it quickly. What is the purpose of looking out to 72 hours' depth? There are many things a corps must do in those hours. They should be used to plan, order and execute those maneuver, fire support and logistical preparations necessary to seize on an opportunity for offensive action.

The presence of any enemy formation in the corps commander's area of influence should trigger a re-evaluation of his long-range plan and generate options for defeating this force along with all others in the area of influence. Several options will probably be retained at this point. However, the range of options narrows as the force approaches and closure time decreases. Almost all options will include attack of the force to inflict delay and disruption. Although distances here are great, the payoff can be considerable, since the critical targets include soft-skinned logistical and command-control elements whose value will be far less when closer to the front-line battle.

As the force closes, its impending impact on the front-line battle will become more apparent, and the relative merits of the various attack options will begin to sharpen. Options at this stage should include deep nuclear strikes with *Lance* or air-delivered weapons. Targets at this stage are far more vulnerable to nuclear effects than at the FLOT. They are still well beyond the danger radius to friendly forces, and the time until closure is realistic enough to allow request, release, and execution to occur.

Of course, the commander must have a strong conventional option in the event nuclear release is not forthcoming. He must identify the critical time at which he must finally commit himself to one course of action. In any event, he seeks to hold the enemy formation out of the division area of influence long enough for division commanders to have sufficient space and time to accomplish their missions and prepare for the next echelon.

When the force enters the division area of influence—about 24 hours' distance from the FLOT—the entire process is triggered again on a lower scale. Here, the importance of real-time target acquisition dominates. Since, at this point, the attacker is committed to specific attack avenues, he has few movement alternatives left to him. The defender can capitalize on that. Again, if tactical nuclear weapons are to be used, they must be used now.

A review has been made of innumerable planning exercises in which assumed enemy penetrations were drawn with great care to reflect that point "beyond which the integrity of the defense is jeopardized." It was found that, if the penetration was allowed to develop as it was drawn in the defended territory, it was *always* too late. If for no other reason, therefore, it is of paramount importance that the planning process begin while that follow-on echelon target is still deep in enemy territory and that nuclear release be requested in sufficient time to allow employment while the target is still 24 to 60 hours from the FLOT.

As in the earlier part of this battle, the commander must integrate the full spectrum of air and land weapons systems. It is, at this point, still an air/land battle, perhaps more air than land, however.

By the time the following echelons close to within about 12 hours of the FLOT, they become the concern of the brigade commander. At the 12-hour line, actions must be taken that not only delay and disrupt the following echelons, but also help to defeat those in contact at the FLOT. Given the right target, and that the enemy has already used chemical weapons, it is here that our use of them can be integrated. They should be used to isolate one part of the battlefield while an attack is launched against another part of the follow-on forces. It is here that the land aspects of the battle predominate—that is, the battle is more land than air.

With a little luck, the outcome will find enemy assault forces destroyed, freedom to maneuver restored, and the initiative captured from the enemy. In the end, this simultaneous attacking of echelons becomes key to the primary objective of the extended battlefield—to win, not just to avert defeat.

Studies show clearly that successful interdiction does result in a degradation of the enemy's massive firepower. It is also clear that successful interdiction results in a reduction of enemy momentum brought on through loss of support, and that it provides the defender time to secure nuclear release if required. Finally, interdiction reduces the attacker's alternatives by disrupting his ability to execute his intended plan.

The conviction that well-planned interdiction can provide these results is based in part on the target value analysis phase of a fire support mission area analysis completed by the US Army Field Artillery School. Part of that analysis was a simulation comparison of 1980 European corps battles, first without interdiction and then with interdiction. While the predicted availability of interdiction means may have been sanguine, some significant trends were, nonetheless, observed.

Each of the potential interdiction effects is highly desirable. But their exact significance is more apparent considering the simulation output over time. Specifically, a look at the effect of interdiction on enemy strength at the close-in battle shows the real value of deep attack.

Without interdiction, the enemy is able to maintain consistent superiority at the FLOT over time. During this period, the defender's strength dwindles, freedom of action deteriorates, and the enemy's grip on the initiative decisively tightens.

With properly employed interdiction, enemy follow-on echelons are held out long enough to create periods of friendly superiority in which the initiative can be seized with enough time to act. The longer and more frequent these windows can be made, the greater the chance of winning, providing we are prepared to identify them and act at the time and in the place where they develop.

We may not be capable of creating windows of such frequency and duration across the entire corps front. However, it is now possible to create such opportunities, and, if aggressively exploited, they could lead to the generation of longer, more extensive opportunities for higher level decisive action building toward a major offensive.

Interdiction Planning

Summarizing, it can be seen that interdiction is key to battlefield success. The enemy's momentum can be altered by attacking high-value, second-echelon targets, reducing his ability to mass and build up momentum. Interdiction is the method whereby we achieve the leverage necessary to slow him down and ultimately stop him from achieving his objectives.

It is interdiction that allows us to focus our attacks on those enemy targets whose damage, destruction, or disruption would help us fight the battle to our advantage. Interdiction has as its main objective that portion of the enemy's force which is moving toward the FLOT or is in staging areas preparing to join that fight.

This interdiction concept does, however, imply some changes in current ways of thinking, especially in command-control. In order to execute the concept, we must recognize the need to learn how to skillfully use resources far beyond those organic to corps and division and to plan their application over a greatly expanded battlefield. Of significance here is the establishment of timely and responsive working relationships with air forces for both target acquisition and attack.

The interdiction battle will be fought at the corps and division level. To do this well, it must be practiced routinely. Interdiction targets at division level are directly linked to tactical objectives. At corps, however, interdiction is a function of controlling target presentation rates and densities. As the enemy's second echelon moves closer to the FLOT, interdiction becomes more closely related to the defensive scheme of maneuver.

Advanced planning is absolutely critical to a successful interdiction battle. It is imperative that such planning be conducted continuously. This will ensure that commanders are aware of courses of action open to the enemy, and the vulnerabilities of each, thus enabling them to attack targets which present the highest payoff at a particular time. Prior to and during initial stages of the battle, the division intelligence officer, applying intelligence preparation of the battlefield techniques, must forecast enemy strength, progress and dispositions at selected times. By assessing these developing vulnerabilities, he can recommend courses of action for interdiction attacks. When blended with the scheme of maneuver, these enemy vulnerabilities can then be exploited.

Following such an interdiction planning process, the intelligence officer can develop an enemy probable event sequence which can be used to predict with some high degree of accuracy which courses of action the enemy is likely to follow. That is, the intelligence officer should be able to forecast what events must occur and in what order to produce the desired disposition of enemy forces at any critical moment. This probable event sequence is simply a template against which to assess the progress of events. It identifies interdiction requirements which will have to be met if friendly commanders are to influence the battle in a desired direction.

Interdiction targeting can be a complex and demanding staff process, particularly at division level. Its effect is to create time and space gaps, not to relieve maneuver forces of having to face second-echelon elements. It is most effective when it is an integrated effort, one which effectively integrates fire support, electronic warfare, deception, and intelligence with maneuver.

Current and Future Capabilities

Having made a case for effective, continuous interdiction, what is the Army doing to achieve such a capability? Considering the weapons, sensors and automation capabilities which will be available through Army 86 efforts, we will be able to do these things quickly and efficiently on the battlefield of the mid-to-late 1980s.

But what about now? The answer is that there is, today, considerable potential to do just what has thus far been described. Since the penalty in terms of battle outcome is too severe to wait to adopt the extended battlefield concept until 1986, our Army must set about seeing how we might get the most from current capabilities.

Even using conservative planning factors, interdiction of critical enemy second-echelon elements is possible within existing means. But, to make that a reality, we must begin transitioning to those concepts now and practice them daily. If we begin that transition with the resources at hand, we will thus be better prepared to fight and win while simultaneously maturing the conceptual notions in the day-to-day work of operational units. Such an approach will also ensure that we have the right capabilities included in the Army 86 force designs.

And so, as in all aspects of our profession, we must practice now what we intend to do in war. We must train as we will fight. Management of sensor assets in peacetime by those who will be expected to use them in war is the only prudent approach.

The same applies to the correlation of data in determining high-value targets. We must get the data into the hands of those who will be expected to use it in the future. We must establish integrated targeting cells in all fire support elements now. It is important that this capability be developed at corps and divisions for nuclear as well as conventional and chemical targeting. It is important that it be done in all US Army units worldwide.

For the present, many of the acquisition means and most of the attacking means will come from air forces. This is particularly true for corps interdiction requirements. Regardless of who owns them, these are the means we need to gain the best battlefield return. Applying them according to the conceptual notions described above is the way to realize their greatest potential.

Recent exercises have demonstrated that the type of targeting information described earlier is available now—with current means. What next needs to be done is to design exercises for corps and divisions which will focus that information at their level. To make the interdiction battle occur properly, and in a timely manner, corps and divisions must also be able to manage the current family of sensors.

We know the tendencies and patterns of threat units when they are deployed as they would be in a second-echelon formation. The task is to make this information available to corps and division commanders for their use in interdiction targeting.

For timely acquisition, we need to ensure that corps have control of sensor systems such as the *OV1D* side-looking airborne radar, *Guardrail, Quicklook*, and the Integrated Test/Evaluation Program. Of equal importance is that there be a direct down-link of this information to divisions.

Data from a number of other supporting means must also be made available. This category includes the *RF4C* and other national and theater systems. Among the most challenging problems is to create the down-links necessary to pass what is already available to corps and divisions in a timely manner.

The Need for Training Target Cells

To begin an adequate effort at fusing this data and developing interdiction targeting, cells must be established in all fire support elements at levels from brigade through echelons above corps. These cells must learn to exploit enemy vulnerabilities by blending the information and expertise available from all-source intelligence centers and electronic warfare support elements. Historically, we have focused all our training efforts on winning the fight in the main battle area. However, we are now entering a new dimension of battle which permits the simultaneous engagement of enemy forces throughout the corps and division area of influence. To accomplish this, we must emphasize training in four basic areas:

- Friendly acquisition capabilities.
- Threat tactical norms.
- Friendly attack systems.
- Specific techniques such as target value analysis and intelligence preparation of the battlefield.

For this to be totally successful, both Army and Air Force targeteers must be trained to work together in these functions. Microcomputers, which are currently available in an off-the-shelf configuration, can provide excellent assistance to this training effort. They can store a multitude of data, from terrain features to fire plans, from friendly weapons systems to likely threat courses of actions. They can perform target analyses and display them in alphanumerics and graphics. If such systems were available in division targeting cells now, and we created the necessary down-links for passing acquisition data, targeteers could train now at their wartime tasks in a realistic manner.

In a notional division fire support element, the operations cell includes target analysts. What needs to be done, and we have embarked on this course, is to establish the targeting cell and staff it with people who are currently performing smaller tasks elsewhere. We must bring the operations types and the targeting types together.

For such a fire support element to be effective, its personnel must train together daily, as a team, using real-time or near-real-time data supplied by an integrated sensor network such as that described earlier. If actual real-time data is not available, then simulated acquisition information could be used, so long as the data base was developed from previously collected actual information.

Through continuous intelligence preparation of the battlefield, a clearer analysis of the area of operations can be developed, one which will facilitate updating interdiction plans and thereby better support operations plans. Such a training activity would contribute greatly to developing confidence and proficiency. By exchanging views and working together, Army and Air Force target cell personnel could establish a credible capability now to deal with any future second-echelon threat.

Remaining Challenges

Like most things of great worth, this capability will not be easily gained. There are many challenges but, in the end, it will be worth all the effort necessary to make it happen. Foremost among the challenges are those which inhibit our ability to blend current operational requirements of sensor means with the need to conduct real-time training at divisions and corps. It will also be difficult, though essential, that appropriate security clearances be acquired for all personnel working in the target cells. This is especially important, for they must have access in peacetime to the data they will be expected to process in war.

Recognizing it is beyond our capability to conduct actual exercises which simulate threat second-echelon patterns so target cells will have something to train against, it is within the state of the art for computer simulations to postulate and portray scenarios which the enemy traditionally follows because they are based on his known tendencies. This would be a useful substitute for targeteers to practice such analytical tasks as event sequencing. Lastly, we must continue to upgrade our communication capability and take advantage of existing commercial facilities. If we do all this, the payoff will be more than worth the investment.

Summary

The challenges notwithstanding, the message of all this is quite clear:

- Attacking deep is essential to winning.
- Attacking deep and the close-in fight are inseparable.
- The extended battlefield concept is the keystone of force modernization.
- We can begin today to practice, learn, and refine the extended battlefield concept.

The ideas of the extended battlefield concept are, in fact, the very same ideas upon which the Army 86 concepts are based—see and attack deep. And, as might be expected, therefore, organizations of Division and Corps 86 correspond in makeup and function to elements of the extended battlefield team.

The question before the Army now is how to implement the concept quickly. While there are yet some questions, it is not likely that man-years of study will clear them up to the satisfaction of all concerned. It is, therefore, time to field and learn to use the concept on the ground with real troops, real equipment, and the real-world problems of field commanders.

The time for implementation is now. This is so because there is, first of all, promise of a major increase in combat effectiveness with current means. There also exists an enhanced capability to exploit new sensors, weapons, and command-control systems as they are fielded. This enhanced capability is even more evident in the field of microprocessors and computers. As a nation, we have a considerable advantage over our potential adversaries in this technological field. If we strive to put that advantage to work for us, it could become a significant combat multiplier. And, finally, of equal importance, there is an opportunity to cause the enemy to wrestle right now with a problem he has traditionally assumed does not exist.

Army leadership is so convinced that a real potential exists now, if current assets are organized correctly, that a four-phase program has been developed. Phase one, already begun, includes conferences at each major command designed to lay down the basic ideas. This article is part of that phase. In phase two, the US Army Training and Doctrine Command and the major Army commands will jointly refine implementation proposals to fit specific priorities and assets.

In phase three, the joint product will be provided to corps and divisions in the field. In phase four, Army service schools and centers will conduct training in the concept and implementing procedures to ensure that officers and noncommissioned officers leaving the training base are ready for their respective roles on the extended battlefield.

Future Battlefields

Armor Conference
Fort Knox, Kentucky
13 May 1981

Recently I was listening to some captains describe for me their busy world—couldn't get any training done, high deadline reports, too many distractions, too little time. I asked them, "Fellows, what are you trying to get done? What are your goals?" One of them responded, "General, my job as a company commander is simply to try to divine which of the 25 or 30 things I have been given to do is first priority today." He whipped his notebook out. In it was a list of some 27 items. He said, "Now, every one of those things listed has been prescribed as number one priority by the battalion, brigade, division, or corps commander. There's no room left for my priorities. My job is to determine which one of those things is first today, because it will be different from what was first yesterday, and different from what will be first tomorrow." By the way, the captain pointed out to me that not one of those 27 items was directly related to training or maintenance!

I mention this conversation because it's important to keep in mind as we talk about readiness in the Armor Force. A major problem for those young officers is that they can't get their company-level units ready for combat because they don't know what the first priority is and they haven't got the time to sort it out.

Certainly, if we are going to be ready for tomorrow and beyond, we must have some framework in which to view the challenges we'll face. What kind of world are we getting ready for today, tomorrow, and beyond? Three things seem to me to represent the challenge. I'll cite those, then try to build the rest in that framework.

The first challenge is the Soviet problem. The Soviets and, through them, their surrogates, allies, and followers will likely continue to gain in military capability relative to the United States and its allies, at least during the next few years.

Second, the growing dependence of many nations on Middle East oil and on critical mineral resources in South Central Africa represents an economic, military, and political vulnerability that will continue to be exploited by the Soviets; it is one that quite likely could precipitate a major world crisis.

Third, the nuclear strategy of the West continues to experience several fundamental dichotomies that will likely be acutely aggravated by adverse trends in the world's military balance and that will, therefore, remain unreconciled for many more years. The difficulty of coping with the nuclear problem simply emphasizes the necessity for an added level of readiness in conventional forces.

Considering these challenges, we must recognize that the battlefield we have traditionally looked on as our primary concern—that in Western Europe—is expanding to one that, though very much like the European one, could be encountered in other regions of the world. Those regions include the Middle East, Northeast Asia, and perhaps even South Central Africa, looking ahead 20 or 25 years.

What are the principal characteristics of that battlefield? First, it will be dense with large amounts of modern equipment. This is so largely because we confront potential enemies who fundamentally believe the road to success is in numbers. I would contend, however, that the quantity of weapons systems is far less important than how that quantity is applied. Most—all—of these weapons systems will be of high quality. Here again, I would contend that quality is less important than how quality is used. Whatever the quality or quantity factors may be, battles, particularly at the point where decision is arrived at, will be very, very intense.

Second, just as the ground battlefield will be dense with large numbers of high-quality systems, so too will be the air over that battlefield. This third dimension of modern war will be dense with high-quality air defense and air systems.

Third, in such a potentially complicated and confusing environment, command and control will be ever more important. But here, too, just when needed most, command and control will be more difficult because of the electronic warfare environment in which these battles will be fought.

Fourth, the variety of weapon systems on that battlefield will be such that it would be impossible to field any one or even a small number of systems capable of coping with the total threat. Therefore it will continue to be an all-arms battle, a combined arms battle, a ground battle, an air battle—an AirLand Battle.

That will be a very demanding world. If a single word best describes the problem of ensuring combat readiness in armor units today, tomorrow, and beyond, it is complexity. The reason that a remotely controlled tank ran out here a moment ago was because, while it represents a considerable complexity, it is really very simple and costs very little. It typifies one of our challenges. We simply must determine how to reduce complexity to manageable proportions and somehow put it to use to our advantage.

There are two approaches to coping with complexity. One approach says that complexity makes our world one in which it is so difficult to operate that we really can't predict what's going to happen. So we continue using the proven patterns—teach soldiers their individual jobs fairly well, send NCOs to NCOES, send officers to branch schools, to Leavenworth, and try to cope with complexity as we face it.

The second approach says that, because of complexity, we can't take the first approach. Despite complexity, there are things we can do to simplify our problems. But we've got to think it out in advance and be better prepared to take the necessary steps to improve our readiness to fight on that battlefield in spite of and indeed because of its complexity. If ever we are to get the Army, and the Armor Force in particular, ready for today, tomorrow, and beyond, we simply have to figure out how to cope with complexity, and we must do it in advance.

Consider some fundamental notions:

First, the history of battle tells us that small units—battalions, squadrons, companies, troops, batteries—well trained and well led, are more often than not what wins in battle. This is especially so in battles where one side outnumbers the other. In these cases the outcome of battle most often defies the measures we have traditionally used to predict battle outcome. The day of the Lanchester Equation is done. War seldom turns out that way. War is not a firepower equation. It is, rather, a contest of human wills. Leadership is not a matter of administrative or managerial ability; on the battlefield, leadership is the ability to harness the courage of human beings into concerted action in a most dangerous and complex undertaking.

Second, battle has become a complex problem that can only be controlled by overcoming those complexities. To get anything done in battle, it is necessary to issue orders. Complex situations lead to complex orders. Complex orders are easy to misunderstand. They take time to deliver, and the mistake level is always high. Therefore leaders must learn to identify complex situations in advance, to find logical answers, and to prescribe regular drills that they, their soldiers, and their units can use to cope with those complex situations. They must synchronize those drills, because they are and will be combined arms drills that integrate soldier tasks, unit tasks, and all-arms tasks.

Third, to win, leaders must think. Thinking takes time. In battle, there is little time to think. We'll fight with weapons similar to those of the enemy. We'll fight on the same terrain, under the same weather conditions, and in the same climate of uncertainty and fear. The only thing that will make a difference will be the brains of the commander. Therefore, commanders must be trained to think, think ahead, think correctly, logically, quickly, precisely, with creativeness and imagination about the complex situations they can expect to encounter.

Fourth, because command and control will be made more difficult in that complex electronic warfare environment, because commanders cannot be everywhere, because critical situations always seem to arise where commanders are not, it is important that noncommissioned officers and soldiers know what must be done. The best way to do this is to drill them to proficiency in battle drills, drills designed to cope with the most likely situations. In this way they will be capable of doing something about right until the boss can get there.

Fifth, the characteristics of modern weapons are challenging our ability to train up to full systems design capability. The gap between systems capabilities and our ability to realize full system potential has grown to the point that it is necessary to look for some completely different ways of training and assigning soldiers, of organizing battle, and for conducting battle.

Sixth, technology provides marvelous potential for reducing the problems of individual and collective skill complexity. Simple, inexpensive electronic training devices of all kinds must be used to advantage if we are to reduce to manageable portions the complexity revolution that faces us and overcome the high cost of training to readiness proficiency.

That remotely controlled tank you saw a moment ago has some great potential. We are not exactly sure what, or where it ought to be used, but it's obvious we've got to consider its use somehow. The machine that sits to your left front is one on which we have duplicated the controls of an infantry fighting vehicle. It's an electronic game like those in a typical arcade. Walk by one of those sometime. You'll note they are always full. The young people in there are volunteers. They've paid a quarter or more to play those games. They are all learning something. In many cases they are learning something that has a military application. If you put all that together, it tells you something—we ought to use that technology and appeal. And so we are. For not much money they've converted that game to use IFV controls and to display a fairly representative sight picture. For not much money we could put these in units and, if soldiers learn the right things, it would be much cheaper than buying more IFVs or an expensive simulator.

Next, in battle the unexpected can be expected. In fact, the unexpected will be routine. To face that truth, we must train our officers and men for it, in advance, with drills that are designed to train them to do something, to do it about right, do it quickly, and do it in consonance with the mission of the force, the aim of the battle.

Finally, that complex and dangerous battlefield will be a place suited only to and for the actions of trained commanders who have prepared in advance to cope with situations they are likely to encounter. The age of the amateur hero is gone forever.

Now, let's sum up. For nearly a decade, our nation has been unwilling to provide its military forces with either the materiel resources or the moral stamina necessary to satisfactorily perform the tasks assigned them. Emotionally mesmerized by Vietnam, morally bankrupt by a pseudo ethic that denied the worth of individual obligation to the nation or anything else, we have cut away at the sinew of our military strength, hollowed out our intelligence community, embraced the notion that training soldiers for war is too expensive in peacetime, and abandoned conscription in favor of joining whatever cohort we could in what was apparently to be a uniformed job corps.

For our Army, we have taught several generations of officers that managerial skills, administrative talents, and an unbounded faith in technology were surrogates for leadership and were indeed more important than a comprehensive understanding of the breadth, depth, and scope of the military art, its history, its practice, its precepts, its basic fundamentals and its enduring values.

The history of the decade just past should be instructive to us—to the nation and to the Army. For we have now embarked on a course to set aright many things cited as having gone wrong in those difficult years. While I applaud that, as we all must, I submit to you that we are an impatient people with a very, very short memory. The euphoria that presently envelops the affirmative side of the "more for defense" debate will soon pale if you and I cannot show what we've done with all that treasure. There is, and always has been, far more that can and must be done by applying some of the notions I've tried to suggest to you this morning than can be had by just spending more money. It is not enough to open the vaults of the bullion depository, if that could be done, to satisfy the apparently insatiable appetites of military folks who keep demanding more of everything without being very able at accounting for what they have or what they've done with it. There's just no free ride. Readiness in the context in which you are addressing yourselves to it these next few days is a lot more than more dollars. Readiness is a state of mind, a way of life, a determination to do right what must be done regardless of how the dollar debate comes out.

So on that complex and most dangerous battlefield of the next war, the determined, prompt, correct, and calculated battle actions of trained commanders, soldiers, and units are what will win—commanders trained to train and trained to lead the soldiers and units entrusted to them, and soldiers trained to follow, trained to fight determinedly and fiercely in units whose survival and success is their life's blood.

And so it is that sometime, somewhere, once again the fate of the nation's aims, goals, programs, ambitions, and perhaps even survival will rest on the determination of a few good soldiers, a few good leaders, and a few good units, trained well in time of peace in order that they might be ready to fight well in time of war.

Use of Phase Lines

Message to Lieutenant General William R. Richardson
Fort Leavenworth, Kansas
4 June 1981

1. I have in hand CAC msg 292250Z May this subject.

2. My concern with phase lines in defense is in several parts:

a. Your msg did not reexamine the definition of phase lines. What are they for?

b. In the Leavenworth problems in which I objected to their use, phase lines were frequently used in place of other control measures which were more appropriate to the situation; frequently used when they were not needed; frequently used in situations in which no control measure at all would have been [a] better solution; frequently used when they did not establish—by definition—the degree of control that was sought and which was required by the situation.

c. To several generations of officers, of which my generation is one, phase lines automatically meant delay. While you contend that not to be the case, I believe it is. In addition, one of our persistent problems with the execution of the active defense is that it tends all too often to take on several aspects of a big delay action on successive positions. I'm talking about how Army brigades tend to do it on the ground. One reason for that is that, when we drew up the battle area and battle position goose eggs in the early days of the active offense, [no one] was comfortable without all those boundaries, primarily for fire coordination. Rather than go against what they considered to be the grain of things, many people turned to phase lines, believing that, if we did not like boundaries, then maybe we would not object to phase lines. Since I was personally responsible for part of that problem, I say again that if the US Army cannot live without its boundaries, then we'll have boundaries. I would much prefer that we go back to boundaries than to adopt something else new—phase lines—and further confuse everyone.

3. While your soothing words were meant to assuage my fears, you know me well enough to know that I listen carefully for the Lorelei song in all that golden prose. Your message therefore only served to further convince me that we are not together on this matter. While I have no objection to the use of phase lines for appropriate purposes in describing control measures for defense, because of their nature their use should be considered the exception rather than the rule. In your problems it is the other way around.

Thermal Imaging Technology

National Infrared Information Symposium
Treasure Island Naval Station, California
8 May 1984

It is a rare privilege to have been asked to chair this session of the 32nd symposium, a session that has been titled "User Experiences with Thermal Imaging Technology." Soon, you will hear from a selected few of today's users who are trying to train troops and crews to take advantage of the vastly improved capabilities provided the AirLand Battle team by the array of night-vision and thermal imaging equipment being made available in the field.

My role in this proceeding is, as I see it, to set the stage for the presentations that follow, and to do so with a perspective drawn from having been the user or user's primary representative for the several years that I commanded, first, the Armor Center and School at Fort Knox, and then later the entire Training and Doctrine Command for more than four years.

In the former role I was, of course, responsible for user requirements, to include night-vision requirements for tanks, ground scout vehicles, aerial scout vehicles and attack helicopters, and ancillary and supporting equipment for all those vehicles. In the latter incarnation I was, of course, responsible for user requirements, to include night-vision requirements for the entire fleet of combat and supporting vehicles with which the United States Army intends to fight the AirLand Battle.

As a result of that experience, it has become my conviction that we are now in possession of, or are about to possess, the capability to fight at night virtually as if it were day. Indeed that was the goal we set for ourselves more than 10 years ago as we summed up our observations of the Yom Kippur War, welcomed the first editions of the M60A3 tank with its vastly improved night kit, and summed up some experimental but highly useful operational experience with night-vision gear in the Vietnam War.

It has also become my conviction that we are in no way ready or able to take advantage of the improved capabilities for fighting the AirLand Battle afforded us by this new equipment. Nor will we be able to ready ourselves and take advantage of what we have unless we are prepared to make some significant and fairly extensive changes in organization, operational concepts and tactics, training, and support systems.

Further it is my contention that if we are not able to find the means to capitalize on the capabilities provided us by thermal imaging systems, we will have wasted the significant treasure we have poured into their development and fielding, and we stand the significant risk of being done in by an adversary who has fielded and trained up in the use of like systems that, while technically inferior in some ways to our own, are nonetheless adequate to support his operational concepts.

If it is now possible to operate at night and under conditions of degraded visibility virtually as if it were day or good visibility, then where are the crews/organizations/manpower to operate the systems over those expanded parameters of time? The Air Force crews some aircraft with crew ratios of two, three, four or more to one. The Army has never crewed its aircraft or its combat crews at more than one to one with some margins for transiting and casual personnel.

If we've spent a considerable sum, and we have, making the AH64/M1 team capable of operating 15 to 18 hours a day, then where are the organizations/crews/personnel to capitalize on that investment? Are the crews to fly at night the same ones who just finished a long grueling day? Experience tells us something like three days and two nights consecutively is about all the best of us can tolerate. If the addition of a night-fighting capability to our air/ground team enables an 18- to 20-hour combat day, then where are the organizations to take advantage of that capability? If the addition of night-vision equipment adds complexity to the soldier's burden in terms of the need for more time to train to proficiency in more tasks, then where are the additional training resources—time, ammunition, cadre, systems—to meet that demand?

If we have provided ourselves a capability to operate the AirLand Battle team 18 to 20 hours a day, who's to support it? Where are the spares? Where are the maintenance, recovery, and evacuation? Where are the mechanics and all the panoply of support obviously required by this new capability?

If the enemy has similarly equipped himself, and he has, then how are we to contend with his improved combat capability in the face of our business-as-usual allocation of training resources, crewing ratios, and support capabilities?

There are not satisfactory answers to most of those questions. To those of us who began asking them 10 or more years ago, it is very much like emptying the slop jar over the windward rail of a sailing vessel—it all blows back in your face. During the Carter years in this country, it was the view that military training was a luxury one could not afford in peacetime. And so the training establishment was shredded. Now we have embarked on a massive buildup in equipment. The training reductions having been imposed, no one is now willing to tackle head-on the problem of restoring a capability to train to the level of capabilities of the equipment that's been provided.

The other day I read a GAO report on the performance of our tank and mechanized battalions at the National Training Center at Fort Irwin. The report cited "systemic" and recurring problems in battalions undergoing training. One of those problems, perhaps the most significant, was the inability of those battalions to conduct sustained operations at night, in bad weather, and during reduced visibility.

You are about to hear from some dedicated folk who have attempted to tackle these problems head-on. I hope you will find their work refreshing, stimulating, ahead. And I hope their work will serve as a spur to us all to knuckle down and begin, at long last, to devote the necessary attention to organization, training, crew ratio, and support, all of which are essential to enabling us to take advantage of this marvelous opportunity that technology has brought to our doorstep.

TRADOC's Analysis of the Yom Kippur War

The Jaffee Center Military Doctrine Joint Conference
Casarea, Israel
16 March 1999

To fully understand TRADOC's analysis of the Yom Kippur War, a brief background is necessary. US involvement in the Vietnam War ended early in 1973. Before planning for the future could begin, there were serious problems to be resolved. There were near-term problems and longer range problems.

Near-term problems were critical and urgent. The nation had used up its Army in the Vietnam War. Soldiers and units deployed in Europe saw themselves as minor speed bumps for Soviet forces en route to the Rhine and beyond. Not only did they not believe they could defend successfully, most did not believe they could win. While there were many reasons for this, three are dominant:

- Resources necessary to sustain a reasonable rate of force modernization had been consumed by operational needs of the Vietnam War. New doctrine and subsequent development of equipment, organizations, and training and education for soldiers and leaders—noncommissioned and commissioned—had stood still for nearly 10 years. This process—doctrine, development of equipment, organization, and training consistent with the doctrine—lies at the heart of an Army; without it an Army stagnates.
- Mobilization on a scale necessary to support, first, large forces deployed to Europe, then to Vietnam as well, was denied by President Lyndon Johnson, determined that nothing—not even the war—should interfere with progress toward his Great Society. It was necessary therefore to use the entire Army, the Army deployed in Europe and the Army stationed in the United States, as the rotation base for Vietnam. Further aggravated by a one-year tour length in Vietnam, this effectively increased personnel turbulence in units far above a level at which reasonable unit combat proficiency could be achieved and sustained.

Both the militant revolt against all authority that characterized the generation of the young in the 1960s and growth of a substantial drug culture in the nation, especially in that same generation, were reflected in the nation's armed forces. Military jails were full. The drug culture pervaded all but the best units, and there were very few best units.

Largely because of this situation, conscription was shut down in 1972, a full year ahead of the expiration of the draft law, without much real confidence that the Army could recruit sufficient volunteers to fill its structure, but finally acknowledging that the draft law as implemented had become totally dysfunctional.

Army soldiers and leaders returned, many of them several times, from a war in which they had won all their battles, only to find that, while they were away, the nation had lost the war. The result was a crisis in confidence—soldier confidence in leaders, leader confidence in themselves, in units, in the Army, most importantly in the political leadership of the nation.

Longer-range planning for the future was beset by complex problems:

- The Soviet Union/NATO standoff remained. Worse yet, the Warsaw Pact had grown considerably in capability in the decade that US forces had been consumed by the Vietnam War.

- In Soviet forces modernization had continued apace. New Soviet doctrine embraced a triad of mass, momentum, and continuous land combat, with or without nuclear weapons, at tactical and operational levels of war, in order to win quickly before NATO could mobilize and before NATO could make the all-important nuclear decision. The Soviets considered that they could probably win both tactical- and operational-level battles with or without, preferably without, the use of nuclear weapons.
- As in the past, NATO/US forces could expect to be substantially outnumbered from the beginning. Due to this traditional disparity in force size, it had long been considered that operational- and tactical-level nuclear weapons would need to be used fairly early in NATO's defense. Thus, for NATO, there had always been an unstated but inevitable linkage between the use of nuclear weapons at theater level and an intercontinental ballistic missile exchange. This was a linkage Soviet doctrine now sought to avoid. For, given the enormity of respective thermonuclear inventories, inevitability of such linkage simply invited Armageddon.
- Militarization of conflict in the Third World, already in train, had continued to increase dramatically, best characterized by the iterative series of Arab-Israeli wars. Inventories of modern, highly lethal weapons of all kinds in the inventories of military forces in the Middle East, and elsewhere as well, had increased an order of magnitude, most dramatically perhaps following the Suez Crisis and the 1956 Arab-Israeli War.
- By end-1972 the US Army had redeployed from Vietnam and demobilized to the point that the only residual (nondeployed) forces were those earmarked for deployment to Europe in the event of a Soviet attack. In the eyes of the force planners, the obvious question loomed large: future planning for what?

Whatever the future held, there was widespread consensus that the Army needed substantial rebuilding. That work commenced with the appointment of General Creighton W. Abrams as Chief of Staff late in 1972. By summer 1973, rebuilding was underway in earnest with reorganization of the Army command structure. Reorganization included dividing the traditional Continental Army Command (CONARC) into a Forces Command (FORSCOM) and a Training and Doctrine Command (TRADOC). The former would command Army forces stationed in the United States and the US Army Reserve, while the latter would be responsible for establishing doctrine and for developing equipment, organization, training, and education for all ranks to implement the doctrine. That work began as TRADOC came into being in July 1973.

No sooner had we developed tentative ideas about what needed to be done than, in October 1973, the Arabs again attacked Israel. Very shortly General Abrams dispatched the Chief of Armor, Major General Donn Starry, and the Ml tank Program Manager, Brigadier General Bob Baer, to Israel. There we walked battlefields with Israeli Defense Force (IDF) commanders who had fought; studied in detail the battles just finished; and carefully examined IDF equipment, organization, training, and leadership that had produced such magnificent performance in battle. Our observations were conditioned by the background already described. We knew what we were looking for. In addition we were armed with two overriding questions put to us by our Chief of Staff. First, what about the tank—were the M1 requirements right or not?—special attention to the need for a new, larger caliber gun (question to both of us). Second, for the Chief of Armor, what does the Chief of Staff of the US Army need to know about the Yom Kippur War?

From the Commander of the IDF Armored Corps, General Musa Peled, and others who fought on both fronts, and from General Israel Tal (Talik), who had even then produced a very early prototype of what became the magnificent Israeli-designed and produced Merkava tank, we learned about tanks and battles at tactical and operational levels.

What did we learn? Operationally we learned that:

- We should expect modern tactical battlefields to be dense with large numbers of weapons systems whose effectiveness at range will surpass previous experience by nearly an order of magnitude. While modern battlefield systems will enjoy greater protection due to increased standoff range and improved armor, it will nonetheless be true that what can be seen can be hit, what can be hit can be killed. The long-range antitank guided missile—Sagger, TOW, HOT, Milan—will add considerably to lethality at range in the battle at the forward line of troops. Direct-fire lethal battlespace will be expanded more than two, perhaps three, orders of magnitude over corresponding battlespace in World War II and the war in Korea.
- Because of numbers and the lethality of modern weapons, the direct-fire battle will be intense; enormous equipment losses can be expected in a relatively short period of time. We noted that total—both sides, tank losses in the critical six or seven days of the Yom Kippur War exceeded the total US Army tank inventory deployed to US Army, Europe, in units and in war reserves combined.
- The air battle over the tactical battlefield will be characterized by large numbers of highly lethal aerial platforms, both fixed and rotary wing, and by large numbers of highly lethal air defense weapons systems. Although there were not large numbers of armed helicopters employed in the Yom Kippur War, it was not difficult to foresee their introduction, postulate their effect on the battle, and analyze their impact on other factors that we could examine firsthand.
- The density-intensity-lethality equation will prevent domination of the tactical battlefield by any single weapons system; to win it will be necessary to employ all tactical battlefield systems in closely synchronized all-arms action. Further, integration of tactical- and operational-level battles will need to be closely coordinated to ensure a synergistic relationship between operations at the forward line of troops and those projected into the extended battle area.
- The intensity of the battle will make command-control at tactical and operational levels ever more difficult. Additionally, effective command-control will be inhibited by the presence of large numbers of radio-electronic battlefield operating systems whose prime purpose it is to interfere with effective command-control.
- The modern battlefield will require commanders to think clearly about some very complex situations; to decide quickly what must be done; and to issue clear-cut, simple instructions about who is to do what, where, and when in order to get done what the commander has decided upon. That kind of clear thinking takes time. In battle there is really no time to think. Therefore it is necessary that, to the extent possible, complex situations be perceived, solutions thought through, and reasonable courses of action postulated in advance in order to foreshorten the decisionmaking cycle—turning it, in time, inside that of the enemy.
- In modern battle, regardless of which side outnumbers the other, and regardless of who attacks whom, the outcome of battle at tactical and operational levels will be decided by factors other than numbers and other than who attacks and who defends. In the end the side that somehow, at some time, somewhere in the course of battle seizes the initiative and holds it to the end will be the side that wins. More often than not the outcome of battle defies the traditional calculus used to predict such outcomes. It is strikingly evident that battles are

yet won by the courage of soldiers, the character of leaders, and the combat excellence of well-trained units—beginning with crews, platoons, companies, battalions, and squadrons and ending with regiments, brigades, divisions, and corps. The best tank on the battlefield is yet the one with the best crew. The best units on the battlefield are yet those that are well trained and well led and those who have trained together to a high level of excellence before battle's onset.

- Extended battlespace characteristic of modern battle demands the ability to "see over the hill," to look deeply enough into the follow-on echelons to find, target, and deliver weapons against them to delay, disrupt, and destroy their ability to load up the battle at the forward line of troops. Surveillance and target acquisition systems and weapons systems for the deep battle need [to] be under control of the operational-level commander.

In a few words, the message was clear: The United States Army had to be ready to fight in the environment just described, fight outnumbered, and fight and win first and succeeding battles without sustaining significant or unnecessary casualties, and without having to resort, at tactical or operational levels, to the use of nuclear weapons to overcome disparities in numbers or other capabilities. After listening to our recitation, that was the task given by General Abrams to TRADOC and those of us responsible for seeing to the details of rebuilding the Army.

For several years after the Yom Kippur War, TRADOC and other agencies in the United States would collect data, conduct analysis, and publish studies. All too often the height and breadth of data and information could be measured in kilometers, the depth of analysis in millimeters. In the end nothing changed significantly the conclusions we had drawn early on from walking the battlegrounds with those who had fought, listening to their descriptions of what had happened, and availing ourselves of the penetrating operational analysis, even at that early time, by Musa Peled and his stalwart lieutenants and by Talik and his tank designers as they built Merkava. That very early visit was the beginning of years of frequent visits, long discussions, and considerable reflection that produced the programs to move the US Army from where we were to where we knew we needed to be. For all that the United States Army stands ever in debt to Musa Peled, the IDF Armored Corps and its intrepid leaders, and to the imagination and engineering genius of Talik and his tank designers.

How about the tank? Having examined several hundred damaged and destroyed tanks—Soviet, US, and British, of many vintages, that had come from Yom Kippur's battlefields, three important facts stood out:

- First, gun/ammunition: a larger caliber gun would eventually be required; more lethal ammunition would have to be designed and engineered; all ammunition should be stowed below the turret ring; combustible-case ammunition, characteristic of the German 120mm gun under consideration, would have to be proven troop safe to the satisfaction of the user—the Chief of Armor; improved fire control—especially fire-on-the-move and night-fighting capability—would be essential.
- Second, armor protection: armor with new geometry, made of new materials and material combinations, would be required; the aim of such armor would be to defeat chemical energy shaped-charge warheads, which had rather rapidly overtaken kinetic energy penetrators as the antiarmor weapon of choice.
- Third, protection from land mine detonations: vulnerability of US-built torsion bar suspension systems to land mine detonations had convinced most of us who fought with armor

units in Vietnam that we should rid ourselves of torsion bars and move to a less-vulnerable bolt-onto-the-hull system.

The Chief of Armor was instructed by the Chief of Staff to revise the requirements documentation for the Ml tank. Improved armor was already planned. Of the remaining requirements, it proved possible only to make provision for mounting of the 120mm gun when it became available, for improvement of ammunition for both l05mm and 120mm guns, and for fire-on-the-move and night-fighting fire control improvements. Ammunition stowage below the turret ring and redesign of the suspension system to defeat the mine threats were both considered too far removed from traditional US tank design philosophy. Fear of delaying the Ml program to implement the desired design changes caused the user to agree to set aside these requirements for another day.

For 17 years after 1973, the US Army labored hard at repairing the ravages of Vietnam and preparing for the next war. How all that was done is a story for another day. But almost all of what was done was set in place in the first eight years after the very early evaluation of the Yom Kippur War. By 1982 our early observations about warfighting at tactical and operational levels had been converted to new doctrine and published as AirLand Battle in the 1982 edition of FM 100-5, *Operations*. That doctrine formed the basis for development of equipment, organization, training, and education of soldiers and leaders at all levels. That doctrine, and everything that resulted from it, was put to test in Desert Shield-Desert Storm in 1990–1991. Following are some brief observations about that test:

- Desert Shield-Desert Storm, 1990–1991, was a limited war on the model of a Korea or a Vietnam. There were, however, some significant differences, differences that reflected our determination to not relearn the hard lessons of those earlier limited wars. In this case the political goal of the operation was made clear at the outset. Success of the operation was contingent on joint operations by US forces as well as extensive participation by coalition allies and on public support among US and allied "people at home." When the agreed-upon political aim had been accomplished, forces were redeployed. Despite some after-the-fact grumbling that more should have been done, the deployed military force did what was set forth at the beginning and withdrew, for better or worse, despite the risk of crystal-clear 20/20 hindsight about what should have been done offered by arms-length strategists.
- The military force fought outnumbered and won, almost without personnel casualties, and without invoking the specter of tactical or operational use of nuclear weapons.
- The part of the force that brought the operation to a successful termination was a corps-sized mechanized force of combined arms employed with lightning speed and devastating lethality. The mechanized force employed deep surveillance and attack systems to the depth of an extended battlefield. Large-scale attack helicopter units were integrated into the tactical scheme of maneuver, as were fixed-wing fighter aircraft. Everything worked—battle tactics, extended battle, Abrams tanks, Bradley fighting vehicles, Apache helicopters, JSTARS (Joint Surveillance and Target Acquisition System), ATACMS (the long-range Army Tactical Missile System). New organizations proved more effective than their predecessors. Soldier and leader training proved more than adequate. In sum, AirLand Battle—doctrine, tactics, equipment, organization, training, and education—was an unqualified success. Indeed, some of it worked better than its designers had hoped. In all, the US Army force that deployed to and fought Desert Storm was in all respects the finest military force the United States has ever fielded. It reflected a three-quarters-of-a-century-long struggle to understand the effects of modern technology on

military operations—opportunities offered by new technical capabilities, beginning with the advent of the tank in the First World War. It also reflected, for the first time in US Army history, a coherent concept of modern battle at tactical and operational levels, missing from US Army thought since the mechanization of war began with the advent of the tank in the First World War.

5. Combined Arms

Page

Observations on the Tank-Antitank Battlefield 228

Combined Arms Team 231

Mechanized Infantry and TOW Employment 232

Infantry Weapons Systems 233

Mechanized Infantry Combat Vehicle 234

Heavy Infantry Fighting Vehicle 236

Infantry Fighting Vehicles (Multiple Addressees) 240

Infantry Fighting Vehicle White Paper 241

Infantry Fighting Vehicles (Williams) 242

Combined Arms 243

Mortar Requirements 245

Combined Arms Operations with the M1 Tank 248

Systems Mobility and Agility 249

Combat Support and Combat Service Support 250

Observations on the Tank-Antitank Battlefield

Armor Magazine
January–February 1974

". . . The antitank defenses were situated on a hill overlooking the valley up which the armor charged. . . . Units which suffered least were the ones who presented almost all their vehicles to the enemy simultaneously. Even this benefit diminished after the infantry leaders gained experience controlling the fires of their weapons. . . ."

"Key leaders were killed early leading the attack . . . cannot emphasize enough the importance of utilizing terrain . . . amazed by the tremendous importance of a small pile of rocks or little knob not even on the map. These features were the difference between survival and death in the attack. . . ."

"The tanks when stationary had a high probability of detecting attacking antiarmor helicopters . . . detection occurred long enough prior to missile launch to permit effective engagement of the helicopters, sometimes even with the main gun. . . . In general moving tanks did not detect or engage the attacking helicopter, and helicopter gunners were able to successfully launch their antitank missiles with good hit results. . . ."

". . . well trained aerial scouts flying nap of the earth operated very close to the enemy . . . one scout aircraft, properly employed, was found able to do the work of several ground scouts. . . ."

". . . well employed antitank weapons with good fields of fire can wreak havoc amongst vehicles in mass armor assaults. The object (of the defense) is to allow the enemy to drain his mobile tank forces against the defenses . . . perhaps even at the cost of those defenses . . . then, depleted and disorganized, the attacking armor forces are crushed with fresh mobile reserves, the attacker's rear is penetrated and, unless he grossly outnumbers the defender, the attacker will have little left with which to counter. . . ."

". . . enemy infantry was organized into antitank teams sited well forward . . . tanks attacking without their accompanying infantry were devastated trying to get at the enemy tanks which were sited at standoff ranges to the rear of the infantry antitank belt and interspersed in and among the secondary belt of antitank guided missiles."

". . . learned again and again the essentiality of integrating all elements of the combined arms team—tanks, infantry, artillery, attack helicopters, tacair. . . ."

Sketches from the October War? Not at all; scraps from various commentaries on our tests and evaluations in the continuing search for the truth about the tank-antitank battlefield.

Historically each new development in antitank weapons has occasioned a surge of commentary about the declining utility of the tank. Recent years have seen all the world's armies proliferating antitank weapons. In some armies infantry organization and tactics have been realigned to take best advantage of the new quantitative and qualitative antitank wealth. In the air, heliborne antitank systems add a new dimension to the already changing character of the battlefield. Quantitatively, more than qualitatively, recent changes in the antitank sector of the battlefield spectrum give reason for pause, reflection, and reevaluation. For it is now truly a tank-antitank battlefield.

So now, possibly more than ever before, hard questions must be asked, objective answers sought, sound tactics developed to fit the changing complexity of the battlefield. The search is not without emotion as old loyalties and new confront one another in the dialogue. It would be comforting to say that our experiments confirm the old teachings, validating tactics, techniques, organizations, equipment.

Such is not the case; indeed the only bit of ancient dogma that remains unscathed is the essentiality of the combined arms team. Having acknowledged that fact, however, it may be that on the tank-antitank battlefield little else recognizable remains. The warp and woof of the whole cloth are not completely discernible yet, but one can at least draw up some hypotheses growing out of what is now known about the tank-antitank battlefield, hypotheses to be tested, investigated, validated, or nullified in the search for more precise definition of the metes and bounds of tank-antitank war.

First is a set of hypotheses concerning the role of the combined arms team and tasks for its component elements. Second, and flowing from the basic task hypotheses, is a series of subsets of hypotheses which address themselves to tactics, techniques, organization, and equipment which are the logical outcome of the basic premises themselves. We have space here only to outline tentative postulations about some basic tasks and to lightly sketch in possible outcomes of investigating those tasks in depth.

The Combined Arms Team. The primary task of the tank-mechanized infantry-attack helicopter-artillery-tacair team is the conduct of *offensive* operations whose purpose is destruction of the enemy—destruction by penetrating into areas of enemy weakness, rolling aside the flanks of the penetrated force, storming into the enemy rear to destroy reserves, trains, lines of communication, command-control. This team is the basic factor in the combat equation; below the level of brigade its role is always offensive; the division may defend—its task forces attack, always attack.

Each team member provides capabilities reinforcing weaknesses of other members. Infantry protects tanks against antitank teams and provides an antitank capability of its own, protected by long-range tank fires. Antiarmor helicopters attack the enemy when he is moving, on his flanks, in his rear, adding high mobility antiarmor fires to the fight. Tanks attack enemy armored vehicle formations, penetrate at weak points, [and] accelerate the destruction exponentially into the enemy's rear.

Infantry. The primary task of infantry on the tank-antitank battlefield is tank killing. Armed with newer and more capable antitank weapons in greater quantities, infantry becomes a formidable threat to the tank's most significant vulnerability—well trained infantry employing effective antitank weapons such as the RPG7, TOW, DRAGON. Infantry's secondary task is a historical task of every unit—reconnaissance to find the enemy and hold him in place long enough for killing firepower to be brought to bear on him.

If reconnaissance and tank killing are infantry's tasks, then is the rifle squad properly organized and equipped? Should not perhaps its fire teams be antitank teams, [or] an antitank team and an automatic weapons team? With what weapons should the squad be equipped? Is an automatic weapon still primary to infantry battle, or should the antitank weapon take its place? With what devices must we equip infantry to perform its reconnaissance task—in weather, at night, at what ranges? What kinds of indirect firepower must infantry have so that, once found and held, the enemy may be destroyed?

Tanks. The basic task of the tank is destruction of the enemy by offensive action. A secondary task is reconnaissance. In executing its basic task the tank's primary targets are enemy armored vehicles and enemy antiarmor-equipped aircraft. If these task hypotheses are assumed correct, what kind of weapons should the tank mount, and what ammunition type(s) should these weapons fire? If the enemy has a significant capability with armor-defeating heliborne weapons, should the tank platoon consist of five tanks, or four tanks and a flak wagon of some kind, three tanks and a flak wagon? Should tank coaxially-mounted weapons continue to be area-fire weapons, or a point target-capable cannon?

Cavalry. The primary task of cavalry—air and ground—is to find enemy weaknesses so that a combined arms team attack directed against the weak area will penetrate, enabling armor task forces to enter the enemy rear and destroy him. Cavalry's secondary task is one of preventing the enemy from finding weak areas in friendly force deployments.

If these task hypotheses are near correct, does it not follow that cavalry should be armed with the same tank-antitank systems that characterize the main force, adding only the enabling communication, surveillance, and target acquisition equipment required by its unique tasks? Should not cavalry be simply heavy cavalry—a few light scout vehicles, perhaps aerial scouts only, to add scope to its vision and range to its finding capability? Is not the role of the infantry part of cavalry primarily one of recon/antitank—the same roles postulated for other infantry forces? What surveillance and target acquisition equipment in what combinations need be added to what part of a heavy cavalry force to give it the required finding capability? Should we continue to insist that every cavalry platoon be a self-contained mini-combined arms team, or could we beget a more flexible force by tailoring a few lighter scout elements (air, perhaps?) with tanks and mechanized infantry and do the same thing—better?

As with the basic task premises, hypotheses may be set forth as tentative answers to each of the suggested questions, and many others as well. One would hope that in the validation or invalidation of the hypotheses some more precise perceptions about the tank-antitank battlefield might be acquired. Tactics, technique, organization, and equipment requirements then flow logically from these hypotheses; alternatives become clearly defined; tentative decisions may be made, set against cost of alternatives, and a rational course of action charted.

It's always upsetting to ask questions whose answers are difficult to divine. Particularly is this true in a bureaucracy, and more particularly is it true when the questions involve a spectrum of systems whose individual proponents are prone to stand on traditional biases. But question we must if we are to be true to the heritage bequeathed us by that small band of farsighted men who thirty years ago were asking the questions of their day and seeking the harder answers.

Combined Arms Team

Letter to General Bruce C. Clarke
Palmyra, Virgina
6 September 1974

Be assured that both he [General DuPuy] and I believe that one of the great lessons the IDF rediscovered in October 1973 was that of the essentiality of the combined arms team. Your informants were right—they lost heavily in tanks the first two days in the Sinai because they attacked with tanks alone. Once they restored balance and went back to the time-proven team idea they did well. We are teaching the combined arms as we always have. Together with Benning we have produced a training circular on the company team, and are working on one on the battalion task force—a real first in service school coordination. Not easy, but done!

Mechanized Infantry and TOW Employment

Letter to Major General Willard Latham
Fort Benning, Georgia
18 May 1976

I was about to write you . . . and provide some first reactions on mechanized infantry in the Army in the field. My first observation is a tactical one, and it relates to the employment of antitank weapons—especially TOW, in conjunction with mech infantry. As you know, our TOWs are assigned two per mechanized company and the rest in antitank platoons of the mechanized battalions. As I have walked through all the battle positions of our mechanized and tank battalions in an on-the-ground review of our general defense plans, I find that there's considerable confusion about how to employ the TOW and the infantry, especially together. What I find is that, when the TOWs are properly emplaced, the accompanying infantry in the wrong place. If the infantry is in the right place, then the TOWs are in the wrong place. In wooded or close terrain, where infantry can best fulfill its role in the combined arms battle, the Dragon is about all that can be used effectively because of restricted fields of fire. The result of all this is that we are not getting full use of the capabilities of our mechanized infantry, of which we have too little to begin with, and we are not fully utilizing our TOWs, of which we have too few to begin with. Part of this problem is organizational—we probably should look seriously at concentrating all the TOW systems in an antitank company in the mechanized battalion and getting them all out of the line mech companies. Then the battalion commander can task organize, putting TOWs from the AT company with his mechanized companies as appropriate.

Further, I can see little or no use for TOW systems in scout platoons of mech and tank battalions. They too should be consolidated somewhere else—in the antitank company. In addition, I can't emphasize strongly enough how badly our mechanized infantry, as well as our scouts in mech and tank battalions and the scouts in the cavalry, need a suppressive weapon better than the .50 caliber machinegun. Therefore, as I have done before, I urge you not to tie up equipping our M113s with a suppressive weapon of some kind with a correlated project to PI the M113—they should be separate programs. The M113 is about OK. What is not OK is the suppressive weapon on it, and we need it badly.

Infantry Weapons Systems

Message to Major General John R. Cleland
8th Infantry Division
1 March 1977

Suggest our most significant infantry issue for mech infantry on defense or offense is the essential incompatibility of the systems with which we have equipped them. We have short-range infantry—Dragons, machineguns, LAWs, and long-range antitank infantry—TOWs. Employment techniques for each are pretty significantly different, and we've not made that distinction clearly enough in tactics and training, and not at all in organization. Believe it's time to bring that issue out in the open so that we can get it resolved as part of the division restructuring effort. I will take necessary steps to coordinate General DePuy's itinerary to provide time for him to listen to and give guidance on your presentation.

Mechanized Infantry Combat Vehicle

Letter to General Bernard W. Rogers
Army Chief of Staff
25 March 1977

Last week I appeared before a subcommittee of the Senate Armed Services Committee, together with Ed Miller and Howard Cooksey. The issue quickly resolved into a discussion about the MICV.

We, of course, supported the MICV and may have done some good. If we did make any impression on Senator Nunn (who was absolutely dead set against the MICV at the outset), it was largely because Dr. Phil Karber of BDM swung over to our side.

Three civilian witnesses were called by the committee, no doubt at the instigation of staffer Record and Senator Hart's assistant, Mr. Lind of 100-5 fame. They were: Karber, Luttwak, and Canby.

Karber had lunch with me before the hearing and we were able to persuade him that, although the MICV isn't perfect, it is the best infantry fighting vehicle at this time and that we couldn't, in good conscience, send American soldiers into war in the M113 in Europe. This sounds like taking the credit for Karber's performance. In fact, he did a better job than we did in pointing out that the M113 is first generation, the BMP and the Marder are second generation, and the MICV is third generation, at least in terms of mobility and firepower. He testified to this effect.

Karber is a very good man, honest and smart. He is concerned about the disparity in protection between the XM1 and the MICV. So is Luttwak, so are the Germans as pertains to the Leopard and the Marder, and so are the Russians as pertains to the T72 and the BMP. So are we concerned.

Phil Karber had originally intended to recommend that we discontinue development of the MICV and turn the XM1 chassis into a MICV. When he found out this would take some time, he reluctantly agreed to support the MICV as the only realistic option open to us at this time. Luttwak testified against the MICV, as did Canby. Luttwak, like Karber, wants more protection. Canby is just plain vague and confused and confusing.

For over a year I have been thinking about the requirement for an Armored Infantry Fighting Vehicle which would have armor protection equal to the XM1. This thinking was prompted strongly by the Israeli solution to the vulnerability of armor by including a mechanized company organic to each tank battalion in wartime (in peacetime they can't afford the manpower). I have also been influenced by the fact that George Blanchard favors a composite battalion because he wants to train the way he plans to fight.

Because of the obvious impact on the MICV program and the nightmarish thought that we could find ourselves stuck with the M113 for another 10 years while we make yet another start on the MICV, I have kept these thoughts within TRADOC.

Day before yesterday I read in the *International Defense Review* that the British Army has resurrected its development program for a MICV, but this time plan to build it with Chobham armor. It seems to me that a combination of the Karber, Luttwak, Nunn, Hollingsworth, Bartlett concern, coupled with the British initiative on a heavily armored MICV, probably signifies the hopelessness of keeping this idea bottled up.

The more I think about this matter the more I believe that we must begin to move in the direction of a truly armored infantry fighting vehicle to be married organizationally with each tank battalion. These vehicles would no doubt be expensive—they certainly would be heavy—and probably be an overkill for the battalion formations of the mechanized infantry for which we would still need the MICV—we certainly don't want the M113.

Another thought which may help is that we are nearing the end of the MICV development—in fact, it would be essentially finished were it not for the revision of the turret and the Bushmaster development. As we look beyond the current set of new equipment which will be available by about 1985, there is very little on the drawing board. In other words, there will be a trough in the development program which we might just as well start to fill now.

The idea of an AIFV using special armor is being quietly examined at the Tank-Automotive Research and Development Command. However, the cat isn't too far out of the bag at this moment.

I suggest that you and Dutch and George Sammet and Shy Meyer and Howard Cooksey and I meet sometime in early April to discuss this issue and plot a course for the future. The sensitivity of it is obvious and the dangers to the MICV program are clear. Thus it would seem to warrant the best thinking of your top people involved in such matters.

Heavy Infantry Fighting Vehicle

Message to Multiple Addressees
16 November 1977

1. The purpose of this message is to set forth TRADOC concepts regarding the heavy infantry fighting vehicle (HIFV) and its relationship to the infantry fighting vehicle (IFV) and to related weapons development programs.

2. First it is necessary to consider operational concepts:

a. When requirements for the mechanized infantry combat vehicle (MICV) were first written, the operational scheme for its employment appears to have included the following:

(1) It was to be a combat vehicle system from which infantry fought, dismounting to assault or to fight dismounted in close terrain or builtup areas. It was *not* simply an armored personnel carrier intended to afford protection from small arms and fragmentation weapons to infantry squads being transported to battle.

(2) Its primary weapon was to be a cannon. The cannon was to provide high volume automatic fire for suppression while attacking mounted and fire support of the rifle squad fighting dismounted. The vehicle rapid fire weapon system (VRFWS), later named Bushmaster, was the developmental candidate. For reasons now not clear, its caliber and other essential characteristics were determined by studies which pitted it primarily against other like vehicles in the threat array. Its comparative effectiveness was set forth in like terms. It was for some time in its early development analyzed in a "war of the MICVs."

(3) The rifle squad aboard the MICV was to be an eleven-man formation; dismounted it fought in two fire teams, like any other infantry squad. In this mode it was to be supported by automatic weapons fire from the MICV cannon.

(4) Antitank weapons for the MICV and its squad were to be those short-range weapons organic to the squad—Dragon and LAW.

(5) It was to accompany the tanks wherever possible—i.e., most of the time. While tanks would normally lead, terrain, builtup areas, and forests could cause infantry to lead, usually dismounted. It was also considered that tanks and infantry would normally attack together on the same axis, but that they would also move on separate axes, or tanks could overwatch by fire when obstacles, terrain, or other conditions dictated.

b. TRADOC's evaluation of the Yom Kippur War and other analyses led to a redefinition of MICV's operating concept. Thus, in early 1974, TRADOC's operational concept for MICV was redefined in terms of requirements to support tank-led combat teams by:

(1) Providing long-range suppressive fires against enemy infantry antitank teams.

(2) Providing suppressive fire while moving cross-country with tanks.

(3) Providing high volume of suppressive overwatch fires to support dismounted infantry attacking enemy infantry who haven't been suppressed.

(4) Defeating the BMP beyond the range of the 73mm smoothbore gun mounted thereon.

(5) Firing antitank guided missiles from an armor-protected station aboard the vehicle.

(6) Protecting the crew against automatic weapons fire of calibers up to 14.5mm.

(7) Dismounting, when required by terrain, enemy, vegetation, or scheme of attack, a fire team of four, five, or six men armed with machineguns and short-range antitank weapons, and overwatching that fire team with fire from the fighting vehicle.

3. Now, as between [these two concepts], what is the same and what changed?

a. The ideal of a fighting, not transporting, vehicle with a cannon system for suppression and protection for the crew remained essentially the same.

b. The fighting mode of mechanized infantry fighting dismounted changed considerably, from a squad with two fire teams supported by the vehicle cannon to a single fire team fighting with its fighting vehicle as the other element of a two-element fire and maneuver team at squad level. This change recognized, first, the reality that a full eleven-man squad would never be present for duty and that a single fire team of four, five, or six men would likely be all that could ever be dismounted for operations. Secondly, the changed concept recognized that, in the initial battle, the density of systems and intensity of the fight would make command-control so difficult at lower levels that it was necessary to reduce echelons to be controlled, in this case from three to two at squad level. This would also make platoon and company commanders' tasks somewhat less demanding.

c. The primary function of the cannon changed somewhat, emphasizing in place of unquantified statements of suppression needs the clear need to stand off the BMP by outranging the 73mm smoothbore and the need to suppress dismounted long-range antitank teams.

d. New was the requirement for a long-range, armor-protected antitank system. This change reflected recognition that survivability of tanks in the central battle needed to, and could be, improved considerably by a long-range antitank capability as part of combat teams fighting there. This was the genesis of the need to add the TOW-Bushmaster armored turret (TBAT) on the MICV.

4. Recognizing that TBAT represented such a change in materiel requirements for the MICV that it would result in a considerable MICV program delay, TRADOC began to search for an interim armor-protected, long-range antitank capability—interim between today's exposed crews, dismounted or in M113s, and MICV's protected weapons and crews. The outcome of that search was the improved TOW vehicle (ITV) program. The ITV, however, was never seen as a competitor for MICV's mission. It was and is a mobile, long-range antitank system with limited armor protection. It was intended to replace the then-fielded TOW vehicle which featured a pedestal-mounted TOW system erected above the M113 deck and fired by a completely exposed firing crew.

5. As time wore on, and our perception of the central battle sharpened, two things became apparent. The first had to do with the employment of TOWs as part of the antitank tactics in the central battle; the second perception was one of questioning the capability of MICV to survive at the same level as the tank in the central battle. The Hunfeld studies at Knox, and subsequent

work in V Corps, clearly demonstrated that thin-skinned M113s and Sheridans did not survive on a dense battlefield in an intense fight. They are simply too vulnerable to too many enemy weapons. This led to a conclusion that, given the level of armor protection we had required on it, MICV was not much more survivable than M113s or M551s. Therefore to survive they all had to stand off—on flanks, overwatching or wherever, especially in the attack. Also it became increasingly clear that our infantry system suffered from a mismatch of short-range and long-range weapons which confronted it with two insurmountable dilemmas in tactics. First, with regard to TOWs with infantry in our present organizational distribution, it became apparent that, where the short-range infantry deployed with properly emplaced TOWs, the infantry was in the wrong place on the battlefield. Contrariwise, when TOWs deployed with properly deployed short-range infantry, the TOWs were in the wrong place. Neither, therefore, could do its job very well if encumbered by the presence of the other, and one team member was always malemployed. Secondly, with regard to TOWs on MICV in an integrated turret, while the long-range antitank capability certainly made MICV a more versatile system, it also brought along a built-in dilemma—the missions for which one could effectively use the cannon against hard targets like BMP did not require TOW, and against soft targets at greater ranges the TOW was obviously not the correct weapon at all. With tanks in the central duel TOW could provide longer range antitank fires than the tanks, standing off even tanks, but it could not survive nearly so well as tanks. TOW on MICV made it a tank killer, but MICV's survivability level made of it a stand-off system. The counterargument to this is, of course, the one reflected in the BMP—that is the Soviet doctrine of complementariness. As the Sagger and 73mm smoothbore fit the Soviet concept of complementary systems in the same vehicle, so do the cannon and TOW on MICV. Which is the correct view cannot be determined for super certain, but it does appear that neither the BMP nor MICV can stand *mixed in* with tanks in the central battle and survive. Soviet BMP employment doctrine reflects this fact.

6. What grew from the perceptions just outlined was the idea of a heavy infantry fighting vehicle. It would have survivability equivalent to XM1. It would fight integrated with XM1 in the central duel. It would carry an assault team which would dismount to pry out an enemy gone to ground, but who for the most part would fight mounted. It need not, indeed could not mount an ATGM, but it did need a good armor-defeating direct and rapid fire weapon of fairly long range which could reach out to destroy lighter skinned vehicles at longer ranges, giving some stand-off, but which could also defeat tanks at the closer ranges of the central duel itself. It could, in short, unstress our tanks in the main battle, first by defeating enemy armor vehicles and antitank systems and second by joining in with the tanks, a target obviously so dangerous as to require the attention of more enemy systems, drawing them away from our tanks. The only weapon on the horizon—that is, under development—which seemed to match this need was the ARPA 75mm automatic cannon. While not everyone agreed at the time on this gun-vehicle match, what was fairly well agreed upon by all was that the HIFV was neither a competitor nor a follow-on for MICV. The two clearly had important, separate, but complementary roles on the battlefield.

a. Survivability was key—HIFV equaled XM1.

b. Direct rapid fire armor-defeating weaponry was essential, but an ATGM system appeared not survivable enough, and not fast-firing enough, and a small caliber cannon not lethal enough.

c. Fighting with the tank in the central battle, it represented a lethal capability to all enemy armor within range. Therefore it would be a target which the enemy could not ignore, all of which spelled a system which could help unstress our tanks.

7. A task force has been appointed to evaluate this concept. While the task force is obviously chartered to examine such conceptual excursions and variations as may be appropriate, it is still essential from TRADOC's view that the original concept be thoroughly wrung out. There appears some confusion between several TRADOC agencies as to precisely what the HIFV concept is; the purpose of this message is to provide that information.

8. In summary:

a. We need the IFV.

b. The HIFV makes sense.

c. The HIFV would have comparable survivability to XM1; capability to kill tanks at shorter ranges; comparable mobility to XM1; long-range suppressive fire capability, especially against armored vehicles other than the tank and against antitank teams.

d. The question remains open as to whether different organizations and different mixes of IFV/HIFV may be needed.

Infantry Fighting Vehicles

Message to Multiple Addressees
9 January 1978

1. I have reviewed the briefing prepared by USAARMS and USAIS. . . . It falls well short of the mark. In fact it's horrible. If we are to satisfy our critics in OSD, the Congress, and elsewhere of the need for the IFV/CFV, then we must put together a very convincing case, which this briefing does not do. We must describe operational concepts which generate the requirement for an IFV with the capabilities we have said we require, e.g., kill BMPs, XM1-like mobility, etc.

2. . . . One of the reasons we have not been successful in articulating our case for the IFV is because our critics view it as just a product-improved 113, which it is not. The 113 is a carrier; the IFV is a fighting vehicle. It is the difference between the two, and why a fighting vehicle is needed, that we have not clearly articulated to our critics. To make the case, we must first describe how tanks, long-range infantry and short-range infantry must fight together on the battlefield, clearly describing the complementary roles of each. Structuring the battlefield by infantry as described in the briefing certainly is not one of them. Then we must show why the M113 is inadequate for operations with the M60 series tanks, given its limited armor protection and limited firepower. When we go to the XM1 the situation just gets worse. Here we must clearly describe the stress on the tank due to the 113's inability to operate with our current and future tanks. Having done all this, the operational concept for employment of the IFV with the XM1 must be shown. But, again, not described as simply a product-improved 113 nor, on the other hand, should it be made to appear as a light tank. It must be shown as a fighting vehicle that can unstress the tank and also carry infantry.

Infantry Fighting Vehicle White Paper

Message to Major General William R. Richardson

9 January 1978

. . .

2. . . . Our argument should lead off with a statement on the critical requirement for infantry on the European battlefield. We should point out the tasks that the infantry perform, i.e., in the defense; occupy dismounted avenues of approach; prepare and maintain strongpoints; and reinforce long- and short-range antitank fires. In the offense, we should emphasize the need for infantry in counterattack operations. Here we can stress the need for infantry to operate with tanks providing overwatching fire, suppressing antitank weapons and engaging in close combat when required. In this discussion limited visibility activities should be highlighted. We should make the point that the infantry is an essential element that is present on the battlefield. When properly equipped, we are going to receive a higher return on this manpower-intensive investment.

3. Following the above comments we suggest you emphasize the complementary nature of the combined arms team. The TOW and 25mm systems on the FVS provide a long-range (3,000 meters) CE defeat capability against tanks and BMPs. This capability assists us in slowing and disrupting the enemy's movement. Its shortcomings are its rate of fire and limited stowed load. However, it is complemented by the tank that has a higher rate of fire and stowed load and an excellent medium-range (2,000 meters) KE defeat capability. The point of emphasis is that the tank and infantry systems complement each other. When a deficiency exists, the other system must attempt to pick up the slack.

4. This leads us into a third major consideration, that of unstressing the tank. When looking at the battlefield today our tank is faced with the "one on many" (what it must defeat) "many on one" (what is attempting to defeat it) situation. The end result is that the Soviets can concentrate their effort on our tank component. The acquisition of the FVS would redress this imbalance by forcing the Soviets to show greater interest in our FVS components. Simply stated, an XM1/FVS force has a much greater degree of balance than an XM1 force equipped with 113s. . . .

5. A further point to consider is the impact on NATO interoperability if we maintain the M113 as our primary infantry carrier. The Leopard II main battle tank possesses many of the same fine performance capabilities of our XM1 (mobility, shoot on the move, etc.). Effective interchange of US infantry units with FRG tankers on a reverse situation will be limited if we retain the M113 as our prime asset. We suggest closing the introduction with a short discussion of recent changes in Soviet force structure in which the BMP has been emphasized.

. . .

7. Another point that we feel must be emphasized is the current status of the program. We are very close to an OT. We should be funded to complete that OT, see where we are with the system and how well it performs. I am confident that if we can get this test under our belt we will be in a position to demonstrate the improvements that the system will give us.

Infantry Fighting Vehicles

Message to Colonel Bruce Williams
United States Defense Attaché Office, Tel Aviv, Israel
19 January 1978

We're having a hell of a time trying to save the infantry fighting vehicle. All R&D and procurement was cut by OMB in budget process. If he hears of it, tell him [Musa Peled] I haven't lost my mind—just surrounded by people who haven't fought a war and whom we can't make understand the battle equation.

Combined Arms

Armor Magazine
September–October 1978

The Combined Arms Team and Armor—frequently they're thought of as synonymous. In some ways they should be—the concept was introduced into our Army at Fort Knox, not by anyone now on active duty to be sure, but by a small group of our distinguished predecessors. The concept has been developed, protected, husbanded, expanded, and even criticized at Fort Knox. In fact, everyone who has served there has participated in at least some of those activities.

So we might say all is well with Armor and the Combined Arms Team; we could mutually congratulate ourselves, smug in the knowledge that Armor has the inside track on all that's necessary to win the critical battles of the next war. That's a tempting security blanket, but not a real one.

The Army is not that homogeneous in its outlook, nor is the world in which it lives. In the Army today there is a parochialism that challenges and sometimes even denies many things about the combined arms idea. That parochialism sometimes may concern leadership, or tactics, or administration, or any of a hundred other things. It is easy to get confused, even discouraged, on finding that everyone doesn't understand the message as do those in Armor. It is also all too easy to join the throng that is quick to point out problems, but offer no solutions.

Not all the question asking is bad. It is a necessary part of Army dialogue—in progress for 200 years—and we probably shouldn't want it any other way. For, despite Field Manuals, How-To-Fight Books, and the other written paraphernalia with which we surround ourselves, approved doctrine on any matter is often the opinion of the senior officer present. Now, while that may give me no small measure of satisfaction, it doesn't help anyone else—nor did it satisfy me when I was younger.

It does, however, point out a strong feature of our system—we can and should argue the merits of operational concepts with which we intend to fight. Operational concepts are important—they set the framework for tactics, organization, equipment development, and for training. They are the guts of our Army; therefore, a consensus about them is important. However, a word of caution. A common starting point is necessary for any intelligent dialogue to proceed. Each discussant must recognize that everything for which the other stands is not inherently wrong. To believe that is folly, a folly that rejects the value of dialogue.

It is this failure to recognize the merits of a dialogue, and its bounds as well, that troubles Armor and the Combined Arms Team. Instead of listening intelligently to one another, we are divided into two or three strident camps. In one, the tank is supreme. In another, it is the armed helicopter. In still another, it is the antitank guided missile (ATGM). There is no room for compromise; rationality is not a virtue in any camp; all draw their best examples from the same source, the Yom Kippur War. Listening carefully, one wonders if in October of 1973 there were several wars or just one.

So while we chorus our huzzahs for the Combined Arms Team, in a quite parochial aside we add "fine but *helicopters/tanks/ATGMs*—insert one of your choice—are the real answer." So

---

Reprinted with permission from *Armor* Magazine.

at this point a summing up seems appropriate, followed by suggestions for a perspective that might help cope with the dilemma in which we find ourselves.

The Armor Combined Arms Team in our Army was created by a few farsighted men—Chaffee, Van Voorhis and others—who persisted against a lot of entrenched tribal wisdom. Their victory was short-lived, but it lasted long enough to win World War II. Then, in a rush to get back to "real soldiering," we disbanded our large Armor formations—all we really needed was a few tanks to support Infantry. Many still believe that. Today, this group would have us believe antitank guided weapons have taken over, and the tank is dead.

The antitank helicopter is a new and attractive dimension in battle. It is so new that those who understand it the least have made it the center of too much attention. Its singular advantage—the ability to move rapidly from one part of the battle to another—has given rise to mistaken notions about what it really can do. Ignoring the limitations of weather, terrain, air defenses, and the inability to occupy ground, enthusiasts raise up the helicopter as the answer to the warrior's prayer. Some would even trade battalions of tanks for squadrons of attack helicopters.

Then there are the tank purists; after cursory study of the Yom Kippur War, they redecided in favor of more tanks to the exclusion, or at least neglect, of other Combined Arms Team members. All we need is an elite, sophisticated, highly proficient tank force.

Versions of these arguments have passed by us all at one time or another. All contain some tempting arguments. Their failing is that they defy everything the Combined Arms Team was designed to be. Most alarming is that they interact most violently in the ranks of Armor. The Armor soldiers of our Army seem unable to speak with one voice. Every one of us who has successfully commanded a unit of tanks, mechanized infantry, cavalry, or attack helicopters is an expert at how those units should be organized, equipped, and employed.

Unable to put aside the nearsightedness of personal experience and embrace a broader Combined Arms Team perspective, we debate endlessly. We continue to talk long after saluting would be a more appropriate gesture.

So my appeal is for perspective, not parochialism; for rationality, not rashness; for teamwork, not lip service.

If the Yom Kippur War demonstrated anything, it strongly affirmed the utility of the Combined Arms Team with strong emphasis on the operative word, *team*, a team which embraces a balanced force of artillery, mechanized infantry, tanks, air defense, engineers, and supporting arms and branches; and a team which draws its effectiveness from balancing the capabilities of these systems and from the synergism of their combined efforts. True, the balance is constantly changing, but it is always interrelated. Armor is part of this interrelationship. As legatees of the Combined Arms Team idea, it seems to me Armor soldiers have a special duty to ensure that imbalances are redressed. We all must be willing to understand and logically examine each proponent's advocacy in terms of what's best for the Combined Arms Team. If we don't, I predict our detractors, aided by some well-meaning voices in Armor itself, will destroy or imbalance the team and ultimately jeopardize our chances for victory. The team—the Combined Arms Team—deserves a better fate.

Mortar Requirements

Message to Lieutenant General Glenn Otis
Deputy Chief of Staff for Operations
22 February 1980

1. Your request for recommendations on mortar structure did not specify what time frame you were talking about. The answers could be different if you want a 1980 solution instead of a 1986 solution. My response addresses the mid-range (1986) solution. If we can get agreement on the future mortar composition, we can then backward plan and determine how to get there.

2. Heavy division. By 1986 the heavy division's indirect firepower capability will be significantly enhanced by: the addition of more 155mm howitzer tubes (72 vs. 54), the addition of more 8" How (16 vs. 8), the introduction of the MLRS (9 SPLL per division), increased stocks of dual-purpose improved conventional munitions, artillery-delivered AT and personnel mines (RAAMS and ADAM), Copperhead, and increased fill of battery computer systems and TACFIRE units. We may also see increasing numbers of FIST vehicles and airborne laser locator designators (ALLD) on mast-mounted sight-equipped scouts. Further significant improvements in the indirect fire capability will be gained through the later introduction of SADARM munitions to our 8' How (and eventually 155s) and precision guided munitions (PGM) for the MLRS.

3. Given this substantial increase in indirect firepower, the basic question is: "What is the role of mortars in the heavy division and how many, if any, are required?" The role of mortars in the heavy division, we believe, is somewhat different than the role of mortars in our light infantry organizations. In heavy divisions the mortars should provide the task force commanders with responsive close-in fires to support task force indirect fire requirements against light armor and dismounted forces. These fires fill in gaps in 155 DS supporting fires. The mortars' higher angle of fire can be brought in closer to our own dismounted units than can the howitzer fire and can get at forces on reverse slopes better. Secondly, the mortars will be able to provide highly responsive smoke and illumination fires to the task forces, relieving the artillery of much of this mission. With the introduction of thermal sights on XM1s, XM2s, and XM3s, the current requirement for illumination will decrease, but the need will probably never be completely eliminated. Therefore heavy divisions need a mortar capability at the task force (battalion) level. Because of the increased indirect fire capability of Heavy Division 86, and because we do not envisage that company teams of the task forces will routinely be operating in non-supportable, independent actions, we do not believe it necessary to have dedicated mortar tubes for each company team. The remaining questions are: "What caliber of mortar and how many do we want at battalion?" The basic versatility, growth potential, and volumetric efficiency aspects argue for the heavy mortar over the medium mortar. Since the tracked carrier will be dedicated for either, weight of the system and manportability are not issues; we therefore believe the requirement to be for the heavy mortar. In the near term that is the 4.2" mortar; for the long range, and since we may need to buy more heavy mortars, we may wish to transition to something like a 120mm smoothbore mortar.

4. In order to provide continuous support during fast-moving situations, mortars must be able to displace by section, with a minimum of two guns per section firing. This argues for a six-gun platoon at the battalion. In order to standardize organizations and simplify training requirements, the mortar organization should be the same for mech and tank battalions in heavy

divisions. We are also providing the reconnaissance squadron with a six-tube mortar platoon since it will habitually operate on independent missions.

5. Ranger and Airborne Infantry Companies. There is general consensus that ranger and airborne units need dedicated man-portable mortars at company level. That's why the lightweight company mortar was developed, and the need on which we based that requirement has not changed. The significant problem associated in providing man-portable mortar support is the transportation of ammunition. Therefore, lightweight mortar requirements in ranger and airborne companies are basically an emergency support capability unless some mechanized means of delivering the ammunition is provided. You simply can't carry much around on the backs of soldiers. The 82d Airborne may not want to take its lightweight company mortars with them on every mission, but it should have the option for contingency areas where the lightweight company mortar would be adequate and transportation problems can be solved (such as the Dominican Republic). Currently airborne division requirements call for three lightweight mortars per rifle company. Two lightweight company mortars are assigned to each ranger company, for a total of six per battalion. We support these requirements. More on mortars at the airborne infantry battalion later.

6. Infantry Divisions (7th, 9th, and 25th Divisions). The role of mortars in the light division has been the subject of much debate recently. Improvements in the division indirect fire support capabilities in the mid-range period will be similar to those provided in the heavy division. There will be no 8" How in the light division, but we are proposing a battalion of MLRS to provide general support or reinforcing fires for the division. If the light infantry division will face a dismounted threat, then there is an increased role for the close-in fire support capability of mortars in the light divisions. The requirement for smoke delivery may be somewhat less than the heavy division, but the requirement for illumination in the light divisions will be greater, since we are less likely to proliferate thermal night sight and image intensifying capabilities to the dismounted forces to the levels these are in heavy units. Retaining mortars at the company level generates a logistics problem, as mentioned earlier; there is also a significant training problem. We do not expect rifle companies to be employed habitually on independent missions or in widely dispersed sectors. Therefore we do not see a tactical requirement for dedicated mortars in each company. Our goal for light division maneuver battalions is to eliminate company mortars and to increase the capability at battalion. The question is, "Which mortar should we use?" The procurement of the high-mobility multipurpose wheeled vehicle (HMMWV) will allow us to carry either the heavy or medium mortar in light divisions. The lethality of the heavy mortar, as compared to the I-81, argues strongly that we select the heavy mortar. As an example, the heavy mortar gives an improvement of over 20 percent in lethal area coverage over the I-81mm for the same investment in people and tubes. Also the potential capability of the heavy mortar to fire precision-guided munitions (such as the follow-on Copperhead seeker) must not be ignored. If light infantry divisions are required to face enemy armor, an antiarmor capability for the heavy mortar is an option which must be seriously evaluated.

7. Since the light infantry battalion consists of three companies (versus four in the mechanized infantry battalion), the provision of six heavy mortars at battalion will provide greater fire support per light division rifle company than that available to the mech infantry company team. This solution further standardizes the mortar platoon for the light infantry divisions with that of the heavy divisions and standardizes the mortar itself.

8. Airmobile Division. The only remaining issues, therefore, are, "What mortars are required in the airmobile division?" and "Is there a requirement for a battalion mortar within the 82d Airborne and, if so, what should it be?" So far we have argued for a force structure that consists of heavy mortars and lightweight company mortars. We must decide either that the 101st and the 82d can operate effectively with the heavy mortar at the battalion level or that there is a requirement for a third mortar system. Given the procurement of the HMMWV and the UH-60, we believe that both divisions should use the heavy mortar at the battalion level. The mortar organization should be standard at six tubes per platoon. The final issue is, "Should there be company mortars within the 101st Airmobile Division?" Training and logistics problems associated with company mortars argue against the inclusion of mortars at company level. Tactics for the airmobile division, and the probability of units operating away from their battalion and possibly outside direct support artillery range, argue for the inclusion of lightweight company mortars in airmobile companies. We believe these mortars should be included at the company level, and the division given the option of using them or leaving them in the Conex.

9. The mortar solution we recommend in the long term for the Army, therefore, consists of two types of mortars and two types of organizations—a six-gun heavy mortar platoon at each maneuver/ reconnaissance battalion and a three-tube lightweight mortar section with the 82d and the 101st rifle companies. We see no reason to change the assignment of two lightweight mortars per ranger company. This solution will provide the Army approximately 4,200 manpower savings and substantial procurement savings by eliminating one mortar system. It will simplify the training tasks and provide adequate close-in responsive mortar support to maneuver units within the modernized divisions that we expect to field from the mid-1980s on.

Combined Arms Operations with the M1 Tank

Message to Multiple Addressees
26 May 1981

The differential in operational capability afforded by M1 is so much greater than anything we've experienced before that we must make a coordinated and determined effort to fold it into our way of doing business as quickly and as thoroughly as possible. If we do not, we will never realize the potential of the machine.

Systems Mobility and Agility

Letter to Lieutenant General Sir Robin Carnegie
5 February 1982

With regard to mobility, I would caution it is still true that systems mobility/agility do not appear to buy much survivability per se—some will say to the contrary, but we've a large body of evidence that says otherwise. Therefore the effort to improve systems mobility/agility will always be encumbered by the need to provide protection on today's most lethal battlefield. One apparently cannot have significant degrees of both at the systems level at the same time. Force mobility, the need to have greater mobility for its effect on the battle as opposed to its effects on system survivability, is quite another matter. For I am convinced that one cannot win, especially outnumbered, without attacking. But to attack on today's battlefield requires craft and cunning—a concept of offense that we've perhaps but only begun to understand. For example, if I had the choice in the face of limited budgets, I'd buy only the advanced attack helicopter, the M1 tank, and some fast SP artillery.

In support of that combination I'd buy some good lift helicopters like our Blackhawk; their purpose—to move the infantry around. If infantry need protection in moving around, I'd use whatever armored personnel carriers are left in the fleet—fix them up and make 'em last a little longer. But only as APCs. Had I some more money, I'd look seriously at a heavy infantry fighting vehicle, one with the same protection levels as the tank. Perhaps the Israelis are right in providing a tank big enough for their infantry to ride around in if necessary. I would not, however, put the infantry in a machine like our infantry fighting vehicle; it has too many disparate weapons systems aboard and it cannot survive at nearly the same level as the tanks and so can't go where the tanks go. Besides, it is not nearly so mobile as the tank—even though it has about the same speed, it just can't go along with the tanks. It just didn't turn out the way we intended for it to!

So be careful with your cavalry, for in the name of mobility you may make them so light that they are not survivable, and we can't afford that. And, in the other part of the force, if the infantry can't go along dismounted or in light-skinned vehicles with the tanks on today's lethal battlefield, then we must find some other mode for them. Obviously there is terrain on which they will be essential, but there they will also be less vulnerable if properly deployed.

If my perception of mobility/agility at systems versus force level is correct, then you might be able to build a mobile enough force with the right survivability, given the correct mode of employment, and that concept could include heliborne or carrierborne infantry.

Combat Support and Combat Service Support

Message to the Joint Chiefs of Staff
21 June 1982

1. The proposed allocation system outlined in Ref A [JCS/J5 MSG SAB, 041928Z May 82] will not resolve the overall shortage of combat service and combat service support (CS/CSS) units, but will at least restore some order to the planning process. Until the ratio between combat forces and CS/CSS units is improved, our overall combat capability will remain limited. Moreover, the impact of insufficient CS/CSS units is far-reaching.

2. As a first step, we must identify the real need, both in numbers and types, of CS/CSS units by thoroughly scrubbing our plans. Once this is done, we will be better able to articulate the true dimensions of our CS/CSS shortfalls. We then should be able to apportion those CS/CSS assets now in the force in a manner that makes sense when considering likely contingencies.

3. As apportionment is considered, planners must keep in mind the vast differences in support required between theaters. For example, SWA may require, in proportion to fighting forces, five times the number of pipeline companies that a European scenario would require.

4. We must also recognize that some CS/CSS units are very scarce, such as engineer port construction units and terminal service units, yet absolutely essential to some operations. Consequently, planners must be provided priorities so that the right mix of CS/CSS forces, particularly the scarce ones, will be apportioned not only in consonance with the numbers of supported combat units but also in terms of the special requirements for a particular theater.

5. Once we have highlighted the shortage of CS/CSS assets, requesting additional support forces in the PPBS process will be easier to justify. We believe the direction should be toward restoring a better balance between combat and CS/CSS units.

6. It is absolutely essential that we have the capability to support any forces we commit to a given contingency. If we do not or cannot, then we must say so to the civilian leadership.

6. Command and Control

Page

Operations and Training—Command, Control and Management 252

Combat Information 255

Command and Control: An Overview 256

The Role of Knowledge-Based Systems in Command and Control 258

Commander's Notes...

no. 2

3 May 1976

OPERATIONS AND TRAINING -- COMMAND, CONTROL AND MANAGEMENT

This Commander's Note sets forth the Corps Commander's concept of command, control and management of operations and training.

The responsibility and authority for command, control, and management of operations and training will be decentralized to the lowest level of command which has, or to which can be made available, the requisite resources to plan for, organize forces and resources for accomplishment of the operational mission or training objective.

The idea of decentralization is but one part of a broader concept; that of the mission-type order. Everyone knows that a mission-type order need have only three fundamental ingredients:

-What is to be done; what the commander issuing the order expects to have accomplished.

-What is not to be done; what controls, limits, or constraints the commander issuing the order considers necessary to impose on what his subordinates do, primarily for the purpose of coordinating their actions.

-What the task is to be done with; what resources the commander issuing the order assigns to subordinates for task accomplishment, and what support subordinates can expect from outside their assigned resources.

Note the absence of the word how. The fundamental difference between what and how is the key to understanding the concept of the mission-type order and idea of decentralization. At each level there is a built-in tendency to prescribe more and more of how the job should be done; to embellish the order as it works its way down to company, troop and battery. All too often a broad mission-type order issued at a higher echelon becomes a detailed "how to do it" by the time it reaches the bottom.

COMMANDER'S NOTES NUMBER 2
Operations and Training -- Command, Control and Management 3 May 1976

Interestingly enough, decentralization, although integral to the mission-type order concept, has historically been plagued in execution by considerable ambiguity in statements of what is to be done, not done, and done with. And so one frequently finds a greater or lesser degree of abrogation of responsibility by commanders who, in the name of decentralization, dump the whole load on the company, battery or troop commander. Many times the same commander who tends to overspecify in his orders can be found bending over backwards to avoid specifying anything in his training directives, because training is decentralized.

Decentralization is delegation of responsibility and authority for executing a mission to the lowest level of command which has, or to which can be made available, the requisite resources to accomplish the mission.

If the mission is training in order to maintain a satisfactory state of readiness, then decentralization means delegation of responsibility and authority for planning, programming and executing training to the lowest level of command having the necessary personnel, equipment, fuel, ammunition, spare parts, and planning, programming and executing expertise to accomplish the training mission.

Planning involves devising ways to achieve specific goals; programming involves rationalizing goal-achieving plans to fit within resource constraints; both describe what is to be done. Executing involves devising how to do what is programmed.

-If a readiness goal is a satisfactory level of tank gunnery proficiency among crews and platoons in a unit;

-Then a plan might prescribe quarterly crew and platoon qualification course firing in order to achieve the goal;

-But, a program might provide for semi-annual crew and platoon firing based on range, ammunition, fuel and other constraints;

-And execution would involve training crews and platoons in an orderly progression so as to enable them to qualify semi-annually.

Generally, battalions and squadrons are the lowest level at which a commander can fully plan and program; both have the requisite staffs, both are full tactical, administrative and logistical entities. The company, battery or troop cannot plan or program in the sense that these terms are used here; they have neither the staff, the resource management

2

COMMANDER'S NOTES NUMBER 2
Operations and Training -- Command, Control and Management 3 May 1976

capability, nor the expertise. The company executes programs. If decentralized training works properly, the company, battery or troop commander need only figure out how he is going to conduct training; what he is to do must be fairly clearly spelled out for him. If decentralized training works properly, there is no need for a training staff at company level. The company commander attends a weekly programming meeting with the battalion S3, at which time S3 and company commanders iron out coordinating details for two or three weeks hence, and make last minute changes based on what the company commanders feel need be added to or deleted from their already laid out training programs. The battalion, not the company, publishes training schedules, schedules describing what is to be done, schedules precise enough about what is to be done that all that remains is for the company commander and his platoon leaders to figure out how they are going to do it.

When the time comes to test platoons and companies, these tests are designed and administered by the battalion. Many will argue that conduct of platoon tests by the battalion commander is not decentralization. On the contrary, this is the essence of decentralization. The battalion commander decides what is to be tested, based on what training he has required in his training program. Then, since only he has the requisite resources to conduct a comprehensive test, and since good testing requires standardization, he provides resources, specifies what is to be done, and scores how it is accomplished. This is his measure of how well the company and platoon commanders have done their part of the job -- the how of training. Properly applied, decentralization breeds better leaders, but at the time requires better leadership on the part of those responsible for their upbringing. It requires of the battalion commander and his staff the utmost in professionalism, planning and programming expertise, good management of limited resources, and a complete willingness to accept mistakes, set them aright and proceed. Of the company, battery and troop commander, decentralized training demands the utmost skill in the details of how to train men. Empathy, perception, initiative, imagination and creativity are his especial purview, for it is through the process of development and application of these qualities that the fine junior leaders of today become the splendid battalion, brigade, division and corps commanders of a few years hence.

DONN A. STARRY
Lieutenant General, USA
Commanding

DISTRIBUTION: "B"

Combat Information

Letter to General Bruce C. Clarke
Palmyra, Virginia
4 October 1976

Our REFORGER experience this year convinced me that my preconceived notions were correct—command and control is our biggest problem. From battalion to corps we are less capable than we were ten or fifteen years ago. This is so because we have added a great deal of sophisticated equipment which has slowed down the flow of information, and we have let our headquarters get so big that they are constipated internally. FM nets, for example, are now equipped with secure gear that has some mechanical reliability problems and cuts the range of the radios. In AM nets the RTT equipment still doesn't work very well, and at battalion the Spanner Study several years ago removed one of the AM sets. From brigade up to corps the AM RTT nets are clogged with too many messages that are not too important, and all reporting is done sequentially. This means that the whole command is at the mercy of those nice young men and women who fiddle around with the messages in the message centers. It may be that the generals should see the message right away, but if SP4 Whatshisorhername doesn't recognize that, it never gets to the general. Not a new problem, but aggravated by the increasing volume of traffic and the size of headquarters.

VHF equipment is a way around, but it is not flexible and can't be moved quickly. While some amount of internal reorganization, and eventually central information processing, might solve some of this, all that is a long way and a lot of bucks off. Meanwhile I've got a problem—how to command V Corps while all the wise men are devising all this sophisticated stuff to do it better.

There's no question that both divisions and corps need TAC CPs. But if the TAC CP simply depends on the main for information, or even if the TAC were to monitor certain information going into the main, the TAC would be so far behind the battle that it would serve no purpose. What the TAC needs is combat information—before it gets dumped into the fusion center and merges indistinctly with all the other information flowing there. There are several historical examples which you will recall—Twelfth Army Group's Phantom Service, Patton's Army Information Service, and Rommel's monitoring the nets of his lead panzer regiments. All were systems designed to bring the commander immediate combat information—information he needs to make timely decisions, decisions about resource allocations, movement of forces, targeting priorities in the second echelon, and so on. Unless the commander has that he can't run the battle. He won't get it from the main tactical operations center under the current system. So what we are doing is identifying what we think the division commanders and I need to know to run the battle. Then we will devise some way of getting that information direct from the units which report it—in many cases battalions. This will eliminate the time delays that come from sequential reporting and volume in the main information flow.

This is the only way I can see to do it. Until all these studies address themselves to that specific problem, they are probably not worth the time and effort, and certainly not the money, we are putting into them.

Command and Control: An Overview

Military Review
November 1981

In the March 1981 issue of *Military Review,* in an article entitled "Extending the Battlefield," the operational concept for what is now called the AirLand Battle was set forth. No element of that concept is more essential to the development of a credible war-fighting capability than command and control. Command and control has ever been an essential element in battle. Also, it has always been a very, very difficult problem to master. It is perhaps more so today than ever before because of recent technical developments in the field of electronic warfare.

The command and control problem goes something like this: In order to fight the battle successfully, the commander has to find out what is going on, decide what to do about it, tell somebody what to do, and then keep track of how the battle is going. He needs to turn that information-decision cycle in time inside the enemy's information-decision cycle so that, instead of simply reacting to what the enemy does, he can seize the initiative. In order to win the battle, holding the initiative is essential. It is even more important that we recognize the need to retain the initiative and, equally important, that we acknowledge detractors in our command and control system which tend to prevent this from being done.

Actually, command and control problems come at us from two directions—one is external and the other internal. External to today's communication systems, there is so much information from so many sources that we are hard put to know how to assimilate all of it.

Two factors aggravate the problem of making useful information of all that data. First, the priority system internal to the message center system is not designed to determine what is truly important. Second, in our operational reporting systems, everything is reported sequentially—platoons report to companies, companies to battalions, battalions to brigades, brigades to divisions, and so on. Every time information passes through one of those sequential reporting nodes, two things happen: it gets slowed up and it gets mixed up.

Historically, the successful commanders—the Pattons, the Montgomerys, and others—have taken extraordinary means to bypass those sequential nodes in order to get at the essential information in a timely fashion. In the Third Army's operations in Europe in 1944–1945, General George S. Patton used a mechanized cavalry group whose forward elements reported directly to Army headquarters from the Army's leading battalion task forces. These elements reported where they were, what they were doing and how the fight was going—just a few essential elements of information. The history of Patton's operations abounds with stories of his terrorizing division and corps commanders because they did not know as much about where their battalions were as he knew through the auspices of his Army Information Service.

General Bernard L. Montgomery's contribution to command and control lore was called the Phantom Service. Somewhat different in organization, it nevertheless provided him the same capability that the Army Information Service provided General Patton.

Internal to our command and control systems, we have equally tough problems. One has been called the self-constipating staff system. The G3/S3 or G2/S2 vans get separated, their occupants do not talk to one another, no one works between them, and the twain never

---

Reprinted with permission from *Military Review*.

meet. There is a further problem sometimes called the tyranny of the message center. Here, administrative procedures regarding generating and processing messages and message center or communications center procedures themselves take over, becoming more important than the message traffic itself. In this system, traffic critical to the command and control of operations can languish literally for days while waiting to be folded, spindled, or mutilated by the message or communications center clerks.

These problems are of long standing; they are also solvable. The fundamental problem is that, all too often, we turn the problem over to the communicator. This stalwart promptly assumes that, because all that information is there, it is his job to bring it to the commander. Not so! In fact, that is exactly backwards. The first order of business is for the commander to set out the elements of information he needs to run the battle. Then he tasks the communications system to bring him that information.

Some years ago, we designed such a system in a corps-size force. The Battlefield Information Reporting System specified what the commander needed to run the battle. Then the communications system was tasked to bring that information to the commander. The system was extremely responsive. It provided more than sufficient information to wage the corps battle successfully. Manual in operation, it was designed so that it could be automated. It was also designed to revert to manual operation when the computers went down.

So, the message is this: command and control is manageable, but it takes a lot of thought and even more hard work, not by communicators alone, but by commanders. If the commander does not take charge in a positive way, all the obstructers outlined above will combine in some way to prevent or inhibit effective command and control.

. . . Without effective command and control, we can never hope to fight successfully and win the AirLand Battle.

The Role of Knowledge-Based Systems in Command and Control

Armed Forces Communications and Electronics Association
Kansas City, Missouri
18 October 1984

Under contract to the Defense Department, a company that shall remain nameless designed a system to assist in military decisionmaking using the latest in knowledge-based or expert systems technology. It was named the Military Expert Strategy System, acronym MESS.

The general was receiving his first demonstration of the product, which had suffered considerable delay and several cost overruns. The system was designed to combine knowledge of the current situation, some analogous military history, geographic, time, distance, deployment, and other relevant data. And, according to the rules and protocols for conducting a commander's estimate of the situation, it was to produce a complete estimate, to include appropriate alternative courses of action.

Seated at his shiny new keyboard, the general tapped in the query, "If the Soviets launch a first-strike missile attack on the United States, will it consist of only land-based ICBMs, or will it include sea-based missile systems as well?" The computer whirred and clicked, burped twice, and up on the screen came the response, "Yes." "Yes, what?" typed the agitated general. "Yes, sir, General!" was the response.

In the beginning may I say that it appears quite likely that knowledge-based systems have a potentially very useful role to play in their general application to command-control. However, may I also express some cautions about artificial intelligence and expert systems as decision aids in the C3I area. My cautions are based on several observations.

First, there is good news. There is an impressive menu of expert systems products being worked on in R&D centers in both industry and the military. The whole idea of artificial intelligence, of which expert systems—knowledge-based systems—is part, seems to be an idea whose time has come. Expert systems of all kinds are beginning to provide a variety of tools for solving a host of specific and rather narrowly focused problems.

There is also some less-than-good news. There is a lot of "hobby shopping." There appcar to be some fairly well-exaggerated claims for what is being done and what we can expect to be done. There is an urgent and growing need for a clear-cut statement of user requirements in order to focus the considerable time, talent, and monies being spent in this field on the most important problems.

What to do about that? Before suggesting some answers, let me provide some definitions so that we're all clear about what I think I'm talking about.

An expert or knowledge-based system is one that combines knowledge and reasoning techniques to solve problems normally requiring the abilities of human experts. It is generally considered a subelement of the general field of artificial intelligence. Artificial intelligence is a discipline devoted to developing and applying computational approaches to intelligent behavior. Intelligence is the ability to respond to any situation in a controlled manner.

Now about user requirements. As with many, indeed most, systems, it may be difficult to find the real user. Logisticians and others are seeking ways to speed up the laborious and time-consuming logistics process, from supply to maintenance to transportation. Information engineers are seeking ways to process the flood of data that has reached deluge proportions in today's information systems. And we have some few folks trying to figure out how to get the commander the vital information he needs to conduct the AirLand Battle.

May I just suggest a few fundamental notions that might form the basis for work proceeding toward development of a relevant set of user requirements. It seems to me that we are, in truth, talking about two related but quite different things when we speak of command-control. First is the command of forces; second is the control of functions. With regard to the former, there appear to me to be serious if not insurmountable obstacles to the application of expert systems in all but a very few aspects of the art of the command of forces.

With regard to the latter—control of functions—there appear to me to be unlimited applications of expert systems capabilities to the solution of most, if not virtually all, of the labor-intensive accountant-type activities so typical of this population of problems, for they are indeed problems to which computational approaches offer promise of fairly straightforward solution.

Why do I believe we have a particular problem with command of forces? Why is it not like the control of functions? Command of forces in battle is an art form. It requires a disciplined mind that has been trained in the complex art of thinking logically about a specific category of very tough problems. The human mind applies intelligence to problem solving. The essence of intelligence is the ability to find ways to solve otherwise intractable problems by limiting the search for solutions. People, in solving problems, use extensive background knowledge about the world's regularities to constrain the solution search. Intelligent behavior, therefore, involves application of a whole lot of background information that, in humans, is taken for granted but, in computers, must be dumped in by the bucketfuls. The human intellect can invoke mathematical theorems, apply rules of thumb, reason by analogy, apply instinct, use intuition, invoke a sixth sense, bet on the outcome, and so on—all of which requires an enormous pool of background knowledge and experience that is almost instinctively used to limit the search for solutions. What is missing, but required, to apply expert systems to the command of forces is a coherent model of what takes place, more often than not, inside living systems as they go about solving problems. So what is called for is considerable additional research in the behavior of living systems—cells, organisms, organs, organizations, at all levels, in order to develop expert systems that behave, in the main, pretty much like the living systems they are designed to emulate.

The history of battle instructs that successful commanders were those who invoked extraordinary means to bring to themselves a relatively limited set of information about the battlefield and who made their decisions based on that information. Read the history of General Marshall's use of pigeons in World War I and of the German, British, and US systems developed by the successful commanders—Guderian, Rommel, Patton, Montgomery, Clark, and others—in World War II to understand what I'm referring to. Many of you have heard me tell this story before, so I'll not belabor the point with you tonight. I simply refer you once again to the lessons of history and to the pioneer work of Dr. James Grier Miller on the "general theory of living systems" and on the applicatory research work done by TRADOC in US Army battalions during the 1977–80 period.

With regard to control of functions, it seems to me the user should set his sights on those problems whose solution can save the most in time, manpower, and other scarce resources, and that do not lend themselves to simple-minded accounting procedures for solution. For example, we are producing an expert system designed to recycle the space shuttle from landing to launch, adjusting the hundreds of thousands of things that must be accomplished as changes to one or more of those activities occur virtually by the hour. This is obviously not a simple accounting problem. This requires a substantial-sized inference engine and a lot of expert input to its construction. Most control of functions problems are amenable to this kind of solution.

The general requirements for command control of the AirLand Battle should form the basis, then, for whatever expert systems development we believe might be necessary and appropriate.

First, commanders at appropriate echelons must identify what information they really need.

Second, the communications architecture and its imbedded protocols must provide for delivery of that information in a disciplined reporting system that will, in turn, provide a disciplined information flow.

Third, interneted databases must be developed and used extensively, especially in the control of functions area, in order to provide the basis for widespread dissemination of large volumes of routine information, eliminating redundant and duplicative reporting requirements.

Fourth, multilevel security gateways must be provided in switching systems in order to integrate security considerations with normal information flow processing.

Fifth, multiple frequency switching procedures must be applied to both low- and high-speed transmissions in order to make enemy jamming and direction finding difficult, if not impossible.

Sixth, the special needs of commanders for command of forces type information, especially information relative to the deep battle, must be recognized and provided for, even at the expense of single-purpose architecture and special systems.

Finally, expert systems should be designed and imbedded first in the control of functions activities and processes and, as possible, to assist the commander in his command of forces function, which has been, is, and always will be an art form rather than a straightforward matter to which one can apply computational power and expect the kind of solution that reflects the best thinking of a disciplined military mind trained and educated over a lifetime to think logically about very, very tough problems.

7. Doctrine

Page

La Guerre Révolutionnaire 264
Doctrine and Materiel 271
Rationale for Changes in Doctrine 272
Defense Doctrine 275
Doctrinal Literature 278
Effecting Change 279
Israeli Relationships 280
New FM 100-5 281
German-American Coordination 282
Battle Book 283
FM 100-5 Defense Philosophy 284
The Corps Battle 285
Helping the Republic of Korea Army 295
Tank Allocation Basis 296
Corps and Division Operations Manuals 297
Doctrine and Studies 298
Active Defense (Benedict) 300
Battlefield Reserves 303
Instruction on Offensive Operations 304
FM 100-5: *Operations* 305
Doctrine Writing 311
The Central Battle 312
Me, the Central Battle, and HQ TRADOC 318
Doctrinal Literature 323
Active Defense (Faith) 324
The Central Battle Again 329
Doctrine Rules 334

	Page
US and Federal Republic of Germany Doctrine (Thurman)	335
Division 86 Operational Concept	336
Division 86	337
Operational Concepts and Doctrine	338
US and Federal Republic of Germany Doctrine (Burns)	340
Battlefield Terminology	342
FM 100-5, *Operations*	343
Models and Methodology	345
Attack of Follow-On Echelons	346
Battle Models	347
The Writing of Doctrine	351
FM 100-5	352
AirLand Battle Operational Concepts	353
Doctrinal Literature Program	356
Allied Cooperation	357
Army 86 Implementation	358
Air/Land Battle 90	359
Ground Forces Issues	362
FM 100-1	364
AirLand Battle I	365
Allied Forces Staff Talks	366
Implementation of AirLand Battle	367
Division 86 Transition Planning	369
Implementing AirLand Battle Concept	370
Active Defense (Haig)	371
AirLand Battle II	372
Joint Operations	382
REDCOM Mission	385
Fighting Outnumbered to Win	388
Heavy Division 86 (Otis)	389

Page

Heavy Division 86 (Meyer) 390
Undcrstanding AirLand Battle 391
Joint Doctrine 392
Third Wave Concepts 393
To Change an Army 394
SACEUR and AirLand Battle 400
Evolution of US Army Operational Doctrine 402
AirLand Battle III 412
Joint Operations 424
American Military Thought: A Perspective 431
The American Way of War 438
AirLand Battle Determinants 448
Welcome to the Counterinsurgency Century 451

La Guerre Révolutionnaire

Military Review
February 1967

On 3 July 1962 France turned over control of Algeria to the Democratic People's Republic of Algeria. The French Empire had withered as a result of spreading anti-colonialism and the mortal debilitation of French strength in the first great war of this century.

At the end of World War II, the French were the first to meet the expanding Communist revolution in a form which has received wide recognition in the rest of the world only recently. For, when most of the world was focused on traditional ideas of general war and its doctrine of massive retaliation, the French were in Indochina struggling with Ho Chi Minh's inheritance from Mao Tse Tung—the war of "national liberation," the Communist revolution.

In the Indochinese War the French military forces, in tragic defeat, felt the loss of their traditional expertise and their professional ability. Following defeat at Dien Bien Phu, they set about to restore their ability to conduct this new form of war. For two years at least, possibly more, military intellectuals struggled with theory and doctrine and evolved a theory of revolutionary war and how to deal with it—*la guerre révolutionnaire* and *action psychologique*.

Scarcely had they come to a satisfactory theory of *la guerre révolutionnaire* and a doctrine for combating it than they were obliged to put it to the test in Algeria. New doctrines and new organization yielded results. As had been the case in Indochina, however, military and political voices were not in harmony. Although the military forces abetted the fall of the Third Republic and the rise of the Gaullist Fifth Republic, their political hopeful never quite communicated to his military leaders. As a result, the professional officer corps revolted. Their revolt a defeat, their generals in disgrace and condemned to death, their secret army in Algeria a failure, the army came home to be reconstituted in a new image.

It is significant that, throughout the empires, republics, restorations, and governments that rose and fell, the army remained loyal to France. The French Army avoided politics and was generally contemptuous of the government, but quite unwilling to take positive action against it. *La guerre révolutionnaire*, however, as they came to understand it, challenged the very core of the profession.

Contributing Factors

In the attempt to restore expertness to the profession in the face of this new challenge of war, the army met its demise. There were many contributing factors:

- General development of the parameters of the Cold War.
- The worldwide fate of colonial empires.
- A political kaleidoscope of changing governments in France.
- Traditional aloofness of the French military services from the rest of the populace.
- Continuing deaf ears in Paris to the voices of military counsel.
- French military defeat in the war in Indochina, and conquest without victory in Algeria.
- Prideful struggle of the French military leaders to restore professional know-how.

Reprinted with permission from *Military Review*.

The war in Indochina had been lost, but the army came out with some hard-won lessons. They conceived that revolutionary wars are fought for the allegiance of people, and that the point of convergence of the effort is largely social and psychological, not military. They saw the need and application of conflicting practices: education and social programs on one hand, propaganda, psychological programs, involuntary population resettlement, and political indoctrination on the other.

Two significant lessons struck French military intellectuals ruminating on their experiences with the Communist revolution in Indochina. The principal lesson they drew was the way that an inferior force could always defeat a modern army—if it had the support of the local populace. They were also impressed by the seeming cohesion of politico-military command action demonstrated by the Viet Minh. They concluded that proper psychological measures could form the cohesive bond between the civilian supporters and fighters that this war required. A kind of formula emerged: revolutionary war = partisan war + psychological war.

Five Stages

Commandant Jacques Hogard, vocal among the military intellectuals, analyzed *la guerre révolutionnaire* in five stages:

- Propagandists and agitators work secretly among the people, sounding out and appealing to resentments against prevailing social and economic conditions.
- Having gained a foothold, agitators and organizers form sympathizers into age, occupation, and other groups. Gradually these groups spread a net of resistance to the regime over the country.
- Some of the groups are organized into armed bands for ambush, raids, sabotage, terror, and assassination.
- The level of terrorism rises, forcing government elements to withdraw from some areas which are promptly seized by the agitators and declared liberated zones.
- A general psychological and political campaign against the government, and a military campaign against its armed forces, is gradually developed.

From this analysis the intellectuals then sought a counterrevolutionary doctrine aimed at the weaknesses of *la guerre révolutionnaire*, which they saw as:

- The superiority of Western ethical value codes should bring liberalism to the top, providing it could demonstrate equal aptitude in the application of psychological techniques.
- Subversion, especially in the early stages, requires a great deal of time to organize. Security forces, properly warned, can intervene during this time.
- The insurgents, especially at the outset, seldom have an adequate logistics base and are, therefore, vulnerable to a starve-out campaign, withering them on the vine.
- The insurgents can draw out the battle indefinitely, but can almost never produce a decisive military blow unless given an unusual opportunity.
- Population control is based on a clandestine politico-military network of cells, groups, and sympathizers covering the country. If this organization can be broken up, the insurgency collapses.

The action program to support the doctrine developed plans to isolate the insurgents from outside sources of supply and help; destroy insurgent regular forces and large guerrilla bands; protect essential communications and economic centers, but not tie down regular military forces to this task; accomplish community resettlement of the population as required to deprive the insurgents of support and a culture in which to germinate; and to reeducate insurgents captured or who surrendered.

The theory then developed a program of action and successful reaction as shown in the chart.

Insurgent	Incumbent
1. Forms agitation and propaganda cells.	Police and intelligence activity increase; strengthens administrative and military controls; propaganda, social, and economic reforms.
2. Expands and coordinates resistance into effective extensive network.	
3. Intensifies propaganda; terror, sabotage, assassination, and other armed action.	Mobilizes state's pollitical and physical resources; attacks the subversive infrastructure; organizes self-defense units; resettles population; regains control over liberated areas; isolates and destroys enemy regulars.
4. Creates liberated areas; installs provisional government; gradual change from irregular to regular war.	
5. Regular war with fully developed diplomatic activity.	Destroys regular forces; continues diplomatic psychological, social, and political action.
4. Regularization fails; reverts to guerrilla war.	
3. Reduced liberated zones; small-scale guerrilla war.	Reduces auxiliary military effort; continues reforms.
2. Opposition and some sabotage continue.	
1. Continues propaganda and agitation.	Continues reforms and reeducation; new order emerges.

Against *la guerre révolutionnaire* the French sought to bring the most touted Communist weapons of psychological action and political indoctrination in the context of a general war common to Marx-Lenin-Mao. To complement their action program, they developed *action psychologique*—a theory of psychopolitical action.

In 1957 the French Army took action to reorganize the staff and command effort, implementing the new concepts of countering *la guerre révolutionnaire.* To their four-part general staff-type organization they added a fifth section, G5, in the army, field army, corps, and division staffs. At the top level there was a third deputy chief of staff specialized in psychological and political operations. Finally, in the French military schools, there were introduced courses of instruction in psychological war and political operations.

At the height of its organizational experience, the G5's responsibilities included psychological action, psychological warfare, morale—civilian and military, troop and public information, chaplain activities, and civil affairs. G5 coordinated with G2 on intelligence bearing on psychological or civil affairs matters, and with G3 on planning matters. With this organization and theory, the French Army went to Algeria.

Source of Revolt

The source of the Algerian revolt was Marxist, chiefly because its theories of revolutionary war were derived from Karl Marx, [Vladimir] Lenin, and Mao Tse-tung. The leaders of the Algerian revolt were quick to realize that the Communist theory of wars of "national liberation" was the context in which many changes were in progress around the world. This helped them describe their view of the nature of their struggle—against imperialism, total and anticolonial.

On the other hand, the political attitudes of the Algerian rebels, their liking for strong centralism in their social-democratic state—almost in the image of 1789—and their preoccupation with the liberty metaphysic represented Jacobin nationalism at its best.

Finally, the rebels apparently felt they could outlast the French, as Ho Chi Minh had done, taking skillful advantage of each opportunity while building a subterranean nation.

Similarities

Analysts of *la guerre révolutionnaire* held the view that the revolution played into the hands of international communism; that it was plotted, channeled, and controlled to a remarkable degree. Much of the French Army eventually believed this. Actually, the Algerian War was much more spontaneous, far more confused, and less cohesively directed than the Indochinese War had been. But, to the French, Algeria became the model into which Indochina, Greece, Malaya, and the Philippines fit with precision. Some of these similarities were:

- Algeria was a war for the control of civilian support and loyalty.
- Algeria was a struggle of psychological and propaganda mechanisms with both sides using a wide variety of psychological and political action techniques.
- There was in Algeria a total allocation of resources in a limited area. It was general war, but general within the borders of a single country—a pattern to be repeated in later wars of "national liberation."
- Ideology became a factor of contest, openly on the Algerian side, more surreptitiously on the French side.

Eventually almost three-fifths of France's active military establishment was committed to Algeria, augmented there by the *gendarmerie*, several civil-defense groups, and Moslem auxiliaries. Their task was to fight the guerrillas but, above all, to administer, control, and convert. This preponderance of French forces and auxiliaries enabled military pacification to proceed until rebel units were reduced and isolated. When this occurred, resistance levels fell back to terrorism and small-scale skirmishes.

The *Sections Administratives Spécialistes* (SAS), a group of reserve officers with a knowledge of the Algerian situation and fluency in Arabic, was called to duty and became the principal administrative unit in the struggle to regain administrative control. The army coordinated its activities with SAS at all levels through its G5, making almost complete immersion in political and psychological action inevitable.

The psychological role of the military services—said to be second only to its social role—appeared in ministerial directives and was the basis for the "war for the crowd" and "war in the crowd" theme in French military writing. In Algeria, the war against *la guerre révolutionnaire* was divided into two areas, rural and urban, each with a different line of attack for the counterrevolutionary forces. The political and psychological effort expended by the French military forces was of great magnitude. The French Army newspaper, *Le Bled*, for example, at one time had an Algerian circulation estimated at 350,000 copies, more than most newspapers in France.

The first great test of *la guerre révolutionnaire* and *action psychologique* came in the Algiers revolts of 1957, fought between the troops of General Jacques Massu and the local element of the National Liberation Front. The rebels had created a liberated zone of Algiers centered on the city's Arab quarter. Into this situation came the 10th Parachute Division of General Massu, fresh from the frustration of Suez. In a repressive year, they overcame the rebels in the city. To the proponents of *la guerre révolutionnaire* and *action psychologique*, this was vindication of their methods and theories and proved the validity of their doctrines and action programs. They seized the circumstances as justification for their years of research and intellectual soul searching.

By 1958, when the army played a hand in the coup that brought down the Republic, the army was deeply involved politically and had developed expert knowledge in political and psychological warfare.

French Understanding

In Charles de Gaulle the French Army felt it had someone who understood it better than all others. The army, using all the resources of *action psychologique* in Algeria, secured better than 60 percent support for the referendum adopting the Fifth Republic in September 1968.

However General de Gaulle soon disavowed the concepts of war in Algeria held by the bulk of the army there. One of the paramount factors that made the army's position so strong was the widespread French feeling for Algeria as part of Metropolitan France. To resident Frenchmen, as to the army's leaders, Algeria was France. The army had demonstrated their know-how in successful containment of the revolution in Algeria.

Then, on 16 September 1959, General de Gaulle made the pronouncement that self-determination was the only method worthy for France to follow. Reaction to this pronouncement by the Algerian French was revolt in Algiers—the week of the barricades in January 1960—which failed. The trials of the participants clearly revealed the deep involvement of the army in the political, psychological, and social fiber of Algeria.

In the aftermath, the G5 system was disbanded in the French Army— *action psychologique* had run its course. General de Gaulle, after touring in Algeria, and sensing the deep involvement of his army in political affairs, had directed a gradual reduction of political activity by the military services.

As 1961 arrived, there was a growing sentiment in the French military forces that Algeria had become another revolutionary struggle lost through civilian control of military affairs. Then came the coup in Algeria on 22 April 1961, followed by the trial of its leaders.

Notes on a Theory

Some writers contend that, had there been no Indochina and no Algeria, *la guerre révolutionnaire-action psychologique* would never have come into being. Its previous colonial war experience around the world, and the bitter lessons the French Army learned in Indochina, created a remarkable background for the military profession to work out its destiny. This school also points to other revolutionary wars of the period—Greece, the Philippines, Malaya—with the observation that no such turbulence came out of them.

On the other hand, it should be remembered that most of the world, the United States in particular, had only recently become aware of the nature of wars of liberation. Only since Chairman Nikita Khrushchev's speech in January 1961, and President John F. Kennedy's reaction to his Vienna meeting with Chairman Khrushchev, had serious attention been given this problem at top levels of government in the United States. With Marshal Lin Piao's confirmation of the methodology of Indochina as the preferred Chinese method, the West is faced with developing some type of doctrine to counter the Communist drive to power.

Further complicating the problem of countering *la guerre révolutionnaire* is the fact that in the contemporary world most such wars will be conducted inside the border of another sovereign state. The French, at least, were in the position of a colonial power with initial access to the area and some degree of administrative control.

The problem we shall face increasingly is how to project aid to those governments faced with insurgency without seeming to interfere in the internal affairs of other states. So the problem of doctrine development is, perhaps, more acute for the West today than it was when the French began their period of meditation after Indochina. Truly the US military profession may be said to be in its period of meditation—searching for theory and doctrine.

It would seem that, to be truly useful, especially in a democracy, a doctrine such as *action psychologique* must be politically viable, part of a balanced national strategy. *Action psychologique* was politically viable from the standpoint that it recognized that one sure way to counter a totalitarian regime was to impose one similarly constructed to counter it. Having done this, who is to say whether in so doing the effect has not been to destroy that which one sought to save in the first place.

Goals and Objectives

Initially *action psychologique* was politically viable in Algeria because, given the world temperament and trends of the time, it was probably the only method that could have succeeded in countering the revolution and saving Algeria for France. But the minute France recognized self-determination for Algeria as the only appropriate goal, the doctrine fell apart.

Doctrine can be politically viable only so long as it is cast in pursuit of objectives of the state on which there is at least common understanding, if not consensus. Changing objectives, especially during an operation, is a hazardous undertaking, as General Douglas MacArthur and others have testified. In a military establishment with as elaborately constructed and dogmatic a doctrine as *la guerre révolutionnaire* and *action psychologique,* it is likely to bring down the whole structure.

Goals and objectives, clearly stated, are difficult to come by in a liberal democracy. It is just not in the democratic way of doing business for a democracy to establish clear, long-range goals and proceed either directly or indirectly from here to there. Democracy's soldier is continually placed in the position of having to divine the intentions of his leaders. He must do this to establish force structure and get on with management of budgets, purchase of hardware, and deployment of troops.

How he is supposed to do it is obviously a subject that needs some study. We have just looked at the case history of an entire military establishment that misunderstood the goals of its government, and of a government that imposed on its military force a national goal inimical to the survival of an elaborate and intricately contrived network of doctrine and theory.

Negativist Premise

Is the doctrine of *la guerre révolutionnaire* and *action psychologique* sound?

Strategically, the theory looks at the world as one continually at war. Just as does Marxism-Leninism-Maoism, this view ignores the possibility of mutual accommodation between nations, of balances of power, and of spheres of influence. It therefore begins from a negativist premise.

While the interpretation of *la guerre révolutionnaire* as a negative, total dogma is quite correct, there is nothing to say that this same philosophy must extend to the counterprogram, regardless of its rationale. To accept the negative case of *la guerre* is to deny the very principles on which the liberal West rests its case. For better or worse, the institutions of Western democracy will survive or fall on the positivist, humanist, individualist values of the liberal ethic. If groups or governments of the West, in pursuit of better methods for dealing with communism, should adopt the philosophical bias of communism itself, then liberalism is doomed.

Action psychologique as a counterrevolutionary doctrine was probably correctly drawn for the colonial experience. Given their long history of colonial wars, the French military leaders probably divined the right method of dealing with what they found in Indochina, even in Algeria. Being broadly objective about *la guerre* and *action*, however, we must conclude that they were, in fact, theories of propaganda and political indoctrination of large groups of people using military power and organization to get the job done. The French extended the idea to activities not normally military.

One cannot help but be struck by the narrow vision of this program. It was aimed at the right target—people, and it dealt with the key problem—public administration. But there are surely other ways of implementing programs like this without turning solely to the military services.

The French Army took the entire task upon itself largely because, in the dogmas of Marx, Lenin, and Mao, such a course is quite appropriate. But again this is a method inimical to the structure of liberal democracy. Armed forces can surely fight against the effects of subversion, but military force in a democracy is not the proper agent to deal with the causes of subversion.

If such a program is to be meaningful and not self-defeating in a democracy, there must be a broader base of disciplines, skills, and resources applied. Economic, political, social, psychological, and educational programs with a nonmilitary bias can all be constructed.

When the world first recognized general war of the nuclear kind, there was much talk of the economic and military coordination necessary to produce victory. With general war in the counterrevolutionary sense, we must realize that economic, military, political, social, psychological, and educational forces must be brought to bear. Victory may not be so total as we once thought it to be. Realization of the totality of war, and the ambiguity of victory in this sense, is simply to admit implicitly that, since war is total, the military services can no longer handle more than part of war.

Doctrine and Materiel

Letter to Lieutenant General Orwin C. Talbott
US Army Training and Doctrine Command
18 March 1974

This is to acknowledge your note of 4 March calling my attention to the report of your trip to Israel, as well as copies of your separate letters to General Abrams on our tanks and the new IDF tank. We are, and have been, working in the areas you highlighted, albeit much more slowly than any of us would like.

On the tactics side, Tom Tarpley and I plan to jointly produce a training circular covering greatly simplified tactics for the combined arms—tanks, mech infantry, armored cavalry. We've both given up on the field manual route; the manuals do have to be changed, but there isn't time to wait the process out, so we're going the training circular route and the manuals will have to follow along.

In tank gunnery we are pursuing the same course, and intend to produce a training circular that will police up our gunnery procedures. Following that, we'll revise FM 17-12, leaving most of it to the -10 manuals where it already is and belongs, simplifying and reducing the bulk of 17-12 thereby.

Our most difficult problems are really in the materiel area. The materiel developers simply don't accept gracefully having anyone raise even the shadow of a doubt about their programs, despite lessons of the October War. Hence we have PM ARSV scrambling around trying to save some part of his program. The AMC team on the ASH task force has built a reduction model showing how the only thing we can do is build a new airplane—despite my positive statements to the world that what they propose is the last thing I'd even entertain. And I find PM TOW Cobra refusing to consider fixing many of the faults identified in OT/DT I and II because neither General DePuy nor I came in with a specific list of things to be fixed, despite the fact that they were set forth in the report of test. PM AAH and I have reviewed the MN for that vehicle, and the development community rises in anguish every time I challenge some part of what they are doing for fear the fact that I even raised the question could impede or cancel the program.

This is a most frustrating business. The hard tests are yet before us. I find myself under increasing pressures from all sides to stop asking difficult questions, quit challenging programs—not because the challenges shouldn't be raised, but because it might upset some fellow's program—and just incidentally his job, money, prestige, and so on. There's less integrity, intellectual and professional, displayed in this whole area than I believe we can countenance. If we can't persist and overcome it, the lessons you so vividly set forth will never be reflected in improved combat gear for our Army. If I sound disturbed, I am, and I realize you weren't soliciting this kind of outburst with your note. But I'm really about fed up with the midgets who won't see the Army for worrying about the Army Air Corps, their private program, or some other parochial interest. And so if you hear that I'm on the attack somewhere, that's probably a correct report; when outnumbered and surrounded, my experience has been that the simplest thing to do is attack and keep on attacking.

Rationale for Changes in Doctrine

Message to General William E. DePuy
Commanding General, US Army Training and Doctrine Command
8 July 1974

1. The genesis of my parting comment to you last Tuesday lies in a growing sense that, to the non-TRADOC world, and even within the family, we are sort of backing into all the changes we're trying to make. Several observations:

- One of my young captains observes that he doesn't understand why the fact that FMs are too fat and bulky requires us to change all our doctrine just to get a manual of a size you and I think more appropriate.
- A 3d Cav major expostulates that the Army's senior infantryman is hell bent on screwing up the cavalry and I am his captive party to the crime.
- A senior ORSA after the ASH IPR stands up and says he still doesn't understand why we need the scout helicopter.

2. And anon. There's no groundswell against us, at least not one I can detect. But there's a lot of opposition. So simplicity, fat manuals, better gunnery, etc., will not suffice as rationale for what we're doing. We have to have a groundswell *for* us, and my guess is we haven't much time to get it swollen. So, pursuant to your affirmative reaction to my suggestion, here's what I'm doing.

3. What follows is a paper I've written outlining what I view as the underlying bias of what we're doing. While I believe that what we're about needed to be done anyway, it is noneless convenient for a number of reasons to tie it at least loosely to the October War. Here it is:

Many observations have been made and lessons drawn from recent battles in the deserts of the Middle East. Several important facts about these battles combine to provide the basis for an intensive reevaluation of the logic underlying tactics, organization, logistics, and equipment development for armored forces. Three critical facts are there:

(1) Long-range high-velocity tank cannon and long-range antiarmor missile systems dominate the modern battlefield. Their range capabilities are such that virtually anything that can be seen can be taken under fire and hit. They are so lethal that virtually anything they can hit can be killed.

(2) Long-range air defense cannon and missile systems dominate the air above the battlefield. They are so effective that they can deny forward fighting elements effective close air support; they severely limit operating profiles of the low, slow-flying platforms of the Army aircraft fleet.

(3) The US Army must learn to fight outnumbered and win. The masses of armor and air defense weapons employed by Israel's foes not only testify to a Soviet ability to supply her allies with vast amounts of first line materiel, but portend the masses of such systems the Soviets themselves would use. The tank ratios on the Golan Heights in October 1973 were not at all unlike those to be expected in a war in Central Europe.

These principal facts indicate a need for a revised set of priorities; they are:

- Detection and identification of the enemy at maximum possible distances from the friendly main body in order to prevent engagement of the main body under adverse conditions—unwarned, poorly deployed, not poised to fight.
- Firing fast first in a tank-antitank battle, especially at antiarmor-capable targets. In a tank duel, accuracy is important—firing first is more important.
- Control and distribution of tank-antitank fires in order to kill targets rapidly and save ammunition to engage the next attacking echelon.
- Battlefield movement only along covered avenues—making maximum use of *terrain* in order to avoid or evade enemy long-range observation and fires.
- Suppressive fires delivered from *overwatching* positions in order to reduce the chance that maneuvering friendly forces can be seen and engaged by the enemy.
- Flight of Army combat aircraft in forward battle areas at nap-of-the-earth, or as close to the ground as navigation aids and pilot training will permit, in order to survive. Scout and attack helicopters must live and operate in the ground battle environment.
- Suppressive fires against enemy air defense systems, especially radar-directed systems, in order to permit scout and attack helicopters to operate more effectively.
- Attack and counterattack on reverse slopes in order to protect attacking forces from long-range enemy observation and fire.
- Operating in darkness or other conditions of reduced visibility in order to reduce range and accuracy of enemy observation and fire.
- Precision, discipline, speed, and security in the directing and reporting of the battle in order to win the battle quickly, unimpeded by enemy countermeasures.
- Adequacy of stowed loads of ammunition and fuel; speed, responsiveness, and security of resupply systems in order to reduce the need to resupply, but to ensure prompt resupply when it is needed.

4. There is more, but you can get the thrust from that. Now what I've done is to include a tailored version of the above into:

Tk Bn ARTEP
TCs for Tk Plat, AC Plat, Co Tm, and Tk Gunnery
ARSV Study Plan
CSAC Report

When the time comes, a suitable version will appear in:

TCs on Bn TF and Bde.

A version will appear in my personal column in *Armor* for September-October. A version will form the basis for an LOI we're writing with regard to definitions, functions, terms of reference, and MOE to be used as a standard in our test and evaluation efforts, task forces, requirements documents; in short, to drive the combat developments process in the right direction. Target 1 August.

A version will be delivered by either Huck or me to every officer and NCO class at Knox starting immediately.

And finally, I'm sending a letter to the scattered armor brotherhood from battalion commander up saying this is what we think and here's what we're doing about it. This will go out within the month.

5. At your convenience I'd appreciate your reaction, suggestions. Also, if you disagree at all with the approach, I'd appreciate knowing it now. We really have to get moving—we had precious little time to begin with; now we may have even less. We just have to plant the conviction that we're right, it's logical, and it deserves support.

Defense Doctrine

Message to General William E. DePuy
Commanding General, US Army Training and Doctrine Command
17 October 1974

1. On the 9th, Ross Franklin and company from Benning briefed me on a concept of defense stylized as the "Strong Point Concept." You will apparently see it soon. While I do not intend to preempt, it is nonetheless imperative that we come to a mutual understanding of the philosophy which must underlie our defensive scheme. Therefore this message is to set forth what we believe to be the basic parameters of defense on the modern battlefield and solicit your views with regard to how wide we may be off the mark that must be set.

2. Fundamental to our outlook on the defense must be the realities of the modern armor battlefield, much as we outlined them as a prelude to our offense tactics discussions. Massed armor, friendlies outnumbered, exposed elements will be destroyed unless the hostiles have been destroyed, suppressed, or their view of the proceedings inhibited by clever use of terrain, weather, smoke, night, and so on.

3. Basic considerations which govern commander's layout of his defenses seem to us to be:

   a. Recognition of technical realities of the threats on the armor battlefield—range and so on, and of how the enemy fights his masses of armor.

   b. The imperative of maximum use of terrain to take advantage of the inherent benefits that accrue to the defender, all enumerated well in your French Soup Letter.

   c. The imperative of making each defensive layout a well-prepared ambush, probably sited along reverse slopes.

   d. The imperative of eventually seizing the single advantage the defender lacks—initiative, in an attack scaled to destroy the attacking echelon before the next echelon can gain the battle. Scaled also to take place on reverse slopes protected by suppressive fires and smoke, as well as terrain.

   e. Inherent in parts of these four elements—threat, terrain, ambush, attack—are several elements preferred as common practice that represent a considerable change in US doctrine regarding defense. They are the use of reverse slopes and the use of limited counterattacks to destroy enemy forces in each echelon before succeeding echelons join the fray.

4. Consider for a moment the Fulda Gap, specifically the Highway 45 approaches down the Kinzig River Valley, and the forward part of the sector which lies north of the Spessart and southwest of Wasserkuppe. Forward of the Fulda you will recall there is a salient to the east, and that in that salient is an important road junction.

5. The covering force is cavalry—the 11th; in our view it should be beefed up by adding:

   a. Divisional squadron of the 3d AD.

   b. Corps attack helicopter company(ies).

   c. M60A2s from the main body if permitted by the defensive scheme.

Its mission is to meet the enemy, make him deploy and organize for battle, thereby determining as much as possible about his strength, dispositions, and intentions, and at the same time delaying his advance as much as possible. There is some confusion about the role of a GOPL and COPL, both prescribed by current doctrine. In our view they probably aren't necessary in this situation, at least not as presently depicted doctrinally. Further there is some difference of opinion about where the main body begins its defense—e.g., forward of or behind the Fulda. This really depends on the terrain, but as a general rule we believe the defense of this area has to begin as far forward as possible and continue in that area as long as possible, and in the process destroy as many of the enemy as possible by ambush and attack.

6. The main defensive area is characterized by a succession of well-devised ambushes using reverse slopes, taking every advantage of terrain, using mines—especially scattermines, using artillery and all other means in a sort of symphonic destruction of the attacking echelon. At an appropriate time the forward task force commander attacks—an armor sweep, covered by terrain, fire, smoke, night—to destroy as much of the attacking echelon as possible—not necessarily to "restore a FEBA," and so timed as to be done and have the ambushers reconstituted in new positions to meet the second echelon as it arrives. The covering force should move aside to cover flanks, move to cover areas where the defenders are deliberately thin, and move to detect and signal the approach of the second echelon. It should not be withdrawn through the main body, told off to "rear area security" and forgotten. The defending task force commander must have prepared his initial ambush plan, his attack plan, his second ambush plan and its accompanying attack plan and a third ambush plan with its attack scheme. At least. His people must have reconnoitered, sited weapons, prepared positions, and so on.

7. Now I can't characterize that as an area or a mobile defense, or anything else. It is what some of us believe the mobile defense should be, but there is considerable difference of opinion about that. I do happen to believe that if we don't do something like what I described, we're going to get hurt—bad. And so I don't like to give it a name lest we get into a contest of semantics. It does, however, make sense to me in the context of the things that drove our offensive schemes, and it recognizes the critical elements of defense—threat, terrain, ambush, attack. However, there's a larger problem with regard to what we say about defense. The school year 1956 was the last time we taught mobile defense by brigade (combat command). That was the year we changed our pitch and said that the mobile defense should only be conducted by divisions, and that this was so because the division commander was the only fellow who had the resources to do the job. We said this largely because we had for the first time a nuclear capability at division level—first HJ, then 8-inch. And so the combination of divisional nuclear fires and divisional-level reserves was to be the decisive element in the defense scheme. The forward deployed elements were simply to delay and canalize the enemy into areas where he could be destroyed by the nukes and the counterattacking forces. Then soon we had a 155 nuclear capability, in DS units, but no one ever said that, now that we did, it was time to see if we should go back to mobile defense by brigades. I thought that was probably a mistake at the time. Now I also believe that the mathematics of the armor brawl make it foolhardy to persist in the belief that we will somehow be able to stage a massed armor attack by a division-level reserve—larger tank forces, in the face of the considerable antiarmor threat present in the enemy second, third, and succeeding echelons. I realize this is Jack Cushman's business, but a clear view of it is sort of essential to what gets done in the forward area. There's no sense in doing what I described in 6 above if the whole purpose of the forward elements is simply to canalize the enemy into places where someone else will attack him. So, fundamental to what Tom Tarpley and I say about defense, is what the Army believes about the subject in the larger

sense. You will, I believe, see this dilemma in the Benning presentation—it gives them a lot of trouble. We believe division commanders should probably have some reserves, but we also believe that its employment—that is, our doctrine about counterattack, needs drastic overhaul to accommodate to the realities and dimensions of the modern battlefield. I have a scheme for this also, but this message is too long now, so I'll save it for later.

8. In all the above I have avoided picking at what Benning has done. This is deliberate. You should see what they've done unencumbered by NITPICKS from me. We are, I believe, in substantial disagreement. So with their briefing in hand and this message as a skeleton of what we've worked out here, you can judge for yourself. Whatever we do has to be credible, especially to those along the Fulda. It seems to me that if we have them even halfway onboard about the offense, and have made a convincing case, that our defensive scheme has to be tied directly to the same driving rationale that brought us to revise the offense. I believe the threat-terrain-ambush-attack idea makes much more sense than anything else I've heard described on the subject.

Doctrinal Literature

Message to General William E. DePuy
Commanding General, US Army Training and Doctrine Command
1 November 1974

1. Your letter concerning the "Small Group of Five or Six in Shirt Sleeves" at AP Hill was most welcome, and I'd like to presume a couple of suggestions for thought before we convene.

2. First, the proliferation of Training Circulars is a matter of concern. They take lots of people, time, money, and fragment effort rather badly. Further, there aren't that many people who can write well, so we've run out of authors. There's no way for us to continue to proliferate TCs and at the same time take a decision to start a crash to produce FMs—just can't be done, we haven't the horses.

3. Second, it strikes me that now that we have what seems to be a general acceptance of our views of the modern battlefield, that the need to produce TCs as a device to spread the word has diminished. [There are] some reservations in the conduct of defensive operations, but we may be able to finesse that. What I'm suggesting is that now may be the time to skip the TC phase and move directly into FMs. We could do this for openers by just consolidating work in progress—combining several TCs, both those already written and being written, into one FM, and devote all the attention and effort to the FM. To me this reinforces what I said above about the workload and our inability to cope with the sheer magnitude of the task.

4. Third, we've reviewed our previous position with regard to what ought to be written. On the combined arms side I now suggest we simply have these basic manuals:

- The Company Team, which would cover tank and mech infantry operations from platoon through company.
- The Battalion Task Force, tank and mech battalion operations.
- The Brigade, armor, and mech.

We have in preparation a manual entitled "Cavalry." It covers all cavalry, armored, air, platoon to regiment. In our view it should be the single manual on that subject. Tom and Dave will have to speak about how to deal with foot and airmobile infantry, airborne, and artillery. The point is that in the combined arms business we should strive for as few manuals as possible, each with as few pages as possible. If we get the right kind of deathless prose we can do it. If we don't we'll never get it done by your deadline and, like you, I'm convinced that's about all the time we've got.

Effecting Change

Letter to Major Wilder M. Snodgrass
Newport, Rhode Island
1 October 1975

If you're looking for change in military systems see my "First Battle of the Next War" for an opening commentary on the volunteer and mobilization systems. The truth is we are now living with a military system which is an anachronism. It's probably outlived its usefulness. But we haven't 50 years to recognize the fact—which is what we took to change from volunteer to mobilization systems. And we have yet to demonstrate that we can effect a change of a system from within—the last time Elihu Root did it for us, over the protests of his generals. Where is our modern day Elihu Root? Or where are the generals who could get done what he did without someone like him?

Finally, if you want to look at a system, look at the horse cavalry in the 20th century—an anachronism from early in the century, but it survived a major war in which it served little purpose, and managed to thrive for the next 24 years.

So much for history. Looking today and at tactics, we could not do what we're doing were it not for the fact that everyone's doing it. In other words it has to be driven from the top—TRADOC. Nothing we've done at Knox could we have done had General Bill DePuy not supported us. And had not he in turn laid the necessary ground work with first Abe and now Fred Weyand. Having laid down the framework and got support, how do we percolate change to the Army? It has to be done in literature—FMs to set the stage, ARTEPs to test units against what the FMs set forth. Literature must be supplemented by courses at Knox and Benning for general officers, SCOCS, trips by school commandants to visit major commands, and all that. It's a sales job. And you have to work at it. And it takes an inordinate amount of time.

Israeli Relationships

Letter to Lieutenant Colonel Michael D. Mahler
Fairfax, Virginia
28 April 1976

It may interest you to know that most of the recent TRADOC literature was stimulated by my visit to Israel shortly after the war and subsequent work with the Israeli leaders.

New FM 100-5

Letter to General Alexander M. Haig Jr.
Supreme Allied Commander, Europe
26 July 1976

Attached is a copy of the final edition of our new Field Manual 100-5, *Operations*. It has just been finished up by TRADOC after more than two years of research and hard work. It is in consonance with the German manual 100/100, *Command in Battle*, as a result of work done between General Bill DePuy and German Lieutenant General von Reichert, the German Vice Chief of Staff. There are some differences in terminology, but none in concept. General DePuy sent me some advance copies; knowing how far you are now away from the US distribution system, I thought perhaps you would enjoy having an early copy.

German-American Coordination

Letter to Lieutenant Colonel Samuel D. Wilder
Fort Monroe, Virginia
23 August 1976

We're doing pretty well with [FM] 100-5. CENTAG is revising the whole defense plan based on the work we have done and the recommendations we have made in concert with VII Corps. We are still having some difficulty stamping out the killing zone but, since the CINC was the corps commander when it was adopted and at least at one time thought it great, it is a little difficult to suggest to him that he simply tell them to fix it right and press on. General Kroesen is in a difficult position, playing it well and very close to the vest. Anyway, soon we will have a new defensive scheme which will fully reflect what we have been writing about the last three years. So we can turn to other things.

Battle Book

Letter to Colonel V. M. Robertson Jr.
Fort Leavenworth, Kansas
27 September 1976

Lieutenant Colonel Jerry Schurtz will have a battle book with him when he arrives on or about 12 October for meetings with General Otis' guys. It is classified, but we have not an unclassified version to send. Obviously a version of it could be used for unclassified work—there's no classified magic about the system itself.

A word of caution. We have been accused of stereotyping the battle, the implication being that everything will be so set piece that commanders will not be able to respond flexibly to anything other than what is set up. I recognize that risk. And, for whatever it may be worth to you, here is the philosophy which underlies the battle book.

It seems to me there are two ways to prepare for war.

In the first way one reasons that war is very complex these days—so complex indeed that there is no way of knowing in advance what may develop. Therefore, we should train soldiers in their respective skills, train commanders to think, and then leave them to solve the battle situation for themselves as they encounter it.

In the second way one reasons that, in spite of the complexity of modern war, there are probably a great many repetitive situations—or nearly repetitive, and so nearly so that thinking out in advance what one might do given a certain set of circumstances has great value. The best solution to a situation can be worked out before the battle and adapted to battle circumstances as they develop. We must teach our commanders and soldiers to think about the battle in an organized way. The hardest thing to teach a young sergeant, lieutenant, captain, or battalion commander is to use all the resources he has and integrate them wisely in pursuing the battle. If he learns—almost by rote—to do this in battle planning, thinking out possible situations in advance, then he is less confused in battle, and in the heat of battle new situations do not cause him to immediately go into a whole new planning process from the very beginning.

Obviously the second method is reflected in the battle book. There may come a time when we will find it no longer necessary. I doubt that, simply because commanders change so rapidly—we are always bringing up a new generation. For the moment, at any rate, it is proving an invaluable tool. The whole exercise of having to develop the battle book and play out the battle using the target servicing ideas has in fact built up a lot of confidence in our ability to do what before seemed impossible because we had not thought about it in an organized way.

FM 100-5 Defense Philosophy

Letter to Major General George S. Patton
Fort Hood, Texas
11 November 1976

As the author of the chapters on offense and defense in 100-5 I can assure you, and anyone else who will listen, that there was no intent on our part to try to say that, because of modern weapons, defense now rules supreme. What we were trying to say was that on the modern battlefield, if you don't defend well in the first battle, you won't be around to fight the second and succeeding battles. Further we recognized that the only way we will ever be able to get the Soviets' attention is to pile on and kill a lot of the bastards early. When we have the Soviet commander to the point that he sees himself being destroyed if he presses the attack, and yet knows that if he goes back the commissar will rip his knickers, then we've got him about where we want him. While he's trying to decide what to do next, then we'll attack! But the only way to get to that situation is to kill (attrite) a lot of him first.

The Corps Battle
Briefing Script
1977

What I would like to do is talk to you about what TRADOC does, and I would like to do it in terms of the tactics, organization, equipment, and training necessary to get the Army ready to fight the next war. If we are going to do that, we need to consider first the central battle; that is, the critical point on the battlefield where all the aspects of firepower and maneuver come together to cause a decision. In order to describe the tactics, organization, equipment, and training that are necessary to put soldiers in organizations and equipment on the battlefield in effective fighting outfits, we have to consider this central battle as a first order of business.

So here we are today. We have some blue systems and some red systems, and I'm going to try to describe for you how those blue and red systems interact in a central battle in 1977. To do that, I'm going to use the US V Corps in Europe in CENTAG as an example, and I'm going to use a description of the battle that relates to the calculus that I'm about to describe.

Consider the probability of victory with an attacker versus defender ratio of one to five, that is one attacking five, and across the spectrum to five attacking one. If we take the traditional calculus of battle that we have used since the days of Lanchester's equations trying to describe the mathematics of aerial combat, we get a curve that tells you that, if you are one attacking five, you don't have much chance of success; your probability of victory is very low. It tells you that, if you are five attacking one, you are in pretty good shape; you are liable to win.

There are two things wrong with this. First, it says that the fellow who has to fight outnumbered really is foredoomed to defeat from the very beginning. Now, if that's true, and we expect ourselves to be outnumbered at the beginning of any conflict that might come our way, then we are in a pretty hopeless situation.

But if we look at the history of a thousand (I say a thousand, I fudged a little, it is 960 plus) tank battles that we have looked at over the period that tanks have been used on the battlefield, we find quite a different picture. What it shows is that the probability of victory varies very little, regardless of the odds, from one to five to five to one. It says that there is something about the battle that we can't describe using the original model. That tells us that there is some hope for us fighting and winning outnumbered, both on the defense as well as on the attack. And so let's see if we can apply that calculus to a battle, the V Corps battle, which we call the corps battle.

Consider the V Corps sector of Central Europe: Frankfurt, Fulda, the autobahn from Eisenach in East Germany into West Germany and down toward Frankfurt and the interzonal border. Traditionally the corps boundary has run parallel to this autobahn and just north of it and ends on the border just east of a hill mass known as the Knullgebirge. More about that in a moment.

On the south, traditionally, the corps boundary has run along the Spessart and ends at the border in the vicinity of a high hill mass known as the Höhe Rohn, the place where the Air Force radar towers are. Recently we moved that corps boundary from 10 to 15 kilometers to the south and southeast, so that now the boundary itself actually runs along the southeastern slopes of the Spessart.

In the middle of the corps sector, there is a high hill mass known as the Vogelsberg. The Vogelsberg is significant, not that it is necessarily key terrain, but if you look at the drainage systems in the corps, you will note that, east of the Vogelsberg, the Fulda, the Haune and along the border, the Werra and farther south, the Ulster River systems all run north and south. This means that, east of the Vogelsberg, the hills and valleys run in a north-south direction, and the terrain is more defensible if you are moving from east to west because you have all these cross compartments across which you can fight.

If you go back to the west of the Vogelsberg, you will notice that the Nidda, the Kinzig and several of the smaller valleys that open out to the northwest are all northwesting or southwesting and, in fact, change the nature of the terrain compartments in the western part of the corps sector. Therefore, if you don't defend this area along the Vogelsberg line, you can't use the good terrain to best advantage to the east of the hill mass. Once the enemy breaches the Vogelsberg line, coming either down the autobahn or down through the Vogelsberg itself or down parallel to the Kinzig and across the Main, you have in effect opened up the area into the Wetteral corridor and the approaches on to the Main crossings at Frankfurt and the Rhine crossings at Mainz. Therefore the corps defense is built on the area to the east of the Vogelsberg.

First avenues of approach. There are three in the eastern part of the sector. One is down the autobahn from Eisenach to Bad Hersfeld to Alsfeld to Giessen and on to Frankfurt. One is astride Highway 84 from Eisenach through the town of Rasdorf on the border and into Hünfeld in the Haune River valley. The third comes from Meiningen in East Germany through the hills to the east of Fulda and into the city of Fulda itself. From Fulda this avenue can open up through Schlitz and Lauterbach on to the Alsfeld approach or turn south down the Kinzig River valley.

Now the terrain. From Bad Hersfeld, looking northeast, the hills in the background are the Knüllgebirge, heavily wooded, badly cut up, steep ravines, narrow valley floors, steep slopes, easily defensible by mechanized infantry. As a matter of fact, a mechanized infantry heavy task force defends this ground. On the Stoppelsberg looking east, with the town of Eiterfeld in the foreground, the hill mass and the woodlands of the Knüllgebirge now open out into this broad, open agricultural land.

Another thing to note is that we have a case of what's popularly called urban sprawl in this area. These little villages are beginning to connect with one another with buildings that are built along the roads that run through the area. This is a problem and we treat it, except for the central parts of the older parts of the town, just like we do terrain. For example, there is a building that, if the enemy occupies it, gives him observation for indirect fire all across this battle position on which we are standing. So you either have to deny him use of the building by occupying it or by doing something else to keep him from getting in it and bringing observed indirect fire to bear on your positions.

Now let's talk about the enemy. There are 12 enemy divisions in the area opposite V Corps. Of those 12 enemy divisions, 9 are Soviet divisions and 3 are the divisions of the East German Army. Of the nine Soviet divisions, five belong to the First Guards Tank Army, headquarters in Dresden, and consist of four tank divisions and one motorized rifle division. The other four Soviet divisions belong to the Eighth Guards Combined Arms Army, headquarters in Nohra. Three of those are motorized rifle divisions, and the 20th Guards Tank Division, with headquarters in Jena, is a tank division. The three East German divisions belong to Military

District 3 of the East German Army with headquarters in Leipzig, and, of the three, two are motorized rifle divisions and one, the 7th in Dresden, is a tank division.

Now how might the enemy attack? Well, just using his description of how he intends to conduct combat operations, he could probably fit as few as four divisions in here in a holding attack. The terrain will hold about six divisions.

We know on the other side that we are the V Corps with headquarters in Frankfurt. The 3d Armored Division has headquarters in Frankfurt and brigades in Kirchgoens, Freidburg, and Gelnhausen. The 8th Mechanized Division has headquarters in Bad Kreuznach with brigades in Mannheim, Mainz, Wiesbaden, and Baumholder—four brigades in the 8th Division. The brigade in Wiesbaden is Brigade 76, the 4th Brigade of the 4th Mechanized Division out of Fort Carson. There are two artillery groups, the 41st in Babenhausen and the 42d in Giessen; the 130th Engineer Brigade is in Hanau; the Corps Support Command has headquarters in Frankfurt and six battalion-sized units scattered through the area; the 11th Cavalry has headquarters and a squadron in Fulda, a squadron in Bad Hersfeld, and a squadron in Bad Kissingen.

Now as to the force ratios. I think the significant thing to note here is the artillery force ratio. The numbers used represent tubes. If we take the amount of ammunition, the weight of ammunition, that those tubes can throw at one another in a 24-hour period, we are outnumbered about five to one instead of about three to one. The difference comes from the multiple rocket launchers, which the Soviets have aplenty. The number usually used is deceiving because it relates only to tubes, so you can see we are outnumbered on the order of two to two and one-half to one in some cases. In some cases, we simply have no counterpart, as is the case of the BMP versus the M113.

The mission is rather prosaic and straightforward, but it's necessary to translate that into some other meaningful words in order to get on with an analysis of the battle, and we have done that. Now, why do we say destroy the first-echelon army? Well in the first place we believe it will be necessary to do that to get their attention. Given the Soviet training and their mode of operation and what they say about the importance of mass, momentum, and speed, their insistence that their commanders make 40 kilometers a day, we believe that it's going to be necessary to cause that first-echelon army commander to face up to the fact that he can't go any farther. He is going to have to go back and tell the commissar that he can't make it. And so he is on the horns of a dilemma. He has to either come ahead and face the sure destruction of his army or go back and tell the commissar that he is not going to make it and face destruction at the hands of the commissar. It's in that sort of uncertain frame of the mind that we would like to meet him out there some morning when the war begins.

The corps battle, or the central battle in this corps, is phased as follows: deployment and covering force and then the main battle area fight. The covering force consists of eight battalions and squadrons: the three squadrons of the 11th Cavalry beefed up by the organic cavalry squadron of the 3d Armored Division, the 3-12th, and the organic squadron of the 8th Division, the 3-8th, both of which have had tank companies and self-propelled artillery batteries added to them so they are in fact configured like the regimental squadrons. There are three tank-heavy task forces. Two of those come from the 3d Armored Division, and one in the Fulda Bowl area comes out of Brigade 76, Task Force 1-70th Armor. The reason that those task forces have been added to that formation is because it takes that kind of a force to service the targets that come at the 11th Cavalry in the covering force. I'll come to that in a moment.

In the main battle area, we have, behind the covering force, the three brigades of the 3d Armored Division and the four brigades of the 8th Mechanized, with the second brigade of the 8th, the Baumholder Brigade, backing up the 4th Brigade of the 4th Mechanized, Brigade 76, along the Fulda avenue of approach. If we look at the brigade battle areas, there are four in each sector, the fourth one in the 3d Armored Division sector being an area into which the headquarters of the 11th Cavalry moves when the covering force fight is over, picks up some fresh battalions, and goes into position along the Hunfeld avenue of approach. So what we have here in the main battle area are battalions stacked up four or five battle positions deep along two of the major avenues of approach into the area, Hunfeld and Fulda.

How might the enemy attack? Let's go back to that for a moment. He says that he has two essential formations, an advance to contact and a breakthrough. Let's take them in turn. Consider a motorized rifle division moving to contact. The width of the formation is about 15 to 25 kilometers. The cells in the attack formation are the echelons as he describes them—his reconnaissance, his advanced guard, his first echelon, and his second-echelon regiments. We can calculate generally the rate at which targets arrive, the type of targets that we can expect to arrive at that rate, and the dimensions, the geometric dimensions, of the area in which that formation will be deployed. That's how we begin target servicing.

Now let's consider a tank division in a breakthrough formation. The frontage is narrower. The reconnaissance elements are missing, the advanced guard is missing; there are some flank security elements on either side, but the frontage is much narrower. There are two echelons, and the density of equipment in each of those echelons is much greater than it was in the other, the march to contact.

Now let's see how the battle might develop. A US division would be deployed across a front of about 40 kilometers. If it is attacked on anything like breakthrough frontages, we say it will be 7 to 15 kilometers because, in trying to analyze this on the ground, we found that it was difficult, if not impossible, to employ breakthrough tactics on our terrain. The reason is that the terrain tends to absorb the gross geometric differences between the two formations. Now that's not to say that some US battalion, brigade, or even division commander is not going to see something coming at him that looks like a breakthrough attack as he sees it. When he docs that, he has to form up to meet the breakthrough attack. He does that by moving battalions around, concentrating so that he has half-way respectable force ratios in the area where he expects the main battle to take place. He covers his flanks, or fills in the areas that he has denuded of ground forces by this concentration, by putting out there his air cavalry and his attack helicopter units. They watch over the flanks, detect the presence of the enemy, bring the enemy into the area, get hold of him, and do something with him until the commander can get something there on the ground.

A brigade battle area would be made up of several battalion battle positions. In this fight we have elected to strongpoint a certain position. The enemy attack comes in the open ground, and we attack him with two battalions, with another battalion holding. We do it in such a way that we kind of tidy up this fight before the next echelon comes along. More about that later.

As an aside, I think it might be appropriate to point out that not all the weather and visibility is predictable. This brigade fight I have just described takes place in battalion battle positions that may have weather that lets you see the whole area. But other times it doesn't. There are many days such as this in Europe.

Target servicing is the problem at battalion level. Let's talk about the concept first, and then let's apply it. Consider a battalion battle position, an actual example and a bad example, which is why I have elected to use it as an example. This was an M60A2 tank battalion of about 30 weapons systems if you count the two tank companies, some TOWs that he had, a mortar battery, and a 155 dedicated artillery battery. They were defending an area in which there are two small hills. The terrain was such that the enemy could come into the area with about 210 systems of the lead two regiments of a motorized rifle division at a rate of march of about 18 kilometers an hour or 300 meters a minute. The terrain was also such that, from an intervisibility standpoint, he could expect to open fire on the enemy at about 2,100 meters.

If the enemy is moving 300 meters a minute and you could open fire at 2,100 meters, he is going to close with you in seven minutes. If you have 30 weapons systems, each firing one round a minute at 210 systems, that's seven also and that says that every one of these has to destroy seven of those with one round a minute. There is no room for error; there's no margin for error because he will close with you in seven minutes. This works only if you assume no losses and that everything goes well. I have taken out the artillery and so on just to simplify the example, but it tells you that that is kind of a dicey fight and we have probably given that fellow more to do than he is capable of handling. He either needs more systems of the kind that he already has or he needs some different kinds of systems, like tanks that fire faster than do the Shillelagh-firing M60A2s. So the problem is obviously to put in the right weapons mix to do the job.

Now let me expand that to another battalion battle position and describe for you what we believe the pace of the battle might be and the way the battle might go. Let's pick up the action at 10 minutes into the battle. We are defending with a battalion with three companies. One company team made up of M60A1 tanks and some TOWs; another company team is made up of M60A1 tanks and some TOWs; and a company team is made up of M60A2 tanks in three platoons and some TOWs. Ten minutes into the fight, the Team A commander has issued some instructions to his platoon leaders about where to go. He has put the TOWs in position. He has detected the presence of the enemy, called for artillery fire on them, and reported to the battalion commander what he has done. He has done a lot of other things, but those are the four major things that he has done.

Team B has done much the same, but Team C is not quite so busy up to this point because he can't see the enemy yet. The battalion commander has been very busy. He has looked at this whole thing and he has called for artillery. He has called for tac air. He has called for some attack helicopters, because it looks like he has a major threat coming at him.

Fifteen minutes into the battle, enemy regiments have closed to within direct-fire range of Teams A and B and later of Team C because he has longer weapons systems at his disposal. If you examine the situation closely, you note that every battalion-sized formation in that enemy array is being fired at by at least two platoons of the friendly force. The attack helicopters are on the way, but we don't know about the close air support yet. We have intensified the artillery and mortar fire.

A number of decisions have been made and acted upon by each of those commanders at this point, 15 minutes into the battle. It's been a busy time for the Team A and B commanders. The battalion commander is beginning to slow down a little bit because he is concentrating his attention on getting the extra artillery, the tac air, and the attack helicopters.

Twenty minutes into it the battle is essentially over. What's happened is that the enemy has closed into a pocket. The attack helicopters have arrived, and we have moved a platoon of tanks down to provide fire suppression so that the helicopters can open fire without lots of air defense systems firing at them. We have begun to attack advancing enemy formations by fire. We have moved a platoon to get better flank shots at an advancing tank regiment. We have begun to use smoke so that we can finish up this part of the battle before the second echelon comes along down the road. Red losses so far are about 192 systems and blue losses about 30 systems.

Again, here are the numbers of decisions that have been made and acted on. The helicopters got there, but the tac air did not, as there simply wasn't time. It has been a busy battle, and I have described it very, very briefly. The point is that the battle is going to be very intense. The battlefield is going to be very dense, and there is not going to be much time to sit around out there contemplating what to do next. We are going to have to act automatically and do it about right the first time.

We can construct a tally, a sort of killer/victim scoreboard against each friendly system: M60A1s, M60A2s, TOWs, and artillery, the number of pieces, weapons, or whatever that were available to fire; the time that was available for them to fire at the enemy based on the range; the kill rate, given some attrition of each one of these in kills permitted; the red losses as a result of those kills; the blue losses as a result of red return fire, 192 to 30. So we conclude that red can probably be defeated before he closes with Team C, but not before he closes with Teams A and B. There is another echelon coming down the road, and so what to do next is the big question.

Now that is target servicing. Let's apply that to a real situation. There are three major avenues of approach, the autobahn, Hunfeld, and Fulda, into the corps sector. The commander of the covering force has looked at his situation and drawn out some subavenues of approach, along with the number of enemy battalions that we think the enemy can put in that particular piece of ground. Now let's focus our attention for a moment, for illustrative purposes, on the sector that runs astride Highway 84 through Rasdorf and Hunfeld. What we have is a sector that's defended in the covering force area by the 1st Squadron of the 11th Cavalry. West of the line the 1st Battalion, 32d Armored, an M60A2 battalion out of the 3d Armored Division, has a sector. There are five company teams in the cavalry squadron; there are five company teams in the 1-32d Armor.

This squadron commander has laid out two squadron-sized battle positions. This battalion commander has laid out about four battalion-sized battle positions: one forward of the Haune, one in the hills just west of the Haune, and a couple more farther to the west.

Let's look at the 1st Squadron of the 11th Cavalry target servicing. We have 89 blue systems against 348 red systems at the beginning. We think the battle might last as much as 74 minutes, and we will have about 33 friendly systems left, while we can reduce the enemy to zero. We can do this, we believe, with a little luck. The 1st Squadron can destroy about nine battalions of the first initial assault echelons, or about five battalion equivalents if you convert those nine battalions to US battalion equivalents. I use this example because this is the place in the corps sector where the force ratios are the greatest. You can appreciate that we are fighting against some pretty stiff odds here, even at the beginning.

Now let's expand that kind of analysis to the total corps covering force area. We picked 454 blue systems against 1,628 red systems, about four divisions. That is because four divisions are all that will fit into that terrain. We reduce that force in about 70 battle minutes; we don't destroy them totally because we run out of systems or ammunition or both. In this particular case, we run out of Sheridans because they are thin-skinned and are very sensitive to artillery fire, so they are destroyed before they get a chance to fire all their missiles. Thus we don't get a chance to destroy all the enemy. We were outnumbered at the beginning about 1 to 3.4. Each US battalion or squadron destroyed about 3½-battalion equivalents of the enemy.

Now let's turn to the main battle area. We had a battle position forward of the Haune. There is one just west of the Haune, another up in the hills, and another in back of that. We are six battalion battle positions deep in this area. Our reduction model pits 78 blues against 423 reds. We reduce to about parity at the end, because here we run out of ammunition and we run out of missiles in the M60A2 tanks. The tanks are not destroyed, but because they have a low basic load, a small basic load of missiles, we run out of missiles before we run out of targets. Now, here again, we had a five to one force ratio at the beginning and about one to one at the end, and we have an ammunition problem.

Let's expand that to a summary of the total problem in the whole main battle area. There are 1,333 friendly systems against 2,894 red. In about 50 battle minutes, we reduce the enemy to about parity and, here again, the problem is that we run out of ammunition. We were outnumbered just over two to one in the beginning. We have him down to about one to one. We ourselves are at about 40-percent strength as against his 19-percent strength. We think, therefore, that we have a high assurance of defeating the enemy. We said 90 percent—take your own number, but a high assurance that the first-echelon red army can, in fact, be destroyed somewhere at or near the third set of battle positions in the main battle area.

The corps I have mentioned will be about half strength. The fight may last as few as three or as many as six days, and we can delay against the second-echelon army. But to defend against it, we will need more ammunition, replacement personnel, and equipment. As a matter of fact, what one really needs to do at this point is to attack. But, here again, you need more ammunition, some replacement personnel, and equipment before that can be done successfully.

At that point we think we will have our friend in a situation something like this: His first-echelon army is at 19-percent strength. He doesn't know how badly he has hurt us. He knows that, if he turns around and goes back, he has to explain to his boss that he can't make it, his 40 kilometers a day. He knows that, if he comes ahead, he risks certain destruction at the hands of all those evil fellows he has been fighting for the last two or three days. So he has a problem. We would like to meet him as he addresses himself to the solution to that problem.

In this description of the central battle, I used the corps battle in V Corps as an example. In the description of the central battle, we have shown some deltas, some deficiencies in our own organization, our own equipment, our own tactics perhaps, and even in training that need to be redressed before we can create a situation that we think we can win. Let me just cite some of those deficiencies to show what kind of an analysis can flow out of this sort of description of the central battle.

The first problem is deployment. What we have done in analyzing the enemy is said that he can move from his kasernes to dispersal areas in about four to six hours. He has about three routes for every division from their assembly areas to the border in no more than about 13 or 14 hours, so from kaserne assembly area and from assembly area to the border adds up to about 24 hours from the time he starts to move. Thus he has the capability of attacking us in this particular area with about six divisions in about 24 hours from the time he begins to move.

How do we know? We watched him do it. We watched him do it in maneuvers. We watched him do it in map exercises. He can move along those roads blacked out at night; he can move along those roads in radio silence. It's a capability, and it's a demonstrated capability.

Now the question we have to face is that, if he can do that, if he can attack us with six divisions 24 hours after he begins to move, what chance do we have of knowing about it, with all the surveillance equipment that we have at our disposal, looking and listening? The intelligence guys tell us that there is probably an 80-percent chance, 8 chances out of 10 that we will know within 12 or 15 hours after he begins to move. Well, if you subtract the 12 or 15 hours, that says to you that you have to be on the border in something like 9 to 12 hours ready to fight. Let's see how well we can do that.

These are the distances: 120 to 150 kilometers for the 3d Armored Division; somewhat longer, up to 250 to 285 kilometers, for the 8th Division's most distant brigade, the Baumholder Brigade. So we have a problem.

The fundamental problem in deployment is ammunition loadup. We have recently loaded the ammunition, the tank gun ammunition, in the tanks of these brigades, and we have cut the deployment time about in half. The fact of the matter is that, as a result of our analysis, we are now able to get the covering force on the border, closed on the border, and ready to fight in something like the prescribed time, and we are working on the forces in the main battle area.

Let's look at offensive air support. You probably noticed that, in target servicing, we had some targets left over, targets we could not service with the ground systems that we have. What do we do about that? Well one possibility is to call on offensive air support. If we simply convert the unserviced targets to aircraft sorties using the Air Force rates, we generate a requirement for about 400 sorties a day for the first three days of the battle and about 300 sorties a day for the next three days of the battle.

Now I'm not saying that we are asking for 400 dedicated sorties or 300 dedicated sorties, we are just saying that that's the scope of the problem. If several of NATO's eight corps are attacked by breakthrough attacks, and the Soviets have that capability, then it may be that there will be several of NATO's corps commanders who are seized with the same problem in fighting their part of the central battle as this corps commander is. Thus this is a picture, a framework at least, around the size of the problem. There are competing demands on close air support, and I'll come to those in a moment.

In order to make air support possible, we have tried to simplify the way in which we go about furnishing close air support. We have done that by drawing a series of target boxes. These target boxes are essentially in battle positions. There are pop-up points shown in each target box, and at each pop-up point there is an identifiable terrain feature. The pop-up points are approached along an azimuth or a heading from a release point, and those are identified. In one case it is a slag heap. A heading and a time of flight to the pop-up point, and a picture of the identifiable feature in the target box, are given to the pilots.

They come across at low level, 700 to 900 feet above ground level, and from that altitude they simply pop up. They can identify the target because it is to the right or the left of the reference point. We have reduced the number of passes these aircraft have to make to one pass per sortie, and we have reduced his exposure to enemy fire by bringing him in low. We have increased the turnaround time and so we have solved part, at least, of the close air support problem. This is an application of what some of you may know as the Creek Braille System to the central battle.

There are other demands, as I pointed out before, on offensive air support—such as air superiority, deep interdiction, aerial surveillance, or do you want him against the first echelon or the second echelon? All of these are decisions the commander has to make, because he has used this part of his close air support to make up the delta in target servicing capability that he does not have on the ground. If he can't count on that kind of support to service the first echelon, and he needs them against the second echelon and the interdiction and air targets, then he may have to change his way of fighting on the ground.

Now let's consider sustaining the corps battle. I will briefly discuss ammunition and command and control. First the ammunition problem. This is what we have, one tank load for every five days. We think we need 1.3 loads per day for the surviving tanks. We have 55 artillery rounds per day and we believe we need 360 rounds per day of 155. All of this comes out at the bottom line to the facts that the tanks run out of ammunition in about six days, or the corps will run out of tank gun ammunition in about six days, out of artillery ammunition in about seven days, and out of TOWs at about noon on the second day. So we have to get some more ammunition, and we have to position it in the right places. Part of the problem, as I pointed out before, is loading it on the tanks and improving our readiness to fight in the central battle.

Directing the corps battle, the command and control task, goes in a cycle. You get the information, you decide what to do, you tell somebody what to do about it, and you keep track of it. The problem is this: over the past years we have cluttered up the nets between battalion, brigade, division, corps, and so on, with complexities that range all the way from something we have done to the equipment, as in the case of the secure equipment on the FM radios, to our own problems with computers and to our problems with what I call internal softwares, the message center.

Let me illustrate. Our FM radios didn't work very well in the beginning; as far as range goes, they weren't designed for long range. We reduced their range by applying secure to them, and so now they don't work very well. They don't work as well as they did in the first place because we tried to make them secure. The Spanner Study took one of the two RTT sets out of the battalion. It took two to keep one running in the first place, and so we really can't depend on AM RTT the way we used to. The same thinking applies up the line. The VHF rig, the multichannel shots particularly, is a beautiful system, but it takes a long time to tune it and it takes an inordinate amount of time to move it.

We have a Corps Support Command with two 360/40 computers in it. This is the front face, or the field face, of the wholesale logistics and administrative systems that come all the way from the CONUS, and there is nothing, or almost nothing, in there that's usable to the practitioner in the field. In other words the retail system is very difficult to pick up off the front face of the wholesale system, and we have an interface problem. There is nothing much in that management information, SAILS [Standard Army Intermediate-Level System], SIDPERS [Standard Installation Division Personnel System], PERDDIMS [Personnel Development and Distribution Management System], STANFINS [Standard Finance System], all those. There is nothing very much in that that is useful to the corps commander in trying to fight the central battle.

How to solve the problem? Well, it relates to these things: the mass of information; the multiple sources; the priority disciplines of the net; the fact that everything has to be reported sequentially from company to battalion, from battalion to brigade, and so on. And, internal to the system, we have what I call the self-constipating staff system. That's where the major has to go see the lieutenant colonel, they both have to go see the colonel, and they all have to go see the general. You can spend days doing that before the message gets out.

The other problem is what I call the tyranny of the message center. That starts with the fact that you have to punch the tape at 25 words a minute. You can only send it at 66 words a minute, and if the young man or young lady who runs that teletypewriter doesn't want it to go out, it's not going out anyway. We have to figure out some way of solving some of those problems, because they add at least 50 percent, if not more, to our problem in command and control.

What we have done in trying to solve that is experimented with a system like the one that was used by the British, like the one that was used by the Germans, and like the one that was used in Third Army in World War II where, in the corps tactical command post [TAC CP], or in the division tactical CP, we monitor the nets of the lead maneuver battalions. It turns out that we can't do that using FM and, of course, they don't have the AM equipment here anymore, so we move that up to the brigade level. These battalions report FM to the brigade, the brigade broadcasts AM simultaneously to the corps and division main and to the corps and division TAC CP, and that works fairly well. It works better if you use a satellite terminal at each brigade level. We have experimented with that and it works very well. But we are working on ways of solving this problem using existing equipment. It is very difficult. We are going to need some new things in this area to totally solve our problem.

The brigade simply takes the battalion battle reports and transmits its own location, what's going on, degree of commitment, the commander's outlook on how the battle might go and something about how things are going, fire support, fuel and ammunition, mobility problems, and the enemy situation. And so that's the central battle for a corps.

Let's go back to the general description for a moment. There is some training in there, some tactics, some equipment, and some organization. Had we taken the time to go through each one of those, we could have described the little deltas that come out of our description of the central battle in 1977 with existing kinds of blue equipment and red equipment.

The next thing that has to happen in old TRADOC is for us to extend into the central battle in 1985, 1987—pick your own time—those developments that are now underway: the T72 tank; the new SP artillery for the Soviets, as well as their current emphasis on counterfire systems; tac air and artillery; XM1; MICV [mechanized infantry combat vehicle] or the infantry combat vehicle; the infantry fighting vehicle, the ICM [improved conventional munition], if we reorganize division structure; roll-on on our own side. Then we replay that battle to try to develop the deltas in 1985, then project from the 1985 battle on out by using a technical assessment and trying to develop out of the technical opportunities that one sees in the laboratories of the country a technical strategy that we might pursue to get us to a description of the central battle in the year 2000.

That's the way we intend to approach the problem. In the end, it's quite apparent to us that battles are won by these three things, the courage of the soldiers, the quality of the leaders, and the excellence of the training. That's what makes force ratios come down into some reasonable proportions, influenced by something in the leadership and something in the training that won the battle, not just the calculus of the battle alone. But the use of the calculus to describe the battle is a way of laying it out so that the leaders can contemplate what has to be done to lead the soldiers and then decide what has to be done to train them correctly in order to fight in the central battle today, 1977, and tomorrow, 1985.

Helping the Republic of Korea Army

Message to Lieutenant General John R. Thurman
Deputy Commanding General, US Army Training and Doctrine Command
19 September 1977

1. This responds to your LVW 1192.

2. I believe we should be willing to help the ROKA [Republic of Korea Army] with whatever TRADOC products we and they determine might be useful to them. The ultimate determination of what we might do, how much we might do, and what bounds might be placed on such activity by the current situation with regard to troop withdrawals will have to be sorted out with Shy [Lieutenant General Edward C.] Meyer.

3. [Lieutenant General John H.] Cushman was here last Thursday. We all gathered to hear him on the simulations done in I Corps Group using first battle as a basis, and on close air support. He made the statement that he did not believe 100-5 applies to the Korean situation. By extrapolation I take that to mean none of the manuals derivative from 100-5 apply either. I understood him to say that he is trying to work out a "whatever seems to work best" solution to the problems they face using simulations, and then at some point incorporate that body of conclusions into plans.

4. If I have correctly understood him, this puts us squarely on the horns of a dilemma in trying to help the ROKA. For if 100-5 does not describe a modus applicable to the Korean situation, then none of our manuals, games, simulations, and other doctrinal material can be of much use to them. For, while we did not begin all our efforts with a very clear perception of the calculus of battle that underlies 100-5 and its derivatives, it is true that more and more our games and simulations are being validated against what we now believe to be a correct calculus.

5. That brings me to two important considerations. First, we have not the resources to undertake development of a new calculus of battle applicable only to the Korean situation. Secondly, I do not agree with Cushman's assessment that 100-5 and its derivatives are not applicable to a war in Korea. It is true that war in Europe and war in Korea have some rather different parameters. Especially is this so with regard to terrain and the type forces one could expect to employ and to encounter in the first battles. However, 100-5 is based on some principles, some calculus, which I find rather equally applicable to first battles in either theater. If one accepts that, then we might be able to help them; if one rejects that, we could find it most difficult to help them. In addition it will be rather awkward to help them with CG I Corps Group holding the view that 100-5 does not apply.

6. Despite all that, I do believe we are obliged to respond to their request. . . . Let's get together amongst ourselves and with Shy to determine if we can and should help, and what limits we might have to operate under. Then we must go to Cushman and resolve what I understood to be a rather fundamental difference of opinion between us, one which could make it difficult if not well nigh impossible to provide much meaningful help to the ROKA. If that difference can be resolved satisfactorily, then I am certain we can be of at least some help to them.

Tank Allocation Basis

Message to Lieutenant General John R. Thurman and Others
17 October 1977

1. In developing the need for a FIST vehicle and for the FAC in a track, we have raised once again the question of the need for the seventeenth tank in the tank company, and for the third tank in the tank battalion headquarters tank section.

2. The issue at hand is the 45-tank cavalry squadron developed in USAREUR and now part of USAREUR's formal MTOE submission. I've got from 36 to 45 by adding tanks at troop and squadron level for FOs and FSOs, just as we've always had in tank units. DCSOPS is balking, and my agreement with Shy [Lieutenant General Edward C.] Meyer is that we will make the case for those extra tanks, if indeed there is a case to be made. However, this also raises the bigger question of all tanks long included in TOE for FOs, FSOs, and FACs.

3. Therefore I'd like for you all to get together and draw up a proposed TRADOC position on the matter under CAC's fine Spenserian hand. Especial consideration must be given to the critical question of the survivability of command and control for our fire support means. In addition, I'm about convinced that we should mount the second in command in tank companies, cav troops, tank battalions, and cav squadrons in tanks. You'll probably have to do some simulations or play Battle several times, but we should get an answer together by late November.

Corps and Division Operations Manuals

Message to Lieutenant General John R. Thurman and Others
28 October 1977

1. I've worked through the pieces of [FM] 100-15 which Lou [Major General Louis Menetrey] gave me some weeks ago, and through several chapters of [FM] 71-100 Ed [Lieutenant Colonel Edward Scribner] provided me. What follows is a reaction to that exercise and some additional instructions for the reiteration, especially of 100-15.

2. First, these two manuals are to explain the application of principles set forth in 100-5 to corps and divisional operations. They are therefore important links in a logic train that starts with the generalizations of 100-5 and ends with target servicing at battalion and brigade levels as described in 71-2 and the brigade parts of 71-100.

3. Second, both 100-15 and the divisional parts of 71-100 must take what is said in 100-5 and describe what division and corps do about each of those items, concentrating on the uniqueness of what is done at the echelon of interest—corps or division, and how it relates to what is done at the next echelon up and next down. In considerable measure the uniqueness of what's done at each echelon will depend on the resources available at that echelon that are not available at the next echelon down or up—not just more resources, but different kinds of resources, different capabilities in other words. Offensive air support, intelligence, and general support maintenance are examples of three pretty widely separated functions to which this principle applies most obviously and probably most clearly. They'll be fairly easy to deal with. Others will be less clear and therefore more difficult to handle, but the principle applies to everything—enemy, weapons, organization, tactics, and so on.

4. Third, what sets corps and divisional activities apart from similar matters at brigade and lower units is the degree of involvement of each echelon in target servicing in the central duel. Brigades, battalions, companies, and platoons service targets and target arrays. Divisions and corps pull together intel, forces, air, log, and so on, and apply resources to the battle based on perceptions about larger forces, areas, and other considerations. Both division and corps commanders direct the battle, but each does so in a little different way than do commanders at brigade or battalion. Indeed, in terms of finding out what's going on, deciding what to do, and doing something about it, directing the battle is probably the corps and division commander's most difficult single task. We must in each book describe the problem from the vantage point of the fellow trying to do it at that level and how he relates to those doing their part of it one level up and one down.

Doctrine and Studies

Letter to Dr. Seth Bonder
Vector Research, Inc.
29 November 1977

Thank you for sending me a copy of your keynote address presented at the 16th AORS. We share some convictions about many of the Army's ops analysis problems.

While we might agree as to what must be done, the harder question is how to get on with it. With every new layer of professional question askers hired at OMB and OSD, it is necessary to devote ever-increasing amounts of talent to answering questions that are at best only vaguely related to the problem. In fact, all too much of our ops analysis talent gets sopped up that way. While it is a phenomenon characteristic of every change in administration, it does seem to have persisted somewhat longer than my recollection of the same saga in times past—perhaps just from my changed vantage point.

In addition, there are sharply eroded resources with which to do anything—ops analysis included. We are confronted by a legacy of an almost complete void in data about human performance factors so essential to good analysis of today's weapons and insights into what to do better in tomorrow's. There is a requirement to begin extensive cost and training effectiveness analyses of the vast array of simulations and training devices we are fielding. The calculus of the forgetting time, (learning) curve, and the transfer function have yet to be defined, even in a tentative way. Aside from all that, everything is going swimmingly.

I'd be happy to discuss these and related matters with you. However, it's only fair to warn you that I know you're in business to do studies; studies take money and time, and we haven't enough of either. Historically, when the Army hires a study effort it gets too little guidance from the guy who wanted the study in the first place, and the contractor goes off on whatever bent his perception of the problem leads him. Therefore we have volumes of wrong answers to right questions, right answers to wrong problems, and in the end we have not much of what we now find ourselves needing desperately to get on with our business. The magnitude of our problem is not of the instant—it is rather the accumulation of many years of the kind of mutual self-delusion I described abovc. And I must say that in that futile encounter the Army is as much at fault as anyone else—we let it happen to ourselves. My problem is how to get organized to get a few good answers to the toughest problems and move on, developing from that beginning whatever refinements time and money may permit. If you can help solve that problem I'd be grateful.

One matter in your paper with which I take exception. Drawing on FM 100-5, you allude to the fact that the active defense uses a light covering force deployed to do certain things, and you go on to say that idea has never been tested in simulation or battle. I've reread that part of 100-5 again, and I can't find the basis for your statement. I wrote both the offense and defense chapters of that book, and I can assure you that nothing was further from my mind than a light covering force—especially in Europe. In fact, in applying the active defense idea set forth in 100-5 to the real world V Corps situation in Europe, I found myself as corps commander with a covering force of eight battalions or squadrons—what it took to service the part of the first echelon we believed it necessary to be accounted for by the covering force. Incidentally, you might be interested in the simulation we developed as a combination training and analytical device.

It's called "Battle"—a high-resolution, low-level battalion simulation—time-sequenced Monte Carlo calculus. Leavenworth has one set up and running. With it we hope to expand our ability to work out tactics on the terrain board in a manner consistent with a calculus representative of the battle itself. It uses a Wang 64K minicomputer.

Active Defense

Letter to Major General C. P. Benedict
1st Infantry Division (Mechanized)
13 March 1978

Recently I read your REFORGER 77 After Action Report. I was somewhat disturbed by one of its summary paragraphs:

> d. Active Defense Doctrine suggests forward placement of combat forces with minimal reserve at Division level. It further suggests that as frontages broaden and forces thin, more of the combat power should be positioned forward at the expense of reserve units. The 1st Infantry Division found tactical success in retaining a strong reserve to counterattack decisively in the destruction of enemy forces on a fluid, porous battlefield. Battalions were moved frequently and placed OPCON to brigades at the point of main attack. Attack helicopter companies were moved with great frequency to blunt enemy thrusts in combat with maneuver forces on the ground. But, in the final analysis, it was the strong, decisive counterattack which destroyed the enemy force, destabilized his offensive, and denied him the initiative that underscored the tactics employed by this Division. In my judgment, we need to rethink active defense doctrine. We need to re-examine the role of a reserve force. We need to review the porosity of forces on the wide frontage battlefield and their vulnerability to penetration in darkness and periods of reduced visibility. With these understandings, the defender can seize the initiative for limited periods of time, thus significantly disrupting the enemy's offensive drive.

That paragraph suggests to me that we've failed to get across to the Big Red One the essence of what we are trying to say in FM 100-5.

In the matter of the reserves, for example, you apparently perceive that we intended that there be no reserves at all. That is not the case. What we were trying to say was that the target servicing problem is likely to be so acute, especially against a breakthrough attack, that the division commander may not be able to afford to hold out a reserve. If he does hold out one, it will likely be a force disposed in depth astride a major avenue of approach, where it can defend in depth, should the main attack indeed develop there, or be used elsewhere by moving quickly to wherever the main attack might be. These are terms familiar in armor-mech doctrine for years; when forces are slim the reserve is the uncommitted or least committed force.

In CENTAG, I would think the corps commanders would want to create a reserve as quickly as possible as reinforcements from the States begin to arrive. But the possibility of maintaining any sizable reserve, under terms other than those described above, is in my mind out of the question—there just aren't enough forces there to begin with. In V Corps there are two brigades echeloned in depth along each of the two main avenues of approach in the MBA; the rearmost brigades are deployed under command of the division commanders, but are not to be employed without approval of the corps commander. This condition will have to obtain until reinforcement units arrive in sufficient strength and the battle stabilizes enough for a reserve to be created.

Even when a reserve is possible, I still believe it not likely that we could employ it using the terms of reference once set forth for the reserve in the mobile defense. For the same reason that we believe it possible to annihilate large numbers of Sovs coming at us in mass formation, it will be possible for them to do likewise unto us. That's what's different about the active defense; that sets it apart from the mobile defense.

When I went to V Corps, I first walked along the GDP positions with every battalion commander and had each explain what he intended to do in the position he'd been given to defend. What I found was a reserve at every echelon—from company to corps. When I tallied up the aggregate, 56 percent of the tanks in the corps were in reserve at one echelon or another. The problem was that no one had ever added it up that way before. I challenge, therefore, the assertion that the forward defenses are any more porous in the Active Defense than they were with over half the tank strength in reserve in the mobile defense. The two ideas are mutually exclusive.

Now it is also true that a lot of counterattacking will be necessary, and to some extent at least that idea didn't come clear in 100-5, although when we wrote it we fully intended for it to stand out. Remember we talk of attacking by fire, and by fire and maneuver, always bearing in mind that when the defender gets up out of the terrain to maneuver he gives up the terrain advantages that are his as a defender. Apparently that's a harder concept to grasp than we realized. But it is also true that the Soviets pay almost no attention to terrain, and therefore it is the battle multiplying effect of terrain that can be one of our greatest advantages if we but learn to use it properly. So counterattack, but do it in smaller battalion increments rather than brigades, as before, and do it carefully—not cautiously, but carefully.

I'm sending you a copy of *The War of Atonement*. In the first seven chapters Herzog describes the battle on the Golan Heights. It is my firm conviction that in a war in Europe or in the Middle East, should we be required to fight there, some brigades of the US Army will fight a battle like that of the 7th Brigade. I've walked that ground and refought that battle with the commanders who were there. I did it right after the war when dead men and dead tanks still littered the battleground. I did it again last May. There's no question in my mind that the outcome of that battle was a miracle—God's hand over His chosen people if you will. It's also quite clear that its outcome was the product of some terribly hard fighting by some superbly trained crews who were incredibly well led. I don't know the Lord's will with regard to our Army—I'd hope we are on His side sufficiently for the miracle to be reworked if something like that ever happens to us. But I also know that to the extent we can we're obliged—you, I, all of us, to do the best we can to provide our soldiers the kind of training and leadership that give us the best possible chance to win—with a little luck and perhaps a miracle or two.

In the weighing out of whether or not we can ever hope to fight and win outnumbered, the miracle of the 7th Brigade loomed large in our considerations. But so also did some other realities, realities about the War in the Sinai, the 1967 War, and the experience of several hundred tank battles since World War I. In the end we could find no other rational way to proceed than the way we set out in 100-5. The Germans agree. I would therefore hope that we could all pull this wagon together by seeking a deeper understanding of what went into the development of 100-5 and what it really says and means. It will serve no purpose at all if any or all of us—especially us senior folk—speak of rethinking the doctrine. We've decided to do it that way—not TRADOC has decided, but the Army has done so. We've gone to considerable lengths to work out some differences with the Germans, who had independently come to most of the same conclusions as had we.

May I hope therefore that the problem is that we were not able to express ourselves clearly enough in 100-5, and it is that lack of clarity in concept which causes the instant problem. FM 71-100, *The Division*, is aborning. In it we have tried very, very hard to provide the requisite balance to redress what has apparently become an unbalance in 100-5 itself. I hope you'll look 71-100 over carefully when it comes to you this summer. We are all available in TRADOC, from me on down the line, to provide whatever assistance we can to help further our mutual understanding of what we were trying to set forth and what we believe more firmly than ever the US Army has to do to win—outnumbered.

Battlefield Reserves

Message to Lieutenant General John R. Thurman
Commander, Combined Arms Center
22 March 1978

Following our discussions of last Friday, I want to be sure that we all understand what we tried to get across in 100-5 and what we must present in our tactical courses of instruction regarding reserves. Some perceive that we intended that there be no reserves at all. That is not the case. What we were trying to say was that the target servicing problem is likely to be so acute, especially against a breakthrough attack, that commanders may not be able to afford to hold out a reserve. If a reserve is held out, it will likely be a force disposed in depth astride a major avenue of approach, where it can defend in depth should the main attack indeed develop there, or be used elsewhere by moving quickly to wherever the main attack might be. These are terms familiar in armor-mech doctrine for years; when forces are slim, the reserve is the uncommitted or least committed force.

Battalions probably will not have a reserve, and brigades may or may not, depending on the division commander's decision. If the division has a reserve, brigades will not have reserves. The corps commander may direct his divisions to echelon brigades in depth along a main avenue of approach in the MBA, these brigades to be deployed under command of the division commander, but not to be employed without approval of the corps commander. These then would be both the corps and division reserve. This condition will have to obtain until reinforcement units arrive in sufficient strength and/or the battle stabilizes enough for a reserve to be created. Even when a reserve is possible, it is not likely that it would be employed using the terms of reference once set forth for the reserve in the mobile defense. For the same reason that we believe it possible to annihilate large numbers of armored forces coming at us in mass formation, it will be possible for them to do likewise unto us. That's what's different about the active defense; that sets it apart from the mobile defense.

When I went to V Corps, I first walked along the GDP positions with every battalion commander and had each explain what he intended to do in the position he'd been given to defend. What I found was a reserve at every echelon—from company to corps. When I tallied up the aggregate, 56 percent of the tanks in the corps were in reserve at one echelon or another. The problem was that no one had ever added it up that way before. As I count battalions, Bob Washer's example still had about 25 percent of the tanks in reserve. Against five divisions, that's still probably too much. I don't want to screw up his method of figuring out how to allocate forces, but believe our problem still is one of too big a force designated as reserve(s).

Instruction on Offensive Operations

Message to Lieutenant General John R. Thurman
Deputy Commanding General, US Army Training and Doctrine Command
29 March 1978

It seems to me shortsighted of us to set the only offensive instruction [at Fort Leavenworth] in the contingency corps scenario. Not that contingency operations don't need covering. But Offense in the Mechanized Battle—Europe, is the thing we explained least well in 100-5, therefore it seems important to me that we resolve that by having an instructional unit on it. Otherwise we are perceived to have skirted the whole issue of how to attack in Europe. And attack we must in order to win. Were we to change, that begs the issue of the contingency corps and how it operates. There may be some other ways of handling that matter without seeming to have skirted it. It's of less concern to me that we underemphasize the contingency corps than it is that we ignore the offense in Europe.

FM 100-5: *Operations*

Inter-University Seminar
Fort Leavenworth, Kansas
30 March 1978

Our Army has regulations that cover many subjects, but not military tactics. Whether this is by design—to protect the ignorant—or by oversight, I'm not sure. Probably, it's because we all consider ourselves experts, and so, we never agree long enough to write a regulation.

But we do set forth operational concepts, tactics, in field manuals. Normally, these manuals live a pretty quiet life, serving as references and as exhibits for various inspectors' checklists. In many cases, they are little read, less often followed, and not the prime topic of Army conversation.

That is, at least until the 1976 edition of Army Field Manual (FM) 100-5, *Operations*, appeared. It's safe to say that no Army manual has ever been so widely commented on and misunderstood. It is quite unusual for a field manual to generate such a wide discussion as this one has. I'm not quite prepared to say why the discussion developed, but develop it did.

So, the first point I want to make this morning is that that's not all bad. About anything important, a dialogue is essential. Operational concepts are the stuff on which armies feed and nurture themselves. And so a dialogue about them is essential to the Army's well-being. My purpose this morning is not to be defensive about the concepts we have espoused for our Army. I don't believe they need all that much defending. But enough of the dialogue has struck recurring themes that I think it appropriate to offer some perspectives as a backdrop for your deliberations here in this seminar.

To understand any description of war and the proposed fighting of it, both complex problems, it is useful and quite necessary to understand the circumstances under which the liturgy was laid down. Let's begin with that.

The time is 1970–73. The US military is withdrawing from Vietnam. The US Army, Europe, is in disarray, rent asunder by its role as part of the rotation base for forces deployed in Vietnam. The Army training base in the continental United States has concentrated almost exclusively for 8 years on providing units and individual replacements to Vietnam. The combat development community has concentrated on Vietnam to the exclusion of work to modernize the Army's ability to fight in other theaters. Doctrinal development is still in the mindset of the 1950s.

In the Pentagon, the Army Staff is trying to restructure the Army, to find for it a size, shape, composition, and operational concept consistent with the needs of the US national military policies, whatever those might be. Three major factors strongly influence this restructuring.

First, there is no well-articulated national military policy. Second, there is the strong feeling that, after every war, armies always set out to figure out how they might have fought that last war better. There is an even stronger determination to avoid that process, and this time to look ahead, not back. Finally, there is the grim realization that modernization of the Army, normally a constant process, has been set aside for Vietnam. No matter what operational concept is derived, the Army is faced with a desperate need for a massive and expensive materiel acquisition program.

So, in the context of 1970–73—times of social, political, and economic upheaval in our society—what did we see for our country and our Army as we tried to look ahead? We saw the possibility of two wars, one a mechanized war such as we might have to fight in NATO Europe or perhaps even in the Middle East. The other war would resemble a Korea, a Vietnam, a Lebanon crisis, a Dominican Republic. Each obviously requires different kinds of forces—mechanized on the one hand, light infantry on the other.

Estimating likelihood of occurrence, war in NATO Europe, while probably least likely, was certainly the most important from the standpoint of our national survival and the well-being of Western civilization. Contingency operations with light forces, while perhaps less important to our national survival, were still probably more likely.

With Nixon doctrine beginning to reaffirm our national interest in Western Europe, our military focus narrowed to NATO. Here, our strongest and most dangerous enemy was much stronger than when we turned our backs on him to go to Vietnam. And he was growing stronger almost daily. So, we decided to begin with developing operational concepts to cope with our most difficult problem, the mechanized war.

What might that battlefield be like? First, and most apparent, was the improvement in quality of modern weapons. Tanks four or five times more effective than in World War II. Artillery 10 times as effective. True, it didn't happen overnight, but gradually, in an evolutionary way, quality had arrived. Quality brought with it some sophistication, demanding more training for the soldiers.

Second, and almost as apparent, were the twin problems of quantity and variety. In the main part of the battle, there were likely to be quantities of weapons systems never seen before and such a variety that no single weapon system could cope. To win, it's going to take a little bit of everything, a well-integrated combined arms team.

A new dimension, radio electronic combat, will make the leader's job harder. It will be more difficult for him to communicate with those he has to talk to in order to get on with the battle. Modern air defenses make the air over the battlefield a very dangerous place. The battlefield may be dense with obstacles of many kinds—blown up equipment, antitank ditches, mines.

What might we have to fight against? Almost every potential threat mechanized army uses Soviet operational concepts, organizations, and equipment. Key concepts are mass, momentum, and continuous land combat. Mass means numbers and concentration of forces. Momentum means sustaining advances of 40 to 50 kilometers a day. Continuous combat is echelonment of forces; when one unit is consumed in the battle, the next one is committed without a pause. Modern vision equipment on both sides enables the fight to continue around the clock, in smoke, in bad weather.

Operationally, an enemy attack might find a whole tank division, more than 300 tanks, employed on a front as narrow as 5 kilometers, deployed in first and second echelons.

Even moving to contact a motorized rifle division might be about 15 to 25 kilometers wide, divided in echelons—reconnaissance, advance guard, first and second echelons. The frontage is wider than the breakthrough, but speed is equally important.

Operational mass, a problem in itself, combines with a larger problem—our historical propensity for being outnumbered, losing the early battles, then mobilizing to outnumber our enemies and so win the war. A look at today's world strongly suggests that logic to be bankrupt. So, we come to the visceral question—can we hope to fight outnumbered and win? What operational concepts would defeat such an enemy? While there is no rule that the outnumbered side loses, mathematicians and statisticians would have us believe it. So, must we admit defeat in such a war, especially in Europe? Do we have almost five divisions in Europe as a token, a hostage force, or have we a chance to win?

Let's explore the phenomenon of fighting outnumbered. If we plot the probability of victory, . . . a traditional calculus develops this curve: if you are 1 attacking 5, you can't win, but if you are 5 attacking 1, you win every time. That's not very encouraging for a country whose forces, even with allied help, can expect to be outnumbered at the outset of a conflict, and in today's world to stay that way for a long time.

But the history of a thousand tank battles tells quite a different story. The probability of victory seems to change very little, regardless of the odds. Apparently, there is something about battle that can't be described using traditional calculus. Military forces can and frequently do fight and win outnumbered.

So, back to the original question: what operational concepts can we employ? Being realistic about NATO Europe and a policy prohibiting first attack, we would begin a war there defending, just by the circumstances. The defense was also an appealing problem because of the natural advantages afforded an outnumbered force—terrain, preparation of the battlefield, and so on.

We considered it essential to use terrain to beat the enemy, exploiting his propensity to mass without regard to the ground. As you know, the defender loses the initiative to the attacker. But we also believed that, to win, we needed to regain the initiative. Even if for short periods, it would be necessary to attack while defending. To avoid suicidal massed assaults of our own, defensive attacks by fire and maneuver at critical times and places would be the rule.

Eventually, an operational concept grew: see deep to find the following echelon, move fast to concentrate forces, strike quickly to attack before the enemy can break our defense, finish the fight quickly before the second echelon closes. All this using the defenders' natural strength—terrain to multiply the strength of the defense.

Many problems intrude between a short, snappy operational concept and a full-blown field manual description. The full liturgy didn't come easily. Ancient precepts had to be rethought, changed, or set aside.

The new operational concept unveiled new and difficult problems. We elected to describe how to defeat the breakthrough attack, believing that if we could do that, we could solve the other problems. I still believe that.

An attacking enemy combined arms army deploys to a depth of about 100 kilometers; its first-echelon divisions are 30 kilometers deep; about 50 to 60 kilometers back are second-echelon divisions of the first-echelon army; 120 kilometers back are reconnaissance elements of the second-echelon army.

According to our operational concept, we had to see deep. How deep? In what detail? To what level? In what timeframe? How and with what? As the tactical concept developed, some answers to these questions began to appear—a scheme of maneuver evolved.

We found the brigade commander must have information of second-echelon regiments of the first-echelon division; the division commander must have information of second-echelon divisions of the first-echelon army; the corps commander must know about the second-echelon army. Generally, the deeper he must see, the more a commander must rely on sources he doesn't control directly. To get the right information, each commander must mount a tough relentless operation using his own and outside resources—target acquisition and surveillance resources.

To defend successfully, the defender needed to deploy forces to find the enemy and force him to start his main attack before coming on the main defending forces. Borrowing an old term, this was called the covering force. The use of an old term didn't help many old tacticians, because they transferred the old mission along with it. This covering force was different. It was meant to fight, not just to gain time but to deceive the enemy as to our main battle area and to draw him out from under his air defenses. It simply could not afford to fire a little, too little, then move back too soon and too far, the traditional modus operandi of covering forces. Now, the covering force was to defend, its battalions and squadrons fighting just as would similar units in a main battle area where the real defensive battle was to be fought.

Where the rear fight is, at battalion and below, the battle is won by destroying enemy systems—servicing targets. A simple explanation of servicing in the defense is to determine how many targets must be destroyed in time and space in order to persuade the enemy to give up the attack. The battalion might have to destroy as many as 250 targets in about 10 minutes, the division more than 2,000 targets in hours or perhaps days, the corps more than 3,000 targets in perhaps 3 to 5 days.

Now, a look at move quickly. If a US division, deployed across about 40 kilometers, is attacked by anything like a breakthrough attack, the division commander will have to concentrate 7 or 8 of his 11 or 12 battalions very quickly. He fills in areas that he has denuded of ground forces using air cavalry and attack helicopter units.

In the main battle area, we can win by servicing targets, concentrating on the flanks of the attack and deploying deep at its front. The battle is fought in a series of successively deeper battle areas and positions until the attack is killed. There is, however, no traditional massive counterattack to eject the enemy. We are outnumbered; the luxury of a large reserve force is beyond our means. No idea in active defense doctrine has been harder to get across than the absence of a large reserve. It means everybody else has to do his job right—the first time.

Some perceive that we intended no reserves at all. That is not the case. What we were trying to say was that the target servicing problem is likely to be so acute, especially against a breakthrough attack, that commanders may not be able to afford to hold out a reserve. If a reserve is held out, it will likely be a force disposed in depth astride a major avenue of approach, where it can defend in depth, should the main attack indeed develop there, or be used elsewhere by moving quickly to wherever the main attack might be. These are terms familiar in armor-mechanized doctrine for years; when forces are slim, the reserve is the uncommitted or least committed force.

Battalions probably will not have a reserve, and brigades may or may not, depending on the division commander's decision. If the division has a reserve, brigades will not have reserves. The corps commander may direct his divisions to echelon brigades in depth along a main avenue of approach in the main battle area, these brigades to be deployed under command of the division commander, but not to be employed without approval of the corps commander. These, then, would be both the corps and division reserve. This condition will have to obtain until reinforcement units arrive in sufficient strength and/or the battle stabilizes enough for a reserve to be created.

Even when a reserve is possible, it is not likely that it would be employed using the terms of reference once set forth for the reserve in the mobile defense. For the same reason that we believe it possible to annihilate large numbers of armored forces coming at us in mass formation, it will be possible for them to do likewise unto us. That's what's different about the active defense that sets it apart from the mobile defense.

The alternative, maintaining a large reserve as a counterattack force, seemed and still seems out of the question. There just aren't enough forces to begin with.

Now, it is also true that a lot of counterattacking will be necessary. Remember, we talk of attacking by fire, and by fire and maneuver, always bearing in mind that, when the defender gets up out of the terrain to maneuver, he gives up the terrain advantages that are his as a defender. Apparently, that's a harder concept to grasp than we realized. But it is also true that our enemy pays much less attention to terrain, and therefore, it is the battle multiplying effect of terrain that can be one of our greatest advantages if we but learn to use it properly. Counterattacks are important, but done in smaller battalion increments rather than brigades, as before, and done carefully—not cautiously, but carefully.

The division is normally the lowest level at which a counterattack plan is formally prepared. A counterattack is by definition a hasty attack and, if it is not successful, then the division may have to conduct a deliberate attack. Here, the old graphics about an assumed penetration that would be attained were even less useful than before. A more sensible way to control the counterattack might be to divide the real estate into recognizable compartments. The commander can direct the most logical subordinate unit to counterattack using the predesignated segments. Thinking about this for awhile, it gets more and more logical. Those who criticize this concept for the lack of a large counterattack force frequently complain about the porous nature of the main battle area—everything is forward. Now, holding out a large reserve can't make a battle area any more dense. Indeed, the reverse occurs. You can't have it both ways.

To attack, we found it appropriate to study how the enemy defends. Echelonment of forces is also characteristic of their attack doctrine. The enemy defense attempts to canalize and misdirect any attack. Massive counterattacks at times and places of his own choosing are also important. The two most singular features are density of troops and the massive size of the counterattack forces—for a Soviet front, an entire tank army.

Again, our operational concept was to see deep, concentrate, suppress enemy fires, and strike into enemy rear areas. The principles of seeing deep already discussed apply once again. This attack concept is not a great change from before. Attacking enemy rear areas is now a key feature of any attack rather than a special operation. This is so because the main fight must be with the second echelon, otherwise we'll be caught in a battle of attrition that could return us to the tactics of World War I. Striking enemy rear areas goes hand in glove with the see deep idea. The rear area facilities of command, control, and communications are the vulnerable parts of an enemy defensive operation. They must be destroyed.

Most observers believe that the Soviet Achilles' heel is in C3 and logistics. If we can destroy these by fire or maneuver or by overrunning them, the cohesion of the defense will be broken.

FM 100-5 has been the target of but one major criticism about offense. Critics say we didn't stress it. Now, page count and graphic aid presentations are irrelevant. Lack of stress is in the eye or mind of the reader. It may be that for nearly 20 years we've been a defensive-minded Army. We need to change that, but printing the offense chapter in blood isn't the way to do it. It has to be instilled in the soldiers' minds. The best defense is still, as in football, a good offense. The active defense is designed as it is to stop the enemy, to destroy his attack force, so that we can go on the offense.

Other aspects of FM 100-5 have drawn comment, but time is closing on your upcoming deliberations. I hope when you treat the subject of operational concepts for the US Army, you consider the perspectives I've offered, and that you disown the role of the critic who offers no viable alternatives.

One final point. We believed in 1970–73 that, by describing the battle, we could develop a package of tactics, force structure, equipment, and training. The operational concept—the tactics—are now guiding force structure and equipment development. But we didn't forget the last part—training. From the tactics, we derived training principles that move our training.

These principles perceive that wars are won by the courage of soldiers, the quality of leaders, and the excellence of their training. Of the courage of our soldiers, there is no doubt, nor should there be. The other two principles are related and require some discussion.

When considering war, it is obvious that, to win, one must think, but thinking takes time; in battle, there is no time. So, we must train all our soldiers, officer and enlisted, to think things through in advance. Only if we are mentally prepared ahead of time can we be successful. The pressure of combat leaves no margin for error.

We and our enemy have similar weapons. We fight on the same terrain, in the same weather, and we both fight fear. Only in the ability of our leaders is there any difference. Therefore, we must train our leaders to think logically about difficult problems, and to do it quickly and under pressure, so the decisions are on time and right—the first time.

Our commanders must know best what to do in complex situations, but the rub is that, in battle, our commanders can't be all over. Complex situations will develop where the commander isn't present or can't get there. Soldiers must be trained to handle complex situations with standard drills, to do something right until the boss arrives.

To get things done in battle, decisions must be made and orders given. A complex battlefield calls for complex orders, and these can easily go wrong. In battle, such difficulties can be fatal, so we must train the whole team through complex situations on a regular basis.

Our training system is built on these principles. From Skill Qualification Tests and Soldiers' Manuals for the individual soldier to Army Training and Evaluation Programs for unit training, they all support concepts set forth in FM 100-5, *Operations*. The cement that bonds together good weapons, sound tactics, and effective organizations into winning battle teams is training—training to develop excellence in the skills of leaders and soldiers to the end that they have both the capability to fight the tough battles and the conviction that they can and must fight hard and well and that, if they do, and have a little luck, they'll win against all odds.

Doctrine Writing

Letter to Major General Paul F. Gorman
8th Infantry Division
18 April 1978

Thanks for your 3 April letter about [FM] 71–100. . . . I wish now that we had published the draft with which I was confronted when I arrived, although the systems business was very, very heavy, and we have in fact had to abandon that method of describing things since, so perhaps it's better. I put the tag on Leavenworth to do it and they bombed out again, so it's in here as before; Scribner and I and a couple of unsung heroes are working on it. Terribly disappointed in our inability to get Leavenworth to produce. No fault of Roy's, I guess—there's just a horrendous bureaucracy out there that apparently defies progress. Not the least of our problems has been a residue of resistance to anything from here—the feeling that we would go to AP Hill with the thing in the end anyway, so why should they labor too hard out there. Still lots of resentment to that method—I have therefore avoided it, although I thought it was super at the time—and still do! Very little on paper with the brigade manual, and having a bitch of a time on the corps book. We'll get it done somehow, but it will take far longer than any of us foresaw at the outset.

The Central Battle

Association of the United States Army
Fort Benning, Georgia
24 April 1978

This morning, you've heard what the other side is capable of doing. I'm sure most of our speakers would agree that the crystal ball we've been gazing into is a little murky. If it's that way as we try to look at the threat, it's equally so when we try to look at ourselves. But look we must. So, what I'd like to do is describe for you the way we look at ourselves and the enemy. The method we've elected to use is one of trying to describe what the battle might look like.

We call this description the Central Battle. It is an attempt to show the interrelationships on both sides of tactics, organizations, weapons, and training. The Central Battle is that critical arena on the battlefield where all these aspects join together to cause a decision.

Here's a rough idea of how we describe that arena. It's 1978. Soviet systems are T64/T72 tanks, BMPs, ZSU 23-4s. Our systems are M60A1s, M113s, TOWs, Vulcans. These systems are run together in battle simulations. The outcome is a description of the capability deltas—pluses and minuses—for both sides.

Next, we check our current development programs to be sure we are working to redress the imbalances and doing it in a timely way. If not, we have to get something started quickly.

Now, let's look ahead—say 10 years. Take systems known to be coming in between now and then, add new capabilities that plausibly will be added between now and then, and run the simulation again. Going back to our example, we'd tot up today's deltas, add systems now under development—T80 tank, the new SP artillery, and lots of armed helicopters, TAC air, and artillery for the Soviets. On our side, add XM-1, the infantry fighting vehicle, Roland, then we replay the battle to develop 1988 deltas. Using a technical assessment, we must project to beyond 1988, trying to foresee in the programs in our laboratories opportunities for a technical strategy we might pursue to move into systems for the Central Battle in the year 2000. Very quickly, that's how we work the problem.

But let's go back and talk our way through Central Battle 1978. For that battle and on toward Central Battle 1988, we see the possibility for two wars. First, mechanized war, such as war in NATO Europe, perhaps even in the Middle East. And the other war—a Korea, a Vietnam, a Lebanon crisis, a Dominican Republic. Each requires different kinds of forces—mechanized on the one hand, light infantry on the other. Estimating likelihood of occurrence, war in NATO Europe, while probably least likely, is certainly the most important from the standpoint of our national survival and the well-being of Western civilization. Contingency operations with light forces, while perhaps less important to our national survival, are probably more likely. The importance of our commitment to NATO dictates that we focus on Europe first, for that's our most difficult problem; solving that will also help solve the "other war" problem.

Now, what might the battlefield in the mechanized-armor war be like? First and most apparent is the quality of modern weapons. Tanks today are 4 or 5 times more effective than in World War II, artillery 10 times as effective. True, it didn't happen overnight, but gradually, in an evolutionary way, quality has arrived. Quality brings with it some sophistication, demanding more training for the soldiers.

Second is the two-sided problem of quantity and variety. The field of the Central Battle will be crowded with large quantities of weapons systems and such a variety of systems that no single weapon system can cope. To win, it's going to take some balance of everything, a well-integrated combined arms team.

Electronic warfare, radio electronic combat, will make the leader's job harder, for it will be more difficult for him to communicate with those he has to talk to in order to get on with the battle. Modern air defenses will make the air over the battlefield a much more dangerous place than ever before. The battlefield may be dense with obstacles of many kinds, blown up equipment, antitank ditches, mines, all of which will make mobility more of a problem than in past times.

How might we expect the enemy to fight? Almost every potential threat mechanized army uses Soviet operational concepts—tactics, organizations, and equipment. Key Soviet concepts are mass, momentum, and continuous land combat. Mass translates to numbers and concentration of forces. Momentum means sustaining advances of 40 to 50 kilometers a day. Continuous combat is echelonment of forces so as to reinforce success in the attack and to ensure defeat of an enemy attack while defending. Modern vision equipment on both sides enables the fight to continue around the clock, in smoke and in bad weather.

Operationally, an enemy attack might see a whole tank division, more than 300 tanks, employed on a front as narrow as 5 kilometers and deployed in first and second echelons.

Even moving to contact, a motorized rifle division might be about 15 to 25 kilometers wide, divided in echelons—reconnaissance, advance guard, first and second echelons. The frontage is wider than the breakthrough. Speed is equally important.

What operational concepts might we adopt to accommodate to that enemy on that battlefield? And what might those operational concepts lead us to in developing tactics, weapons systems, organizations, and training systems? Recall, we said NATO Europe represented our most difficult problem. Our job there is to cope with an enemy first attack, so let's look first at the defense.

To counteract the Soviet operational concepts I have already discussed, we have developed some operational concepts of our own. Simply put, those concepts tell us we must see deep to find the second echelon, move fast to concentrate forces, strike quickly to attack before he can break into our defense, and finish the fight quickly before the second echelon closes. All the while, we must take advantage of the defender's natural strength, the opportunity to use terrain and weapons to best advantage, multiplying the strength of the defense.

It is essential to use terrain to beat the enemy, exploiting his propensity to mass without regard to the ground. It is also true that the defender gives up the initiative to the attacker. And we believe that, to win, we need to regain the initiative. Therefore, even if for short periods, it will be necessary to attack, attack while defending. But it will also be necessary to avoid suicidal massed assaults of our own. Defensive attacks by fire, or by fire and maneuver, at critical times and places must be the rule. Now, let's look at what all that means.

Consider a schematic of a combined arms army making a main attack for a front. We find the army deployed in a depth of about 100 kilometers, with its first-echelon divisions about 30 kilometers deep. About 50 to 60 kilometers back, we have the second-echelon divisions of the first-echelon army. At about 120 kilometers to the rear of the front line, we begin to meet the reconnaissance elements of the second-echelon army. This geometry poses some problems.

The problems are twofold. One has to do with how far we can see, the other with how far we can fire. Consider the depth of surveillance provided by the target acquisition and surveillance systems available to the corps commander. Notice that the bulk of the target acquisition and surveillance means reach not quite to the second-echelon regiments of the first-echelon divisions. The SLAR and a few sensors are the only things we have that can see those regiments. Nothing under the immediate control of the corps commander can see the second-echelon divisions of the first-echelon army.

With the exception of the nonnuclear Lance, notice that the corps commander with his organic fire support can fire only about as far as he can see. His fire support means of primarily cannon artillery cannot reach the second-echelon regiments of the first-echelon divisions.

To see deep, we believe the brigade commander must have information of second-echelon regiments of the first-echelon divisions; the division commander must have information of second-echelon divisions of the first-echelon army; the corps commander must know about the second-echelon army. Generally, the deeper he must see, the more a commander must rely on sources he doesn't control directly. To get the right information, each commander must mount a very aggressive operation using his own and outside resources—target acquisition and surveillance resources.

To defend successfully, the defender must first find the enemy and force him to develop his main attack before coming on the main defending forces. This is done by the covering force. The covering force *fights* to gain time, to deceive the enemy as to the whereabouts of the main battle area, and to draw the enemy out from under his air defenses. Covering force battalions and squadrons fight just as would similar units in a main battle area.

To fight in the main battle area, forces must move quickly. If a US division, deployed across about 40 kilometers, is attacked by anything like a breakthrough attack, the division commander will have to concentrate 7 or 8 of his 11 or 12 battalions very quickly. Areas denuded of ground forces are filled with air cavalry and attack helicopter units.

In the main battle area, the fight is won by servicing targets, concentrating on the flanks of the attack, and deploying deep at its front. The battle is fought in a series of successively deeper battle areas and positions until the attack is killed. There is, however, no traditional massive counterattack to eject the enemy. We are outnumbered; the luxury of a large reserve force is beyond our means.

At battalion and below, the battle is won by destroying enemy systems—servicing targets. A simple explanation of servicing in the defense is to determine how many targets must be destroyed in time and space in order to destroy the enemy attack. A battalion might have to destroy as many as 250 targets in about 10 minutes; a division more than 2,000 targets in hours or perhaps days; a corps more than 3,000 targets in perhaps 3 to 5 days.

To attack, we first must study how the enemy defends. Echelonment of forces is also characteristic of his defense tactics. In the defense, he attempts to canalize and misdirect any attack. Massive counterattacks at times and places of his own choosing are also important. The two distinguishing features of his defense are density of troops and the size of the counterattack forces—for a Soviet front, an entire tank army.

Our operational concept in the attack is to see deep, concentrate, suppress enemy fires, and strike into enemy rear areas. We've already discussed seeing deep. Concentration is necessary to fracture his first-echelon defenses. To move anywhere at all, suppression will be a must. Striking deep into enemy rear areas, we seek to destroy command-control, communications, and logistics, the most vulnerable parts of an enemy defense. If we can destroy these by fire or maneuver or by overrunning them, the cohesion of the defense will be broken.

So much for operational concepts for the Central Battle. From battle simulations and tests, we have observed some imbalances, some trends. Since you're primarily interested in materiel developments, let me highlight some conclusions of our Central Battle analysis.

First is the obvious need for fielding the XM-1 tank. While it's still part of the combined arms team, the tank is the central actor in the Central Battle. Because this is so, the tank is the prime target for almost all enemy systems. It is, therefore, the most stressed system on the battlefield. It is at once required to destroy more, yet to survive attacks from more antagonists than ever before. This is precisely why the XM-1 tank is such an important development. Its armor protection, survivability, is a quantum improvement, an improvement that will affect every other weapons system in the Central Battle.

The next obvious need is for a fighting vehicle for our mechanized infantry. The US M113 is an armored personnel carrier. It contributes little to the Central Battle. The troops it carries and their antitank guided missiles do contribute when deployed, but compared to Soviet BMPs and ATGM carriers, the M113 is a zero. It has little firepower and less survivability. It is obvious that a mechanized infantry fighting vehicle with armament that can stand off tanks with its ATGM and other armored vehicles with its gun is, after the XM-1 tank, the major US requirement. Its prime role is to reduce stress on the tank.

Now, a word about counterfire weapons. First, air defense—we have the distinguished, but elderly, Vulcan/Chaparral mix. Even when Roland is deployed, there is still an imbalance, both short and long term. This brings us to the requirement for a DIVAD gun. A secure air defense umbrella over our own forces will, in turn, free more weapons systems such as Air Force fighters and Army helicopters to be used offensively against the enemy. Thus, other systems become less stressed.

No facet of our Central Battle methodology is more important than its use to evaluate the ability of a system to stress the enemy and to take stress off other friendly systems. So, if we can improve our air defense, we will have a more offensive Air Force and helicopters. If we can improve our mechanized infantry, attack helicopters, and antitank guided missiles, we will have more effective tanks.

Central Battle analysis also shows imbalances in combat support and combat service support. On the modern battlefield, resupply will be an awesome problem. The ability to resupply quickly and frequently is indeed a combat multiplier. One obvious need is for an armored resupply vehicle that can move into and resupply systems fighting in the Central Battle and survive while doing so.

One of the most difficult things to analyze is our ability to control forces in the Central Battle. In recent years, we've seen a vast increase in the amount of information available, the number of sources that provide it, and the variety of means to move the information around. However, all this is a mixed blessing. We find ourselves with so much information that we cannot identify the parts that are important and necessary. We are, therefore, confronted with an information overload, a situation in which, as information inputs increase, outputs decrease as the overload threshold is reached and passed. Now, systems have ways of coping with overload.

Not understanding this phenomenon too well, we find historically those who have coped with the problem have used multiple channels more than any other technique. Most frequently, the multiple channel has been a separate one, one that in fact bypasses the established information flow system and proceeds with some ad hoc system to move critical information around. In both World Wars, several clever systems were used by military leaders on both sides. Commanding the Central Battle calls for real-time information and a speedy discrimination process, something that will key up the important information in a usable format and in time to be used. This is a tall order, but it too will reduce stress in the system and, hopefully, obviate the present need for considerable duplication of effort.

One final word. In the Central Battle, organizations, tactics, and equipment are about a wash. Their tanks are better than ours in some ways and ours better than theirs in some ways. For every tactic, there can be developed an effective countertactic. All organizations are fairly effective and fairly equal. It's how we put all this together that counts. Remember, I said that training was an important part of the Central Battle. Let's talk about that for a moment.

First of all, there is the problem of training to fight outnumbered. Now, there is no hard and fast rule that says if you're outnumbered, you lose. But a lot of people, particularly well-meaning mathematicians and statisticians, would have us believe so.

If we plot the probability of victory with attacker versus defender ratios of 1 attacking 5 to 5 attacking 1, a traditional calculus develops a curve indicating that, if you are 1 attacking 5, you can't win, but if you are 5 attacking 1, you win every time. This is not very encouraging for a country whose forces, even with allied help, can expect to be outnumbered at the outset of a conflict and, in today's world, to stay that way for a long time.

But the history of a thousand tank battles tells quite a different story—the probability of victory seems to change very little, regardless of the odds. Apparently, there is something about battle that can't be described using traditional calculus. Military forces can and frequently do fight and win outnumbered. That difference between the curves can be explained in the courage of soldiers, the quality of leadership, and the excellence of training. Most important, it's training that ties the other two together.

We've done a series of Armywide evaluations of proficiency. These have measured the difference between the capability of the weapons systems with high-performance crews and low-performance crews. That performance gap can be explained in one word—training. If you test a unit right at or after its annual gunnery, you get a high curve, but then training decay sets in. A proficiency test given 6 months later results in a low curve.

We war-gamed these levels in some of our computer simulations and found that the final outcomes changed significantly. They showed that two high-performance companies destroyed one extra enemy tank company, plus the tank companies that would normally be destroyed by low-performance companies. In addition, the well-trained soldiers survived better by losing fewer of their own systems.

This difference becomes more dramatic when you project it to a corps—six additional tank battalions, two-thirds of a Soviet tank division, are destroyed. That's the kind of edge in the odds that training can produce.

I mentioned that courage of soldiers, quality of the leaders, and excellence of training are the keys to winning. Of the soldiers' courage, there is no doubt. The leaders' quality can be developed by good training. Our training system is built on that principle.

So, this has been a quick look at requirements for our Army. As we develop our Central Battle methodology, some of these may change. In the briefings that follow, both today and tomorrow, you will be provided more details about equipment and systems requirements and developments. May I leave you with two thoughts.

First, understanding the operational concepts with which we intend to fight is essential, for they are the driving force behind our development of tactics, organization, materiel, and training. However, having grasped operational concepts and drawn from them a coherent scheme embracing tactics, organization, and equipment, one must recognize that this is just the price of admission. All that just gets you on the field and gives you a capability about equal to the other guy.

The more important thing to know is that the cement that bonds together good weapons, sound tactics, and effective organizations into winning battle teams is training—training to develop excellence in the skills of leaders and soldiers to the end that they have both the capability to fight the tough battles and the conviction that they can and must fight hard and well and that, if they do and have a little luck, they'll win against all odds.

Commander's Notes...

no. 2

2 May 1978

ME, THE CENTRAL BATTLE, AND HQ TRADOC

The formal TRADOC mission could be summarized as, "Get the Army ready for the next war." That's sufficient as a general statement of our responsibilities. Our real missions derive from a description of what I call "the Central Battle." Our tactics, organization, weapons systems, and training must all be clearly related to the challenges implied in the scenario of the Central Battle.

Time passes. Systems come on line. Ours and theirs. So the concept of the Central Battle needs constant revision. As programs and systems begin to produce, and even as ideas develop into studies, all must be integrated into the Central Battle concept. Once integrated, then we can judge impacts and criticality to the total Central Battle concept. Knowing this, how to assign priorities and resources becomes somewhat easier.

* * * *

Half the job of the TRADOC staff is to bring me information--when I need it--to keep updating constantly the Central Battle concept. I need *all* the pertinent information available. Thorough, conscientious staff work by the TRADOC staff will dig out the information required. But it's all useless unless I get it assimilated. It's even more useless if it comes too late.

I can get more information from the TRADOC staff if each of you gives me less. Here's why. In a week, about 110 staff actions show up in my in-box. I could handle that in a week if all I did was work the in-box. About 70% of my time in the Headquarters goes not to the in-box, but to briefings. I can handle that dilemma, too--by listening to briefings and thinking about staff papers at the same time. I don't. Most of the information I need to command TRADOC is in the "field." Much of my time must go there. In February, for example, I was here six days.

COMMANDER'S NOTES
ME THE CENTRAL BATTLE, AND HQ TRADOC

NUMBER 2
2 May 1978

Within those six days, add 15-20 office calls, a dozen or so visitors, seven social engagements, two or three ceremonies, and 32 telephone calls. These are the realities.

To work the problems of the Central Battle within the restrictions of the realities, I need less information. But every piece of the "less" has to be pure. Every piece must go through that old filter of: need to know, good to know, nice to know. I need the need part, not the rest. You need to synthesize, condense, strip-out, boil down, distill, abstract--like a good newspaper editor.

Here's your challenge: reduce six months work to a 10-page package, or a package to a page, or a page to a paragraph, or a paragraph to a sentence, or a sentence to a few words, or a few words to a model or diagram. I need concepts, bottom lines, central themes, summaries, abstracts. Any action officer who can condense accurately is worth ten who run out poopsheets by the pound. My measure of "completed staff work" is less paper, not more.

The TRADOC staff is an information processing system--and the action officer is the foundation of that system. I expect a good action officer to drive for breadth and depth in a given area. This makes him a subject matter expert. His expertise is the source data of the information system. This is why I prefer to deal with him directly.

Working directly with source data greatly reduces the exponential distortion that occurs each time information changes hands (or heads). Working directly with source data also enables me to get a specific answer to a specific question. I can state directly to the source what it is I "need to know." When I can't do this, each intermediate link between me and the source invariably adds some "good" and some "nice"--in both directions. Then I've got to sort.

Branch chief positions in the TRADOC staff were designed mainly to coordinate and control staff work. That's a good purpose. But timeliness and source data contact are so important to me that I'm willing to give up some degree of coordination and control and permit action officers to operate on general guidance from the primary staff and division chiefs. We don't really have a need for branch chiefs.

We don't really have a need for "pre-briefs," either. Every pre-brief takes *time*--preparation time, and presentation time, and the time required waiting to "get on the calendar." Furthermore, pre-briefs invariably change the source data content. I need the boss's broad

2

perspective, but I also need the action officer's source data--concepts and facts from the subject matter expert. Pre-briefs distort source data content.

In my view, the intermediaries between me and the subject matter experts have four main jobs: (1) know my priorities and transmit them to the right subject matter experts; (2) support the experts; (3) distill and synthesize the information flowing upward; and (4) expedite all three processes. These four tasks capitalize on attributes that, in my mind, characterize the intermediary: rank, experience, judgment. I want the intermediaries to be "Total Systems Officers"--to know the details of the *systems* that work through their directorate, not the details of staff action papers. An intermediary with detailed knowledge of each action officer's paper is, to me, an intermediary who is putting his time to the wrong purpose. And an intermediary who is spending his time (and mine) giving me a long and detailed briefing is not spending his time doing the four things I want him to do.

I obviously cannot deal personally with every action officer on the TRADOC staff. The triage is determined by my assessment of our task priorities. For me, as I explained at first, this assessment comes from the concept of the Central Battle. My priorities give the primary staff the basis for aligning *their* priorities, and so on down through connecting links to source data level. Let me give you one example of how this works.

As I study the Central Battle, I see an enormous demand for a wide and complex array of trained skills--a need for soldiers so thoroughly trained that their skills not only meet the technological challenges, but are also automatic even under the stress and crisis of the first battle. There is, in my view, a critical shortfall here. But this is the difference between victory and defeat in a "come as you are" war. So I turn next to the system whose output is trained skills--the Army Training System.

I assess that system, and its components, and its concepts . . . and I conclude that, while the system design is excellent, its operation is not smooth. I look at two of the system's most critical parts: the organization and functioning of the schools in accordance with Instructional Systems Design (ISD) principles; and the training systems in the hands of soldiers in the field. I see shortfall in both respects. And two priority tasks emerge: (1) the whole Army, and particularly the schools, must understand and implement the Army Training System; (2) the production of soldiers' manuals, how to fight manuals, SQTs, and similar training materials for the field must proceed at all possible speed.

COMMANDER'S NOTES — NUMBER 2

MF, THE CENTRAL BATTLE, AND HQ TRADOC — 2 May 1978

I transmit these priorities to my primary staff. Hopefully, they are transmitted down to the source data level--to the four-five action officers who have the knowledge of how to get these things done.

When I similarly derive 8-10 more priority tasks in other areas of TRADOC's concern, there are perhaps 50 subject matter experts on the TRADOC staff who are my source data. I expect to deal primarily with these experts as I work the priority tasks. And thus the sort is made. Bear in mind, however, that the Central Battle changes, and my priorities will not remain constant over time.

* * * *

The other half of the job of the TRADOC staff is to *help* the hundreds of TRADOC organizations external to Headquarters, TRADOC. The TRADOC staff is part owner of *any* problem that *any* TRADOC organization is working. These organizations are trying to do either or both of two things: (1) give me the information I need to work the Central Battle concept, or, (2) give the Army the TRADOC share of specific capabilities (skills, systems, ways of operating) it will need to fight the Central Battle. I expect the TRADOC staff to pass on my priorities, guidance, and requirements; to *help* the organizations meet these requirements; and to keep me informed of the outputs.

The critical man is, again, the action officer. The intermediary's tasks are, again, synthesize, abstract, expedite. The only difference, in this case, is that the customer is not me, but our subordinate organizations. The critical task is, again, communication.

The effectiveness of any communication is dependent in large part upon the accuracy of the content. However, no matter how accurate the content, it is worthless if the source is not credible. The credibility of the TRADOC staff is highly dependent upon the degree to which we meet the expectations we create. Every failure or postponement of payoff on a promise operates to the detriment of effective communications between HQ TRADOC and our subordinate organizations.

Accurate, credible communications are still worthless if they don't fit into some common context or framework understood jointly by the source and the recipient. For the work of the TRADOC staff and our subordinate organizations, this common framework is a system of some sort.

COMMANDER'S NOTES NUMBER 2

ME, THE CENTRAL BATTLE, AND HQ TRADOC 2 May 1978

Several large systems link the TRADOC staff and subordinate organizations. There's one for tactical doctrine, another for weapons systems, another for resource management, another for training, and so on. To work together on a given requirement, both the TRADOC staff and the subordinate organization must have a common understanding of the system involved--who does what, what impacts on what, inputs, outputs, linkages, relationships, feedback.

Let me illustrate the need for common understanding of a system. I can tell you how to determine if common understanding exists.

The Army Training System is the most complex TRADOC system. It involves the most resources, time, and people. If the TRADOC DCST staff and a subordinate organization like a school, for example, both understood the system, at least two desirable conditions would obtain: (1) any action officer at either end could explain clearly how his area of expertise contributed to the Army Training System; (2) any action officer in the school could quickly locate the right subject matter expert on the DCST staff. Every shortfall in these desired performance criteria, in all TRADOC systems, reduces effectiveness of communication between the TRADOC staff and subordinate organizations.

* * * *

SUMMARY

DON'T DO ANYTHING DUMB! I have reiterated this "summary guidance" several times. I recognize that it is an idealistic charge to my TRADOC staff, but I keep thinking about that damned horseshoe nail. Every error that you and I make, and every bit of time that you and I waste, will exact a price on the day of the Central Battle.

* * * *

DONN A. STARRY
General, United States Army
Commanding

DISTRIBUTION:
H1

5

Doctrinal Literature

Message to Lieutenant General John R. Thurman
Fort Leavenworth, Kansas
18 May 1978

The problem we are now encountering in writing doctrinal manuals is there is little institutional memory of concepts that are the thread of our doctrine. Manuals now go off in directions not originally intended because authors approach many problems as new ones when they are not new. Today's authors should not only consult, but strive to thoroughly understand, the early manuals—especially 100-5, 71-1, and 71-2. To be sure, the dynamics of doctrinal development, changing perceptions of the threat, technical advances, and many other factors will act to change, modify, or expand on what has been written. And manuals written today must accommodate any or all of those factors. It is the role of Leavenworth, in coordination with TDO, HQ TRADOC, and appropriate TRADOC schools and centers to ensure that doctrinal development is indeed dynamic, that its dynamics are properly directed, and that all interested parties are informed as developments occur.

Active Defense

Letter to Major General John C. Faith
Office of the Deputy Chief of Staff for Operations
19 May 1978

I read your letter with great interest and appreciate your concern about the active defense. There are, without doubt, many difficult problems to be overcome in executing such a concept, including, as you suggested, the business of communicating, collecting information and producing intelligence, training our battalion and brigade commanders, and a general recognition by all of the impact of terrain and force ratios on the successful outcome of any battle.

However, I do not believe these problems are unique to the active defense, nor do they imply flaws in the concept. These and other problems exist in every tactical setting, certainly so to no less a degree in the conduct of a position or mobile defense. Let me briefly discuss each of your concerns.

First, force ratios. The fact is we probably will be outnumbered from the very outset of the next war. The question then becomes: How do we defend successfully while outnumbered? As you know, I conducted a study of close to a thousand tank battles while at Fort Knox and found that, contrary to the outcomes predicted by Lanchesterian mathematics, actual battle results were not necessarily dependent upon force ratios. The outnumbered force does have a chance for success, provided the commander expertly uses the many advantages accruing to the defender.

One such advantage is terrain. We often forget that, in any given area, all of the available terrain cannot be used for maneuver by mounted forces. A mechanized enemy will be forced by the terrain to use primarily those avenues of approach that permit fire and movement by combat vehicles.

So, regardless of what the total force ratios are, the terrain will limit just how much combat power the attacker can mass at any given point. This is a marked advantage to the defender who takes the time to carefully study the ground upon which he must fight. A thorough analysis can reveal much about the size force an attacking enemy can employ along each avenue of approach and the number and type of combat vehicles that must be serviced over time and in a given space. Armed with this information, the defending commander can better judge the number of defending weapon systems required to service targets as they appear on the battlefield and where the defender's systems should best be located. In addition, Soviet tactics tend to emphasize formation geometry almost to the exclusion of terrain considerations. If this is so, we should capitalize on it.

Specific intelligence about the enemy is another matter, and one that certainly needs further attention. It is possible that our information collection and reporting systems will be so overloaded on the next battlefield that we won't be able to produce and disseminate the intelligence we need to fight. The problem isn't gathering information so much as developing a discriminating process that provides the critical information in time to be used. Efforts are underway to find a solution to this difficult problem. In the meantime, commanders must devote some of their time to aggressively organizing their information collection systems rather than sitting back, waiting for the right information to come to them. They must direct the attention of intelligence officers and staffs towards those areas where information central to decisionmaking can be found rather than allow the intelligence community to wander aimlessly around the battlefield in search of "what the boss wants."

Good communications are indeed important to the success of the active defense. However, to suggest that an active defense won't work in the absence of good communication begs the question. We must anticipate degraded communications and develop alternative command-control arrangements. Such arrangements include greater reliance on wire, messengers, and hand signals at lower levels, redundant command-control systems throughout the chain of command, and a clear understanding of the commander's concept for fighting the battle. The use of battle books for each position can greatly assist commanders in conveying their concept of the operation to subordinates.

Until we are able to improve our communications systems, several alternatives are available to the commander. He can locate his tactical command post well forward in order to be at the scene of the critical action. By doing so he reduces the number of communication links that must be used to pass information. He can develop a reporting scheme which allows him to continually monitor the actions of units in contact; for example, cavalry may be positioned with forward units to report combat information directly to the commander. As you know, Rommel and Patton used a similar technique during World War II. The best solution to the communication problem is training so that all understand exactly what must be done.

The active defense was based on the conviction that, during the early stages of the next war, US forces could expect to be outnumbered, outgunned, and probably defending across a wide front. It is to be expected that the Soviets will concentrate for a breakthrough when forced to do so, compounding the opposing force ratio problem at that point on the battlefield. It is obvious that under these circumstances a position or area defense will not work, simply because sufficient combat force cannot be generated fast enough to kill all the targets appearing on each part of the battlefield. The mobile defense compounds this problem by unnecessarily subjecting a large, unwieldy reserve to the same kind of destruction we expect to inflict on Soviet formations when massed to attack.

Under these conditions, there are never enough units to be strong everywhere and still provide depth to the defense. So units must be maneuvered from elsewhere once the main effort is located. Maneuver, then, becomes the key to generating sufficient combat power at the point where the enemy main effort is going in.

The problem of concentration is twofold. First, there is the business of physically rounding up tanks or APCs, organizing them for the move (probably on multiple routes), moving them from point A to point B, then rapidly dispersing them into battle positions. The ability to do this is a function of training. The second facet is more complex, as it pertains to the conceptual requirement for maneuver in order to concentrate sufficient combat power *at the place where it can do the most good.*

There are those who suggest that maneuver to concentrate will be impossible because the main attack cannot be identified early enough to overcome time-distance factors, that communications will be disrupted, and that lateral movement will be difficult because of problems with traffic control, enemy artillery fire, and so on.

The counter to this argument suggests that minimum movement will be necessary if the defending commander has properly analyzed the terrain in his operational area, determined with a reasonable degree of assurance where the enemy main effort can be expected and what the terrain will hold there, then positioned sufficient forces to service the targets expected to appear along that avenue of approach. Of course, it will be necessary to take some risks elsewhere by economizing forces in less-threatened areas.

Movement that is required can be facilitated by designating before the battle those units most likely to be moved, by pre-positioning units near the place in which they are most likely to be employed, and by onsite reconnaissance of routes and positions. Of course, each commander must have thought through the process of feeding new units into the fight long before the action begins.

Perhaps feeding forces into the fight as contemplated in an active defense could be described as piecemealing. However, such operations are not new to the US Army, nor is it necessarily difficult to control. There are several examples of successful tactical operations conducted in this manner. The battle of St. Vith comes immediately to mind. As you will recall, Bruce Clarke [General Clarke, then a brigadier] successfully defended for over five days against overwhelming odds by employing an "active defense." The key to his success was rapid lateral and forward maneuver by hastily organized task forces, independent small unit action, and skillful use of the terrain, this done under the most adverse of weather and enemy conditions. Communications and detailed knowledge of the enemy were certainly not our strong suits in the Ardennes during this period.

St. Vith is a textbook example of the application of the principles suggested in FM 100-5 and other How-To-Fight manuals. Clarke demonstrated that a skillful, innovative commander can defend, delay, and attack in a small geographical area with a hodgepodge of units thrown together quickly to defeat a numerically superior force. Similar operations were conducted by the Israeli 7th Brigade against the Syrians along the Golan Heights, 6–9 October 1973. We can surely expect at least as much from our commanders today.

Deciding who is least committed is a difficult problem which has often confronted commanders at least since the days of our own Civil War. There are no easy answers to this question, but we are working on some things which may help. First, we are teaching soldiers and commanders all we can about Soviet tactics and equipment. Hopefully, by recognizing numbers and type of equipment and tactical formations, our commanders can more accurately report what they are up against. Secondly, we are trying to improve our electronic and visual surveillance devices to give the commander a better picture of the battlefield, where the enemy is in strength, and where we can best economize forces. We are also working on ways to better and more quickly report on the tactical situation. As suggested earlier, some commanders may find it necessary to have surrogate eyes operating with each maneuver battalion to report to the commander exactly what is occurring.

Regarding boundaries, I have been concerned about that problem for a number of years, primarily because commanders seem uncomfortable without them. The fact is we must change our thinking in regard to boundaries. We must learn to operate on the battlefield *against enemy formations* rather than *within imaginary lines* on the ground. In the traditional sense, boundaries *restrict* fire and maneuver. In an active defense, boundaries must *enhance* fire and maneuver. We cannot conduct an active defense using graphics designed for a position defense.

You mentioned problems associated with lateral movement and the probability of misunderstandings at unit boundaries. Your fears represent what I call a "boundary fixation syndrome" shared by many. While boundaries do keep the battlefield tidy by clearly assigning responsibility for fires and maneuver, they can restrict our ability to concentrate fires. While in some cases boundaries may be quite appropriate, they most often are simply a convenient tool for commanders to assign responsibility for an area without due regard for the tactical importance of terrain. At St. Vith, Clarke was less concerned with structured, well-coordinated internal boundaries than with the ability to rapidly maneuver and place effective direct fire on the enemy, the essence of an active defense. You may recall that neither of his flanks was tied in with adjacent units.

An active defense does not suggest that operations will not be coordinated between adjacent units. In fact, the concept does suggest that commanders need be even more familiar with adjacent unit operations than ever before. Greater reliance must be placed on personal contact, liaison officers, and perhaps even small elements of one sort or another to maintain contact between units. Problems between adjacent units can be minimized by assigning areas of responsibility based on a thorough terrain analysis previously suggested. The next higher commander must play a more important role in ensuring that potentially dangerous gaps are handled in some fashion or other. Where gaps do exist, commanders may designate specific units to monitor or control them. Cavalry, ground or air, may be most often used, but even remote sensors can be used to detect enemy operating between defending units. The wise commander will use all control measures available to him in order to best control his forces.

The amount of authority a given commander has to fire his weapons and to maneuver is based primarily on the scheme of maneuver and resulting control measures specified by his commander, as is true of any operation. For example, given the requirement to defend from a battle position, a battalion commander may reposition his weapons to better service targets, although his freedom to maneuver may be limited. In such a case, the brigade commander may choose to maneuver the battalion from one battle position to another. On the other hand, if the same battalion commander is given a battle area within which to operate, he may maneuver and fire as he sees fit within the area so designated. It is true, however, that the next higher headquarters has a much greater burden in coordinating maneuver between its subordinate elements due to the nature of an active defense. There are times, for example, when a brigade commander may require the concerted efforts of three or more battalions in a relatively narrow area. Here close coordination will be required. In other circumstances, such as in economy of force operations over wide areas, each battalion must be given greater freedom to maneuver.

The How-To-Fight manuals clearly state that boundaries of any sort will not restrict direct fires. In those cases where our planning is faulty, gunners must have the freedom to engage enemy targets with direct fire wherever they appear, regardless of boundaries.

Control of indirect fires has not substantially changed. One unit may not fire into another unit's area of responsibility without prior coordination. Special coordination may be required if it becomes necessary to maneuver friendly forces in those areas not the specific responsibility of any unit, as is the case between battle areas.

An active defense will not work if units do not train for its execution. The entire thrust of How-To-Fight manuals and films and TRADOC Bulletins is to change the thinking on how to successfully defend on a complex, lethal battlefield when greatly outnumbered. Our training system, from skill qualification tests and soldier's manuals for the individual soldier to Army Training and Evaluation Programs for units, is designed to support the concepts set forth in FM 100-5. The concept is sound. Unfortunately, we have not successfully convinced all brigade, division, and corps commanders to try it.

A word about ARTEPs. The ARTEP is a practical way to find out if units have acquired the basic skills necessary to be successful in combat. It's not practical to conduct a brigade or division field training exercise to determine if each battalion can function effectively. Provisions are established in the ARTEP to ensure that battle positions are ". . . located to take advantage of natural obstacles, have good observation, fields of fire, cover and concealment, and are *mutually supporting*." ARTEPs also specify coordination to be effected by the controlling headquarters. To be sure, we have provided the necessary tools to the field to prepare for defensive operations. It's time now to train using them.

Regarding German acceptance of the active defense and general apprehension among Europeans. The Germans, in fact, participated in the development of FM 100-5. There will, of course, be some disagreements between any commanders on how best to conduct a defense. But, on the whole, the Germans accept the concept of an active defense. In fact, they employed similar tactics on the Eastern Front in 1943–1944 when severely outnumbered by the Russians. The concept, which they call "forward defense," is embodied in their HDv 100/100, a counterpart of our FM 100-5. I have personally walked the ground with Franz Poeschl when he commanded III Corps and studied his plans. From that I conclude we have no fundamental differences in concept or in employment between our two armies.

You suggested that we need to reestablish a wide range of types of combat activity including the delay, the forward defense with large reserves, the active defense, attack, counterattack, and meeting engagements. Except for a thin defense with large reserves, which I believe would be totally ineffective against Soviet offensive formations, each can be found in FM 100-5, FM 71-100, and other HTF manuals. What is important to understand is that an active defense is a potpourri of several different types of operations. While some units may be disengaging to move elsewhere, others may be defending and still others attacking. Sometimes it may even be necessary for some units to economize forces and conduct a delay while others are attacking elsewhere on the battlefield. We have not attempted to limit imagination and innovation, only direct it in a sensible direction. In short, *do what makes sense*!

Let me close by saying that, unlike any of our previous manuals, FM 100-5 and those that follow are pragmatic approaches to solving complex problems the like of which we have never experienced before. These books attempt to illustrate what is possible under adverse conditions. They are based on political, economic, technological, and tactical realities—not philosophical abstraction.

I believe many of your concerns can be mitigated if we train using the tactics described in the various How-To-Fight manuals. Training is the cement that bonds together good weapons, sound tactics, and effective organizations into winning teams. Training develops excellence in the skills of leaders and soldiers. It's time for us to get our heads out of the sand and realize that tactics developed following World War II just won't work on the battlefield we face in 1980. Those of us in positions of responsibility must see to it that the doctrine so laboriously developed is practiced or a better alternative proposed.

The Central Battle Again

US Armor Association
Fort Knox, Kentucky
24 May 1978

This morning, I'd like to talk about the Central Battle again. No new wisdom, but let's talk about some aspects of it in a different way perhaps than before. In TRADOC, we are trying very hard to develop operational concepts from which we might derive tactics, weapons systems, organization, and force structure. We try, then, to put all that together with a training system for the Army.

First, let me review the concepts and say a few words about their interdependence. First are the overriding Soviet concepts: mass, momentum, continuous land combat, and the notion that defense is but a temporary measure. Alongside that are counterconcepts, those we've adopted in attack and defense doctrine. We speak of seeing deep into the battlefield, moving fast to concentrate forces, suppressing enemy fires—especially artillery—with counterfire systems, striking quickly to kill many systems, then finishing the fight rapidly before the second echelon can close.

On both sides, the concepts are interdependent; one is not sufficient to a military system without the others, for they all relate to one another. Let's look at that interdependence as a capability, a systems capability—interdependence of concept, interdependence of weapons systems, of tactics, and of organization and structure.

Let's look first at force interdependence of forces that operate against one another in the Central Battle. Direct-fire red systems—BMPs, tanks, antitank guided missiles, and short-range infantry are arrayed against blue APCs, tanks, antitank guided missiles, and short-range infantry. Contributing on both sides to the battle are red TAC air, artillery, and helicopters operating against blue, and blue TAC air, artillery, and helicopters operating against red. Now, if we put on these systems numbers that represent the contribution of each system to killing systems on the other side, we would get a revealing relationship.

Blue tanks kill about 48 percent of enemy systems destroyed in the Central Battle. Blue antitank guided missiles kill about 20 percent of enemy systems that perish in the battle. So, 68 percent of enemy systems killed by friendly systems in the duel are killed by those two weapons alone, tanks and antitank guided missiles. For the enemy, red tanks account for but 31 percent of the total blue system kills, and antitank guided missiles only 7 percent of the total blue system kills, for a 38-percent contribution by red tanks and antitank guided missiles.

Why the difference? The reason is simple. Blue tanks and antitank guided missiles must kill more enemy systems because we don't have an infantry fighting vehicle and we haven't enough artillery. On the red side, the BMP kills about 34 percent of all blue systems destroyed in battle, most of those being APCs, dismounted infantry, weapons crews, and antitank guided weapons crews.

This clearly says our combined arms team is out of balance. To redress that imbalance, we need many things, but foremost among the many is an infantry fighting vehicle—one that with tanks, antitank guided missiles, short-range infantry, and artillery kills its share of enemy systems in those unique tasks for which it is designed. So, the problem is more than one of balance—

it's interdependence, it's the need to restore effective interdependence in the combined arms team.

Because of imbalances in the interdependence equation, blue artillery is overstressed. We have required artillery to make up for our gross shortage in firepower by adding to it new shells capable of doing many new things. There are cannon-launched guided projectiles, smoke rounds, mines delivered by artillery, nuclear artillery munitions. Someone showed me a scheme recently for a cannon projectile with a television camera in the nose of it. We hadn't enough artillery tubes to begin with. Now, we tie up a lot of our tube capability with these special weapons. In addition, we tie up our forward observer system. The FOs are guiding guided weapons onto targets instead of adjusting HE and smoke. So, we have overloaded the total artillery system with all these new weapons, trying to make up for the shortage of firepower. Part of that shortage in firepower can be measured in terms of the absence of an infantry fighting vehicle; many of the systems it would kill now must be killed by artillery.

Looking at counterfire systems, the Central Battle tells us that red artillery is 30 percent more effective than blue artillery. This is so because there is more red artillery to begin with, and red artillery can deliver more throw weight—more ammunition delivered—than can our own artillery. Red is not encumbered by lots of special mission artillery. So, it isn't just that some magic balance has to be maintained in the combined arms team, it is that we must create a satisfactorily interdependent combined arms team in order to take advantage of the synergism that that team offers in battle.

Now, why is interdependence important? Let's talk about the calculus of battle, considering attacker versus defender ratios, ranging from 1 attacking 5 to 5 attacking 1. Using the traditional calculus of battle derived from Lanchester's Law, we get a curve that says that, if you're 1 attacking 5, you haven't a chance, but if you're 5 attacking 1, you nearly always win. That's bad news for the fellow who believes he may have to fight outnumbered. It says he is foredoomed to defeat!

However, the outcomes of about 1,000 tank battles tell a different story. Empirical data tells that there is apparently more to battle than the traditional calculus would admit. Winning in battle seems to require something more than the calculus tells us. What is that?

Let's look at a battalion battle again as an example. Time is zero. A tank battalion task force is defending in a conveniently U-shaped piece of terrain. The task force is being attacked by elements of a motorized rifle division—the commander doesn't know how many. He has three companies, two M60A1 tank companies (Team A and Team B), an M60A2 tank company (Team C), and, in support, a mortar platoon and a direct-support artillery battalion. The terrain is such that the enemy can deploy here at about 6 to 1—6 attackers to 1 defender.

Using the traditional calculus, this battalion is foredoomed to defeat. But let's see if there isn't something we can do to get closer to the red line that history tells us is possible.

Here comes the enemy with about three reduced strength regiments—about 250 systems. Ten minutes into the battle, the team commanders have made a number of decisions. Commander, Team A, has put three platoons into position and positioned some TOWs. The Team B commander has put his platoons into position, also with TOWs. The Team C commander has taken up good firing positions on the reverse slope of a hill, with TOWs firing from the military crest.

The enemy is within range of supporting artillery, so blue has opened fire with artillery on these two motorized rifle regiments. The artillery was firing madly when out of the smoke came some tanks. Blue quickly shifted some artillery over and began to fire mortars.

So, at 10 minutes into the battle, here are the decisions that have been made and acted on—positioning the platoons, positioning the TOWs, identifying the enemy, calling for fire, and reporting what's been done. The battalion commander has responded, asked his direct-support artillery battalion commander to get some more fire, and called for attack helicopters and TAC air.

Fifteen minutes into the battles, the fight is brisk. The battalion commander has distributed the fires of his teams in such a way that each battalion in the enemy formation is being fired on by two platoons of the defenders. He did this by outlining a terrain mosaic in the center of this U-shaped area so that he and the team commanders might accurately direct fire or maneuver into those areas.

The decisionmaking pace has picked up. The Team A commander is directing the fires of his two northernmost platoons on a motorized rifle regiment. At the same time, he has directed the fires of the TOWs and the third platoon on parts of that regiment that seem to be turning in the general direction of Team B. He's firing into the flanks of that outfit. Later, he shifts the fires of the third platoon into the flanks of the middle motorized rifle regiment. The TOWs have been shifted to fire on the tank regiment that is moving more rapidly than the other two, having just come into the battle.

Team B is firing all its TOW fires into the flanks and rear of the attacking tank regiment. The second and third platoons are firing into the flanks of that regiment, and the first platoon is delivering crossfire onto other elements of the motorized rifle regiments that, because of the lay of the ground, cannot be reached by Team A. Initially, most of the fires of Team C were directed against elements of the tank regiment. All TOWs and the second and third platoons are firing there. The first platoon is firing on lead elements of the motorized rifle regiment.

Twenty minutes into the battle, the battle is essentially over. The attack helicopters have arrived. In order to clean up this battle area before the second echelon comes along, the battalion commander has artillery fire a mixture—HE and smoke—behind the fight in progress. The second platoon of Team A is still firing at some trail elements of the motorized rifle regiment in the middle. TOWs and the third platoon of Team A are working over remnants of the two motorized rifle regiments in the center and on the left.

The first platoon, since its fires are masked in the engagement area, has moved down in the throat to fire against enemy air defense systems, permitting the attack helicopters to come in. They are on station with some aircraft firing, some in the forward area rearm and refuel point, about 30 kilometers away, and some in transit at low level, dropping to nap of the earth as they come into the forward air defense envelope. They recycle about every 10 minutes so we can keep missile-firing helicopters on station for some time. The close air support aircraft have not arrived yet and are not expected before we have to have this battle tidied up. Team B is attacking the tank regiment remnants with two platoons, first and second. Its TOWs were destroyed by hostile fire. Third platoon is moving to join the counterattack, overwatching the two attacking platoons.

Now, let's talk battle outcome, the killer-victim scoreboard—a battle analysis. You all know that tank crews perform better some times of the year than others, usually when they have just finished annual gunnery qualification. "Forgetting" takes place as time passes and proficiency drops off until it's time to qualify again. So, we have a standard sawtooth curve with which analysts are so familiar. That curve represents a lot of things, but in this case, it represents learning and forgetting. What I'm going to do now is show you some battle analyses of our battalion battle that were conducted with low-performance crews, with high-performance crews, and then with high-performance crews that were members of fully trained outfits in which all of the interdependent factors in the battle worked together and worked well.

First, consider a low-performance killer-victim scoreboard. Here are the systems, two companies of A-1s with 15 tanks each, a company of A-2s with 12 tanks, 6 TOWs, the artillery, and the attack helicopters. There were 250 red systems in the attacking array; about 150 of them have been killed; 100 of them are left. It is quite likely that most of those 100 are tanks. Out of 71 systems of our own, we've lost 37. On balance, this fight has not been very successful. In fact, this commander is probably going to have to get out of here. He can't fight successfully against the remaining 100 tanks, certainly not with 34 systems left.

Now, here is another battle analysis, different kill rates, different kill ratios resulting from the work of high-performance crews as opposed to low-performance crews. This outfit kills at the rate of about 6½ to 1; the low-performance crews killed at the rate of about 4¼ to 1, so this represents a rather dramatic improvement in performance just based on crew performance.

Now, let's take some data representing well-trained crews in fully trained units where the interdependence of all systems working together is brought to bear. The outcome is dramatic. This outfit kills at the rate of 8¾ to 1. Now, what made the differences in that battle? Let me see if I can summarize it for you. Blue was out of balance. With no infantry fighting vehicles, his tanks and attack helicopters had to kill more than the enemy's. Red did not employ antitank helicopters. He used a lot of artillery, but blue was mostly tanks and survived fairly well. So, all things considered, the advantage provided by technology may have gained blue something like 3 to 5 percent, taking full credit for the sophistication of guided missile guidance systems, tank fire control systems, and so on. Crew performance, the difference between the performance of the low crews and the high crews, added about 15 percent to battle outcome. The difference between the organizational performance—the high-performance crews—and the fully trained unit added about 25 percent to the battle outcome.

Organizational performance, using interdependence, is what made the difference. That's leadership. Leadership is a function of training—training the crews, training the organization. Leadership is what puts it all together. I hesitate to use the word leadership to describe organizational performance, because leadership in our Army has long been looked upon as an individual thing. An individual is a leader. He has charisma or he doesn't have charisma. He makes good decisions or he makes bad decisions. We tend always to focus on the individual aspects of leadership—what made General So and So or Colonel So and So a great leader. What caused Private What's His Name to dash from his squad and suddenly become the de facto leader when the squad leader was killed or wounded and lead the squad to victory? Personal individual leadership, therefore, has traditionally been our bag.

Admirable as that may be, it is truly the exception. Individual leadership has its place, but what is more important is that leaders must also lead platoons, companies, battalions, and these are organizations. The kind of leadership that will get the synergism—the plus—from the interdependent effects of a well-trained organization is a leadership that comes from the organization itself. It can give each individual the strength he draws on as a result of being part of a well-trained, highly motivated outfit that is doing good things. That kind of an outfit can take high-performance crews, squads, platoons; put them together; and capitalize on their interdependence to achieve phenomenal things in battle.

Organizational leadership comes from training. It's much more than flags, guidons, and esprit. It's the simple sure knowledge that each man and every crew are highly trained and that they all belong to a solid, firm, confident, well-trained organization that knows where it's going and what it has to do.

That's the kind of leadership we have to strive for at every level, from E-Nothing on up, if we are really to make the combined arms team work. If we train and lead our units in that manner, we'll go a long way toward capitalizing on the interdependence potential of the combined arms team. There is no other course, for we can and must win the battles of the next war. Our soldiers, our Army, our country, depend on it.

Doctrine Rules

Message to General Bernard W. Rogers
Army Chief of Staff
7 September 1978

This responds to your 281610Z August 1978 message concerning the 1978 Army Commanders' Conference. . . . Suggest we talk first about how to fight, then about how to train and how to go to war—your tentative agenda . . . has training first. We really can't talk about that until we know what we need to be training for—that's how to fight.

US and Federal Republic of Germany Doctrine

Letter to Lieutenant General John R. Thurman
Deputy Commanding General, US Army Training and Doctrine Command
27 September 1978

This responds to your letter to me enclosing General Doctor Kohlmann's letter to you concerning apparent differences in our concepts of operation, particularly the active defense. I believe the differences cited by General Doctor Kohlmann are more apparent than real. The doctrine in the two manuals 100/100 and 100-5 is compatible.

As one of those responsible for making it so in writing, I can attest to the fact that, at least in the minds of the authors of both documents, there were no inconsistencies.

Also I can say that as a corps commander trying to apply 100-5 in concert with German corps and division commanders on my left and right respectively trying to apply 100/100, I found no real inconsistencies.

In the beginning I do believe the Germans emphasized more the idea of killing the enemy forward of the main battle area, or at least in the forward part of that battle area. This meant that the big battle would have to be fought in the security zone. In the minds of my German commander friends this stems in part from a concern that they not give up any of their country. This is a sentiment which I share with them; even though I'm not German, I've spent 10 long years of my life out on that border—I have an emotional investment in it which, while not as deep as a German national might have, is nonetheless every bit as earnest. I too want to defend forward—as far forward as possible.

However, it is not prudent to assume that a scheme of defending and defeating the enemy forward of the main battle area will always be possible. Certainly is this the case when confronted by a breakthrough attack. Therefore it will sometimes be necessary to use the terrain cleverly, judiciously, and in depth to accomplish defeat of the enemy force, playing him against one defensive position after another until he is dead in his tracks. It is this latter realization that led us to emphasize the organization of battle positions in depth. By forcing our commanders to organize in depth we do several things: we make them look at a broader array of terrain than they might otherwise, and make them study it in detail; we provide in advance as much coordination as is possible in depth, so that coordinating arrangements for fires, maneuver, and so on, do not have to be worked out at the last minute and in the heat of battle; third, we provide a planning basis for reserves to be deployed in defense positions which are defensible with minimum preparation, and yet which are sited so as to provide every opportunity to counterattack with the reserve.

Now it is true that not all American officers understand this matter just as I have described it. The process of educating both our armies on the concepts and execution intended by 100/100 and 100-5 is yet under way. My experience is that it is essential for commanders from corps to the lowest level to walk the ground together, talk through the battle together, and together agree upon how the doctrine set forth in the manuals really applies to the piece of ground they have to defend. Only by doing this can we make the apparent differences disappear and arrive at a truly interoperable understanding of how we must fight together in order to win.

Division 86 Operational Concept

Letter to TRADOC Installation Commanders
13 November 1978

The complexity and variety of modern weapons systems demand that we, who represent the "user," know how these means of war are to be employed. More than that, we must state in clear language what we so often describe in stereotyped jargon. Decision makers and legislators, responsible for the arming of our forces, must be better able to understand what we are trying to do with all the equipment and gadgets that we say we need or that we will soon have.

Clarity of thought and of language is required to describe the needs of the Central Battle and of Force Generation. That clarity is difficult to achieve in a discipline which involves not only thousands of givens, but also countless imponderables. Nevertheless, in the future we have to do better in describing the interactions of weapons, organizations, and tactics. The soundest way to accomplish our objective is to develop operational concepts that lead to doctrine, organizations, and weapons systems integration.

The Operational Concept for Division 86 is a step in that direction. It may well become a model for the dynamic process needed to achieve a meaningful evolution in our military theory. As such, it is open to improvement and change, and I request your continuing efforts and participation in its refinement.

Division 86

Letter to Lieutenant General E. C. Meyer
Deputy Chief of Staff for Operations and Plans
30 January 1979

Received your letter . . . reference the Operational Concept for Division 86. . . .

The edition you refer to was admittedly weak offensively. We applied a few fixes to it and issued a revised edition following the 30 November General Officer Workshop. I thoroughly agree that we've got to get out of a defensive mind-set and restore the "spirit of the offense" to our thinking. But the task is not that simple. We need to reaffirm that the strength of the active defense comes from small unit initiatives which are offensive in nature (ambushes, flank engagements, attacks from the rear, etc.). Second, we need to think through when larger units (brigades, divisions, the corps) can best transition into more coordinated offensive operations properly supported with effective joint offensive suppression and containment of second echelons. It makes no sense to jump off on an offense, to penetrate various defensive lines of, let's say, a first-echelon Soviet army, and then find ourselves staring into the face of a fresh, unattrited, mobile second-echelon army or even front. Answers to these theater aspects of the offense will help us to better define the offensive role of Division 86. . . .

Your concern for the effects of urbanization and natural obstacles to armor movement is also shared. While we do not anticipate large-scale fighting in urban or forested areas, which can be bypassed, the frequency at which these areas occur in Central Europe certainly requires our consideration. Looking at urbanization as an extension of terrain, we have got to determine how to best use it to our advantage. Our commanders must be impressed with the potential value of built-up areas in both offensive and defensive operations. Properly positioned forces can slow an attacker's momentum and, if he chooses to bypass, require him to leave a portion of his force behind to isolate the area or suffer the consequences of attack from his flanks or rear. In the offense, by striking to an enemy's rear area and seizing key urbanized terrain along his LOC we can isolate his forward units and cut his logistical lifelines. The rapid urbanization of Central Europe will continue. We all have to give it serious thought.

In closing, let me add that we do plan to follow our Division 86 effort with studies of both a European Corps and a light division. Our work on the forward-deployed corps will get us deeper into the theater-level considerations which you mention.

Operational Concepts and Doctrine

TRADOC Commander's Notes No. 3
20 February 1979

1. Concepts.

All professions have vocabularies of professional terms. Over time, many such terms become establishment "in-words" and are so ill-used that their original meaning is lost. Often it is only necessary to use the words to evoke affirmative head nodding; even though no meaning is conveyed, everyone professes to understand what is meant.

The US Army has its "in-words." Among them is the word *concepts*. There are concepts for many things; there are concept papers on many subjects; there is visible nodding of heads when the word *concept* is used. However, it is apparent that the word means different things to all too many of us.

When the Central Battle idea was first conceived, I made the point that operational concepts had to be the driving force for describing interactions that were to occur on the most intense part of the battlefield. While everyone seemed to agree, more and more I have the impression that concepts are being created not to describe the Central Battle, but to justify some individual weapon or other system or piece of equipment. If this is true, we have got the *concept* of concepts just exactly backwards.

Since we now have signed or are working on concept papers with several US Army agencies and allied nations, it's past time we agreed on what a concept is and what it's supposed to do.

A concept is an idea, a thought, a general notion. In its broadest sense a concept describes what is to be done; in its more specific sense it can be used to describe how something is done. The Soviet notion that numbers win is an example of the former; the implementing notions of mass, momentum, and continuous land combat are examples of the latter. Concepts are therefore rather broad and general; they can also be more specific.

In the sense that the term is used here, concepts are military in nature, relating to ideas, thoughts, general notions about the conduct of military affairs—the operational concepts alluded to earlier. How might the term operational concept be defined? Here is a definition:

> Operational Concept—A description of military combat, combat support and combat service support systems, organizations, tactical and training systems necessary to achieve a desired goal.

Concepts are and must be the first agreed upon part of any project. They must also be dynamic—changing as perceptions and circumstances change. For our purposes, TRADOC concepts must be tested in the Central Battle; effects on and interactions with other concepts—tactical, organizational, training and materiel concepts—must be defined.

2. Doctrine.

Concepts are <u>not</u> doctrine until tested, approved, and accepted. Not all concepts will eventuate in doctrine. This is why concepts are dynamic, not fixed; this is why they are not tied to a specific piece of materiel or a system. Rather they address themselves to the needs—requirements that flow from the Central Battle.

Doctrine is what is written, approved by an appropriate authority and published concerning the conduct of military affairs. Doctrine generally describes *how* the Army fights tactically; *how* tactics and weapon systems are integrated; *how* command-control and combat service support are provided; *how* forces are mobilized, trained, deployed and employed.

3. Concept Formulation.

As a general rule the need for an operational concept may be generated by:

a. Recognition of a problem for which no doctrine exists.

b. Recognition or assignment of a mission for which no doctrine exists.

c. New or improved technology having military application, not yet exploited.

An operational concept, then, is a statement of a way of doing business. Once formed, an operational concept is evaluated. This can be done in a variety of ways: by one of several battlefield simulation processes; by field testing and evaluation; or perhaps by routine staffing.

How are concepts born, and how do they become TRADOC products? Concepts should derive from some perceived imbalance in our ability either to fight the Central Battle or to generate forces to fight the Central Battle. The Tactical Doctrine Office is the keeper of concepts—the clearing house and catalyst for evaluation and infusing of new concepts into all our work.

Anyone can formulate a concept. Concept testing and approval, however, are the business of TRADOC. We should solicit concepts from many sources. We must not drive away new ideas about the Central Battle and force generation simply because they were not invented here.

. . .

US and Federal Republic of Germany Doctrine

Letter to Colonel William F. Burns
TRADOC Liaison Officer to Federal Republic of Germany
30 April 1979

Thanks for sending me John Weckerling's report on the maneuver at Munster.

While I don't question the quality of the observer—John is one of my favorite guys—his report adds fuel to fires kindled by those who say that the US and German Armies are far apart and growing further apart than ever before tactically. This is alarming to me because, while there are some differences between us yet, I do not believe the gap is as wide as our critics would like us to believe. My experience has been that the biggest part of the apparent difference lies in the lack of understanding of what is intended on both sides by words in field manuals and by practice in the field. Of all those who observe, John Weckerling should be least guilty of misunderstanding either side. While his report does not portend to me that things are all that bad, to those who want to believe that they are, it will ice their cake.

If there are differences between our armies, they appear most readily in concepts of what is to be done in the covering force area, in the main battle area, and with regard to reserves.

When working our differences on paper in 100-5 and 100/100, we found the Germans saying the battle had to be won in the covering force area. They couldn't visualize giving an inch in the main battle area. We finally convinced them this couldn't be. In the first place they never put enough force in their security zone-covering force area to slow, stop, destroy, or do anything much else to the enemy. It was just that they weren't willing to give up any more of West German soil than was absolutely necessary. We finally reconciled our differences in the manuals, and while there is still a perception that we are too willing to give ground too quickly, they have conceded that there will be a fight in the main battle area as well as in the security zone. And so in the end we aren't all that far apart. Our perceived willingness to give ground is a matter of our own misunderstanding of what the doctrine is saying, and the tendency therefore to give ground in maneuvers too quickly. It was not the intent in 100-5 to prescribe that practice as a normal one.

Somewhat the same situation obtains with regard to reserves, their constitution and use. We haven't enough force—either of us, to tell off substantial parts of it to constitute a reserve. Certainly is this true at the outset of a short-warning war, but it is also a condition that will continue for some time. When I took command of V Corps, I found 56 percent of the tanks in the corps in reserve at one echelon or the other—clearly not sound tactics. The fourth element in the German battalion and brigade is a skeleton organization—it would not [be] present in strength in a short-warning war. Therefore, while they use it as a reserve in maneuvers, and write about it, the truth is that it won't be there for some time. It is a wise concept—for it provides redundancy that the force structure needs badly. But when we look at it and try to compare it to our own perceived lack of reserves, we must consider it in the context in which it exists—some time will elapse before the Germans can create that reserve, just as some time will elapse before the US forces in Europe can marshal reserves from units deploying from the CONUS. So at the outset of battle, and for some time thereafter in a short-warning war, neither of us will have any reserves to speak of—we simply won't be able to afford them. Like the covering force-main battle area debate, I find in the argument about reserves that we are frequently too willing to jump to conclusions without really understanding the full context in which each side is setting whatever is being observed or talked about.

I do not in any way want to change the impressionist quality of reports from our liaison net. But it is essential that we not let grow, any more than it has, the argument that we and the Germans are so far apart tactically. My personal experience is that we are not, and that, if we but take the time to thoroughly understand our own tactics and theirs, the differences will become more apparent than real.

Battlefield Terminology

Letter to Lieutenant General J. R. Thurman
Combined Arms Center
14 May 1979

This letter responds to the concern you expressed . . . about the lack of a common battlefield terminology. You are right that there is confusion and that we are trying to turn this around.

The critical tasks in the BDP were developed to define the functions on the battlefield. The CAPCATs and now the mission areas which have replaced them provide unique, resource-related categories to which materiel can be assigned. As a result, common terms for both functional analysis and resource management may not be feasible. However, a crosswalk is essential if our functional analyses are going to influence resource decisions.

BDP I critical tasks accomplished two objectives. They provided definitions of the functions being performed on the battlefield and they were combined to describe the concepts of force generation and central battle. Our focus on fighting the second echelon and on reconstituting the force demonstrates the success of the effort. Mistakenly, many people view the critical tasks as restricted to the central battle and force generation concepts rather than being the independent set of functions which can be used to describe any battlefield activity like covering force activities, rear area security operations, or fighting the second echelon.

We need to complete our review of the critical tasks and the interface between them and the mission areas quite soon, not only for the reasons you mentioned but to establish a sound basis for examining Corps 86 and preparing BDP II. Our work to date suggests the addition of one task, altering the battlefield environment, to better represent the primary engineer functions which are currently subsumed in several other critical tasks, and the change of surveillance/fusion from a task to a subtask of intelligence which would incorporate surveillance/fusion, target acquisition and battlefield information collection and analysis. . . .

Using this revised list of critical tasks, the mission areas and critical tasks have been formatted as an initial basis for describing the crosswalk. We have included one possible change in our mission areas, the addition of combat support (engineering and mine warfare), to correspond to the change in critical tasks and improve the interface. Again, we will never get a perfect match when comparing functions to materiel categories as long as certain items can perform multiple functions and certain functions require different types of items from multiple mission areas. However, we should be able to describe the crosswalk.

FM 100-5, *Operations*

Letter to General E. C. Meyer
Army Chief of Staff
26 June 1979

This responds to your letter . . . on the subject of doctrinal matters related to FM 100-5. Written in your other incarnation [as Army DCSOPS], the subject is nonetheless one on which you and I should reach some agreement, then either convey that agreement to the rest of the commanders of the Army—or seek some consensus if necessary. So let me treat the points you made seriatim, describing what I plan to do about them.

First is the charge that FM 100-5 should be applicable across the board. You cite a doctrinal deficiency with regard to how to fight in Northeast Asia. I would add the problem of how to fight wars against surrogate forces, insurgents, revolutionaries of one kind or another—a problem that we have recognized from the outset. [It is] one that was created in part by our desire to point the US Army clearly toward its most challenging problem. Further we took that tack because, with it, we were able to reach agreement with the Germans rationalizing with HdV 100/100. While we might somehow cope with the former problem in a subsequent edition, we cannot generalize too much and still be compatible with the Germans, whose only mission is to fight in Europe and whose doctrine will always reflect that. It is for that reason that I've held off starting a rewrite of 100-5, even though this and other frailties might persuade us that it's necessary. In due course we must revise it, but for the life of me I can't see how we are going to cover all the contingencies—especially those involving surrogate forces, wars of national liberation, and so on. I believe I know how to handle the Middle East and Korea, and it is my intent to do so, but the others I'm not at all sanguine about. In any event, we should not tackle that until we have the Corps manual on the street; that will probably be mid-year next year. In September, I'll have a new counterpart in the Staff Talks with the Germans—LTG Burandt. I know him, but not well. I will however approach him early on with a proposal to undertake a revision of the two manuals. They will be reluctant to do so, so it may take some time.

Second point has to do with expansion of 100-5 to include the corps and theater battlefield. I'm not sure I understand exactly what you mean by that, but believe that much of it—especially the part regarding the corps—will be adequately dealt with in the Corps manual.

Third point concerns 100-5's concentration on the breakthrough. We treated the threat that way simply because that was in so many ways the most difficult problem to solve. I do not agree with Phil Karber's thesis that the Daring Thrust represents a tactical revolution in the Soviet lexicon. The idea is really about forty years old, and they've written about it extensively. We didn't ignore it, although I do believe none of us were sensitive enough to its possibilities when we were writing 100-5. So we should make mention of it in the revision. It too will be covered in the Corps manual.

Defensive orientation is the next charge. You realize that was not our intent, although I do believe Bill DePuy was much more defensive-minded than was I when the final lines were written. The problem we couldn't solve at the time was what to do about the second echelon. We know what needs to be done about that now, and so can write coherent doctrine that will cover both the offense and the defense, and the fight against the first echelons and second echelons as well. However our critics, who trumpet that we should steer away from a firepower-based strategy, aren't all correct—we're going to have to kill a whole lot of them just to get their attention. And we should make no mistake about that! So to say it's all a war of maneuver, and that maneuver will solve all, is to ignore the very real problems with space and depth, especially in Europe, and with logistical support of highly mobile operations. For certain we must find some prose that's more golden than that we used last time. Will try it in the Corps manuals. If that floats it will pave the way for a new approach in 100-5.

In addition to all that, we must address ourselves to the question of the next generation of doctrinal literature as we move toward whatever doctrinal changes are demanded by our decisions with regard to Division 86. There will be many, and my judgment is that they will be significant. When we start, how long will it take, and what it will cost us are critical questions. We're just now closing in on what we started out to do nearly six years ago in the "How to Fight" series. We haven't even touched the "How to Support" series, which is to be its companion set. The drawdowns on TRADOC in the last couple of years have badly eroded our ability to handle a rewrite effort in anything like a reasonable time. In addition there's the problem of who will do it. If it did nothing else, the succession of command here from Bill DePuy to me gave us some institutional memory at the top—and only at the top, that we simply haven't been able to preserve anywhere else. How we can provide at least some of that as we move into the next cycle is a real problem with which we should wrestle.

In due course we need to talk about all that. However Bill Lind is coming to see me week after next, so I'll be much more doctrinally enlightened after that I'm sure. Will keep you posted.

Models and Methodology

Message to Lieutenant General William R. Richardson
Fort Leavenworth, Kansas
5 November 1979

. . .

2. One of the issues raised at the commanders' conference had to do with our inability to show tradeoffs between and among systems with disparate capabilities in terms of their contribution to the total corps or division battle. This is the matter I tried to get at a couple of years ago with the "Central Duel" methodology. That was too crude, too firepower score-driven, and we could not explain why, when we relieved the stress on the tanks by putting some other tank killers on the battlefield, the tanks contributed less to the battle. It's logical in a way, but it makes less of a case for the tanks in the end. Same is true of other systems. I used tanks as an example. In a division- or corps-level simulation in which we try to demonstrate contribution of any system to the total battle we will always be faced with this dilemma.

3. Therefore it seems to me we should look seriously at skipping over the firepower score-driven models which have been our bag in the past and begin to experiment with other methods. I understand you have a division-level simulation under development. Let's get that out and look at it. Sam Parry has STAR fairly well along, and although he's played it at brigade level only, it can be played at division.

4. Dave Hardison and I have agreed to press on with some dispatch to try and solve the division-corps-level simulation problem. I'm waiting now for a proposal from him with regard to time lines, goals, and work assignments to get on with the project. I will convey that all to you in due course.

5. Meantime it seems to me prudent for us to look carefully to ensure we are not starting to eat the elephant from the wrong end. We keep trying to build models from the bottom up. If we look at systems interface and systems tradeoffs as the goal of the simulation, would we not perhaps develop a different thing at the outset, especially if we begin at the division level, say?

6. Ponder that and let's talk.

7. With regard to air: while I realize we have to use whatever the tribal wisdom says is about right for aircraft kill rates and so on, I believe it would be both unwise and self-deluding of us were we to use the exaggerated kill rates per sortie that the Air Force has used to justify the purchase of their aircraft. This is particularly true of the A10. I'm not trying to sell airpower short—no one has touted our need for it more than I have in recent years, but we just must be realistic. We have to be fairly confident in our own minds that we have portrayed the battle in a fairly faithful and representative fashion. We just aren't doing our job if we don't.

Attack of Follow-On Echelons

Message to Lieutenant General William R. Richardson
Fort Leavenworth, Kansas
19 November 1979

[Re your message on] TRADOC proponent for artillery-delivered mines. Hold up the issuance of further instructions. The problem here is not just one of artillery-delivered mines. It is rather one of munitions to be delivered against follow-on echelons, regardless of the munitions effect. Our decision in this situation is therefore concerned with the broader issue of the rationale on the basis of which we assign proponency. To this time it appears to me we have more or less assigned proponencies on the basis of delivery systems—the weapon and delivery means, without regard to the terminal effects of weapons so delivered. That probably was not a conscious decision, but it is the pattern. If we are to reverse that, we open up a Pandora's box of similar decisions. While I don't want a big study of the matter, I do want to consider the instant problem in light of other proponency decisions and the pattern thereof.

Battle Models

Message to General E. C. Meyer
Army Chief of Staff
25 January 1980

1. On 23 January LTCs Dave Marshall and Art Richards were kind enough to expose me to Omnibus 79. Good job by both—my thanks.

2. Permit me some observations on: warning time scenario, force build-up and replacement scenarios, warfighting scenario, the integrated battlefield, loss rates, and the model.

3. Warning Time Scenario. The study assumes 10/15. I am fully aware of the rationalizations, reconciliations, and prevarications that have surrounded this issue for years. And I do not care to enter the lists with another biased judgment. It does seem to me important, however, that the Army concentrate on enemy capabilities, setting aside perceived intentions and presumed likelihood of an enemy build-up that can be detected, correctly interpreted, and acted upon in any postulated number of days. For it is true that, as between highly ready forces which confront one another across armed borders, the signals which portend attack are and can be so deeply buried in the noise levels of day to day activity as to make their detection very difficult and their interpretation if detected even more difficult. One has only to recall the Israeli experience in the 10 months or so before the Yom Kippur War to understand the point. We must therefore contend with capabilities and prepare ourselves accordingly. My numbers may be a little off but, as I recall, SACEUR gives the WP the capability of attacking NATO with about 39 divisions within 24 hours of the time WP forces begin to move. In 96 hours that number becomes about 58 divisions. While some build-up might follow, this is the scenario in which the Sovs strike quickly, with as little warning as possible, and seek a prompt win with swift rates of advance using nuclear, chemical, and conventional weaponry and forces. Doctrinally they write about this; in the field they practice it in exercises. They clearly are capable of attacking NATO in this fashion; if for some reason it doesn't work, they are further capable of protracted conventional conflict. There might then be a delay of some time while forces are brought over from the east and up from the south in order to carry on the protracted war. We too might then have time for reinforcement; how much is problematical. In any event, I would urge that the Army hang its helmet on the kinds of capabilities outlined above. If other scenarios, driven by assumptions differing substantially from these capabilities, are required, then we should respond with an appropriate analysis. But we must always make it clear that our considered judgment, based on capabilities—not on presumed intentions or postulated warning times, is quite different. In a capabilities scenario we might have 48 hours of warning—if we read the signals correctly; and we might have something less than 48 hours of preparation time if we respond with precision and alacrity. If we believe anything like the above, we must make the case, emphasizing the need for quick decision and reaction if we are to survive.

4. Force Build-Up Scenario. On both sides the rate at which forces build up is a function of three things: warning time scenario, pace of battle scenario, rate with which both sides can reinforce scenario. Here again our case must rest on capabilities. On the WP side the scenario postulated in 3 above would build WP forces up faster than in Omnibus 79; without Chinese intervention or threat thereof, there is a capability to reinforce from as far away as the eastern front in the 180 days of interest to Omnibus 79. Therefore the WP forces brought to bear initially can be centered into battle much more quickly than in Omnibus 79, and by the 180th day there

is a capability to reinforce both from southern and eastern USSR. The pace of battle postulated in Omnibus 79 is far short of Soviet capabilities, doctrinal guidelines, maneuver practices, and stated intentions. Their doctrinal rates of advance need not be recited to you. If one follows the warning time scenario set out above, then applies Soviet doctrinal advance rates and presumes they will make every effort to achieve those, then a battle develops in which NATO can possibly defeat the first echelon armies and do it in about a week or 10 days. However, capabilities to destroy second and follow-on echelon armies then are a function of how quickly we can put our own reserves in motion—both US and NATO reserves. It therefore could well follow that the first phase of that war—the one in which WP seeks to overrun NATO with surprise, mass, and speed, is a war of no more than 15 days' duration. In other words, if we can stop that first devastating attack, and defeat it soundly, time for build-up on both sides might be available. If that be the case, then our own mobilization plans and those of our allies must foresee the need for quick reinforcement with combat forces and the logistical sinews of war essential thereto—ammo, fuel, medical, transport, repair parts, and replacement equipment. In sum, the first part of the war may be done in the first couple of weeks. If we survive, a more protracted build-up and warfighting scenario may be appropriate. Finally, the Omnibus 79 methodology for estimating shortfalls in replacement capabilities is sound. However the assumptions going in are probably not within our capability. For example, I seriously question the capability of our combined resources to produce 159,000 trained replacements in 180 days. The driver here is the assumption concerning the IRR—how many of what skills would answer the call. Given that we don't even know where they all are at the moment, and given that in most cases some training would be required, our ability to recover more than 20–25 percent of the IRR in the skills required is questionable at best. This might give something less than 50,000 replacements; we might get another 40,000 from the training base—trainees and trainers. Beyond that our capability is highly problematical.

5. Warfighting Scenario. Omnibus 79 presumes the main WP attack across the North German Main—into NORTHAG. No question they have that capability, and indeed in CPXs and other exercises play out that scenario. They do, however, have the capability to put in simultaneous army-level breakthrough-type attacks against four of the eight NATO corps. It is probably quite imprudent of us to set aside any situation in which they do not put in at least two, and more likely three, of those attacks. In any event, using the Omnibus 79 scenario or one in which we are required to cope with a more impressive range of offensive action, there is the question of the critically thin area in NORTHAG. Omnibus 79 and other studies beef up NORTHAG with forces deployed during the early build-up, pre-positioned equipment sets, or a combination of those. A more robust range of military options could be had were we to also consider what role CENTAG and other forces might play in solving the NORTHAG dilemma. For example, if CENTAG were to attack deep into the rear and on the flanks of the main WP attack against NORTHAG, it might be possible to develop a much more flexible strategy—one that would avoid many of the problems accompanying reinforcement of NORTHAG by coming in behind (west) of the border. In addition, such a scheme would take advantage of the existing US/GE support base in the CENTAG area. With a major allied force astride his support lines north of the Harz Mountains, for example, it would seem highly unlikely that the enemy could press west very vigorously for very long. And so aid to NORTHAG can be had in several ways, at least one of which is in many ways more appealing than simply falling in more forces behind those already in place. That might be necessary too, but only as part of a grander and more flexible scheme.

6. Loss Rates. Long a bone of much contention, loss rates can be postulated in many ways. The point that needs to be made here is that the Omnibus 79 loss rates are different from those in 4102 and from those in the latest TAA and other studies. Our most serious problem in trying to lay out the admin/log system program review is the considerable disarray we are in with regard to an agreed-upon loss model. We must very soon develop and lay out in a common model a range of loss estimates with which we can work our studies, simulations, and plans. Obviously the loss model is determined in large measure by the pace of battle and warfighting scenarios. This makes even more imperative an agreed-upon Army position as to what those might reasonably be. Until we do that, we can't move ahead very realistically in loss model development.

7. The Integrated Battlefield. The Sovs have the capability to conduct integrated nuclear, chemical, and conventional battle from the outset. They plan to do so—evidence is in their doctrine, exercises, training, equipment characteristics, and organization. Whether they intend to do so or not can't be determined—the point is, "They plan to do so," and they exercise the plans. We took a decision at the Nuclear Systems Program Review that we would begin work to put the Army back on the nuclear battlefield. The chemical SPR scheduled for May will produce a similar decision regarding chemical warfare. TRADOC is setting to work to revise the doctrine from the top—FM 100-5. It therefore is imperative that our major study efforts reflect operations on the integrated battlefield—a battle arena that reflects the enemy's capabilities and plans. Here again, if a conventional excursion is required, an Omnibus 79-like study can be done. It should, however, be proffered only as an excursion. The Army position must be that planning for operations on an integrated battlefield is essential and the first order of business. We must set forth with some urgency the case that enemy capabilities have changed dramatically in the past 5 to 10 years. Against an enemy whose strategic and theater nuclear capabilities exceed our own, and whose conventional capabilities—quantitative and qualitative, outweigh our own, we have no reasonable choice but to "plan" for an integrated battle from the outset.

8. The Model. Omnibus 79 uses the CEM model. Its deficiencies need not be recounted. Nor is it necessary to recite again the marginal adequacy of models which deal with FEBA movement rates which are determined largely by manipulation of firepower scores. What this highlights is the imperative of developing a better model. Some months ago I offered to link up with the DUSA(OR) in his effort to produce such a model. Work in pursuit of that goal is not underway. It must be energized as a matter of some urgency.

9. In sum:

a. Army warfighting studies, plans, doctrine, organization, equipment development, and training must all be developed in sobering contemplation of enemy capabilities.

b. There will always be pressures from the nonprofessional echelons above the Army to rationalize and equivocate with those capabilities with which we cannot cope, and for which resources would be required to redress the force imbalance.

c. Soviet capabilities have changed dramatically in recent years—from inferiority to superiority across the spectrum of conflict. Khrushchev's post-Cuban missile crisis statement to Kennedy, "You'll never do that to us again," has indeed become a reality. All our warfighting considerations must reflect that fact.

d. The capabilities philosophy must be the overriding consideration in all our: warning-time scenarios, force build-up and replacement scenarios, and warfighting scenarios.

e. Our warfighting scenario must be built on realistic strategies which seek to offset the enemy's primary operational capability in a war of maneuver—of forces and fires—nuclear, chemical, and conventional.

f. Based on capabilities-driven warning time, force build-up and replacement scenarios, and on flexible warfighting scenarios, we must set to work with some urgency to devise a loss rate model in terms of which we can display an acceptable range of battle outcomes.

g. Given that our conventional operations can no longer be conducted under the umbrella of strategic and theater nuclear superiority, and given Soviet capabilities and plans for operations on an integrated nuclear, chemical, conventional battlefield, we have no reasonable choice but to plan to operate on that integrated battlefield from the outset.

h. Our ability to simulate theater-level war is inadequate at best. This situation is aggravated by our inability to model nuclear and chemical battle outcomes. Work must be accelerated to bring about soon an acceptable theater-level battle model which can also model the integrated battlefield.

10. When you've had time to digest this, send me a signal. If anything like a consensus can be arrived at, we need a game plan.

The Writing of Doctrine

Message to Multiple Addressees
2 May 1980

1. Separately I am sending you all a longer and more complex message that has to do with accounting for the manpower which we dedicate to writing training literature and doctrinal literature. The purpose of the instant message is to explain to you personally what is behind that longer message, and to tell you that it will require your personal attention to bring about the change represented by the content of that message.

2. Because of the way the Army is represented in the military manpower training report—the famous MMTR—many, many people who do not actually teach anything have traditionally been carried against an instructor to platform-hour ratio derived from the MMTR that OSD analysts use to compare services performance in this functional area. For example, traditionally Army combat and training developments personnel have been carried as instructors. This means that many people who never taught anything were counted as instructors. Result: the Army showed a much lower instructor-platform hour ratio than the other services. It was for this reason that, early in TRADOC's struggle with this problem, we separated the CD'ers and TD'ers in the manner in which we now have them arranged.

3. Separation of those who create doctrine and write about it from those who teach it is a mistake—I've thought that from the beginning. My recent investigations into the matter confirm my earlier perception, and indeed indicate the problem is much worse than I had thought. Our instructors are no longer the intellectual masters of the subjects they teach; they are sitting around their coffee cup-cluttered desks waiting for the doctrine to come in from CD and TD. In one school they are waiting for it to "come down from the hill," that being where the TD and CD guys are physically located, much on the order, I suppose, of the Israelites waiting at the foot of Mount Sinai for Moses to come down with the tablets. I'm not criticizing our instructors—they are victims of a situation which we created for them. But they must be the intellectual masters of their subjects! In no other way can the doctrine be vital, alive, and responsive to the Army's needs. They are the critical link in the only continuous feedback system we have with the Army in the field.

4. Therefore it is the purpose of the earlier and longer message to change the situation I've just described. What that tells you to do is merge your doctrine writers and your instructor groups so that those who teach the subjects are those who write about the same subjects. The accounting procedures to accomplish that are set forth in the longer message. Some few of you are doing just what we now intend for everyone to do. In that case, little if any change in your de facto arrangements will be required. In other cases, substantial changes will be required. That's why I signal it for your personal attention.

5. Service schools are for instructors. Students come and go; they retain more or less of what we expose them to based on their level of interest and intelligence. But those who really learn their subjects in depth are those who have to teach every day or so, defend, argue, answer questions, and in the end become the intellectual masters of what they teach. We must get the instructor groups in TRADOC back in command of their subjects and do it quickly.

FM 100-5

Letter to Colonel Robert E. Wagner
2d Armored Cavalry Regiment
19 June 1980

For reasons that I will never be able to understand the defense described in FM 100-5 and related FMs has been taken as being too much defense and not enough active. Having written most of the words about the subject in those manuals myself, I can assure you that nothing was further from our minds. Somehow we just didn't come through clearly.

We have begun a rewrite of the whole series of How to Fight manuals which will, among other things, put the US Army squarely back on the integrated battlefield—chemical, nuclear, conventional. In the rewrite we will have another go at saying clearly enough so as not to be misunderstood that if you want to win you have to attack, whether you're defending or attacking. . . .

AirLand Battle Operational Concepts

Note to Brigadier General Donald Morelli
4 August 1980

The paper you gave me has some good ideas in it, but they are pretty badly fuzzed up. Let me sketch in some notions around which we might build such a paper.

1. World Perspective—1990s:

a. Larger numbers of smaller countries with substantial numbers of relatively sophisticated weapons.

b. Likelihood of many smaller countries with at least a limited nuclear capability; increased military capability; also increased likelihood of nuclear blackmail, sabotage, adventurism.

c. Continuation of Soviet expansionist drive into Middle East; then through the Middle East into eventual domination of the vast resources of Africa, especially those of South Africa.

d. Likelihood that US forces might have to do battle with increasing numbers of surrogate forces equipped and trained by the Soviets.

e. Increasing likelihood that Soviets themselves will risk direct conflict with US forces as they become more and more confident that we have lost our resolve.

f. Increased likelihood of military conflict over scarce resources, wherever they may be found, but especially in the Middle East and, in the longer term, through all of Africa, eventually into the tremendously rich mineral wealth of South Africa.

2. Battlefield Perspective—1990s:

a. The battlefield will be dense, with large numbers of high quality weapons systems.

b. The battle will therefore be very lethal.

c. The battle will also be very intense, especially at the point where the decision is reached.

d. It will be an all arms/all systems battle; Soviets will export to their surrogates the products of their impressive military production rates.

e. Firepower and maneuver forces will be more mobile, especially helicopters and increasing numbers of highly sophisticated armored vehicles.

f. Mobility, numbers, and Soviet-style echelonment tactics will make it imperative that a wide variety of long-range sensor systems be employed to locate enemy forces in sufficient time and at sufficient distance to retain freedom of action and thus the initiative.

g. Mobility, numbers, and intensity will create a pace of battle that will place a high premium on command-control—the ability to direct the battle in all its aspects.

h. Increasing numbers of highly sophisticated electronic warfare systems will make command-control and sensor operation very, very difficult, just at a time when it will be necessary to rely more heavily on both for successful operations.

i. The threat of nuclear weapons employed by surrogate forces or forces of smaller countries makes it imperative that all operations be conducted as if in a nuclear environment, and that the option for US use of small-yield nuclear weapons be kept open and immediately available with quick release provisions at all times.

j. The threat of chemical weapons employment, especially in local situations, to achieve quick success without loss of facilities or resources, other than soldiers and civilians in the area, makes it imperative that all operations be conducted as if in a chemical threat environment, and that US forces have the capability to use chemicals as appropriate at any time. From a national policy standpoint the development of a comprehensive incapacitating but nonlethal chemical capability will be very important.

3. What It Will Take to Fight under Such Conditions:

a. It will be necessary that US forces have a high degree of strategic mobility of an order necessary to reinforce NATO Europe and sustain a fight there, but also of an order necessary to deploy to and sustain a fight in other areas of the world as well—especially to the Middle East, Africa, and Northeast Asia.

b. It will be necessary for US forces to develop a high degree of tactical mobility, so that the disadvantage expected in numbers can be compensated for by the mobility of firepower and limited numbers of forces.

c. Survivability of US forces should be ensured by their high tactical mobility and by emphasis on technological advances in lightweight armor.

d. The operational capabilities of US forces must be soundly based on comprehensive battlefield recon, surveillance, and target acquisition systems (RSTA) which can be fielded quickly, and which can provide adequate information of the enemy far enough away in time and space for US forces to retain freedom of maneuver and therefore the battlefield initiative. This system is based on air cavalry.

e. A much improved and more capable command-control system must be provided US forces at all levels.

f. Longer range air and ground delivered weaponry must be fielded to enable the ground force commander to counter enemy reserves and follow-on forces in timely fashion so that they do not overload the battle against the assault echelons; in this way the commander can provide himself with freedom of maneuver and still retain the initiative, even though engaged. These weapons should include large numbers of "smart" weapons—fire and forget weapons with at least terminal homing, if not full homing sensors.

g. Command-control and sensor systems must be protected from the enemy's highly sophisticated and comprehensive electronic warfare capability.

h. US forces must be provided with an EW capability capable of attacking and disrupting critical enemy command-control and sensor systems.

i. Development of enhanced radiation weapons for employment by surface-to-surface delivery systems of longer range than is now possible is an imperative.

j. Development of non-lethal incapacitating chemical capability for US forces is an imperative; meantime the capability to employ lethal chemicals on a limited scale in localized situations is essential.

k. Tactically it will be increasingly necessary that US forces employ the "indirect approach" rather than attempt to face head-on large forces, well equipped with modern weapons. Mobility, command-control, RSTA, EW, and an integrated all-arms capability featuring highly mobile long-range weaponry with "smart" weapons capabilities [are the] keys to battle success in the 1990s.

Now obviously I've left out a lot; you had some of what I've said in the paper; but think on the above and see if you can't flesh out a little tighter paper with more meat and less avoirdupois.

Doctrinal Literature Program

Message to Lieutenant General William R. Richardson
Fort Leavenworth, Kansas
15 September 1980

1. We are taking action—for the severalth time since I've been here, to try and square away the MMTR, which is the report people use to screw up the relationships between those who teach and those who write. It's been a disaster for years. The Army has the short end because of the way we count things—honestly, and so the other services are hard to convince that anything needs changing. I nearly had Paul Phillips persuaded that he could fix it when he up and retired. We are trying again.

2. It is true that SMEs, if they are in the midst of producing a Soldier's Manual or whatever, have precious little time for teaching. Also true that teachers with a course in session have precious little time for being SMEs on manuals. But the driving thrust of what we do must reflect the notion that the instructor must be the intellectual master of his subject. To do that he has to teach and write about it. After all, service schools are primarily for instructors—students come and go, but the guys who teach and face that critical audience week after week are the ones who really learn. It's a fine line we walk, I know, but we simply can't let some dumb accounting system in the Pentagon screw up the best military school system in the world.

Allied Cooperation

Message to Dr. Walter LaBerge
Deputy Under Secretary of Defense (R&E)
9 October 1980

With both the Germans and the British we've been working for nearly two years now on operational concepts for the AirLand Battle of the 90s. With each partner we are now very close to having an agreed concept. The concepts are essentially identical; whatever differences there might be result more from the separate staff talks environment in which each was produced than from substantive differences in concept. With the British we have agreed to produce a concept paper, materiel requirements, and a development strategy for the tank program, and have all that in hand by end calendar 1982. We hope that work can in the end be tripartite UK/US/GE. The GE/FR tank program is seen by many as an obstacle to a tripartite program. We don't believe that to be the case. The FR/GE program is clearly nearer vice longer term, and should not therefore be in conflict with our longer-term objectives. The Germans, FuHa-level, agree with this; the view at MOD Bonn may be somewhat different. My judgment is that we can win out in the end.

Army 86 Implementation

Letter to General Wilbur L. Creech
Tactical Air Command
10 November 1980

As you are well aware, over the past two years TRADOC has been embarked on a rather ambitious force structure and force design exercise called Army 86. The majority of that work has been completed and the product delivered to Department of the Army for implementation. The key operational notion of Army 86 is that, in Central Europe, to fight and win we must fight concurrent battles against assault and follow-on echelons. You and your staff have been most helpful in promoting a better understanding of that conceptual notion in the Air Force. As a matter of fact, your recent letter concerning battlefield air interdiction will go a long way in clearing up some potential misunderstandings concerning Army/Air Force conduct of the air/land battle.

General Meyer and I feel, however, the penalty in terms of battle outcome is too severe for us to wait until 1986 or thereabouts to implement the concept. He therefore asked me to present some ideas on how we might make that concept work with today's force designs and hardware. I gave the briefing last week at his Army Commanders' Conference. The concept and idea of doing it now were enthusiastically embraced by all the commanders present. As a result, I have set in motion an action plan to assist our commanders in the field in understanding the concept and applying it on today's battlefield. Our first priority will be to the force in Europe.

. . . I understand our Army people working the concept and your TAFIG folks working the Ground Attack Control Center concept will meet later this week to discuss items of mutual concern about targeting cells and attack of the second echelon. I appreciate your assistance and will be receptive to any ideas you and your staff might have on how we can make this concept an operational air/land reality with today's forces. I do not think we can win the battle any other way.

Air/Land Battle 90

Message to General Edward C. Meyer and Others
Army Chief of Staff
28 November 1980

1. Some days ago I sent each of you, and some others, a letter enclosing a draft concept paper on Air/Land Battle of the 90s. It is a start; while I'm not completely happy with it, it does move us in the right direction. With some little more work it should be about what we need to get started on the next phase of the continuing modernization of the Army toward the year 2000. Some of it applies to our world today and belongs in the menu of principal matters on which we must concentrate with the new administration if we are to begin to move in a rational way toward our goals for the future Army.

2. This message is to commend to you that concept paper, reading thereof and reactions thereto, and to highlight some of the 'tween the lines and italics not in the paper or not immediately apparent if one must read it hurriedly, and you all are in that category.

3. The concept treats primarily confrontations, political and military, in areas of the world which trends indicate most likely to be of concern and most contentious for the US to deal with in the years ahead. They are: NATO Europe, the Middle East, northeast Asia, places where there are large numbers of fairly well-equipped forces who use Soviet operational concepts/tactics.

4. Underlying the concept is the notion that the purpose of military forces in peacetime is to reduce to a minimum whatever incentives enemy leaders might perceive as favoring seeking military solutions to political problems. Deterrence is not a word in the Russian language. A clear war-fighting capability is necessary to effectively reduce their incentives—as they view them.

5. In those critical areas our "defensive" strategy must extend beyond simply denying victory to the other side; it must instead set forth definable, recognizable (although perhaps limited) victory as a goal for the defender. There are two important reasons for this:

- First, sometime very soon our political leaders must be made to understand that they cannot commit military forces in pursuit of political aims without setting forth clear-cut objectives which see the military forces as winning something. If this is not done it is most unlikely that the military action can produce anything of substance that will allow a political "win." The purpose of our military actions is not just to somehow avert defeat; it is rather to win.
- Second, enemy leaders must be given to understand clearly that, if they elect to move militarily, no longer is there a status quo ante bellum to be restored; rather they must understand that the situation they themselves have created is one to be resolved on new terms.

6. As the strategic nuclear balance evens (teeters) (shifts), so grows the enemy perception of his own increasing freedom of action at theater levels—conventional and nuclear, and so grows the importance of a balanced modernization of our own strategic and theater forces, now so out of balance on the strategic side. As the possibility of theater nuclear capabilities in theaters other than NATO grows, so grows the urgency of modernization programs that emphasize our general purpose integrated battlefield capability—chemical, theater nuclear, conventional.

7. Soviet-style strategy/tactics embrace(s) two fundamental concepts:

- In the first, mass—momentum—continuous land combat are the operative tactics. Breakthrough (somewhere) is sought as the initiator of collapse of the defender's system of defense.
- In the alternative, surprise is substituted for mass in a tactic called daring thrust. It is a preemptive strategy which seeks to substitute surprise for mass. In NATO it involves the attack of thirty or more BMP regiments of the MRDs [motorized rifle divisions] which, attacking without warning, would seek to deny defending forces the chance to get set forward.

8. Both strategies are essentially maneuver-based schemes whose purpose is to disrupt the operational tactics of the defender, albeit by different methods in each case.

9. Defense strategies in key areas must therefore be designed to preserve the territory/resources/facilities of the defended area for the defender. In none of the critical areas suggested above is there sufficient maneuver room to accommodate a classic defense in depth strategy. The defense must therefore begin well forward and proceed aggressively from there to destroy enemy assault echelons in order to seize quickly the initiative by going on the offense.

. . .

The Soviet enemy generally enjoys a short term advantage in his ability to mobilize and deploy equipped forces quickly; one purpose of the battle concept must be to try and foreclose the possibility of prolonged military operations.

11. Further, the operative tactics seek simultaneously to:

- Deny the enemy the objectives he seeks; prevent the enemy from loading up the assault forces with reinforcing echelons and thus achieving by continuous combat what might be denied by a stout forward defense.
- Create an opportunity to seize the initiative to attack to destroy the integrity of the enemy operational scheme, forcing him to break off the attack or risk resounding defeat.

12. Because of the enemy advantage in numbers, attack of follow-on forces must always begin when those echelons are relatively deep in enemy territory. If an outnumbered defender waits until his numerically superior foe has penetrated the defended territory to mount a counterattack it is always too late to bring to bear effective fires and so to defeat the incursion. Especially true if theater nuclear forces are considered necessary to defeat the penetration.

13. Therefore the operative tactics necessary to support an operational concept of an integrated defense well forward are:

- See deep and begin early to disrupt, delay/destroy follow-on/reinforcing echelons.
- Move fast to mass against the assault echelons.
- Strike the assault echelons quickly so as to prevent them from attaining their objectives.
- Finish the immediate fight against assault echelons and their follow-on echelons rapidly so as to go on the attack and finish the battle against the assault army(ies) before follow-on armies can join the battle.

14. Systems, organizations, and tactics designed to defeat enemy assault echelons, to disrupt follow-on forces, and to seize the initiative by attack must be able to target, move against, fire with nuclear/chemical/conventional fires across the spectrum of the battle—throughout the depth of the battlefield.

15. Key to a credible war-fighting capability on the integrated battlefield—now and for the Air/Land Battle 90s—are:

- Sensor/surveillance systems to prevent surprise attack in peace or war and to provide necessary targeting/surveillance in wartime to permit actions against the follow-on echelons.
- Delivery systems—dual capable, with sufficient range, accuracy, and lethality to hold enemy follow-on echelons at risk in peacetime and attack them successfully in wartime.
- Command-control sufficient to integrate all-source intelligence in near real time in peacetime and in wartime, and to turn that intelligence and targeting information into information for maneuver for employments in near real time as well.
- Maneuver forces organized, equipped, trained, and employed to execute the tactical schemes outlined above.

16. For a moment, consider NATO. While I have long believed that the provisions of 14/3 are permissive enough to accommodate our integrated battle schemes, it is also true that the 14/3 image is much less ambitious than required by the integrated battle concept. In our own JCS documents we are more restrictive than is 14/3 itself, or than 14/3 requires. Both Germans and Brits see the battle scheme under 14/3 as one far more restrictive than integrated battle foresees. This is so largely because they have not the capability to fight the integrated battle now, nor do they see how they could get the resources to develop the requisite capability in anything like a reasonable time. All this suggests that what we really need to do as we further develop Air/Land 90s and integrated battle is seek a 14/4—our initiative to provide a rational, modern strategy for NATO and other regions of the world growing daily more critical. Wouldn't it be refreshing if that initiative came from inside our military establishment instead of from some Tuesday lunch, Wednesday breakfast, or some other of those awesome strategy forums that have become the rule rather than the exception since the Bay of Pigs?

Ground Forces Issues

Letter to David Chu
Congressional Budget Office
31 December 1980

Thank you for the advance copy of your ground forces study. . . . Not having specifics concerning how your cost figures were developed, I'll not comment on them in detail, however some general notions about the budget are set forth at the end.

As for structure, any proposal which would develop a 1.4:1 force ratio has to be attractive to any professional military man. However, there are two major stumbling blocks to trying to resolve the imbalance simply based on the relativity of numbers in a force ratio equation. One has to do with the simultaneity of the so-called war and a half strategy. The other has to do with the internal dynamics of the NATO equation. Let's take the latter first. The NATO strategy, as outlined in your paper, posits an "elastic defense." This implies there is sufficient terrain to trade away for time while mobilization and reinforcement take place. In central Europe there is not sufficient space to trade for time. Too much of Western Europe's lifeblood, in terms of population and urban growth, lies close to NATO's borders with Eastern Europe. Any strategy which calls for space to be given up early in the war automatically is not acceptable to our NATO partners in northwest and central Europe. A close look at our active defense doctrine will show it is elastic, but only within those forward battle areas.

Essential to successful defense in Europe is a capability to target, and attack, with conventional, nuclear, and chemical weapons, follow-on echelons. "Time to mobilize and deploy" implies that somehow the alliance would be able to muster men and equipment in sufficient quantities to stabilize the situation, take the initiative from the enemy, go on the offense and win. That is not possible—we and the Germans quickly run out of equipment at modest force levels long before we run out of personnel. The industrial plant in either country simply cannot begin to produce quantities of equipment in anything like the times necessary. Therefore we will fight with what we have in units and in stockpiles when the war begins, and whatever resolution is arrived at on the battlefield will be determined by those amounts and how they are used. This does not agree that a short war strategy is the correct strategy; it simply argues that we've resourced ourselves into a short war strategy. Like it or not, that's all there is left for us to do. Especially important are logistical and command-control elements, whose disruption could delay, interrupt and create time gaps in the enemy's force echelonment. This would buy us additional time to deal with assaulting echelons without letting follow-on elements bunch up, join the assault echelons, and so overload the defender trying to defeat the assault echelons. We don't have a substantial capability right now to target and deliver against the second echelons, but the capability is vital to a successful defense in any event. Therefore some bucks should be put toward that end—even at the expense of equipping more structure. No question about the need to lessen the threat to central Europe. We should do all we can to encourage Warsaw Pact countries to pursue policies that preclude first strikes against NATO. The ground may indeed be fertile for this, but it needs to be plowed by others, in concert with the Department of Defense. However, the proposal to substitute a West German Corps with a French Corps is just not appropriate. Nothing against the French, but that course of action is just too much bending to meet the Pact's perceived desires or anxieties, and so is not realistic from a number of viewpoints.

Now the non-NATO contingency. First we must realize that we haven't the capability to cope with the war and a half simultaneously. That may or may not be bad, depending on one's point of view—but it is, and that fact must be clearly understood. For it is a risk attendant upon the current budgetary strategy. Secondly, we must admit that the Rapid Deployment Force is no more rapidly deployable today than it was say 25 years ago. The USMC has always been a deployment force. The Army has had such a force designated since 1957. So what's new about the RDF—just that it was thought up by this administration—completely out of context with the long history of the need for and our efforts to provide ourselves with such a force. What we need is not another JTF headquarters, but more fast deployment ships and strategic lift aircraft. Until we have those the RDF is and will remain a political gambit.

So, summing up the NATO problem from a structure standpoint:

- While we do need more divisions, we can't equip or man them in the near term.
- It would take several years—perhaps eight or ten—to buy, at a reasonable rate, the equipment necessary to match our ability to provide manpower fairly quickly and so build a larger force to reinforce NATO.
- The war will therefore be of shorter rather than longer duration and it will quite likely be fought almost entirely with the equipment and forces available or subject to mobilization when the war begins.
- More than just additional divisions, we need to invest in the sensors and delivery systems to provide a credible capability to attack the follow-on forces early on and continuously.

While the Marines are clearly part of our deployment force capability, they cannot and should not be mechanized. It would be entirely too difficult and time-consuming to retrain them to operate in a mechanized environment. It's even more mind boggling to think that we could create the combat support and service support structure necessary for them to carry the fight beyond the Beachhead Line. That's a job for Army forces—a fact which reinforces even more the need to build the roll-on roll-off transport.

While it's not possible to comment on details of your budget strategy, two things need be said. First, the circumstances in which we now find ourselves are the result of nearly ten years of what has amounted to unilateral disarmament. It isn't what the Carters have done—they just came along to ice a cake that had already been pretty well eaten up. Therefore the imbalances you're seeking to redress are of long standing, and will not be redressed in a short period of time.

Secondly, given the first point, it is important that we consider all US forces in the context of a balanced force strategy. It makes no sense, for example, to spend the kind of money programmed for MX, for intratheater transport, and for strategic systems in general unless one looks carefully at the imbalance in spending for general purpose forces. No question the strategic force needs upgrade, so do we need an intratheater transport—but in due course. What we do need is a balanced force development strategy that brings the general purpose forces along to modernization much more quickly than is now programmed. So even with a fixed budget our spending ratios are out of balance. If we spend more pursuing the same strategy, it too will be folly.

FM 100-1

Letter to General Edward C. Meyer
Army Chief of Staff
12 January 1981

Enclosed for your approval is the final draft of FM 100-1, *The Army.* In this revision you will find several significant changes from previous drafts. . . . The most significant of these changes are:

- A new chapter which addresses the Army's organization and functions. Included in this chapter is a discussion of the Army's role in strategy formulation.
- Public Support has been laid down as a tenth principle of war.
- Each principle of war has been discussed in terms of both its strategic and tactical applications.

The result, FM 100-1, represents the kind of product which sets forth in unmistakable terms the philosophical basis for the Army's family of "How to Fight" manuals. You should also know that the book was written at Leavenworth by a task force at CGSC. My judgment is that they have done a first-rate and most commendable job. Help from Glenn's staff and from Jack Merritt's [Major General (later General) Jack N. Merritt, Commandant of the Army War College at Carlisle Barracks] staff at Carlisle has been most helpful. But for the first time since 1973 we have written an important piece of Army doctrine at Leavenworth. We both just have to say *hurrah* to that!

AirLand Battle I

Message to Multiple Addressees
29 January 1981

1. Now that the integrated battlefield and extended battlefield concepts are being assimilated by the TRADOC community, a need exists to tie these concepts into one overarching descriptive term. Though the words "integrated" and "extended" convey concepts, they do not accurately describe the battlefield in its totality. They do, however, have utility when we talk academically or doctrinally about subsets of the battlefield. The term "AirLand Battle" best describes and ties these two concepts, and all others, together. Henceforth, when we talk about the total battlefield, the term "AirLand Battle" will be used. Combat developers working on future systems also need a term to use as a conceptual basis for the development of requirements. That term will be "AirLand Battle of the 90s."

2. I'd like you all to take the actions necessary to ensure that our instructors, trainers, and doctrine writers clearly understand these terms, and that the term AirLand Battle is incorporated into our current and future literature rewrites. All of us must emphasize this term as we go about our daily tasks.

Allied Forces Staff Talks

Message to General E. C. Meyer
Army Chief of Staff
5 March 1981

1. This is to give you a quick report on our third round of talks with the French Army staff, completed 26 February. Results were gratifying, principally because of unexpected initiatives by our FR counterpart, the Deputy Chief of Staff for Plans and Studies, MG Ebert.

2. As you know, Carl Vuono heads the delegation for us. This time around we increased the US delegation size and bit into some tough issues. In this regard, we got an unusually frank assessment from MG Ebert of our draft AirLand Battle concept which provides valuable insights into allied thinking. For the most part, they agree with our logic and conclusions. In addition, our top attack briefing, which stressed emerging technologies, was well received, and our target value analysis from the fire support MAA evoked some genuine interest despite its complexity. As in the past, MG Ebert cautioned that overreliance on high technology risks paralysis on the battlefield (i.e., commanders substitute technical analysis for military judgment or equipment becomes too sophisticated for the operator).

3. From the French delegation, we received some excellent presentations. It became clear they take MOUT training very seriously. For example, French FTXs and CPXs are more realistic than our own, and they have a mature concept for optimizing fire support in built-up areas. Moreover, we saw a superb MOUT film, narrated in English, which we are evaluating for US troop training.

4. In the fire support coordination realm, we saw some clear differences and similarities. Though moving toward automation with Atila (Battalion TACFIRE), the French still rely heavily on LOs. However, their perspective on the complementary roles of artillery and 120mm mortars gave us food for thought.

5. With respect to FR involvement in the 9ID HTTB, MG Ebert came on board enthusiastically, although without specific commitments yet. We will try to keep FR involvement in the HTTB under the aegis of the staff talks. Both heads of delegations know each other and, if the FR CSA sees it our way, both will have proponency.

6. The fourth round of talks in late September this year will probably move to one of the FR schools. Themes under consideration for future discussion include: nuclear/chemical operations; mobility/countermobility; and air defense. With regard to the latter, I hope to use some of the excellent work which has come out of the J-SEAD and J-CAAD studies. In the interval between talks, we are exchanging concepts related to these themes and providing informal comment on respective national concept papers.

7. This is exactly the direction I had hoped these talks would move. The FR/US dialogue is the most sensitive of all the bilateral staff talks and we have laid foundations carefully. This fall we hope to begin exploring ways in which we could harmonize how to fight.

Implementation of AirLand Battle

Message to Lieutenant General William R. Richardson
Fort Leavenworth, Kansas
20 May 1981

. . .

2. VII Corps concerns provide an indication of the range of issues which must be addressed quickly. We cannot tell them to wait and be patient, all will be well in five or six years. Thus, think our answer to them should go something like the following:

a. Corps Weapons Systems. It is true we are currently restricted by weapons systems available to the corps. Scarce acquisition and strike assets highlight our need to get on with the intelligence preparation of the battlefield (IPB) for both planning and exercises and to identify those high-value targets whose attack will produce the required effects on follow-on echelons. Actions taken now to prioritize, quantify, and show the payoff of attacking high-value targets will assist our efforts to develop and field a corps support weapons system.

b. Corps Decision Window. The VII Corps message demonstrates the danger in putting numbers and lines on our charts. If our analysis is right, the point to be made is that the corps fights first-echelon armies. The second-echelon divisions of those armies could well be within 48 hours at the initiation of hostilities. If that should be the case, it is imperative that we be capable of taking advantage of enemy vulnerabilities thus created. The corps cannot limit its planning horizon to 48 hours while we develop corps systems to find and disrupt follow-on echelons out to 72 hours. We must make the point that, if we plan, train, and exercise our forces in accordance with the division of responsibilities established, we put the second echelon at some risk by reducing the enemy's flexibility, initiative, and momentum necessary to mass at a point and time of his choosing.

c. Corps Air Allocation. The alternative of preventing disaster in the close-in battle today as opposed to being overwhelmed by follow-on forces later is not an attractive choice. The entire thrust of AirLand Battle is not to avert defeat but rather to win by creating opportunities for decisive offensive action. We must overcome the notion of a two-part battle. Central to the AirLand Battle is the requirement to tightly coordinate scarce acquisition and strike assets. Given today's environment, the answer is to develop adequate IPB, plans, targeting cells, and links to our Air Force. Until such time as we demonstrate the payoff of attacking high-value targets in our exercises we will not generate the requirements or hard numbers to drive our Air Force in the direction desired. Two steps are necessary now in addition to those mentioned earlier. First, we must develop targeting cells which will identify the high-payoff targets by type and location whose attack will, if timed correctly, produce the required effects. Secondly, we must align Army and Air Force efforts in the one battle. We are in the process of developing a mechanism called the "battle coordination element" to align OAS/BAI decisions at echelons above the corps to ensure the needs of the corps are accurately reflected in the decision process. At corps and division, the targeting cell will link the Army with the Air Force in the planning process.

d. Tactical Communications. In terms of implementing the AirLand Battle there are a number of actions we can take now. First, we must clearly identify and limit information to what is necessary by echelon. Secondly, we must manage our current sensors more effectively.

For example, Guardrail can be down-linked to the division today if tasked. Today we do not routinely make use of that capability in our training or exercises. While we do not have the same capability with Quicklook, it appears we could develop that capability quickly if we choose to do so.

e. Personnel. Manning requirements for the targeting cells at division and corps have not been refined. A full-up targeting cell will be demonstrated at the June 1981 Field Artillery SPR. The impact on spaces, location, and organization will be addressed as part of that demonstration. We will be getting back with the corps and divisions to discuss implementing procedures following the demonstration.

3. Try the above in answer to VII Corps. I want to maintain the excellent momentum EBCT has created. We must continue to attack the problems within today's resources.

Division 86 Transition Planning

Message to General John W. Vessey Jr.
Army Vice Chief of Staff
1 June 1981

1. I understand there is a session with CSA on 11 June regarding the sorting out of the mechanics for the management of Division 86 transition planning efforts, the timing of Division 86 conversion, and the issuance of HQDA guidance. Recommend you consider the following thoughts prior to your session with CSA.

2. The most important elements of the Division 86 transition process are to transition to the new Division 86 organizational designs with current equipment (interim organization) by type unit and convert as many as we can as soon as we can starting in FY83. Our goal should be to convert type units to the interim Division 86 organization within a six-to-eighteen-month period and shoot for the final Division 86 organization with modern equipment as procurement allows. In no way should we revisit the issue of making organizational change dependent upon new equipment. Such a method dictates conversion over an extended period of years with the force plagued by prolonged organizational dichotomy and attendant doctrinal problems. Despite the challenges, if we put our best efforts forward we can proceed in 1983 and wind up conversion to interim design in FY84; maintaining the present strategy by conversion of type units at a specified e-date window is vital to an orderly transition.

3. My final thought deals with the milestone schedules which I am informed CSA will be considering in the plan. Although they do meet the FY84–88 POM cycle windows, there are some close calls. If any delays occur in the dissemination of additional HQDA guidance, or if the recipients falter in developing transition plans for DA approval, these windows will quickly close and at least a year will be lost. The bottom line is [that] decision and execution are urgently needed on this transition proposal. HQDA guidance is critical, as is stringent compliance by MACOMs with the HQDA schedule of required inputs to meet FY84–88 POM development.

4. TRADOC is fully committed to putting Division 86 on the ground as quickly as possible. The Army is on the verge of accomplishing one of the most comprehensive force enhancements ever attempted. I assure you that TRADOC will continue to move full speed ahead to accomplish this goal.

Implementing AirLand Battle Concept

Message to Lieutenant General William R. Richardson
Fort Leavenworth, Kansas
17 June 1981

1. Momentum is building to carry out the decision we took at the 1980 Army Commanders' Conference to implement the AirLand Battle Concept. The concept emphasizes the requirement to interdict second-echelon forces. Army 86 force design efforts build in this interdiction capability. We have achieved a consensus amongst the MACOMs to get on with it and we are moving rapidly toward gaining allied endorsement.

2. As we view the pace of modernization in our Army, it comes as no surprise that our NATO allies achieve differing capabilities over the same time. Given their current status of force planning and programming, our allies will not achieve an equivalent interdiction capability in the mid to late eighties. While we are coming to grips with implementing a war-winning concept, we continue to see the traditional approach to NATO reinforcement—forces deployed during the early build-up, pre-positioned equipment sets, or a combination of those. For example, Omnibus and other studies continue to point up the critically thin area in NORTHAG and our force planners continue by means of TPFDL and POMCUS actions to restrict programming actions to division packages into NORTHAG. However, the primary NATO deficiency in the mid to late eighties appears to be its ability to interdict the second echelon. When one considers AirLand Battle in a NATO coalition context, the best contribution the US could make to NORTHAG in the mid to late eighties might well be a second-echelon-interdiction-capable package.

3. The time is right to challenge the traditional approach and to determine the payoff in introducing second-echelon interdiction packages rather than simply more divisions. Should you agree, we will relook our Army 86 effort and design second-echelon-capable packages. Such an effort would send a strong message as to our commitment to implement the AirLand Battle while providing a wider range of military options in the development of a flexible strategy.

Active Defense

Letter to Secretary of State Alexander M. Haig Jr.
25 June 1981

Originally I had wanted to come talk with you. Our staffs finally agreed the sensitivity of such an event was too much for the current climate to bear. Hence this letter.

Attached are two papers. The first is a reprint of an article from *Military Review* in which we lay out what must be done to win on a modern battlefield on which we're outnumbered. It is the culmination of eight years of work on what Bill DePuy and the rest of us first styled the "active defense." While we thought we had it fairly right then, the spate of criticism that has been leveled at our 1976 version of FM 100-5 tells us that's not the case. So the concept has evolved. We're fairly certain of it now and are using it, with Bernie's [General Bernard W. Rogers, SACEUR] staff and in our staff talks with the Germans and the Brits, to try to get the alliance out of the passive defense doldrums which characterize so much of NATO's defense planning.

We can never defend Western Europe simply by defensive battles to restore the interzonal boundary—there just isn't room. Nor is it an appropriate deterrent gesture to announce that to be our goal. The same logic applies to the employment of contingency forces elsewhere in the world, perhaps in a little different way—but not all that different. And so the concept of the AirLand Battle as it's now styled—*Extending the Battlefield* it's called in the article—has, we believe, sufficient versatility and flexibility to provide the framework for US military operations in pursuit of national political aims in many areas against many adversaries.

. . .

As you may know, I am apparently going to take REDCOM, with a mission of policing up the remains of the Warner-Kelly fracas and getting matters on track. Listening to Pentagon noises, I am increasingly concerned that national policy is being constructed by second-rate analysts in that building as opposed to development by some first-rate people with enough breadth of vision and depth of knowledge to understand what our country can and should do in foreign affairs and to understand the role military forces play, and cannot play, as instruments of national policy. Without that kind of understanding we're in deep trouble, especially in the Middle East where there are more opportunities for Soviet mischief than even the Sovs themselves have divined.

I'm trying desperately to understand what can and must be done to keep the contingency force operation relevant and in bounds. The attached papers are a beginning. I send them to you because of a growing conviction that we need—deserve, better than to have the country get into some impossible circumstance brought on by the narrow and shallow work of people in the Pentagon who never have to pay the price for their mistakes. Shy [Army Chief of Staff General Edward C. Meyer] and I are determined to try and prevent this. We will, soon or late, need help.

AirLand Battle II

Armed Forces Staff College
Norfolk, Virginia
22 September 1981

Today, I'm going to talk about the AirLand Battle as viewed by the US Readiness Command and the Joint Deployment Agency. This includes the joint operation of air-land and naval forces not only from the standpoint of their employment but from the standpoint of their deployment and sustainment as well.

Those of you that have seen the AirLand Battle before will note that REDCOM is coordinating the joint tactics, techniques, concepts, and procedures developed by TRADOC and TAC. The similarity is an indication that we are finally working from the same sheet of AirLand Battle music. You'll remember that the AirLand presentation began with a discussion of the whole war and the half war. I'm sure it was pointed out to you that we have long considered Europe to be our primary problem, and this, the contingency war, is a secondary problem for which we would use those other forces or our light forces. That equation has changed very little today.

It was also pointed out, in terms of likelihood of importance, that while Europe was certainly the most important from the standpoint of the realities of the world in which we live, it is probably the least likely war, while the opposite is true of contingencies in other parts of the world. I would point out that that equation has perhaps changed somewhat today. Certainly, it's changed from what it was 5 or 10 years ago, but I would also say that the notion that one can employ light forces and only those light forces in contingency operations is kind of a myth.

Today, in the Middle East—and it doesn't make any difference how you define the Middle East—there are more tanks than there are in all the Western part of the friendly side of NATO Europe. Any war in the Middle East, or in other parts of the world as well, will require mechanized or at least antimechanized forces and light infantry. Regardless of the fact that light infantry has some utility on the battlefield, it is simply not the force to cope with the heavily mechanized threats of the Middle East.

Our business is employment, deployment, and sustainment. I'd like to make a few points concerning this whole cycle of events, because you have to understand not only our commitment to NATO but our commitments to other parts of the world as well—and they must be viewed in the framework of employment/deployment and sustainment. First, let me make the point that, before you commit military forces to any kind of operation, for any purpose, anywhere in the world, you better damn well decide what you want them to do when they get there. What are we trying to do in the Middle East? Why are we going there? Why would we deploy forces to central Africa, South Africa, Central America? What are we trying to do? One of the recurring and persistent problems in our form of government is that it is never very clear what the military forces are supposed to do when they get there.

So, I find my planners, particularly in the RDJTF, using words like "deter," "dissuade," "delay," and in the end "defer," although we don't specifically use that word. You try to deter the enemy, and if that doesn't work, you try to dissuade; if that doesn't work, you try to delay him; if that doesn't work, what is left but to defer to him? Nobody wants to talk about attacking, defending, holding, seizing, killing, or winning, which are the operative words of military operations, and it is the military fellow's obligation to put those words back into the equation and to restore some sense of reality and balance to the fuzzy words that the political scientists and the politicians like to use with regard to goals and aims. That is our business—yours and mine—and unless we do that, we are not fulfilling the terms of our charter. I'll say this again to you before I'm through today, because it is very important. People tend to get things all fuzzed up with a bunch of words you and I can't translate into meaningful action programs and when we do—somebody upstairs ends up saying, "Oh, my God, that is not what I intended."

First, let's talk about threat, the Soviet threat in Western Europe, the Soviet threat in southwest Asia, or the threat from forces that use Soviet tactics. That, incidentally, is almost everybody in the world except our NATO partners and the Israelis. The Soviets operate in the context of the old triune of mass, momentum, and continuous land combat. They say that on the conventional, chemical, or nuclear battlefield, mass is to outnumber your enemy at the outset; momentum is putting some velocity with that mass and building up momentum and the momentum of the follow-on echelons to build up a continuous combat capability along a particular front or in a given area. They also use another tactic that the Soviets call the "daring thrust," which is a preemptive maneuver. There are those who write about daring thrust, Phil Karber, for example, and I'm not criticizing Phil at all. He is a friend of mine. We just happen to disagree about the problem. He claims this is a brand new thing. In fact, he wrote an article a year or so ago, in one of the strategic review magazines, that said "This is a revolution in tactics." It isn't a revolution at all. Like mass, momentum, and continuous land combat, it's about 45 years old. The Russians learned it from the Germans who ran a tank school for them in Russia from about 1928 to 1934, and this is simply a version, as are mass, momentum, and continuous land combat, of the famous blitzkrieg.

Surprise is another scenario that the Soviets learned from the Germans. It is an attempt to substitute surprise for mass, and today we see considerable emphasis on this in the Soviet system. They have looked at all the antitank defenses that we have purchased and deployed over the years, and believe that if they commit their tank divisions against that kind of defense, they are going to get blown away. They know that they will lose a lot of tanks, so they try to overcome it with surprise. As you probably know from studying them, they are paranoid about surprise anyway, secrecy and surprise. The surprise maneuvers are conducted with the BMP regiments, now in almost all of their motorized rifle divisions. They plan a regimental-level attack on multiple axes, against defenses that are not set. Every one of those is a meeting engagement, which they then try to exploit by pouring on second echelons. Essentially, you have the first-echelon battle problem as you had with mass, momentum, and continuous land combat, and you have the continuous combat problem that you had with the daring thrust. They just happen to be set in a little different framework.

The genesis of the AirLand Battle centered around the US V Corps in Germany, and I'm not going to rework that for you. However, since so many countries subscribe to the Soviet theories, let me extrapolate from that battle and show you a grander scheme of combat that addresses the whole of NATO Europe. In fact, I will talk, in a moment, about the application of these concepts to the conflict in southwest Asia and throughout the world. First, however, let's review the bidding in Europe. The Soviets have the Group of Soviet Forces in Germany and the Warsaw Pact, and in them, they have the capability of putting in about four breakthrough attacks. These are army-size breakthrough attacks against the eight NATO corps simultaneously and can follow that up with their succeeding echelons, which I'll talk about in more detail in just a moment.

Using the daring thrust tactic, they have the capability of putting in 32 of those individual BMP regimental-size attacks. They'll use this in an attempt to achieve surprise, and then they will take advantage of the surprise by moving in their follow-on echelons. These forces are set to exploit those successes gained by the independent regiments. In many ways, this is a more difficult problem to cope with than is the four-fronted breakthrough. It is certainly a more difficult command and control problem, and it has many other aspects that are equally as difficult to manage. So, that's the threat in Western Europe. So, might we cope with these tactics?

What I've discussed here, just to refresh your memory, is the first-echelon regiments of those first-echelon divisions that are coming at you. The second echelon—or follow-on echelon—of those first-echelon divisions consists of the second-echelon divisions of the first-echelon army and the second-echelon army of the first-echelon front. There are four of those fronts stacked up between the border in western Germany and the border of the Soviet Union, so in terms of the dimensions of the problem, numbers, and so on, this is the problem. If you don't do something about these second echelons—whatever that echelon may be—they will quickly overload the first-echelon battle to the point that there is no way to cope with it. You must begin the fight against the first echelon and the fight against the second echelon almost simultaneously. Now this, like the Soviet tactics, is not new. We've always done a little bit of it and called it interdiction. The point I'm trying to make is that interdiction today is an essential part of the AirLand Battle—a most essential part of the battle. It is critical to the extent that it has never been before and, if you don't do it, you can't win. And it's true not only in Western Europe, but it is equally as true in most areas of southwest Asia.

Okay, we know we are outnumbered, but can we fight and win outnumbered? I won't talk you through this thing, but if you used the traditional calculus that we use to describe battle outcomes, . . . if you look at this as 1 attacking 5, don't bother. You have a low probability of success. However, 5 attacking 1—press on, you almost always win. If, however, you examine a thousand battles, it turns out that numbers help, but numbers aren't everything. The history of those thousand battles tells you that, more often than not, the outcome of the battle is determined by something other than the numbers, so don't be deceived by force ratio. This is one problem with our force simulations, particularly theater-level force simulations that depend on weapons count methodologies. Those of you who recognize, or who are so inclined mentally, know that the Sciamachy Curve represents the Lanchester Inverse Squares equation and says that the side that is outnumbered at the outset is foredoomed to defeat. That, however, is seldom the case in battle for a whole lot of reasons. Numbers are important. Don't miss the point! But they are not everything, and there are a whole lot of other things you need to do in order to help yourself win other than just outnumber the other guy.

You need an operational concept most of all. You need to figure out how the battle is going to be fought and you need to determine the force requirements for fighting the battle. You need to move fast and strike quickly against those first echelons. At the same time, you need to look deeply into those enemy formations and begin the interdiction battle at the same time you begin the close-in battle. You need to tidy that whole thing up rather quickly, because there is another echelon coming down the road behind it and you are going to have to fight that battle over again in just a few hours if you are not careful. So, you don't want to delay or let it drag on too long. You don't want to dissuade, deter, defer, delay; you want to kill, win, attack, hold, seize, defend—all those neat things, out in the forward area. Throughout this, you have to keep two things in mind; in the first place, war is no fun, and in the second place, if you don't stop them, you are not going to win and not winning is no fun either. Two other cautionary thoughts must also be addressed. I won't belabor them, but first, terrain and its use is critical, and second, the enemy has the capability and has demonstrated his intent to fight in the conventional, nuclear, and chemical environment—we must be prepared to do likewise.

The friendly brigades fight those first-echelon regiments—but they have to pay attention to the second-echelon or follow-on echelon regiments. How far away? Well, for argument, we said 12 hours and, keep in mind, the time is more important than the distance. In discussing the echelons stacking up, I had some distances. This is because the Soviets lay the echelons out in terms of distances in their field manuals. But I'm trying to convince you that distance isn't nearly as important as time, because it is these forces—that are within 12 hours of being able to impact on the battle—that are the property of or the proper concern of the brigade commander. Because of terrain, and this is laid out to show the effects of terrain on the time sequencing, he can't get the information to and from those forces. It is dependent on distance, but the critical question remains—when will they engage or have a direct impact on the engagement? That is why I say that time is more important than distance.

Eventually, of course, one has to convert these back to distances so you can tell the I&W systems (the intelligence and warning people) where to look and what to look for, and you have to be able to tell the delivery systems where to deliver. You can tell the airplanes where to go and you can tell the artillery where to shoot, but they too are terrain and movement dependent. But in the end, it is time that is the common denominator. We've given the friendly division commander the responsibility for the parts that extend into the whole arena of the first-echelon divisions. He fights divisions and looks beyond. We have given the friendly corps commander the part that has to do with first-echelon armies. In this case, the division commander's line of interest is 24 hours away, and the corps commander's line of interest is 72 hours away. Now, that begins to frame in for you the requirements. If that's the concept that drives the employment of our forces on the battlefield, then from that concept for employment, you can begin to derive the requirements for the forces you need, the kinds of I&W systems, the kind of platforms they need to be on, their range, and so on. We can establish the kinds of delivery systems we need to cope with the close-in problems as we start the interdiction battle. This is imperative if we are to be successful.

That was the land part of the AirLand Battle. Here is the air part of the same battle—we have overlaid the close air support, the interdiction, the offensive air support, and added the recce systems for the overall picture. This includes all aspects of reconnaissance—satellites, fixed wing, remotely piloted vehicles, GLCMs, or whatever you've got. We have shown the strategic air—not just B-52s, but again ALCMs, GLCMs, F-111s, whatever is available to conduct the deeper campaign. This is an AirLand Battle. You cannot fight the interdiction battle without airpower. You cannot fight the close-in battle without airpower. However, airpower can't fight the close-in battle alone, nor can airpower fight the deep battle alone—although the deeper it goes, the more it is an air battle as opposed to an AirLand Battle. What I'm saying is we need to stop arguing about whose roles and missions we are treading on and get on with a joint operation. There isn't enough of anything to do everything that needs to be done. And if we are going to win, we had better get together and figure out how to use what we've got instead of sitting around arguing about whose turf we're on.

To recap, the deep battle—we start at about the 72-hour line where it is primarily an air battle—even at this close range, but certainly beyond that. The purpose of the attack is to hit the command and control facilities, chokepoints, the control mechanisms, the sinews of war, the service support, the softer targets, because they are easier to hit and destroy at the deeper ranges. This is somewhat contrary to the popular notion in Washington that, to win, you have to be able to go out there and destroy the enemy, tank by tank. That would be nice, but I think I could prove to you, in short order, that it is cheaper in terms of resources to go after the things at the 72-hour point than it is to go out there and try to kill him tank for tank for tank. If you get some tanks in the process, okay, so much the better. But if you don't, that's also fine. There will be something for the guys up front to shoot at, and you wouldn't want the fellows in front to lack for targets.

As the enemy closes, and I arbitrarily picked 60 hours—it is still an AirLand Battle, but I picked 60 hours because it is at this point where we can get ourselves most deeply in trouble. Those of you who have been G3s or S3s, operations officers, will remember that when you get your ops maps up on the board, and you get all the symbols on the acetate, and the sergeants are running around with a mouthful of pins, and the ops officer has his grease pencil behind his ear, and the general comes in and says, "Where are your counterattack plans?" Everybody says, "Oh, Christ, we forgot about them." The old ops officer takes his grease pencil and squints at the map, like he knew what he was doing, and he makes a big swipe like that—and says, "That's penetration—we're going to have a counterattack against them, general." The general says, "Right, where's the plan?" The ops officer says, "Well, Major So and So is working on that. It will take him a little while." The fact of the matter is they hadn't thought about it until the general came in, and it's this mentality that we are going to wait until the enemy creates a bulge in our lines before we attack him that will cause us to lose. Some years ago, we took that little line, that all those smart ops officers—and I've done this myself, in fact, I've probably done it more than most people in this room put together. So, I'm not picking on anybody—we took that bulge and said—okay, if we are going to attack him, "This is as far as we are going to let him get before he endangers the integrity of our defenses," or words to that effect. Okay, if you were one of the ones who said that, where was the enemy when you had to start moving forces and assembling fire and getting airplanes and all the stuff necessary to attack him? Where was he? The answer—he was back here at the 60-hour point or beyond. Always—always back here some place. The point is that what you must figure out is how to attack deep, because if you wait until he establishes the bulge in your lines, it's too late. It is always too late, but most particularly so if you are considering using nuclear weapons. This is true for a whole lot of reasons that we can talk about in more detail later, if you want to. There are many reasons it is too late, so we must back the engagement up to this depth—the 60-hour point. That's where you must engage him. You can't wait until they get down in here. You can wait if you want to, but don't invite me to the battle, because it is too late, and the next one I go to I want to win.

Now, to the 24-hour point—this is the division battle now—it is now an AirLand Battle more than just an air battle. We have stated our requirements for real-time I&W, but now real-time target acquisition is a must, and right now we haven't the capability, at least not to the extent that we need it. We are beginning to develop the stand-off target acquisition systems, which will give us what we need. It will give to the divisional commander—even though the enemy is still far away—terrain and timewise—it will give the division the time necessary to plan, execute, and maneuver. If we have done anything to the enemy, if we have been able to do the interdiction job properly in the deep arena, the maneuver force, the attacking force, has few maneuver alternatives at this point. Keep in mind, you may still use nukes if you want to, but the battle is an integrated, a flat-out integrated AirLand Battle—it has to be.

On the close side, we wind up with something like this. You can't just attack him with air alone, for a whole lot of reasons. We must attack him with ground forces, so here we consider the joint air attack team. In some areas, it is just the ground forces, and we are going to attack him on his own ground. If we decide to use chemicals, this is the place. You can use them to facilitate your advance to the targets you want to get to. It is primarily a land AirLand Battle at this point, but it is still an integrated land-air battle. It has to be—there's no way for either force to do the job alone. That's the brigade battle picture, and in the end, what you want is a piece of the enemy's ground.

What you really want is the initiative. The purpose of this whole operation is to wrest the initiative from the enemy. You cannot fight successfully and win if you don't seize the initiative at some point in the battle. There are a lot of ways of doing that, but the history of battle tells us that unless you do it you are not going to win. You can't just sit there and fire away at him in an attrition kind of battle. We all too often have been accused of being an attrition-minded Army, of thinking of every tank as being right on the FEBA, and that's just not true. I say that because I wrote most of the books we are currently using, and I'll just tell you I ain't attrition-minded. It is not our intent. I will, in the same breath, tell you that when there are that many of them out there, you have to kill a whole lot of them just to get their attention. And once you've got their attention, you are going to have to kill a few more to make them understand that you are serious about it. Now, if that's attrition, then we, in a way, are attrition-minded. But we are not—repeat *not*—saying that we are just going to sit there and try to knock them off one at a time as they come over the hill and let that be the end of the battle. We have to seize the initiative from him, we have to "attrit" those tanks en route to the FEBA, particularly if we have given him the opportunity and the advantage of attacking first, and as such have given him the initiative.

All right, here's the message. First, the integrated attack of all targets throughout the spectrum—from the relatively soft, yet highly lucrative and critical, targets out deep, to the tank versus tank, man versus man, arena in the close-in environment—is indispensable. All too frequently, in the past, we have worried service parochialisms over missions and roles. That battle must be set aside to ensure our readiness in the future, because the AirLand Battle concept is the key to stopping the Soviets' mass momentum and continuous land combat theory—cold. To this extent, it is imperative that we get about the business of modernizing our force to complement each other, to do our jobs to ensure the success of the AirLand Battle. The joint attack techniques of the A-10 and the attack helicopter; the coordination of recce, communications, and I&W assets; the development of dialogue and coordination between the infantry, the armor, the air, and all the very necessary and complementary forces supporting them must be enhanced and exercised to the peak of readiness.

Now, as promised, let's expand this to cover our missions and capabilities throughout the world. Harkening back to the earlier "half war/whole war" discussion, this is the whole war as it relates to our forces in Europe today. Although somewhat stylized, this shows our in-place capabilities in the US V and VII Corps and those Warsaw Pact forces facing them across the border. Behind the US forces, some 3,500 miles away in the States, are the remainder of our NATO commitments, 18 divisions, 59 fighter squadrons, and a million people. The Soviets, on the other hand, are backed up by their deep echelon forces and the masses remaining in the Soviet Union. This has, historically, been the whole war!

Okay, to get the whole picture, let's look at our best case. Let's take the old 4L02 Plan, which is the general war plan for Europe. Let's consider divisions—Army divisions and the days from some time at which we begin to deploy. Here's our capability, sealift and airlift. Here's our requirement laid out in the DOD consolidated guidance. We are tasked to build from the 4-1/3 divisions in place to 10 divisions in 10 days—can't do it. It's our shortfall that we are trying to build into the new budget programs. The sealift and airlift improvements in the current administration's attempts to increase the defense budget are a large step in the right direction.

Next, consider tactical fighter squadrons, and I use them as an example only. We have similar problems in moving recce forces, airlifters, etc. This is typical. We must keep in mind that these squadrons are using the same lift as the Army, so the shortfalls will be similar. In this case, we see a need for some 30 squadrons in the first 5 days. Can't do it! Maybe 8 or so—but . . .?

The situation is even worse when you look at sustainment—these are cargo requirements, short tons, requirement to move units as units—battalions, brigades, companies, divisions, squadrons, whatever. Then, there are the nonunit requirements—that is, the individual personnel fillers and the parts supply system that must continue after war begins. Obviously, we have some enormous shortfalls, and not until well downstream do we start to dent our requirements. There are a few ways of coping with the shortfalls—one is to reduce the requirements; the other is to increase the capability. The systems analysts like to reduce the requirements down to equal the capability. That makes everything work out, and obviously, if requirements and capability match, it isn't as bad as you guys said it was. But at the same time, as I said earlier, our business—your business and mine, as military folks—is to state the requirements as objectively and honestly and straightforwardly as we know how to do it—the requirements for employment, the requirements for deployment, and the requirements for sustainment. Because unless you operate in that framework of an honest appraisal of what it takes to do whatever we say we want to do—defend Western Europe in this case—then we are not doing our job.

Now, keeping this best-case scenario in mind, let's shift our focus to southwest Asia, and I use this only as an example, as it allows a different perspective on the problems. The problem may be oil or may not. It may be national stability, it may be our desire to keep the Soviets from extending their influence over the area. Beyond this, I would like to point out that there are three big problems over there. One, naturally, is the Persian Gulf, one is the eastern littoral of the Mediterranean, and the other is the Horn of Africa. More on these in a moment.

Let us put the half war in perspective. You saw the threat and our capabilities in Germany. If it is complicated in Germany, double the distance, take away all reception facilities, all in-place assets, and host nation support, and you have the southwest Asia scenario. But the big problem with people who talk about the Middle East is one of map scale—they look at a map of the Middle East, which is about 1:5,000,000, and they assume that it is a 1:250,000 or 1:500,000, something like that. But the scale is 10 to 20 times larger. Consider Germany compared to the area over Iran. The United States V and VII Corps, part of NATO, cover very little. You can lay the US part down in the middle of Oman and never notice it. We've over 200,000 US soldiers and airmen in that little wedge that you can lose in the middle Oman and lose several times over in Saudi Arabia, Iraq, Iran—anyplace else you want to go. In one plan, concerning southwest Asia, which I saw, we are going in in two different places. It's like putting the landing forces ashore in Augusta, Maine, and landing the Army troops in Norfolk, Virginia. And we are going to control that from one headquarters and it's all going to come off all right. The hell it is! The first thing you want to do when somebody talks about southwest Asia is look to see where his map scale selector lever is turned.

We have spent a lot of time recently worrying about the problem of oil—because this is where the oil comes from—most of the oil, that is. Keep one thing in mind, something like 64 percent of Japan's oil comes out of this area, half of Europe's oil, about 25 percent of US oil comes out of someplace in this area. Iran today produces little and only refines enough oil for its own domestic purposes. They are not a world supplier of oil at the moment, so the problem is—if oil is your problem—Saudi Arabia and the area adjacent thereto—Oman, Kuwait, etc. As to situations—let's just take the Persian Gulf as an example, but remember that the situation could apply to a lot of other areas of the world. Under what circumstances might we want to go there? Well, first, they might have an intratheater of war wherein the Iranians attack Kuwait because the Kuwaitis are allowing Soviet ships to offload gear and give it to Iraq. That is going on today. That's a possibility. The Iraqis attack the Iranians or vice versa—that is going on—that is a possibility. Blockade—somebody wants to prevent movement—they can blockade the Straits of Hormuz, which is the example most frequently used. But, as you can see, there is an excellent opportunity for blockade at the southern exits to the Red Sea. We could have an intrusion from outside by Soviet proxy forces, Cuban divisions for example—they have more than one in the area, or by the Soviets themselves. Of all those things, a Soviet incursion is probably the least likely to occur, at least until the Afghanistan situation clears up and they have decided what to do with the Solidarity Movement in Poland. But anyway, that is a menu of possible events that could cause us to make a decision to go into southwest Asia.

There is, however, far more opportunity for mischief in the situation that exists in the littoral of the eastern Mediterranean at the moment than there is in the Persian Gulf, even though the potential for mischief there is great also. This guy has 2,600 tanks—first-line combat Soviet gear. He can only man 760 of those tanks—that's all. So, if you see more than 760 tanks coming at the western borders of Egypt, and they are coming from Libya, you know damn well that there is somebody in those crews other than Libyans. Who are they? Russians, Cubans? It's a stockpile, a Soviet version of POMCUS. There are something like 2,300 tanks in Egypt. There are almost 4,000 in Israel and 4,200 in Syria. There are more damn tanks in that whole area over there—tanks alone—than there are in all of the tank fleets of all the NATO allies. Now, if you look around at this problem, at the situation in Chad, in the Sudan, and our mad friend is trying to work that equation against the Egyptians—there is more opportunity for mischief in here than one cares to contemplate seriously. Furthermore, if you are worried about access through the Suez Canal, forget it. The Soviets don't care about the Suez Canal. There's already a Cuban division right there, a full-strength Cuban combat division. The banks of the Red Sea and that area are lined with air defenses, first-line, up-to-date Soviet air defenses. On that little island out in the Straits, whose name I've forgotten at the moment, he has taken some T-54 tank turrets, put them in concrete emplacements, and those tanks with flat gunfire can shoot the width of the Strait and obviously control access. So, he has guaranteed himself access to an egress from the southern end of the Red Sea, and at the same time, he is in a position to deny it to us. That's capability. So, it serves no purpose for us to stand around preparing plans with regard to things in the Persian Gulf unless we also consider the circumstances that are likely to occur in the eastern Mediterranean, as well as along the Horn of Africa, and most probably at the same time. This is one enormous, bloody problem at the moment. It will continue to be for some time, and we had better figure out how to cope with it.

Let's shift back to the Persian Gulf now. What do you want to do? What do you want to do there? Do you want to hold on to something—terrain, oil facilities, guard the pipelines or refineries? Do you simply want to control the flow of crude? Part of the problem in Saudi Arabia is the vast, uninhabitable distances. What kind of presence do we have in mind, and what is the mission? Are we going to restore something that has been destroyed, operate a system, guarantee safe passage through the Straits of Hormuz, around the Horn of Africa?

Whatever you say that you are going to do, the force you commit to do it has got to be credible in the view of the Nation. The cost and/or reasonable probability of success equation has got to balance, be understood and accepted—before we go in—wherever we go. Don't forget that! If we are going to secure or restore refinery areas, our light, highly mobile infantry is probably the answer. If we are going to face armored war against forces greater in size than those facing US forces in Germany, we need heavier forces. The question remains—what's the job and what's the threat? Once that has been established, we can work the force structure mix. Once we have the appropriate force, can we move it? Can we move it in time? If the answers to these questions are "yes," then we must next question our capability to sustain the force. If we can do this, we must ask about capabilities to react to another area or an extension of the conflict. If, however, the answer to any one of these questions is "no," then we need to start over—at the top. Finally, all the questions hinge on our agreements with others. Overflight, refueling, host nation support—the entire conflict could hinge on these facets. Don't forget the previously discussed deployment and sustainment shortfalls in support of NATO. Contrast that with southwest Asia—you have all of the movement problems to cope with in an area where you have to haul everything in you use, including fresh water. We lose sight all too often of the enormous infrastructure we built up, of the staging capabilities that we have—not only in Europe, but en route to Europe, that affect our capability to deploy and our capabilities to sustain. What happens to your deployment schedule if you are denied some of the en route bases, or even some en route refueling? How do you balance the deployment means against shortfalls in basing or en route staging, and so on? It's one of the problems the Joint Deployment Agency is trying to work today. We need to be able to mix deployment means. We can't assume that everything is going by air, nor can we assume that everything is going to go by ship. We have to have some balance, some rational balance, of those two movement means—deployment means, and we have to be able to change, and we have to be able to do it quickly.

On the probable/least-probable scale, this may have been a half war. For argument's sake, let's say that the least-probable part of this least-probable scenario comes to pass and we are tasked to fight an invasion of Iran by Soviets. It really makes no difference if we are going to fight the Soviets or restore a pipeline—our force, right now, is the same. These are the forces that have been given to the Rapid Deployment Joint Task Force. This is the Soviet capability. We have light divisions 7,300 miles away, and they are tasked to deter, dissuade, delay, and in the end defer, although again we haven't used the word specifically. But that is what it amounts to. This is a case where the force you have selected to put on the ground cannot, cannot reasonably, be expected to cope with the threat. That force does have some capabilities. There are many things it can do. It can help hold, restore, or operate alone, and it is an indispensable part of a combined arms team, but it clearly cannot go in there alone and fight against those 18 motorized rifle divisions and have any chance of success. This is the problem.

Any way you look at it, however you do whatever it is you do, however you fight against those guys, you have got to do something about this problem. You have got to deal with this problem, because you have the problem whether you are dealing with a corps-size force, a marine landing force, light or heavy, or whatever you are dealing with. You have to have the capability to fight the interdiction battle and the close-in battle at the same time. It's got to be a joint battle, an air battle as well as a land battle. In the southwest Asia, Persian Gulf, Mediterranean area, the joint battle has to include naval forces throughout the spectrum. The AirLand Battle is further complicated by the deployment and sustainment problems. We have got to figure out how we are going to sustain the force after we get the forces deployed, assuming that we can get them deployed. The objective of whatever you do out there and that AirLand Battle, that conceptual notion, is not just to avoid losing something, but it's to win. I ain't going to war again unless we are going to win. I've had enough of that other stuff. Enough of that deterring, dissuading, delaying, and deferring. The next time, we are going out and kill somebody, and to do that, we have got to attack and seize the initiative.

Now, some of you have heard my speeches before about how the winning of wars is the function, as someone said, of "the blood of the soldiers, the guts of their leaders, and the disciplined way their units fight in battle." There's no way you or I can expect the courage of our soldiers to make up for our lack of ability to lead them intelligently in battle. There is no way for us to ask that the courage of our soldiers make up for our inability to get them there on time or our inability to support them properly or, worse yet, our inability to describe for them and their leaders and their units clearly what it is we want them to do after they get there. That's our business—yours and mine—that's the challenge that faces you as planners and as commanders, and it will face you for the rest of your career.

Joint Operations

AUSA Convention
Washington, DC
21 October 1981

It is an honor to be here today . . . and to share the podium with the Commander of the Tactical Air Command, General Bill Creech. For the past two days discussions here have emphasized Army issues. Today we will expand that horizon to joint operations, the fundamental requirements for successful conduct of the Airland Battle now and in the years ahead.

The challenging world and the national strategy have been described for you. Let me emphasize two aspects of the challenge that seem to be most important for our purposes. First is the ever-increasing Soviet problem. The Soviets and, through them, their surrogates, allies and followers will, over the next few years, continue to gain in military capability relative to the United States, at least until current budget programs are given time to come to fruition.

Second, the dependence of many nations on Middle East oil and their increasing and possibly, over the long run, even more critical dependence on scarce mineral resources of Africa, represent economic, political, and military vulnerabilities that will continue to be exploited by the Soviets. These vulnerabilities are ones through which a major world crisis will likely be precipitated if the Soviets can contrive to do so.

Beyond this are other obvious challenges worldwide. In times past, we have forecast the need to fight one and a half wars—the whole war in the high-density, firepower-intensive, chemical, or possibly nuclear environment of Europe; the half war in other areas, the Mideast, Africa, Latin America, or Northeast Asia. This, the half war, we have traditionally seen as a circumstance requiring forces not so robust, heavy, or firepower and mobility dominant as forces for the whole war.

Today, in the Warsaw Pact area, we see the massive Soviet war machine in full array. Today we can also stand on the Golan Heights or in Riyadh and look around at strikingly similar formations and force ratios in a totally different environment. In Europe today, nearly 3,000 tanks and more than 1,000 tactical aircraft confront the United States V and VII Corps alone. In the Middle East, in one of the so-called half-war arenas, there are more than 15,000 tanks and 3,600 combat aircraft total in the fleets of nations of the region.

And so it is that yesterday's half war may well, full-blown, be today's and tomorrow's whole war. A similar analogy would be appropriate in describing the situation in Northeast Asia. This is not to say the role traditionally foreseen for light forces has diminished a whit. Indeed the contrary is the case. But it is to say that we must regard, with some more precision than in the past, circumstances into which we commit US forces, the purposes for which they are committed, and most important the adequacy of the force in terms of what it's expected to accomplish.

May I suggest some guidelines—priorities—that seem to me to be essential in that regard?

First, whatever may be the course of action on which we embark, there must be a reasonable probability of success, and at costs that are bearable in terms of the importance of what's being attempted. A truly good cause may just not be worth the cost. This, in fact, may be the single greatest lesson of Vietnam. Having never clearly set forth the requirements, at least not in terms of the real threat, we waited too long to add up the costs of a reasonable probability of success.

Second, if they are to be employed, it is imperative that it be decided early on what the military forces are to do. Is there a clearly spelled out operational concept, a mission, well-defined tasks, all within the framework of the cost/reasonable probability of success equation? We may wish to send brigades to deter or dissuade something or someone, but deter and dissuade are not missions to be given the brigade, division, or force commander. Next, it must be decided whether or not we have sufficient forces to employ and to accomplish the mission. Are there sufficient backup forces available should something go awry? No one of us is sufficiently adept at the precise tailoring of forces that we can predict with certainty how much is enough.

Third, are deployment means available to move the forces to the area in time to accomplish the mission? Is there sufficient redundancy in the deployment means to provide a reasonable probability of success? If we lose some ships or airplanes, for whatever reason, can we get there anyway?

Fourth, can we sustain the forces we've deployed after they're employed? Can we provide the sinews of war in sufficient quantities, in time, at the places needed to keep the deployed forces operating—doing whatever they were sent to do and succeeding?

Finally, having satisfied one commitment with whatever forces, deployment means, and sustainment resources may be required, we have a need to cover other commitments as well. Can that be done? With a force deployed in the Persian Gulf, for example, could we then respond to a crisis in Europe? Elsewhere?

As we contemplate our national commitments and interests across the world, and the challenges with which we're confronted, it is more than apparent that we have not the resources, nor could we reasonably expect to provide in peacetime the means to cope with our most contentious problems with all up, all ready active forces and do so within the guidelines I've just set forth. What force capabilities do we then require as we approach the year 2000? First, in all the whole wars and the half wars as well, we may find ourselves in environments ranging from terrorism to nuclear conflict. To meet that ultimate military challenge we need:

- A military strategy that embraces coherent operational concepts for employment of air, land, and possibly sea or seaborne forces.
- Rapidly deployable, lighter, appropriately equipped forces that can begin immediately critical actions of the AirLand Battle and that can be reinforced relatively quickly by heavier forces.
- Forces that can operate successfully, if need be, under massive chemical, nuclear, and electronic warfare conditions.
- Balanced AirLand forces that can fight the deep AirLand Battle as well as the lethal, intense close-in LandAir Battle as part of a coherent, relevant concept of operations.
- Forces capable of system and procedural integration hitherto unknown, able to identify, coordinate, and destroy targets and forces, throughout the entire AirLand environment, rapidly and efficiently.
- Forces equipped with communications and information handling technology to permit adequate command and control of joint operations.
- Airlift and sealift forces adequate to deploy AirLand forces quickly, intra- as well as inter-theater.

- Airlift and sealift adequate to sustain the AirLand force once employed, operationally as well as logistically.
- Forces capable of operations in a wide variety of climates, terrain, and geography, as well as against a variety of threats.
- Forces with adequate joint operational tactics, techniques, and procedures worked out and exercised to proficiency in advance.
- A military structure that fully integrates active duty and Reserve components in a set of rational schemes for employment, deployment, and sustainment of joint AirLand forces. A structure capable of responding, wherever the challenge exists, in consonance with national priorities.

For only ready forces with clear-cut, achievable objectives, missions, and rational employment concepts; only ready and quickly deployable forces; only forces that can be sustained once deployed and employed; only a total force with sufficient redundancy, backup, reserves, depth, and training to meet a full range of national commitments; only those forces and the public support that bring them into being and sustain them to serve as a reasonable deterrent; only those forces—those total AirLand forces and the national resolve they represent—provide the means to win should that deterrence fail.

REDCOM Mission

Newcomers' Orientation
MacDill Air Force Base, Florida
8 December 1981

The first question a lot of people ask when they come here any more is why we need a Readiness Command and an RDJTF. If you followed the newspapers last year, that was a big argument, and my predecessor and General Kingston's predecessor in the RDJTF had a lot to say about that in public. All that has gone away now because there is in fact a mission for both commands, and I'll try to explain to you for a few minutes why I think that's the case. There really isn't much of a conflict. It's just that things changed a little bit and we have a little trouble apparently adjusting to the change as we went on down here.

The first thing you have to look at, I think, is what the product is that REDCOM is responsible for—what we are trying to do with the armed services of our country—Army, Navy, Air Force, Marine Corps or the Joint Chiefs of Staff who charter us for the Secretary of Defense, for whom I work, and the President and so on. And the answer to that is that our business is to describe for people how we conduct joint operations, because there is no other agency in the system that is responsible for doing that. Now some people call that doctrine for the operation of joint forces and, when you start talking about doctrine, the services get all agitated—service staffs in Washington begin to bounce up and down in place, the majors start running up and down the hall, telling the lieutenant colonels, who in turn tell the colonels, who in turn tell the generals that those guys down there at MacDill are trying to take over our business for us. That's not true at all. Remember I said joint operations. No service is responsible for that, and it is in the charter of this command to do that. So we call that—to get all out of the doctrine business—joint tactics, techniques, and procedures. It is, in fact, I suppose, joint doctrine, but there are a lot of places, a lot of situations, and a lot of operations where there is no other way to operate together. There is no way to operate together unless someone lays down the procedures you are going to use—tactics you are going to use, the interfaces and the procedural things that have to be done in order to get the job done. Let me just mention a couple. You have an operation with the Army and the Air Force on land, the joint suppression of the enemy's air defenses is important for close air support, reconnaissance and surveillance, and a whole lot of other things that have to go on over the battlefield—some conducted by Air Force aircraft and some by Army aircraft and some by boat. If you are operating on or near a shoreline, or naval forces are engaged in that same sort of business of suppression of enemy air defenses, it is a three-way joint kind of consideration and there's no place except in the JCS publication—which is too general for our purposes—there's no place where that's described, techniques and procedures and the interfaces that are necessary to make that work.

How does the communications system work from one service to the other? How does it work from all the services to the joint commander and so on? Well, that's the thing that has to be described. That's our responsibility—it always has been. In times gone by we have devoted sometimes more and sometimes less attention to that. It so happens that now we are going to devote more rather than less attention to that.

How do we do it? We write plans. We have 40 some-odd plans. Some of them are simple little things involving the deployment of a few people here and there, a communications element, a battalion task force to do something, or whole divisions or corps-sized forces for the reinforcement of Europe, for example. All those plans are our responsibility. All those plans involve joint operations. All those joint operations require some statement of how the forces are going to work together to do a whole lot of things. I mentioned one—suppression of enemy air defenses. Joint air attack team tactics is another one. Joint attack of the second-echelon forces is another one, because it can't be done by one service alone. It has to be done by more than one, so somebody has to lay down how that gets done.

We investigate and write about this on the basis of exercises. Some of those are field exercises, some of them are command post exercises (CPXs), some of them we schedule, some of them the other unified commands schedule. For example, the REFORGER exercise every year in which we exercise a part of the troop list for the reinforcement of Europe. It is a joint exercise. It involves forces of the Army and the Air Force primarily, but the Navy as well—sealift, particularly. It involves all the transportation operating agencies, which are the Sealift Command, Military Airlift Command, and the Military Traffic Management Command. Therefore it is a joint operation, so somebody has to prescribe how that is going to be done. Some of it we can do by simulation. That is, some of what we have to do to develop joint tactics, techniques, and procedures can be done by simulation. We decide how it ought to work. We put it in a computer and simulate it. Does it work? Doesn't work? Well, let's try something else. If it does work—well, let's write about that and then everybody do it that way and put out a little publication that says, "All right, fellows, if you come to war with us, this is the way we do business, and so you have to learn how to do it this way in the exercises, and you have to write about it that way in plans." To the extent possible we do copy what the commands in Europe and Korea and elsewhere have laid down for themselves as operating procedures and try to accommodate to those as we write up our own tactics and operating procedures and so on.

Now those plans that I talked about, which we exercise, which we simulate, which we command post exercise and whatever, involve a wide range of responsibilities. Let me just tick off some of the things we are responsible for planning for—land defense of the United States, the joint land defense of Canada and the United States with the Canadian Forces Mobile Command, the land defense of Alaska, the employment of forces to do whatever is required of us in areas of the world where there is no other unified commander at the moment with geographic responsibility. At the moment that is most of Africa, especially the part of Africa that is south of the Sahara. We are responsible for the deployment of forces that are in the United States—the nine divisions, 40 or 50 air squadrons of the Tactical Air Command, and other forces to support plans of those other commanders in chief—one in Europe, one in the Pacific, one in the Atlantic, one in the Southern Command, and so on. That's why we have with us here, as a part of the Readiness Command family, the Joint Deployment Agency, because that is what the joint deployment guys do—they work up the force list, they describe how the forces are going to be deployed to support the plans and, when exercise time comes, or when for-real time comes, they execute the plans they worked up for the deployment of those forces. And in places where we deploy the forces, whether they are ours or somebody else's, we are responsible for sustaining them once they get there. Fuel, ammunition, food, uniforms, spare parts, you name it—that's our responsibility, to figure out how to get that stuff there, so we have to mix the means and modes of transportation in order to make it possible for the services to move the supplies to support the forces to wherever they may be deployed and to whomever they may come under for command as they are deployed.

So, if you look at it in that framework, our business is really the operation of joint forces and describing how that's to be done, writing about that in plans, exercising it in exercises, simulating it in simulations, and writing up the results of that and making sure it is current so everybody can take advantage of that. If you look at our planning responsibilities—the United States, the United States and Canada, Alaska, most of Africa, and the support of all those other commanders in chief all over the world, that's a sizable job.

Now what's happened here is that, in recent years, the oil situation in the Mideast has first of all caused the Carter administration to say, "Well, we need some more emphasis on our ability to do something in the Middle East." They didn't quite say what they wanted done over there—that's one of our problems, but in any event it was perceived that there was a bigger requirement now than ever before for us to be able to deploy forces to the Middle East. When that became apparent, the mission was given to one of the task forces that was already in this command. Subsequently, on 1 October of this year, we separated that task force out and made it respond directly to the Joint Chiefs of Staff. Don't forget the RDJTF in its present configuration, and however it might grow and whatever geography in the Middle East it might take over, simply becomes another supported command as far as we are concerned. It has its specific piece of geography. It does not relate to the whole world. It will have a series of plans with regard to the geography for which it is responsible, and we will have certain responsibilities with regard to those plans—deployment and sustainment, primarily, just as we do for the plans drawn up by the commander in chief in Europe and the commander in chief in the Pacific and the commander in chief in the Southern Command before employment of forces in their areas of responsibility. So on balance I would say that we've got more than enough to do, whether or not there is a RDJTF, and in my view, at least, that calls for a command the size and with the responsibilities of REDCOM. If I didn't believe that, I wouldn't have come here to take the job.

So to all those guys who say they don't understand why we have a REDCOM and a RDJTF—and why that is all confused, I think that if you remember everything I told you, which I'm sure you don't, you will have an adequate explanation for that.

Fighting Outnumbered to Win

Message to Dr. Sam Parry
Naval Postgraduate School
5 January 1982

I've been telling my "fight outnumbered and win" story using that famous all-purpose cost effectiveness curve and the thousand-battle curve for several years. Some time ago you asked me where that curve crossed the fifty-fifty line; we agreed it was not at zero-zero. Off and on I've tried to sort that out. In so doing, I've discovered something else. The curve that shows the outcomes of a thousand battles is apparently only relevant for force ratios in the one-to-six to six-to-one range. Outside that range of attacker versus defender ratios it seems that more often than not Lanchester wins—the guy with the most "emerges triumphant." Just leafing through Richardson and the more recent studies seems to confirm that judgment. That says that getting yourself to the one-to-six or six-to-one threshold is the price of admission—it's the ante. What happens after that is due in large measure to all the things I've been talking about so eloquently. If true, that's a very important bit of knowledge from a strategy standpoint, for it determines what we can and can't hope to do as we look to generating force at great distances from our bases and sustaining the force through whatever action is intended. You will be able to answer that on the back of an envelope, I'm sure. I haven't time to pursue it in detail, nor have I anyone at hand smart enough to do the calculus. If it seems worthwhile, you might look around for some resident wisdom on the subject or for someone who may have already researched it. There seems to be enough evidence to make it worth looking into.

Heavy Division 86

Message to General Glenn K. Otis
US Army Training and Doctrine Command
4 March 1982

The little bird who tells all has revealed that someone has caused revisitation of the size of heavy Division 86. I object. Could go on for hours, but I object. That's the best restructuring effort we've ever done in recent years. The fact that most of the actors who did it are now elsewhere does not affect the soundness of what we did. The division strength requirement unconstrained stood at close to 25,000. Obviously you can have any size division you want, but we developed the 20,000-man division with a lot of consensus building—as opposed to shoving it down everyone's throat, which was the mode. I suggested many ways to get those spaces. Now, institutional memory being what it is, all that has been forgotten. I just tell you that if we start slicing away at that thing we've got nothing and nearly eight years of effort is down the drain. Surely we can't be that stupid, or are we? What need I do about this?

Heavy Division 86

Message to General Edward C. Meyer
Army Chief of Staff
4 March 1982

A little bird has told me that we're considering reducing manpower authorizations in Heavy Division 86 to something like 18.5K. I just have to protest a reduction in the requirements we finally produced for that division. It's the best and most thoroughgoing reorganization effort the Army has done since World War II. It had the participation and support of all the commandants, the Army Staff, you yourself, and many others. We could have had a 25,000-man division; we could have had a 15,000-man division. The consensus was that the 20,000-man division was about right. We acknowledged stationing problems; we recognized end strength problems; we even went so far as to suggest alternative ways in which the manpower demands could be realized. In the face of all that, if we change now, the leadership of the Army will just appear incredible to everyone who had a hand in that organization and left the effort believing that we had done something right and good for a change, and that's a lot of folks. I am told that Stone is to brief you tomorrow. If so, you will hear what he thinks you want to hear, not what's right in the context in which the organization was finally developed, or any contrary arguments. It took nearly eight years to develop that organization, and don't ever forget that you and I rescued it from the arbitrary shove-it-down-their-throats course it was on under the previous regime. The Army is comfortable with it; let's not make ourselves look dumb!

Understanding AirLand Battle

Message to General Glenn Otis
US Army Training and Doctrine Command
7 April 1982

Jack Vessey's [General John W. Vessy Jr., Army Vice Chief of Staff] 071022Z Apr 82 on AirLand 2000 and Bernie Rogers [General Bernard W. Rogers, SACEUR] reached me. Despite Bernie's reassurances to Jack, my debrief of Morelli's meeting with Bernie confirms my belief that Bernie does not understand, nor does he want to understand, either A/L 2000 or A/L now. The only solution to that is to wait. Meantime, to send Morelli [Brigadier General Donald Morelli, TRADOC Deputy Chief of Staff for Doctrine] to duty at SHAPE, in view of Bernie's comments to Jack, might be something we should reconsider. Not exactly my business, but Morelli has been so valuable to us that it seems short-sighted to put him in a position for Bernie to suppress him—which he's sure to do. We would then be without Morelli's services in any context. Better to keep him doing what he's doing for us here and wait Bernie out.

Joint Doctrine

Letter to General John W. Vessey Jr.
Chairman, Joint Chiefs of Staff
13 August 1982

Continuing our dialogue on joint matters, I now believe I can recommend to you a course of action to get a handle on developing joint doctrine. The recent upsurge in emphasis on the importance of joint operations may make this an opportune time to sort out responsibilities. As we have discussed before, this is an area which needs fixing but has been resistant to previous attempts at repair.

There is no question that doctrine, both tactical and joint, is closely tied to materiel development, what is taught in the service schools, and the perceptions of commanders. However, joint doctrine suffers when it is merely an extension of any of these three or is monitored by four different masters. What I am saying is that the services (and the unified commands) must fully participate, debate and evaluate proposed joint doctrine, but one authoritative agency must manage the program. This is not to limit anyone's right to appeal and speak up, but to make sure we have an efficient system which focuses on the unique requirements of joint doctrine. My recommendation is that USREDCOM be appointed as the executive agent to develop joint doctrine for approval by the JCS.

I believe we are the logical command since we are charged with the responsibility of being prepared to reinforce all the other unified commands. I do not think we are overstepping our intended charter, but we must have the backing of the JCS and the Chairman if our work is to have credibility and proceed smoothly.

I view this effort as independent of any action on JCS reorganization. Of course its goal is the same—making the system work, but I would like to get moving on this now. We need to make a solid step forward in rationalizing joint doctrinal development.

First, I believe it is essential that the Joint Chiefs clearly restate their responsibility for overall joint doctrine policy, approval, and implementation. Although this authority is imbedded in law and regulation, it is clouded by numerous modifications and specific joint doctrinal responsibilities delegated to the services. Then a single agency should be assigned responsibility for direction of the joint doctrine program. . . .

A point which must be confronted is USREDCOM's paucity of Navy and Marine representation. We cannot win if we do not include them and develop truly joint doctrine. We have begun work with Navy and Marine Corps representatives in our joint tactics, techniques and procedures meetings. Furthermore, we hope to produce and get agreement on a four-service doctrinal manual in the near future. I do not want to move on adding Navy and Marine representation to USREDCOM until after you have considered my proposal, but it is a necessary part of its implementation. I feel that proceeding with this modest action would be a step in the right direction.

Third Wave Concepts

Letter to Alvin Toffler
New York City
31 August 1982

Although *The Third Wave* did not specifically address the military as an element of second or third wave societies, the examples, the principles, the very concepts are reforming the American military, just as they are the economic, political and family facets of our society. Reading *The Third Wave* in 1980, I was struck by the many similarities between the then-modern Army and the potential for the Army of the third wave society.

Historically, we in the United States have applied the concepts of the second wave as the foundation for our military thought. In the first half of the century, our industrial capabilities and geography allowed us to enter both World Wars grossly unprepared, then gradually overwhelm the enemy with technology and masses of men and equipment. Since World War II we have lived under "the nuclear umbrella," again a reflection of our superior technology and industrial capability. But, by the 1960s, it had become obvious that our potential antagonists had caught up with or surpassed us in many technical fields, and for certain they outnumbered us. It was time to rethink our basic philosophies. Our military forces could no longer entrench on the hillside hoping to destroy the enemy by sheer attrition as he approached. We were outnumbered at the outset, and likely to remain so for the duration; our potential foes were better equipped than ever before; we had been too long bound by the cultural biases of the second wave—standardization, specialization, synchronization, concentration, maximization and centralization. During the era of the second wave, those were the very biases that secured our military success, just as they had ensured success for industry. In third wave culture, they become constraints.

The more we pondered the third wave concepts, the more we realize that we could no longer fight the enemy, man for man, bullet for bullet, or tank for tank. We had to develop some third wave principles of war which would allow us to fight outnumbered and win. We had to shuck off the second wave cocoon and develop new tactics, doctrine and procedures—Third Wave procedures. These ideas are coming of age with the acceptance of the AirLand Battle and AirLand 2000 concepts, concepts which are guiding the structure, development, and modernization of our Army as we move ahead into the years of the Third Wave.

To Change an Army

Military Review
March 1983

Reform of an institution as large as our Army is problematic under the best of circumstances. The recent history of change in military systems of the world is instructive. Let us examine the story of Sir Ernest D. Swinton's invention—the tank—as well as the history of the development of concepts for mobile all-arms warfare to illustrate the challenges that would-be reformers face in trying to introduce new ideas.

In the British Army, where the idea had its genesis and was the subject of much early development and experimentation, a succession of single-minded tank and mobility enthusiasts persisted in developing the concept of mobile all-arms warfare built around the tank striking force. They did so in the face of persistent opposition by most of their less imaginative peers and superiors. Most of these reformers were "loners." For the most part, they were argumentative, assertive and hardly ever in agreement—even with one another.

Despite support from Winston Churchill, they were forced to work around an organizational system which abhorred change. In frustration, many went public with their arguments and, by doing so, incurred enmity among their superiors sufficient either to bring on their early retirement from the active ranks or to relegate them to some inconsequential posting.

Although field trials were held to demonstrate the new concepts, those who benefited most from the trials were the Germans. They spawned the blitzkrieg based largely on their own study as well as their study of the writings of the British reformers, J.F.C. Fuller and B.H. Liddell Hart, and the record of the trials on the Salisbury Plain.

As war came to Europe in 1939, the British Army found itself with an imperfectly developed concept of all-arms combat based on the tank, to include inadequate tactics, organizations, equipment and training to implement a state of warfare they themselves had invented.[1]

In the US Army, the pioneers were fewer in number, and the institution proved considerably more resistant to change than even the British Army. Therefore, the development of a concept of mobile warfare fared even less well. A succession of Army chiefs of staff rejected the idea out of hand. Even such future practitioners of maneuver warfare as General of the Army Douglas MacArthur testified before the Congress that one should not buy too many tanks, for they were terribly expensive and quickly became obsolete. Strongest among the opposition was that bastion of mobile thinking—the US cavalry. Its last chief, Major General John K. Herr, was the most strident, outspoken opponent of the idea of all-arms warfare which was built around the tank.

There were really only two heroes of this drama in our Army: Major General Adna Chaffee and Lieutenant General Daniel Van Voorhis. Without Chaffee, the US Army quite likely would have had no tanks at all in 1940. And, without Van Voorhis, there would not have been an operational concept for armored formations in World War II. As Edward Katzenbach concludes in his fascinating paper, "The Horse Cavalry in the 20th Century," the Army of the most mechanized nation on earth came to the threshold of World War II firmly wedded to strategy, operational art, and tactics deeply rooted in the 19th century.

Reprinted with permission from *Military Review*.

On the other hand, the Germans seemed to have developed, in what retired Colonel Trevor N. Dupuy calls their "genius for war," a much more impressive willingness and ability to adapt to change. Captain Timothy T. Lupfer describes well the German Army's ability to change operational concepts and tactical schemes in a matter of months in World War I.[2]

Heinz Guderian, reading reports of the armored force trials on the Salisbury Plain, demonstrated the concept with a small force for Adolf Hitler at Kummersdorf in 1934. Kenneth Macksey describes well how the German tank pioneers seized on and matured the preliminary British work on all-arms warfare built around the tank.[3]

With Hitler's blessing of the concept, Guderian, in 18 short months, produced an all-arms panzer division. The division operated within a fairly well-spelled-out doctrinal framework. It included the strategy for mobile warfare; a general operational scheme for how the larger forces would fight; and the organization, tactics, and at least a preliminary array of the type of equipment needed to bring the concept from theory to reality. In his new book, *The German Army, 1933–45,* Albert Seaton describes the German Army's remarkable ability to adapt to change in those very turbulent years.

How did they do it? How were the Germans different from the British or the Americans? Several facts stand out which frame the answer and outline a set of requirements necessary to effect change.

First, the Germans had a general staff element whose primary function was to examine the need for change and, when change was decided on, to draw up the necessary programs to make it happen. True, this capability became diffused as Hitler fragmented his army command into the *OKW* (Armed Forces High Command) and the *OKH* (Army High Command), an overshadowed army headquarters. Indeed, some of the bitter antagonisms that arose between those two organizations in World War II survived until recently even in the *Bundeswehr.* Nonetheless, for the critical developmental years, there existed an institutionalized framework for examining the need for changing doctrine—strategy, operational art, tactics; describing the equipment, organizational training, and other changes needed; and producing the impetus for change through the office of the *Inspekteur.*

Second, the German mavericks were all products of the enormously demanding and rigorous officer selection and training system characteristic of the German Army to this day. Mavericks they may have been, but all had been taught to think logically about tough problems. They were all taught in the same way, in the same schools. Compelling logic to one was, therefore, equally compelling to all. This made arriving at a consensus much easier. And change simply cannot be effected without a consensus achieved by some means.

Third, the principal instigators of reform remained for years in positions related to implementation of the changes they espoused. For example, follow Guderian through the evolution of the blitzkrieg in Macksey's book on Guderian.[4] Change was further facilitated because the senior leadership, to include most importantly Hitler himself, was quick to seize on the strategic advantages Germany could gain over its potential foes by changing the basic ingredients of its military system.

Finally, trials had been conducted—by the Germans in Russia, by the British on the Salisbury Plain and by the Germans and the Russians in the Spanish Civil War. And these closely observed lessons were fed back into the system for the further refinement of their mobile striking forces. Recounting, then, we have a set of generalized requirements for effecting change.

- There must be an institution or mechanism to identify the need for change, to draw up parameters for change, and to describe clearly what is to be done and how that differs from what has been done before.
- The educational background of the principal staff and command personalities responsible for change must be sufficiently rigorous, demanding, and relevant to bring a common cultural bias to the solution of problems.
- There must be a spokesman for change. The spokesman can be a person, one of the mavericks; an institution such as a staff college; or a staff agency.
- Whoever or whatever it may be, the spokesman must build a consensus that will give the new ideas, and the need to adopt them, a wider audience of converts and believers.
- There must be continuity among the architects of change so that consistency of effort is brought to bear on the process.
- Someone at or near the top of the institution must be willing to hear out arguments for change, agree to the need, embrace the new operational concepts, and become at least a supporter, if not a champion, of the cause for change.
- Changes proposed must be subjected to trials. Their relevance must be convincingly demonstrated to a wide audience by experiment and experience, and necessary modifications must be made as a result of such trial outcomes.

This framework is necessary to bring to bear clearly focused intellectual activity in the matter of any change, whether in concepts for fighting, equipment, training, or manning the force. Such a framework was recently institutionalized in the US Army. Let us briefly describe how this came about.

The Army reorganization of 1973 was aimed, in part at least, at the institutional side of the problem we are examining. In those years, the Army needed many changes. Some were purely managerial, reflecting our apprehension of a lot of structure and too little manpower. More importantly, however, the Army realized it needed to change its concepts of warfighting. It addressed the strategic problems of fighting outnumbered and winning; the matter of the operations of larger units, which units perforce would be fewer in number; and the revision of tactics, organizations, equipment, and training to bring the Army out of the Vietnam trauma and to make it an effective fighting force in the last quarter of this century.

The Army found itself confronted by principle antagonists, who were almost always sure to outnumber it, and by a growing militarization and modernization of conflict in the Third World. The Soviets, impelled by their obsession with numbers, were obviously in possession of a maturing operational concept embracing mass, momentum, and continuous land combat in a nuclear, chemical, or conventional environment. Convinced by the realities of our then current and impending resource constraints, we could not afford a like concept. We set about to look for ways to win even though fighting outnumbered. This was a crucial first step. (Russell F. Weigley might argue that that was more of a radical departure from our antecedents than others might agree.)

However, some analysts suggested history clearly endorsed the idea, and the 1973 Arab-Israeli War provided a fortuitous field trial of useful concepts. The lessons drawn from that conflict, as well as other analytical study, led to the Army's conclusion about the requisite strategy, operational concepts, tactics, organizations, equipment, and training. The outcome of this intellectual activity and theoretical study was set forth in what became the 1976 edition of Field Manual (FM) 100-5, *Operations*. Its primary emphasis, at least as viewed by its critics, was on an operational concept the Army called the "active defense."

However well or not so well that work may have been done, it met with considerable criticism from within the Army and without. Some of this simply reflected institutional resistance to the notion of change. Some of the criticism, however, reflected unresolved intellectual and theoretical concerns. But the experience demonstrated that all too little consensus building had been done and that the concepts set forth in the 1976 edition of FM 100-5 needed additional maturing. The results of that realization were several-fold.

First, the Army re-examined and revised its principles of war and published them in a new book, FM 100-1, *The Army*. An early criticism of the 1976 edition of FM 100-5 was that it was not firmly founded on enduring principles and did not even recount our principles of war. This new book began to build that theoretical foundation. The principles of war, as set forth in FM 100-1, spell out fundamental principles on which we must base our military strategy, operations, and tactics in order to be successful today and to meet tomorrow's needs.

While that development was under way, the Army's operational concepts evolved through a succession of changes known as the Corps Battle, the Central Battle, the Integrated Battle, the Extended Battle, and, finally, the AirLand Battle.

One lesson of that experience was that we had imperfectly designed the institutional framework to accomplish change. In 1973, the US Army Training and Doctrine Command (TRADOC) absorbed the old US Army Combat Developments Command. There were several good reasons for that amalgamation—some related to resources and others related to perceived shortcomings with the output of that command. In any event, while strong on equipment development and organizational matters, the new combat developments directorate of the TRADOC staff was weak on conceptual work. Therefore, the bulk of the concept work reflected in the 1976 edition of FM 100-5 was done by a handful of people, none of whom was assigned to the combat development staff at TRADOC Headquarters itself or in the schools.

The realization of this omission in our original concept of how TRADOC was to do its business caused us to create a principal doctrinal development staff element at TRADOC—a deputy chief of staff for doctrine. This officer was responsible for identifying the need for change and for describing the conceptual framework of the change itself. Without that orderly process at the beginning and without one agency directly responsible for it, the need for change would always be ill-defined, and the conceptual direction of change would be cloudy at best.

Now, back to the beginning. The post-1973 reforms were presented to then Chief of Staff of the Army General Creighton W. Abrams. He made many amendments, but supported the general direction of the changes. After Abrams' untimely death in 1974, General Frederick C. Weyand gave his support. That support from the top has continued with both of their successors, General Bernard W. Rogers and General Edward C. Meyer.

The reformers then set about designing tactics, organizations, equipment, and training systems to support the new concept. This resulted in, among other things, the division restructuring study and field trials of resulting organizations and tactics at Fort Hood, Texas. Because the concept was not yet mature, and because, in the trials, an attempt was made to measure performance differentials at the margin with an instrumentation system and a test scheme not adequate to that degree of precision, the trial outcomes were much too ambiguous to gain widespread acceptance.

At this point, it was apparent that the reformers had to begin anew. It became apparent that considerable internal consensus building would be necessary as organizational development proceeded. So, for two and one-half years, school commandants, representatives of the Army staff, major commands, supporting organizations, and other services were gathered at frequent intervals, and what we now know as Division 86 was hammered out at Fort Leavenworth, Kansas.

Consensus building in the Army was difficult for several reasons. In the process of bringing about change, there must first be a conceptual notion of what must be done to fight successfully in the battle environments of today and tomorrow. That conceptual thinking can only result from close, detailed, and reflective study of a wide spectrum of technology, threat, history, world setting, and trends. That kind of thinking can only be done by imaginative people who have trained themselves or have been trained to think logically about tough problems. That kind of intellectual development is one of the most important functions of our Army school system, especially at the staff college level.

It is perhaps here that we have not yet fully equipped ourselves with the requisite means to achieve change. The US Army lacked that great strength of the German system—the intellectual prowess and staff brilliance of its general staff officer corps. US Army officers lacked the cultural commonality that was brought to bear through the process of the German General Staff system, and that was the most impressive, if not the most effective, catalyst in making it possible for them to change quickly—even under the pressures of wartime.

Even though our Army has begun working on this dimension of the problem at the US Army Command and General Staff College (USACGSC), in both the long course and the course now styled as CAS^3 (Combined Arms and Services Staff School), some years will be required before the results of this effort bear fruit. The question has been raised as to whether we should consider a second year at Fort Leavenworth for selected officers to learn more about how we should prepare and plan for war and to hone the military judgment necessary to fight and win.

The USACGSC was a two-year course from 1929 to 1936, during which time some of our most brilliant staff officers and commanders in World War II were produced. The need to train more officers more quickly caused us to reduce the course to one year. Since then, subject matter related to fighting has been reduced to fill the many demands of our increasingly complex world environment. The time to logically think through tough military problems and to develop logical thought patterns was greatly reduced. But the complexities of war have increased greatly, and it is time to give the matter a new hearing.

While much remains to be done, the US Army does have in place today most of the ingredients which history suggests are necessary to effect orderly change. And we are in the throes of changes produced by that system—changes designed to move us into the last two decades of this century. We would be well served in the future if that process could include more sound thinkers in uniform and fewer academic and amateur military strategic gadflies.

We would be better served as the process matures if we could somehow focus the intellectual prowess of the operations analysis community on our fundamental rather than our peripheral needs. We would be much better served, in the long run, if we could learn how to change our institutions from within instead of creating the circumstances in which change is forced on us by civilian secretaries of war, defense, or whatever.

We would be much better served, in the end, if we could develop and refine, in our institution, the cultural commonality of intellectual endeavor and the ability to think logically about tough problems. These are necessary to develop new ideas, mature them quickly, and chart relevant action programs which effect change in an efficient, orderly way.

In short, we need institutional leadership as well as individual leadership. Without a requisite combination of both, history instructs us that the need for change is difficult to define. What is to be done—the goalset of change—is virtually impossible to circumscribe, and the whole process takes so long that not much ever happens. In today's and tomorrow's worlds, we simply cannot afford the luxury of that kind of inefficiency.

The need to change will ever be with us. We may have analyzed the process, framed its essential parameters, and made some considerable progress toward arming ourselves with systemic mechanisms to permit change to take place. But that in no way ensures either that change will occur or that it will be an easy, orderly process. And so the intellectual search, the exchange of ideas, and the conceptual maturation must continue and be ever in motion.

Notes

1. Kenneth Macksey, *The Tank Pioneers,* Jane's Publishing Co., N.Y., 1981, gives a first-rate account of this whole train of events.
2. Captain Timothy T. Lupfer, *The Dynamics of Doctrine: The Changes in German Tactical Doctrine During the First World War, Leavenworth Papers*, Number 4, Combat Studies Institute, US Army Command and General Staff College, Fort Leavenworth, Kan., July 1981.
3. See Kenneth Macksey, *Guderian: Creator of the Blitzkrieg*, Stein & Day Publishers, Briarcliff Manor, N.Y., 1976.
4. *Ibid.*

SACEUR and AirLand Battle

Message for General John W. Vessey Jr.
Chairman, Joint Chiefs of Staff
and
General E. C. Meyer
Army Chief of Staff
9 May 1983

1. Last week, 25–26 April, I testified before the House Armed Services R&D Subcommittee concerning the need for the development of systems included under the JSTARS, JTACMS and JTFS programs. Other witnesses included Jim Wade, Don Latham, Jim Tegnalia, Jim Hollingsworth, and others, to include General Rogers. Some of General Rogers' statements you need to know about, hence this message.

2. You will both recall that about a year ago he tackled Don Morelli at a briefing at SHAPE, saying that Allied Command Europe simply would not sign up for AirLand or AirLand 2000. In the ensuing flurry you talked with him, Jack [Vessey], and elicited an assurance that he didn't really mean what he said. Subsequently, of course, his staff has produced for him the briefing he now uses to show how ACE is going to tackle the follow-on forces. It is, of course, pure AirLand Battle. He himself, however, continues to state that ACE can't apply AirLand Battle doctrine. "We're not going to fight that way," says he. In the last month or so he has made the statement at the Naval War College, the Army War College, and he said it again last week at the hearing. Fortunately, it went by the subcommittee so quickly that no one called us on it.

3. After his testimony he cornered me. The gist of his statement to me was that what he objects to in AirLand and in AirLand 2000, as it relates to ACE, is that we show attacks across the FLOT with both weapons and troops. He claims that this upsets everyone in Europe so that he can't adopt that doctrine for ACE. He further states that 14/3 prohibits that sort of thing; NATO is a defensive alliance; and so on. He has said to several people words to the effect that now that the Army has adopted the strategy he has worked out for ACE he doesn't want us to screw it up.

4. There are several possibilities:

   a. He doesn't understand the concept;

   b. He doesn't want to understand the concept;

   c. He understands, likes it, and wants to be known as the author of the whole thing;

   d. Some combination of the above.

5. Whatever his motive, the Germans and Brits have reported to me that they do not understand what we are trying to do. Through the staff talks mechanism, we have persuaded both nations to adopt AirLand as a concept. Papers have been signed in some instances. The Germans are about to come out with a revised HVD 100-100, which is AirLand at its best. NATO Panel 11 is reported to be in disarray because of the confusing signals they believe they are receiving from us in this country and from the SACEUR himself. In the case of the Germans, it was precisely the aspects to which he objects that proved so attractive to them and was the opening wedge by means of which we were able to proceed. The whole idea of a campaign plan which carried the fight to the enemy with forces and weapons on his soil, especially to include nuclear weapons

on his soil, was the part that persuaded Rudi Reichenberger and Horst Hildebrandt to sign up. We all fully understood at the time that there would be practical limits on where and in what context we might speak loudly about that aspect of the concept, but nonetheless it was the key that opened the door. We even went over 14/3 together, with the Germans pointing out to me that we were the only nation to read into the guidelines a strict prohibition against cross-border operations. They also acknowledged that, for that time anyway, they themselves were seized with an administration that would probably choke at public conversation about the matter. So we agreed to proceed quietly, but with the clear understanding that we could develop such a battle concept in a generic sense without raising the issue of how it might have to be applied for a specific theater campaign.

6. It will of course serve no purpose to try and argue with the SACEUR about the substance of paragraph 5. However, the doctrine now laid out in our FM 100-5, about to be published by the Germans in HVD 100-100, adopted by the Brits, and all the work we've done over the last five or ten years to bring it to this juncture can come to naught if he persists. It would help a great deal if he could just endorse the doctrine as being appropriate in a generic sense with a footnote that it in no way inhibits specific theater campaign applications which must be worked out by CINCs, SACEURs, and others. There's not much sense in trying to reason with him. The only one of us that seems to have any clout with him at all is you, Jack. I'll leave it to you as to how to handle this, but from many sources there are uneasy soundings. He could well cause us to lose much of what we've worked so hard to accomplish for the Army and the country. If that happens, it will be nothing short of tragic.

Evolution of US Army Operational Doctrine

Swedish National Defense Research Institute
Stockholm, Sweden
5 June 1984

The task assigned me is to set forth the evolution of US Army operational doctrine—active defense, AirLand Battle, and future trends. That is indeed a banquet menu. In order to serve it in the modest time apportioned, it will be necessary to simply sketch in critical factors in the equation.

First, some notes about doctrine. The subject of this symposium is military doctrine for Central Europe. More than is the case with other Alliance partners, US Army doctrine is set forth in the broader framework of the need for US forces to be deployed to, and employed in, many theaters as well as in Central Europe. Therefore, there will ever be disparities between what is written for more universal application and what is specified in theater war plans.

One test of sound doctrine, however, is that, in its broader context, it must embrace principles and a logic obviously applicable to and attractive in more narrow, theater-specific frameworks. This test is essentially one of relevance. Failure to meet it has important consequences for, lacking sound operational concepts—doctrine perceived by all to be broadly relevant—it is difficult, if not impossible, for any military force to properly organize, equip, and train itself for battle, much less fight successfully once employed. Therefore, I shall concentrate on US Army doctrine set forth in field manuals and its application to the circumstances in which our Army, and the armies of our allied partners, find themselves in Central Europe.

Corollary to that is the fact that it is impossible to consider doctrine for Central Europe without accounting for the essential conventional force/theater nuclear force linkages. While this may or may not be a generic problem for US forces in a wider spectrum of environments, it clearly has the most important and vital implications for operations in Europe.

Next, we must consider the matter of the threat against which military doctrine is postulated. US Army doctrine, because of requirements for worldwide commitment, must consider many threats. It is, however, true that at its core US Army doctrine takes as the primary threat forces employed under the banner of Soviet-style operational concepts—mass, momentum, and continuous land combat, all potentially in chemical and/or nuclear environments.

I will, therefore, begin by briefly characterizing the threat in Central Europe. Since the end of the Great Patriotic War, Soviet strategy, doctrine, force structures, and capabilities, and those of the Warsaw Pact, have passed through several identifiable stages in arriving at their present circumstance. Each stage has reflected a host of conflicting political, military, and budgetary considerations, and to some extent at least perceptions of the Soviet leadership with regard to the capabilities and intentions of the Western Alliance—their "threat." Post-War II Soviet force capabilities clearly reflected outcomes of their early evaluation of War II operations that, according to Stalin, were to form the basis for Soviet military postwar planning.

Nuclear weapons were considered. They were seen as a new form of firepower, indeed of massive firepower. They were not, however, considered to be the decisive weapons of warfare. Combined efforts of all branches of the armed forces would be the only method by which strategic objectives could be attained.

In 1953, with Stalin gone, there began an important intra-Soviet debate with regard to the inevitability of war (nuclear) with the West. At its core, this debate recognized the awesome power of thermonuclear weapons and the question of whether or not their possession by the Soviets could be counted upon to prevent an attack by the capitalist West. Chairman Malenkov apparently concluded the risk that thermonuclear weapons exchanged between the two super sides would bring about the end of mankind was so substantial that, in effect, these weapons made war between rival world social systems impossible and that even the capitalist warmongers were intellectually bright enough to understand that fact. Khrushchev demurred. His doctrine called for large conventional forces that were also capable of fighting nuclear war. Balanced armed forces were essential, and continued development of heavy industry was yet the cornerstone of the country's defensive strength. Nuclear war was possible, had to be prepared for, and would be devastating to both sides, but the Soviets could and would survive it, as they had already survived the devastations of wars and internal liquidations since the 1917 Revolution. Later in the Khrushchev era, this pattern was to change dramatically, as we shall see.

Broadly viewed, three dominant themes characterized Soviet strategic thought before 1960:

- War could only be won by destroying enemy military forces.
- Nuclear weapons were not the decisive weapons of modern warfare.
- Success could be guaranteed by decisive defeat of the enemy in theaters of operations.

About 1960, there appeared a dramatic change in Soviet strategic thinking. This, in turn, led to major changes in force structures and capabilities reflecting the new doctrine. At its core, this doctrine espoused the view that nuclear weapons systems were to be the primary weapons of the Soviet armed forces; nuclear-tipped missiles were to be the primary means of nuclear delivery, setting aside the long-range bomber force; and the Strategic Rocket Forces were created as the primary service of the Soviet armed forces, displacing the ground forces. Soviet offensive operations in Europe were expected to begin with nuclear missile strikes, followed closely by conventional ground force operations. Apparently, NATO's growing nuclearization of the decade before led the Soviets to conclude that nuclear weapons must be at the forefront of any strategy, and thus, the first role for their own growing nuclear capability was elimination of NATO's theater nuclear forces—air bases, intermediate-range missile systems, tactical airpower, tactical missiles, and nuclear-capable artillery. Any theater conflict in which the enemy had nuclear capabilities would call for an immediate Soviet preemptive strike.

As had the death of Stalin a decade earlier, Khrushchev's fall presaged a renewal of the strategic/doctrinal debate. How much of the change that ensued was precipitated by the shift, in the West, from massive retaliation of the 1950s to flexible response of the 1960s would be difficult to assess. By whatever means, the Soviets in the 1960s embraced several important notions. Whereas Soviet doctrine had previously placed the conventional phases of war after a massive nuclear exchange, this order was now reversed. Some writers even put forth the view that rough parity in conventional capabilities offered fewer incentives for either side to move to use nuclear weapons. There also appeared to be considerable doubt in the minds of Soviet writers about what patterns of escalation one might expect following the initial use of nuclear weapons. In addition, some questions were raised about the relative unpredictability of levels of damage in the theater as a result of theater nuclear exchanges, levels of damage that could well prove disruptive to overall theater operations. Out of discussions of this latter problem, one infers a growing lack of Soviet confidence as to the ability of either side to manage a full-blown theater nuclear war. None of this is to suggest that the Soviets abandoned an interest in or capability for theater nuclear war. It does, however, acknowledge their recognition that it might well be to their own advantage to fight conventional force with conventional force at the outset, seeking to attain Soviet political and military goals without resort to nuclear weapons and the risks that course of action might entail.

One significant outcome of the post-Khrushchev changes was a doctrine that saw the preventive destruction of NATO theater nuclear forces as the primary strategic goal of Soviet conventional forces. With Soviet theater nuclear forces deployed considerably farther to the rear than NATO's theater nuclear forces, the balance of power in theater nuclear force capabilities could be expected to shift to the Soviet advantage with war's unfolding. With ground forces once again preeminent in Soviet force structure, what emerged was, and remains, a strategy and doctrine in which the primary contingency was a war that begins with conventional operations in the theater. Soviet air and missile forces, using conventional munitions, conduct preventive strikes against NATO's theater nuclear forces. At the same time, Soviet theater nuclear forces act to dissuade NATO from first use of nuclear weapons. However, should war begin with a nuclear exchange or should NATO begin preparations to employ theater nuclear forces, then preemptive nuclear strikes would go in against NATO's theater nuclear forces, followed by longer range strategic theater nuclear strikes against deeper NATO targets.

Doctrine for employment of Soviet ground forces embraces the operational concept of mass, momentum, and continuous land combat, all against the backdrop of the theater nuclear equation just set forth, and considering as well the possible use of toxic chemicals. Fundamentally, the Soviets believe that, to win, one must outnumber the enemy sufficiently to overwhelm his forward forces while at the same time striking deep to disrupt his rear, preventing employment of theater nuclear weapons, reinforcements, logistical support, and command and control. Mass in the form of echelon after echelon of forces pressing into the forward battle is given additional force through velocity, which builds the relentless momentum foreseen as the key to operational success.

In the last several years, Soviet military developments have included a revitalization of maneuver group doctrine, part of earlier Soviet ground force operational concepts. Today's Operational Maneuver Group as an independent action force, not part of the follow-on echelons, can be employed in several force sizes at several different levels of command and has the mission to strike deep; disrupt the rear; and attack logistics, command-control, reserves, theater nuclear stockpiles, and delivery systems. There is considerable evidence that OMG doctrine was laid down in the late 1970s in direct response to the active defense doctrine embraced by the United States in 1976 and in principle adopted by NATO in the so-called forward defense.

At this point in Soviet doctrinal development, then, we see Alliance forces in Central Europe confronted by a substantial and growing conventional force threat and by a full spectrum of theater nuclear forces and chemical weapons systems. Mass, momentum, and continuous land combat are the ways of war; NATO's TNF is to be attacked early on by conventional weapons and forces; there is a fully developed capability for first nuclear use, either directly or in anticipation of NATO's first use; and the threat of toxic chemical attack to restore or ensure momentum of the attack is ever present.

With what does the Alliance confront this threat? When NATO was created, the United States controlled an overwhelming strategic nuclear capability. At the time and for a substantial time after, NATO, under the US nuclear umbrella, was likely to prevail in a nuclear war. Certainly was this the case in a first-strike situation; it was true for an even longer time should the Alliance find itself in a strike-second mode. This was so, of course, because the United States had the capability to destroy Soviet strategic forces and/or reduce a Soviet counterstrike to manageable proportions. Since the mid-1960s, Soviet nuclear capabilities have grown dramatically in all dimensions, from intercontinental ballistic missiles to submarine-launched ballistic missiles to theater nuclear missile forces, to the end that now and for the foreseeable future the United States no longer has the capability to reduce a Soviet counterstrike to reasonably tolerable levels. It was early perception of this developing reality that led the Kennedy Administration in the United States to abandon massive retaliation and embrace the flexible response doctrine in its stead.

However, one pervasive problem has beset the Alliance through all the changes from massive retaliation through flexible response. That is the problem of numbers. In the beginning, NATO was grossly outnumbered by the sheer weight of the large Soviet ground force establishment that survived the war and remained in place.

With Western Europe prostrate economically and in various stages of military demobilization, the United States and the Alliance early on sought refuge in the rationalization that numbers on the Soviet side could be offset by firepower—nuclear firepower—on the Alliance side. This notion led to the strategic doctrine of massive retaliation; it persisted at operational and tactical levels as the United States pursued development of and deployed theater nuclear weapons. With the nuclear situation dramatically altered from that extant at the beginning and with conventional force capabilities dramatically altered in every capability from those earlier times, the Alliance today finds itself at a juncture in its affairs at which the doctrines invoked in that long-ago time are largely irrelevant.

Historians will note that this dramatic role reversal—especially in strategic nuclear capabilities, but in theater nuclear and conventional forces as well—has come to pass without any significant effort by the United States or by the Alliance to rectify the state of affairs. It is against that backdrop of changing force capabilities that US and Alliance military doctrines for operations in Central Europe have been postulated

Following World War II, US tactical and operational doctrines were drawn largely from our own experiences in the war just finished and our preliminary evaluations of operations of Soviet forces, but more particularly of operations of the Wehrmacht. With the panoply of massive retaliation, and later of theater nuclear weapons, as a backdrop, conventional force operations for the Alliance were to be defensive in nature and were to rely on a mobile defense. While this term had various interpretations unique to national forces of the Alliance, it did have several fairly well-defined and agreed-upon precepts.

First was a reconnaissance and security force deployed well forward to detect and report enemy movements and, if possible, to locate the enemy main attack. In US parlance, this was called the covering force. It was almost always composed of armored cavalry units, reinforced as necessitated by terrain and threat. Second was a force deployed along the forward edge of the battle area. The general idea was that this force was to be fairly light but capable of fighting sufficiently well to canalize the enemy onto selected avenues of approach. It was almost always composed of mechanized infantry reinforced by tanks. Third was a large mobile reserve or striking force, mostly tanks, that was to attack and destroy reentrants developing the forward battle area as a result of enemy thrusts into and through the forces deployed forward.

At first, US doctrine called for mobile defense tactics at every echelon from combat command or battle group up to corps and field army. Later, as tactical nuclear weapons became available, particularly cannon artillery-delivered weapons, it was considered that the division was the lowest level of command with the capability of successfully employing both a large maneuver striking force and nuclear firepower as well and that the combination of the two could with certainty provide the means to destroy any reentrant in the forward defenses.

While the mobile defense was, and perhaps still is, a sound operational concept, it ignored several aspects of the military circumstances in Western Europe that made it, virtually from the beginning, less than an ideal concept. First was the matter of distances necessary to allow a penetration to develop to a size sufficient to make it an attractive target for counterattack by a large force. Transposed onto a map of Western Europe, the deep thrusts of attacking and defending forces in Western Russia during War II, for example, quickly ran out of maneuver ground in the West of Europe. Second was the matter of political boundaries and the natural urgency not to surrender any of that precious national soil or the population, facilities, industry, and infrastructure set down thereon to an invader—particularly to a Soviet invader. This latter consideration urged for a defense well forward. It also recognized the very real military consideration that defense anywhere is not very realistic if one simply considers some artificial line drawn on a map, especially if that line is a political boundary drawn from other than military considerations. To be the basis for a reasonable success in defensive operations, such a line must correspond to, or be relevant in terms of, some natural obstacle—a river, mountains, a defile. If it is not, it can only be defended either forward of the line or behind it. Defending forward along a political boundary, then, has as an implicit condition the fact that the real defense must be conducted according to terrain forward or to the rear, wherever the best of it may lie.

These circumstances also created in Alliance military plans an understandable tendency to postulate rather shallow penetrations as targets for counterattack. This, in turn, resulted in a need to commit reserves long before the enemy main attack was likely to be evident with certainty. It required that allied commanders somehow be prescient enough to begin to move forces, marshal fire support, and summon airpower with the enemy yet far to the rear in his own follow-on echelon territory—and quite likely long before solid indicators of where the penetration might evolve were apparent. It was also against these very penetrations that doctrine called for the use of tactical nuclear weapons.

Over NATO's 35 years, with those conditions ever present, several other developments have taken place, developments that highlight the need for revision in the Alliance doctrine. Foremost has been the dramatic change in the strategic nuclear balance. Second has been the advent of theater nuclear weapons on both sides. Third is the political import of the military asymmetry reflected in the fact that, from the beginning, Soviet theater nuclear forces have been targeted against Alliance theater nuclear force structure while, on the Alliance side, theater nuclear forces have been primarily regarded as a numbers balancer, a way of offsetting the disparity in numbers of forces between the two sides. Almost from the beginning, this asymmetry with regard to the two theater nuclear forces has existed. It has grown in recent years, as the Soviets have turned away from their intense concentration on nuclear weapons and begun to concentrate more on conventional force attack.

For now, the first targets of Soviet theater nuclear forces, of theater nuclear launch vehicles with conventional warheads, and of Operational Maneuver Groups are NATO's theater nuclear delivery capabilities. This asymmetry does now, and has for nearly 10 years, made NATO's traditional theater nuclear strategy essentially bankrupt. In addition, the growing inability of either side to reason with conviction how one might confine nuclear war to a theater of operations has made a rational nuclear doctrine even more difficult to define.

With these developments unfolding, the United States became deeply embroiled in the Vietnam War. Unwilling to mobilize [Reserve forces] and pay the attendant political consequences, we expanded our force structure by creating new units and, in the case of the US Army, used the entire force, to include the force deployed in Europe, as a rotation base for Vietnam. Thus it was that, for nearly 10 years in the late 1960s and into the 1970s, the capabilities of US units deployed to NATO were at an all-time low. Our Alliance partners, whether they realized it or not, simply stood ground for us under the shadow of US and their own nuclear umbrellas, all of which were quickly growing more and more tattered.

It therefore seemed to those responsible for putting the US Army back together after Vietnam that the time had come to look anew at operational concepts and at the force structure, equipment, and training necessary for their implementation, especially against what remained our most contentious threat—the Soviet/Pact confrontation in Central Europe.

The fundamental requirements for a successful allied defense in Central Europe remained politically to collapse the Warsaw Pact's will to pursue military action and to avoid, if at all possible, the use of nuclear weapons—theater nuclear forces to be sure but, more important, strategic or central nuclear systems, the intercontinental ballistic and submarine-launched ballistic systems. With acknowledgment that massive retaliation was no longer viable, and especially in the face of a dramatically changing strategic nuclear balance, the latter of those two requirements loomed ever larger. It was in at least partial recognition of that fact that the Alliance embraced flexible response, formalizing the strategy in NATO Guideline 14/3. Unfortunately, the Alliance had not addressed itself to the question of what, if anything, was needed in terms of operational concepts, force structure, equipment, and training to make forces of the Alliance fully capable of executing a credible flexible response. In addition, laid out for all to see was the panoply of modern battle in the lessons of the Yom Kippur War: the staggering density of the battlefield at critical points; the presence on both sides of large numbers of modern weapons systems; the presence in the air over the battle of large numbers of sophisticated air and air defense weapon systems; the increased criticality of command-control; the increased likelihood of interrupted command-control due to large numbers of sophisticated electronic warfare means; the inability of any single weapon system to prevail, reaffirming the essentiality of all arms combat; and the outcome of battle reflecting, more often than not, factors other than numbers.

Armed with those realizations, how then might the US Army and the Alliance become more capable of implementing the 14/3 Guidelines? That was the basic question.

A first subordinate question to that larger query is: Can one fight and win outnumbered? The answer to that is a qualified "yes." Numbers, it turns out, are important. But within a reasonable range of force ratios—say 1 attacking 6 to 6 attacking 1, history instructs that, more often than not, the outcome of battle turns on factors other than the force ratios extant at battle's onset. What are those factors? Primary among them, in the history of nearly a thousand battles in which the outnumbered force won out, is that the winner was, in every case, the side that seized and held the initiative, whether attacking or defending at the outset. Seizing and holding the initiative means maneuver; the ability to maneuver means that force ratios between the forces in contact must be held to reasonable levels in order for one side or the other to take the initiative and maneuver.

Now, maneuver, you must realize, is not the traditional US Army way of war. With some notable exceptions, we are essentially a country and an Army wedded to the idea that mass wins in the end, for such was our experience in the dominant theaters in the American Civil War, in both World Wars in Europe, and in the Korean War, with one notable exception. Of the great names in United States military history, but a handful understood and were masters of maneuver: Washington in the Revolution, Lee and Jackson in the Civil War, Patton in War II, MacArthur in World War II and in Korea just about exhaust the first rank.

Nonetheless, with nuclear strategy bankrupt, faced with the imperative of finding some way to win outnumbered, and win without the imminent use of nuclear weapons, it became necessary to describe how one might go about winning although outnumbered, and do so against mass, momentum, and continuous land combat, all in a potential chemical/nuclear environment.

Fundamental to the ability to fight successfully against Soviet-style operational concepts is the ability to contend with two important matters at one time. First is the attack against the forward echelons, assuming we consider NATO defending and the Pact attacking. Second is the ability to begin to attack follow-on echelons in order to disrupt, delay, and destroy sufficient of their velocity and mass to prevent them from building the momentum necessary to success in Soviet-style operations. If the forward echelons can fight successfully, if the follow-on echelons and the OMGs can be prevented from loading up on the front-end battle, then the commander fighting the forward battle has half a chance to seize the initiative and maneuver to the attack.

At the front part of the battle, then, there is a need for a stout defense—a mobile defense, but not of the style brought to mind by that phrase. Let's call that the active defense. It is designed to defend in depth against leading echelons of a Soviet-style attack. Lacking large reserves initially, that defense must succeed until forces can be mobilized and deployed to provide the reserves necessary for large-scale attack and counterattack. The defense is sited well forward for reasons to which we have already alluded. Situations in which the ideal defensible terrain is to the east of the inter-German border should be cause for taking the initiative in order to secure key or critical terrain on which a more favorable battle can be fought. In the case of NATO's defending corps, divisions, and brigades, it means that some will be maldeployed at the outset due to restrictions of the political boundary or because of terrain circumstances in the sector to which they are assigned. That fact should not preclude readjustment quickly to a more favorable posture once the political barrier has been breached by Pact forces.

As the active defense begins against first-echelon forces, the deeper campaign to delay, disrupt, and destroy follow-on forces and OMG is brought to bear. Recce, surveillance, and target acquisition means are directed so as to detect, follow, and target follow-on forces attempting to close on the forward battle. In Central Europe, allied corps commanders should, as a general rule, be cognizant of, and following, enemy forces that are within about 72 hours of being able to intrude into the corps-level battle. On Central European terrain, that time equates to as much as 150 kilometers. Time is, however, the important feature, for of prime concern is the time of closure, the time within which the enemy is able to intrude himself into the corps battle, requiring the marshaling of forces and fires or posing a threat to corps operations designed to seize the initiative through maneuver.

The missions at that depth are to delay, disrupt, and destroy; attacks are made against chokepoints, command control nodes, logistical support, routes of movement, means of deployment, and troop concentrations. It is essentially an air and long-range missile battle. As follow-on echelons close to within about 24 hours of the forward battle, perhaps as much as 75 to 100 kilometers away, depending on the terrain, the divisional commanders begin to follow their progress and to attack as appropriate. Here, the mission is to disrupt, delay, and destroy as possible. Real-time and near-real-time target acquisition is most desirable. The attacking force is beginning to have fewer maneuver alternatives. The battle is an AirLand Battle with close air support, battlefield air interdiction, and ground maneuver forces, as appropriate, being applied against the follow-on forces.

It is against this 72- to 24-hour deep enemy battle array that theater nuclear weapons should be used, if indeed they are to be used at all. For, with delay times inherent in the Alliance decisionmaking process, it must be clearly established at the outset at what point in the battle it is necessary, prudent, and possible to use tactical nuclear weapons. This carries the nuclear fight to the enemy on his ground and against his forces. It avoids their use in territory to be defended with resultant risks to the infrastructure we want so desperately to preserve.

Finally, follow-on forces closing to within 12 hours of the forward battle are the responsibility of forward-deployed brigade commanders. The mission is to destroy and disrupt those forces, defeat the echelon in contact, and go in against follow-on echelons as appropriate to destroy or render the force ineffective before the next follow-on echelon can appear. It is without question a land forces battle, with a full spectrum of all-arms capabilities brought to bear to defeat the enemy as quickly as possible. It is the central and critical duel of the AirLand Battle.

That is the essence of AirLand Battle. Its lessons are severalfold:

- Deep attack and the close-in battle are inseparable.
- Seizing and holding the initiative through maneuver of forces and fires are essential to success.
- The objective of the battle is to win, not just to avert defeat.
- Successfully conducted, the AirLand Battle can, with conventional means alone, considerably postpone the time at which the Alliance must consider the first use of nuclear weapons, thus raising dramatically the nuclear threshold in Central Europe.

AirLand Battle with special adaptation to Central Europe is seen in the SACEUR's Follow-On Forces Attack concept. Here, Alliance doctrine differs from conceptual AirLand Battle in the following ways: Follow-On Force Attack is an effort to bring under central control the necessary recce, surveillance, target acquisition, and delivery assets to find, follow, and attack deeper echelons as they attempt to make their way toward the main battle area. However, this effort is not necessarily seen as a function of the defending force commander's scheme of maneuver. This is so largely because of the need to maneuver and the fact that maneuver implicitly pretty much ignores political boundaries. In recent years, any suggestion that the inter-German border might be violated in the East, even though it has been, or is about to be, crossed by the Pact moving West, has been viewed with an alarm close to paranoia. This disconnect is unfortunate for, from a strategic standpoint, the Soviets must understand that, if they elect to violate the border, then the outcome of that violation is to be decided on the basis of rules that have yet to be set forth. In other words, under those circumstances, it is not necessarily the goal of the Alliance simply to restore the status quo antebellum. That fact in itself adds deterrent value to the strategy that includes AirLand Battle as its cornerpiece.

Second, there are some apparent discontinuities between AirLand Battle and NATO doctrine with regard to the allocation and apportionment of airpower. Attack of follow-on echelons demands a rigorous disciplining of the allocation process. There are limited resources and many contenders for their use. AirLand Battle as a concept, because it focuses on combat operations of corps and their subordinate elements, seeking to win battles and through them wars, of necessity involves corps-level commanders in the process of establishing priorities for employment of airpower. That being the case, NATO's convoluted organization of its allied air forces makes the allocation and apportionment process somewhat more difficult but far from impossible.

Some contend that, while the AirLand Battle cannot be faulted conceptually, the means for its implementation are not at hand, especially surveillance, target acquisition, and reconnaissance means; longer range delivery systems; and command-control capabilities necessary to fight near and deep battles simultaneously. To some extent, at least, this may be true. What is also true, however, is that never until Follow-On Forces Attack has the Alliance made a concerted effort to establish the requisite command-control linkages between and among the growing menu of deep surveillance and targeting systems being fielded. It is sufficient to note here that it is demonstrably possible to provide NATO's corps-level commanders today with near-real-time information of Pact force deployments and other activities to depths well within those required to fight the deep battle and the close-in battle at the same time. The doctrine offers far more opportunities than theater war plans have, to the moment at any rate, been able to assimilate.

To succeed in its offensive, a Soviet-style attack, especially in Central Europe, must achieve success in the air, air defense, and ground campaigns that are its integral parts. Success on the ground is contingent upon momentum built up by follow-on echelons and deep and destructive attacks by Operational Maneuver Groups that fix NATO corps and army groups in place and that deny by conventional means, or nuclear, NATO's theater nuclear force capability. The means to deny to a Soviet-style attack these fundamental goals are at hand. As technology, by a variety of forms, continues to improve the Alliance capabilities, we must be quick to seize the advantages proffered, match our doctrine to the threat, then match the doctrine with relevant force capabilities. This is the genesis of AirLand 2000, an attempt to describe battle a decade or so hence. For, with today's ever longer development times, we are today describing weapons systems to be used 15 years hence. Therefore, we should strive to define as best we can what kind of war they might have to fight.

Never in history has a nation achieved superiority in a wide range of national military capabilities without, at some point, seeking to translate that superiority into some gain or advantage in the power balances that in the end shape the outcomes of international confrontations. Yet, by an unfortunate combination of rationalizations and asymmetries, the United States and its Alliance partners have for too many years given over to the Soviets the initiatives in this vital context.

It is, therefore, quite irrelevant to speculate about whether or not there exists some Soviet master plan, some closely held table of time and circumstance upon which Soviet forces will move upon the rest of us. Nor is it in the least useful to debate whether or not the aging collective czardom in Moscow is benign or bellicose. The fact is that, in a world of complexity, of turmoil, of rapid change, there will more than likely be ample critical circumstances in which the determinant elements will be the relative capabilities of the two sides, their relative willingness to face up to issues of their respective national interests, and their willingness and ability to act with force on behalf of those interests considered vital.

The doctrinal evolution styled as AirLand Battle and its logical extension through technical evolutions now underway can provide the military means. What remains is the political will and sense of purpose necessary to capitalize on this capability.

AirLand Battle III

American Institute of Aeronautics and Astronautics, Inc.
Command, Control, Communications, and Intelligence Workshop
Detroit, Michigan
5 September 1984

Since the early 1950s, the United States and its principal allies in many theaters have elected to rely on technology in the form of nuclear weapons to offset a superiority in numbers enjoyed by their potential antagonists.

Nowhere is this reliance more dramatically demonstrated than in NATO Europe. The hard facts, then and now, were and are that the countries of NATO cannot realistically support the large numbers of traditional divisions and air wings required for conventional defense of NATO's Central Region. As Supreme Allied Commander in Europe, General Dwight Eisenhower proposed 96 divisions and 9,000 aircraft for the conventional defense of NATO's Central Region. This proposed requirement was forthrightly rejected by the NATO ministers in 1951; it was politically infeasible and economically unaffordable.

Later, President Eisenhower embraced the notion that NATO would rely on 26 divisions (12 to be provided by the West Germans), 1,400 tactical aircraft, and 15,000 tactical nuclear weapons. While nearly all of these weapons were built—nuclear artillery shells, nuclear warheads for surface-to-surface and surface-to-air missiles, and nuclear bombs for tactical aircraft delivery—deployments to Europe were limited by President Kennedy's Secretary of Defense, Robert S. McNamara, to about 7,000 weapons. The presence of these weapons allowed enormous reductions in the originally proposed conventional force levels, and the threat of their use against massed attack effectively denied the Soviets a meaningful conventional offensive capability for more than two decades.

Now Soviet advances in nuclear weaponry—strategic, operational, and tactical—have overtaken the Eisenhower solution. It is no longer credible for allied forces in Europe, or elsewhere for that matter, to threaten early use of nuclear weapons in response to a conventional attack. In sum, we now need another solution to the problem of how to cope with a Soviet-style tank/infantry attack, whether delivered by Soviet or Soviet-surrogate forces. It is our contention that, once again, technology can be made to provide the means to reconstitute our superiority over our potential foes, especially at the operational and tactical levels of warfare. For the purpose of this discussion, then, we shall defer strategic considerations of intercontinental ballistic systems and concentrate on weapons systems and command-control systems, especially the latter, at operational and tactical levels.

Operational- and tactical-level realities in every likely theater of war—Europe, Korea, the Middle East—indicate quite clearly that our joint and combined commands will be confronted by at least an initial, if not a continuing, enemy force superiority. These forces, employing Soviet-style operational concepts, will be echeloned in depth so as to build up momentum—the product of mass and velocity—in order to maintain a continuous attack designed to overwhelm and defeat friendly forces with conventional or a combination of conventional, chemical, and tactical nuclear weapons. Given the modern conventional weapons systems in the hands of the Soviets and Soviet surrogates around the world, US nuclear retaliation may provide the only alternative to conventional defeat. It is our conviction that this is an unacceptable alternative and that our warfighting joint commanders must have the conventional force and weapons systems capability to counter initial overwhelming superiority in the air and on the ground without breaching the nuclear threshold—beyond which no one can know the consequences. We further believe that, applying today's technology to the problem, such a group of weapons and supporting systems can and must be developed. To place the need for these new capabilities in a proper perspective, what follows is an operational concept for joint operations on the modern battlefield.

The AirLand Battle concept deals primarily with war in areas of the world where we can be expected to encounter large numbers of relatively modern, well-equipped forces employing Soviet-style operational concepts. Quite naturally, therefore, the threat against which the concept is designed is typified by the Warsaw Pact in Central Europe, the larger aggregations of mechanized forces in the Middle East, and the threat from the north in Korea.

The concept emphasizes one of the frequently ignored lessons of history; that is, when political authorities commit military forces in pursuit of political aims, military forces must win something, else there will be no basis from which political authorities can bargain to win politically. Therefore, the purpose of military operations cannot be simply to avert defeat—rather, it must be to win.

New and radical ways to win are not proposed here. Rather, we will describe the extension of the battlefield, which it is possible to accomplish now and which, if applied, will reinforce the prospects for winning.

AirLand Battle is not a new concept. It is, however, a more descriptive term for indicating the full potential we must now seek to realize from our acquisition, targeting, intelligence fusion, and weapons systems.

The battlefield and the battle in AirLand Battle are extended three ways:

- The battlefield is extended in depth. Enemy units not yet in contact are to be engaged to disrupt their momentum—break up mass, slow down velocity, and complicate command control.
- The battle is extended forward in time to the point that current actions, attack of follow-on echelons, logistical preparation, and maneuver plans are interrelated to maximize the likelihood of winning the close-in battle as time goes on.
- The range of assets figuring in the battle is extended toward more emphasis on high levels of command using joint service acquisition means and attack resources.

What emerges is a perception of the battlefield in which the goal of collapsing the enemy's ability to fight drives us to unified employment of a wide range of systems and organizations on a battlefield which, for corps and wings, is much deeper than that foreseen by previous doctrinal concepts. The word "doctrine" is used advisedly. Our purpose here is less to suggest innovation than it is to pull together many good ideas for making extended attack an integral feature of our combat capability by all services in all theaters.

To ensure that the extended battlefield concept is understood in the full context of the integrated conventional-nuclear-chemical battlefield, I will first review major aspects of the concept, then describe how, by attacking assaulting and follow-on echelons simultaneously, the prospects for winning increase dramatically.

The Concept

In peacetime, the purpose of military forces, especially in the context of operations in areas critical to US interests, is to reduce to a minimum the enemy leadership's incentive to seek military solutions to political problems. In NATO, the Middle East, and Korea, our defensive strategy must extend beyond simply denying victory to the other side. It must, instead, postulate a definable, recognizable (although perhaps limited) victory for the defender. Enemy leaders must be made to understand clearly that, if they choose to move militarily, no longer will there be a status quo antebellum. Rather, the situation they themselves have created will be one that is to be resolved on new terms.

As the strategic nuclear balance teeters, so grows the enemy's perception of his own freedom of action at theater levels, both conventional and nuclear. Theater forces should not be considered solely as a bridge to strategic nuclear war. They are weapons that must be considered in the context of a total warfighting capability.

These considerations dictate that, in Europe for example, NATO strategy must, from the outset, be designed to cope with the Soviet integrated battlefield threat. This includes conventional, nuclear, and chemical combined arms capabilities. The growing threat of nuclear capabilities elsewhere suggests this strategy to be appropriate in other theaters as well.

The Warsaw Pact/Soviet-style strategy embraces two fundamental concepts. In the first, mass, momentum, and continuous combat are the operative tactics. Breakthrough is sought as the initiator of collapse in the defender's defensive system.

In the alternative, surprise is substituted for mass. In NATO, this could involve a number of divisions formed into Operational Maneuver Groups (OMGs) conducting independent attacks. Without warning, they would seek to deny to defending forces the opportunity to get set forward. Both tactics are essentially maneuver-based schemes with a single purpose to disrupt the operational tactics of the defender, albeit by different methods.

The need for deep attack emerges from the nature of our potential enemies, their doctrine, and their numerically superior forces. Whether or not our enemy is stylistically echeloned is not really critical. What is important is that superiority in numbers permits him to keep a significant portion of his force out of the fight, with freedom to commit it either to overwhelm or to bypass the friendly force. The existence of these follow-on echelons gives the enemy a strong grip on the initiative, which we must wrest from him, and retain, in order to win.

Our defensive strategies in key areas of the world must be designed to preserve, for the defender, the territory, resources, and facilities of the defended area. In virtually none of the critical areas of the world to which US forces are likely to be committed is there sufficient maneuver room to accommodate a classic defense-in-depth strategy. The defense must, therefore, begin well forward and proceed aggressively from there to destroy enemy assault echelons and, at the same time, to slow, disrupt, break up, disperse, or destroy follow-on echelons in order to seize the initiative quickly and go on the offense. The operative tactics by which US forces seek to implement the operational concept just set forth must provide for quick resolution of the battle under circumstances that will allow political authorities to negotiate with their adversaries from a position of strength. This is so because the enemy generally enjoys a short-term advantage in ability to mobilize additional forces quickly. Clearly, then, one goal of the AirLand Battle concept must be to lessen the probability of prolonged military operations. Further, these operative tactics should seek simultaneously to—

- Deny the enemy access to objectives he seeks.
- Prevent enemy forces from loading up the assault force fight with reinforcing assault echelons and thus achieving by continuous combat what might be denied by a stiff forward defense.
- Find the opportunity to seize the initiative, to attack and destroy the integrity of the enemy operational scheme, forcing him to break off the attack or risk resounding defeat.

Because of the enemy's advantage in numbers, attack of his follow-on echelons must always begin when those echelons are relatively deep in enemy territory. If an outnumbered defender waits to mount a counterattack until his numerically superior foe has penetrated the defender's territory, it is always too late to bring effective forces and fires to bear to defeat the incursion. This would especially be the case if theater nuclear weapons are considered necessary to defeat the penetration.

Therefore, on an integrated battlefield, systems designed to defeat enemy assault elements, to disrupt follow-on forces, and to seize the initiative by attack must be able to deliver conventional and/or nuclear fires throughout the depth of the battlefield.

Keys to a credible warfighting capability on an integrated battlefield are:

- Sensor and surveillance systems to prevent a surprise attack in peacetime and provide necessary targeting/surveillance information in wartime.
- Delivery systems that are triple capable—conventional, chemical, and nuclear, with sufficient range, accuracy, and lethality to hold enemy follow-on echelons at risk in peacetime and to attack them successfully in wartime.
- Command-control sufficient to integrate all-source intelligence in near-real-time in peace and in war. Near-real-time intelligence targeting information must be provided to maneuver force commanders to permit the timely employment of forces and fires in the maneuver environment.

The operative tactics that support such an operational concept, one of an integrated defense well forward, are:

- See deep and begin early to disrupt, delay, and destroy follow-on reinforcing echelons.
- Move quickly to generate sufficient force against the assault echelons to prevent them from achieving their objectives.
- Finish the opening fight against assault echelons rapidly before follow-on forces can join the battle.

Areas of Interest and Influence

In the execution of such a set of operative tactics, there must be a division of responsibilities among commanders. Just as the means with which commanders see and fight the battlefield vary, so should their primary areas of interest vary.

Each level of command has a dual responsibility. Each must attack one of the enemy's echelons and must see so as to determine the capabilities and intentions of a follow-on echelon. Doctrinally, we say that the enemy's first-echelon divisions, the regiments in front of the assault divisions, as well as the follow-on regiments, are the responsibility of the defending division.

In an attack, those same echelons would also be the division commander's responsibility. The brigade commander fights first-echelon assault regiments. The division commander fights the first-echelon assault divisions. The corps commander fights first-echelon armies. It is the corps commander's responsibility to find and disrupt the advance of second-echelon divisions of first-echelon armies before they become a part of the first-echelon problem.

At the same time, the corps commander is very interested in where the second-echelon army of the front is deploying. At corps and theater levels, the commander must tie into national target acquisition systems and other surveillance means to get information concerning where that army is and what it is doing. Above division level, the commander's responsibility in battle fighting has to do with the follow-on echelons.

Attacking the Follow-On Echelons

It is essential to realize that, as the enemy achieves the echelonment so necessary for his success, he inherently creates vulnerabilities—targets. These same vulnerabilities provide us with the opportunity to put threat second-echelon forces at great risk. But only through repetitive training and tailored systems can we capitalize on his vulnerabilities. Systems that see deep and effectively attack these targets are vital to our success.

What we must do is practice acquiring and targeting Warsaw Pact units now, during peacetime, so we will be prepared to attack them in wartime. In addition, we can do careful intelligence preparation of the battlefield and thus be prepared to attack high-value targets. Such targets include bridges, defiles, and other bottlenecks that will cause threat follow-on echelons to bunch up and present themselves as attractive targets. Additionally, attacking other high-value targets such as combat service support facilities, which must exist to support rolling forces, or selected command posts will also generate delay. Attacks directed in this manner will provide friendly forces time to finish the battle at the forward line of own troops (FLOT).

If the battle is fought with no directed interdiction, enemy follow-on echelons have a "free ride" until they enter the close-in battle. When follow-on echelons are ignored and allowed to stack up behind assaulting forces at the FLOT, they will achieve a breakthrough. The enemy retains the flexibility, initiative, and momentum to apply his mass at a point and time of his choice. Deep attacks seek to deprive him of this freedom.

There are three primary tools for deep attack:

- Interdiction using air, artillery, missiles, and special operations forces.
- Offensive electronic warfare.
- Deception.

In practical current terms, interdiction—principally battlefield air interdiction—is the primary tool of deep attack. At present, the range of jammers precludes effective use against follow-on echelons. However, jamming can be used in the close-in battle as a nonlethal substitute for fires and battlefield air interdiction sorties, which can then be freed to attack deep.

We would like deep attack to destroy enemy forces before they enter the close-in battle, but in today's terms, and in all probability tomorrow's as well, expense and scarcity of assets will limit the practically achievable effects to delay and disruption. Delay and disruption, however, must be aimed at more ambitious goals than just fractional attrition or harassment.

The real goal of the deep attack is to create opportunities for friendly action—attack, counterattack, or reconstitution of the defense—on favorable ground well forward in the battle area. This can be done by avoiding piecemeal employment of acquisition means and attack resources. These resources must be concentrated on critical targets that offer the highest payoff in upsetting enemy plans and to create situations wherein the friendly forces can seize the initiative and win.

It is important to stress here that the deep attack is not just a tool of the defense. It is, if anything, even more critical in the offense. It is essential to winning because it creates opportunities to seize and retain the initiative. It is equally important that corps and theater commanders fight this deep battle at the same time and in close coordination with the close-in battles.

It is also essential, then, that attack means not be applied indiscriminately. Limited strike and acquisition means must be applied in a planned, well-organized, and well-directed scheme to support the plan for winning. Piecemealing long-range target acquisition and attack resources is a luxury that cannot be allowed.

The commander's choice of when to use deep attack means must be taken in such a way that it will create a window for offensive action some hours in the future. That choice must be based on a single unified scheme of maneuver and a plan of fires for the whole of the extended battle. The expected window for decisive action must be created in an area where previous plans have assured the availability of sufficient logistical and fire support as well as maneuver forces.

This demand for careful coordination of present and future action throughout the depth of the battlefield dictates that the plan stem from the concept of a single commander. Separation of the close-in and follow-on battles invites the risk that windows will not be generated or that, if generated, units will be ill-prepared to identify and exploit them.

What emerges from this requirement for unity of command across the near and far components of the fight is a view of an extended battlefield with well-defined depth and width in which the commander is fighting not several separate battles but one well-integrated battle with several parts highly interrelated over time. The depth of this battlefield beyond the FLOT is really a function of the commander's planning horizon expressed in hours.

Attacking Follow-On Echelons: An Example

The following scenario describes an integrated battle situation in which it would be greatly to the commander's advantage to fight assault and follow-on echelons simultaneously. From the outset it is acknowledged that, in this scenario, it would be advantageous to use tactical nuclear and chemical weapons at an early stage and in enemy territory. It is also fully realized, however, that authorization to do this may not be granted in a timely fashion or may be politically unacceptable. That being the case, the battle would have to be fought with conventional systems. Even though this somewhat reduces defensive combat power, the concept described here maximizes conventional weapons capabilities.

The corps commander's concerns in the deep battle are those enemy forces that are within 72 hours of the close-in battle. The corps commander needs to have a well-laid-out, flexible plan extending 72 hours into the future in order to fight both close-in and extended battles, gain the initiative, win the fight, and do it quickly. What is the purpose of looking out to 72 hours' depth? There are many things a corps must do in those hours. It should be used to plan, order, and execute those maneuver, fire support, and logistical preparations necessary to seize an opportunity for offensive action.

The presence of any enemy formation in the corps commander's area of influence should trigger a reevaluation of his long-range plan and generate options for defeating this force, along with all others in the area of influence. Several options will probably be retained at this point. However, the range of options narrows as the force approaches and closure time decreases. Almost all options will include attack of the backup force to inflict delay and disruption. Although distances here are great, the payoff can be considerable since the critical targets include soft-skinned logistical and command-control elements whose value will be far less when closer to the frontline battle.

As the force closes, its impending impact on the frontline battle will become more apparent, and focus on the relative merits of the various attack options will begin to sharpen. Options at this stage should include deep nuclear strikes with missiles on air-delivered weapons. Targets at this stage are far more vulnerable to nuclear effects than at the FLOT. They are still well beyond the danger radius to friendly forces, and the time until closure is realistic enough to allow request, release, and execution to occur.

Of course, the commander must have a strong conventional option in the event nuclear release is not forthcoming. He must identify the critical time at which he must finally commit himself to one or another course of action. In any event, he seeks to hold the enemy formation out of the division area of influence long enough for division commanders to have sufficient space and time to accomplish their missions and prepare to fight the next echelon.

When the force enters the division area of influence—about 24 hours' distance from the FLOT—the entire process is triggered again on a lower scale.

Here, the importance of real-time target acquisition dominates. Since, at this point, the attacker is committed to specific attack avenues, he has few movement alternatives left to him. The defender can capitalize on that. Again, if tactical nuclear weapons are to be used, they must be used now.

In analyzing the history of actual operations, and of planning exercises in which assumed enemy penetrations were drawn with great care to specify that line "beyond which an enemy incursion would jeopardize the integrity of the defense," is always true that, if the penetration is allowed to develop as postulated, it is *always* too late to do anything about it.

If for no other reason, therefore, it is of paramount importance that the planning process begin while follow-on echelon targets are still deep in enemy territory, and that nuclear release, if appropriate, be requested in sufficient time to allow weapons employment while the target is still 24 to 60 hours from the FLOT.

As in the earlier part of this battle, the joint commander must integrate the full spectrum of air and land weapons systems. It is at this point still an AirLand Battle—perhaps more air than land, however.

By the time the follow-on echelons close to within about 12 hours of the FLOT, they become the concern of the brigade commander. At the 12-hour line, actions must be taken that not only delay and disrupt the follow-on echelons but also help to defeat those in contact at the FLOT. Given the right target, and assuming that the enemy has already used chemical weapons, it is here that our use of chemicals can be integrated. They should be used to isolate one part of the battlefield while an attack is launched against another part of the follow-on forces. Here, the land aspects of the battle predominate—that is, the battle is more land than air.

With a little luck, the outcome will find enemy assault forces destroyed, freedom for friendly force maneuver restored, and the initiative wrested from the enemy. In the end, this simultaneous attacking of echelons becomes key to our primary objective for the AirLand Battlefield—to win.

Studies show clearly that successful interdiction does result in a degradation of the enemy's massive firepower. It is also clear that successful interdiction results in a reduction of enemy momentum brought on through an inability to mass quickly, loss in velocity, and loss of support, and that it provides the defender time to secure nuclear release if indeed release is required. Finally, interdiction reduces the attacker's alternatives by disrupting his ability to execute his intended plan.

The conviction that well-planned interdiction can provide these results is based in part on target value analysis. In simulations in which we compared several European corps-level battles, first without interdiction and then with interdiction, several significant facts stood out:

- The enemy was able to mount fewer regimental-size or larger attacks.
- Enemy first echelons were defeated earlier.
- Friendly reserves were not needed so early.
- Enemy penetrations were far less extensive.

Each of these interdiction effects is highly desirable, but their exact significance is more apparent considering the simulation output over time. Specifically, a look at the effect of interdiction on enemy strength in the close-in battle shows the real value of deep attack. Without interdiction, the enemy is able to maintain consistent superiority at the FLOT over time. During this period, the defender's strength dwindles, freedom of action deteriorates, and the enemy's grip on the initiative decisively tightens.

Properly employed interdiction can hold out enemy follow-on echelons long enough to create periods of friendly superiority in which the initiative can be seized with enough time to act. The longer their duration and the more frequently these windows can be created, the greater the chance of winning, providing we are prepared to identify them and act at the time and in the place where they develop.

We may not be capable of creating windows of such frequency and duration across the entire front. However, it is now possible to create such opportunities, and if aggressively exploited, they could lead to the generation of longer, more extensive opportunities for higher level decision action building toward a major offensive.

By now, it must be clear that interdiction is key to battlefield success. The enemy's momentum can be altered by attacking high-value second-echelon targets, reducing his ability to mass and build up momentum. Interdiction is the method whereby we achieve the leverage necessary to slow him down and ultimately prevent him from achieving his objectives.

It is interdiction that allows us to focus our attacks on those enemy targets whose damage, destruction, or disruption would help us fight the battle to our advantage. Interdiction has as its main objective that portion of the enemy's force that is moving toward the FLOT or is in staging areas preparing to join the fight at the FLOT.

This concept does, however, imply some changes in current ways of thinking, especially about command control. In order to execute the concept, we must recognize the need to learn how to skillfully use resources far beyond those organic to corps and division and to plan their application over a greatly extended battlefield. Of significance here is the establishment of timely and responsive working relationships with air forces for both target acquisition and attack.

The interdiction battle will be fought at the theater and corps levels. To do this well, it must be practiced routinely. Interdiction targets at division level are directly linked to tactical objectives. At corps and higher, however, interdiction is a function of controlling target presentation rates and densities. As the enemy's second echelon moves closer to the FLOT, interdiction becomes more closely related to the defensive scheme of maneuver.

Advanced planning is absolutely critical to a successful interdiction battle. It is imperative that such planning be conducted continuously. This will ensure that commanders are aware of courses of action open to the enemy and the vulnerabilities of each, thus enabling them to attack targets that present the highest payoff at a particular time. Prior to and during initial stages of the battle, intelligence preparation of the battlefield must forecast enemy strength, progress, and dispositions at selected times. By assessing these developing vulnerabilities, we can draw up feasible courses of action for interdiction attacks. When blended with the scheme of maneuver, these enemy vulnerabilities can then be exploited.

Following such an interdiction planning process, the intelligence system can develop an enemy probable event sequence that can be used to predict with a high degree of accuracy which courses of action the enemy is likely to follow. That is, the intelligence system should be able to forecast when events must occur and in what order to produce the desired disposition of enemy forces at any critical moment. This probable event sequence is simply a template against which to assess the progress of events. It identifies interdiction requirements that will have to be met if friendly commanders are to influence the battle in a desired direction.

Interdiction targeting can be a complex and demanding staff process. Its effect is to create time and space gaps, not to relieve maneuver forces of having to face second-echelon elements.

Command-Control and AirLand Battle

It should be quite clear by this time that, of the significant challenges inherent in the AirLand Battle concept, none is perhaps more significant than that of command-control. Since that is the subject of this conference, a few words are in order concerning the requirements for command-control from the viewpoint of the operational commander charged with executing AirLand Battle.

In order to fight the AirLand Battle successfully, commanders must find out what is going on, decide what to do about it, tell someone what to do, and then keep track of how the battle is going. At the same time, they must anticipate what comes next, either in terms of what they may do or what is planned for friendly forces to do. History instructs us that commanders who do that will shorten the information/decisionmaking cycle in order that they might seize the initiative. In short, they turn their decisionmaking cycle inside that of their foes—in time, in order to seize and hold the initiative.

However, there are many detractors inherent to command and control systems, both external and internal. Externally, there is the volume of information made available by modern communications systems. There is no real priority discipline; routine reports flow with important ones. Sequential reporting systems move information from company to battalion to brigade to division in the Army, in the Air Force from recce pilot to wing intel to numbered Air Force, then back down the same chain to a fighter pilot, sometimes on the same air base. By the time the information arrives, it is out of date—good for history but not for operations. Internally, there are the procedures, protocols, communications, and message centers we have created to pass information. What has been called our self-constipating staff system requires that many staff elements mull over every bit of information before it gets to the commander. The result is that everything is submerged into a normative point of view, and truly important things are lost in the background noise. The tyranny of the message center frequently turns what should be information fusion into confusion.

The information explosion is a new experience. In today's military world, it compounds some already serious shortcomings. Our forces have never enjoyed high marks for excellence in collecting and processing information in war. As war grew more complex after the turn of the century, the strain on our information gathering and processing system, even in the static environment of World War I, was overwhelming. Commanders often found they knew as little about the location and activities of their own troops as they did of the enemy. Information they did have lagged hours, sometimes days, behind the real situation. Orders issued under these conditions often served only to confuse. The most reliable method of communication of the time, the telephone, depended on vulnerable wires laid on the ground. In an attack, when troops moved, telephones became impractical and information flow slowed or dried up completely.

In the AEF, the American First Army Chief of Operations, Colonel George C. Marshall, recognizing the problem, sent staff officers along with lead attacking battalions. These officers were supplied with some very brave homing pigeons that were periodically released carrying location and situation reports. When the birds arrived at Army headquarters, Colonel Marshall phoned the divisional and regimental commanders to report on where their battalions were and what they were doing—information not available at division or regimental headquarters.

During World War II, once again, battlefield information procurement and processing challenged both sides. And on both sides, alert commanders solved the problem in remarkably similar fashion. The British Eighth Army used liaison teams placed with most lower command headquarters. The British First Army used radio monitoring units to monitor communications of their own leading units. Finally, the British organized Phantom Service to speed up reporting essential friendly and enemy information.

The German Afrika Korps placed a general staff officer with lead elements to report critical information to the corps commander. From the beginning, German *Panzergruppen* Headquarters monitored the radio nets of their lead panzer regiments. The Americans imitated their British Allies in the US Fifth and Seventh Armies with radio monitoring units called SIAM. Originally, these units were intended to monitor enemy radio traffic; however, we quickly learned that we knew even less about our own actions than we did about those of the enemy. Radio intercept units offered a ready-made solution to the problem. These units monitored communications of leading US battalions, providing that information immediately to higher headquarters.

General George S. Patton, commanding the US Third Army, borrowed the Phantom Service idea from his British Allies, using an entire US mechanized cavalry group—the 6th—as an information network. He called the system the Army Information Service.

In Vietnam, heliborne commanders literally hovered over the battle, gathering information firsthand. All too frequently, commanders hovered over one another, all asking like questions or talking at the same time.

Our information requirements today are much more demanding than in times past. Today's commanders must know about today's encounters and be able to visualize tomorrow's battles and those that will occur in the next 3 to 4 days.

Today, all armies have or are developing a host of the most sophisticated and marvelous information systems; the mind is boggled with their capabilities. However, it is quite apparent that all the old problems of information collection and dissemination are still with us. Today, there is more information. Sometimes, it moves faster, and it's somewhat more secure, but humans who must use it still have the message centers and other staff mechanisms constructed according to the shopworn principles of long ago. So much information is coming in that the system is in a state of overload, and we still can't find out what we need to know.

There is an urgent need to recognize that there is a limit to what a command-control system can handle. As organizations approach overload, output increases with input for a time, levels off, then quickly decays to collapse. Research tells us that organizations react to information overload in several ways, to include queuing —waiting in line; filtering—leaving some things out; approximation—getting things about right; and errors—making even more mistakes than were present at the outset. Frequently, multiple channels are sought and used. If these don't work, then errors or omissions take over and system collapse is at hand. Historically, as already indicated, smart military commanders have used multiple channels—bypass channels—to speed the flow of critical information. None of the systems in design or development today take this multiple channel phenomenon into account.

What then are the requirements for command-control of the AirLand Battle? First, commanders at appropriate echelons must identify what information they need. Second, the architecture must provide for delivery of that information in a disciplined reporting system that will, in turn, allow a disciplined information flow.

Third, interneted databases must be developed and used extensively in each of several functional areas to provide the basis for widespread dissemination of large volumes of routine information, eliminating redundant and duplicative reporting requirements. Fourth, multilevel security gateways must be provided in switching systems in order to integrate security considerations with normal information flow processing.

Fifth, multiple frequency switching procedures must be applied to both low- and high-speed transmissions in order to make enemy jamming and direction finding difficult if not impossible. Sixth, the special needs of commanders responsible for the deep battle must be recognized and provided for, even at the expense of single-purpose special architectures and systems.

For the ugly truth is that, with burgeoning technology sprawling at our feet, we are less capable today of providing the kind of command-control devised by Patton, Montgomery, Patch, and Clark, Rommel, Guderian, than were those distinguished soldiers more than 40 years ago. It is almost as if our capability to command-control has proceeded inversely as the square of our ability to apply new technology to communications systems that provide command-control for the AirLand Battle.

And that is the fundamental and practical challenge that faces us today: how to harness today's technology to provide us the tactical and operational advantages we have so long enjoyed from the application of yesterday's technology in the form of nuclear weapons.

Joint Operations

US Army War College
Carlisle Barracks, Pennsylvania
14 October 1986

It is with considerable interest that I note you are beginning a study of joint operations at the Army War College, for there is no more complex, frustrating, or difficult business to try to define, analyze, and come to grips with. That also says that there's likely no more fertile field for whatever seeds of improvement one might sow, for almost anything you can get done is an improvement over what is being done. But one of the stark realities of our time is the significant unlikelihood that US Army forces will ever again take the field independently without the need to operate with the forces of at least one other service, likely more, and without having to take account of the forces of another nation or nations in the process. That fact makes your studies even more relevant, necessary, and in fact long overdue.

What I might add to your study, except to recite the frustrations and pessimisms of an old soldier who struggled mightily, with but limited success, in these same vineyards, I know not. But that's what you've asked me to do, so let's try.

Your course study will lay out for you the details, acronyms, procedures, and processes. What might be useful for me to do for you is to set forth a perspective about the critical elements in the joint process, in its broadest context, to serve as a backdrop for whatever may follow in your course.

Your study should begin with a thoroughgoing review of the authorities and responsibilities set forth in the Goldwater-Nichols DOD Reorganization Act of 1986 and how these differ from authorities and responsibilities defined by earlier legislation. H.R. 3622, the Conference Report on H.R. 3622 published in the 12 September *Congressional Record*, and the most recent revision of JCS Publication 2 (UNAAF), forwarded to the SECDEF by JCSM 165-86 on 25 June of this year, are the key documents. In the new legislation are some substantive changes to joint functions, especially in the JCS itself, and especially in headquarters of the military departments. About those, we can talk in more detail later. It does remain true, however, that no matter who is in charge or what lines of authority are established, the real problems in the joint process in our system remain relatively unchanged. And so, I will begin with those.

The first critical element that deserves attention I will call the political element. Nothing you will read or study about at the national level will make very clear what any joint force commander is supposed to do in a given circumstance. Just where does one go to find the national strategy or the national military strategy written down? It is not written. And that's the trouble with it. Even in deliberate, but especially in crisis situations, there is no quick reference to which you can turn to find out what the national goals and objectives are so that you might structure a relevant course or courses of action in pursuit thereof. Since military force is committed in pursuit of some ultimate political goal, it is important that the joint force commander have a fairly respectable idea of what political outcome is desired before, not during, and certainly not after, the force is committed. The new legislation attempts to cope with this problem. How effectively that may be done has yet to be seen.

A couple of real world stories will illustrate. In the fall of 1966, Secretary of Defense McNamara visited Vietnam. In a briefing at MACV headquarters, General Westmoreland cited for the

Secretary the plethora of missions and tasks he had been assigned by the various agencies and commands empowered to issue instructions to him. He concluded by saying that he had no way of coping with this proliferation of instructions. In response, Mr. McNamara asked that MACV draw up a list or lists of the instructions, cite by whom they had been issued, and tell him what might be the cost of doing all the things MACV had been tasked to do. A small task force was appointed to look into the matter. After lots of cutting and pasting, we were able to put together a list of the considerable number of tasks assigned the command. Major tasks alone numbered almost a hundred, as I recall, and with minor, ancillary, and implied tasks, that number multiplied to the hundreds very quickly. Then, we tried to estimate what it would take to do the job. After several sleepless nights, we finally produced force structure and time estimates that we considered relevant. We then reported, first on the lists themselves and on their redundancies, inconsistencies, and other anomalies. We then estimated that it would require a million and a half men, a million South Vietnamese and half a million Americans, and 10 years to get all that done, and that we would have to invade Cambodia and Laos and at least the southern half of North Vietnam in order to get it done. That was too much. General Westmoreland refused to approve our message to Mr. McNamara, saying, "That's politically unacceptable." Eventually, responding to repeated queries from the McNamara staff, the message was sent, but with a large disclaimer at the end saying that this was what a task force had drawn up in response to the Secretary's questions, although General Westmoreland himself could not endorse it.

Later, I was to learn that this message, unofficial though it was, provided substantial impetus for the Secretary's growing concern about our goals in Vietnam and the long-term cost of pursuing those goals to logical conclusions. In the near term, it had the effect of an imposed manpower ceiling on the US force in Vietnam. Originally set at 210,000 (when the force deployed or scheduled for deployment was about 160,000), this November 1966 McNamara edict provided the basis for space-by-space manpower control of the deployed force. Even with that control, authorized force levels grew to 549,500, and when we began redeployment planning in early 1969, we were just shy of that number in the theater. With that earlier escapade as a background, I was again present when redeployment planning began in 1969 and, in fact, headed the task force to draw up the plans for Vietnamization of the war and for redeployment of US forces. With General Abrams in command in Saigon and General Westmoreland in Washington as Army Chief of Staff, we now had political direction to turn the war over to our Vietnamese allies. Among the first questions we were asked concerned the size of a proposed residual US force. Our answer rounded out to about 250,000 Army, Air Force, and Marines. Before too long, it became quite apparent that the political direction of the new administration—with Nixon in office—was to get out completely and, according to some in Washington, as quickly as possible without regard to the outcome in South Vietnam.

General Abrams understood this situation quite clearly and acted in complete accord with his understanding and with his responsibility as COMUSMACV, a tough line to walk. It was, however, virtually impossible to convince his field commanders of the stark realities we saw in the Washington-Saigon exchange and so it was that, long after it had become obvious to the theater commander that we were on the way out, for better or worse, operational commanders were still in the mode of "look out, let the Americans do it." In some measure, however large, this lack of full comprehension and/or unwillingness to believe and act on it retarded the process of convincing the South Vietnamese that we were really on the way out and that, if they wished to survive, they had better begin quickly to pick up an increasing and, very soon, a full share of the combat burden.

Harry Summers has written about much of this in more detail and more eloquently; you all should be conversant with that writing. The point to be made here hangs on the question of why the theater commander did not have a better grasp of the political realities of his situation long before the McNamara-Westmoreland exchange cited above and, armed with that understanding, why was the theater commander unwilling to lay out for his political masters what he honestly estimated would be required to do the jobs he'd been tasked to do? General Westermoreland said repeatedly during the war, and made the same statement in his memoirs, that he believed that he would be given all that he felt he required to prosecute the war. Although he has never, to my knowledge, said so, I doubt that he ever had any positive assurance from any of his political masters to that end. He just assumed it. Indeed, in 1968, on the heels of Tet 1968, General Wheeler apparently suggested to General Westmoreland that, if he thought he needed more forces, this was the time to say so. The result of that exchange was the famous "206,000 more men" message that brought on the change of command in Saigon, precipitated Mr. McNamara's departure from the Pentagon, led to the appointment of Clark Clifford as Secretary of Defense, and changed the political mood in Washington, as dramatically reflected in the early and following work of the next administration to get US forces out of the war. To those of us who were there at the time, and to anyone with Pentagon experience with that administration, the November 1966 message imposing a force ceiling was a clear signal that resources were not unlimited, that sooner rather than later we had to cope with that reality and decide what we could afford to do with what the Nation was likely to be willing to spend. In the end, the only person in this very, very complex equation who clearly understood the political situation at home and acted both in concert with it and in keeping with his responsibilities for prosecution of the war on the ground was General Abrams.

Looking further back, and very briefly only, at the Korean War, whatever else came out of the MacArthur hearings in the Congress after his relief from command, it is quite clear that at least three times, from Pusan to the Yalu and back, the national political objectives with regard to that conflict changed dramatically without anyone really bothering to make that fact quite clear to the theater commander.

Given General MacArthur, there is some question as to whether or not any advice to that end would have made any difference. That is, however, beside the point; neither he nor his military peers in the JCS were sensitive enough to the political facts around them, and none, therefore, acted in accord with reality.

There is not time to discuss it here, but I commend to your attention the history of the experience of the French Army after the withdrawal from Indochina. Nowhere in modern times is there, to my knowledge anyway, a better example of military leadership trying so hard to conform to what they saw as their political responsibilities, and yet failing so miserably to do so—betrayed, as they would have it, by one of their own kind. The preparation, intellectual and physical, of the French Army for Algeria, for coping with what they called *La Guerre Revolutionnaire*, and the political decision by Charles de Gaulle to pull the rug from under them, is a case study in political-military relations that deserves the closest scrutiny by all of us and lots of contemplation after the scrutiny. It virtually destroyed the French Army for nearly 20 years. I commend it to you.

When I served as CINCRED, one of my frequent callers was Dave Jones, Chairman of the JCS. Inevitably, he would say something to the effect that, "We need a plan to send a division to such and such a place." My immediate and persistent response was ever, "All right, Dave, what would you like to have them do when they get there? More importantly, what do you want to have happen politically as a result of this deployment?" That question never received an answer, either from my friend the Chairman or from the Secretary of Defense, who was equally imprecise about the political goals whose concise statement and understanding by all just had to be the basis for any force deployment planning.

So, you all are afield on fertile ground. Obviously, you could spend the entire time set aside for this course pursuing some of the political-military nuances I've suggested here. No doubt, you will spend much less time on this aspect of the matter than on the mechanics of joint planning and all the intricate niceties of that terribly convoluted system. But at the root of things, the questions I've tried to raise will ever be the most perplexing, challenging, and difficult of solution. Of that, I will guarantee you. They are, therefore, deserving of your most careful attention.

The second critical element worthy of your attention is doctrine for joint force employment. In a word, it is fair to say there isn't very much, if any. You should begin by reviewing the JCS publications that are reputed to be the source of all joint wisdom on the matter. You will find them much too general and sweeping to suit the needs of a joint force commander. As CINCRED, I proposed that the JCS hand off the responsibility for joint force doctrine development, testing, and training to CINCRED. There were several cogent reasons for such a course:

- CINCRED is responsible for training the CONUS-based Army and Tactical Air Forces for operations in the theaters in which they may ultimately be employed.
- CINCRED is further responsible for their deployment to those theaters.
- At the time the recommendation was made, CINCRED was responsible for deployment and employment of joint forces in areas of the world not assigned to another unified command. Of course, a subsequent change to the Unified Command Plan has altered that circumstance.
- With deployed forces operating under rules laid down by their respective theater commanders and with a need to train the deploying forces in employment procedures so that they might be effective immediately upon deployment, CINCRED was faced with a practically impossible task. Therefore, there was, and is, a vital need for a mechanism for rationalizing conflicting employment procedures, rules of engagement, and other matters unique to the several theaters, commands, and services so that the deploying forces might be properly trained.

The Joint Staff, of course, rejected this notion out of hand. Joint force doctrine is the sacred purview of the JCS, and no one from outside that hallowed conclave is quite capable of handling such a matter. Of course, the Joint Staff traditionally has done nothing about it themselves; they just did not want anyone else doing it.

As a compromise, we then suggested a program for the rationalization of joint tactics, techniques, and procedures (JTTP) and, pursuant to an agreement between me and General Jack Vessey, by then the Chairman, we actually began a program to define what we believed to be the procedures most in need of rationalization or standardization and to draw up JTTP publications that would provide the basis for joint force training for forces scheduled to deploy to one or another theater under the current series of deployment plans.

May I just conclude this little commentary by saying that I have some faith in the ability of theater-level commanders to work out suitable joint and allied tactics, techniques, and procedures, for they have done it. The joint force employment system in Korea, especially that between the Army and Air Forces, is pretty much a model, thanks to the efforts of General Jack Vessey when he was commander there. The situation in Europe is, however, quite the opposite, for there are too many nations and too many services involved, and every attempt to straighten things out procedurally inevitably results in another layer or agency empowered to direct or coordinate something that already has several layers or agencies doing the same thing. If there is any staff agency in the world that provides less value added to the forces it is there to serve than the Joint Staff in Washington, it is the extended staff in NATO. The fact that anything at all gets done, ever, is nothing short of a miracle.

Closer to home, I have no faith at all, in the US forces context alone, in our ability to overcome service bias and prejudice and get down to the real business of working together.

Incidentally, in the CINCRED/JCS episode cited above, I learned in the process that General Paul Adams, the first CINCSTRIKE, and so one of my predecessors at MacDill, had tried to tackle the doctrine problem during his tenure in the 1960s and met with the same resistance. Armed with that knowledge, I went to see General Adams to seek his advice. After several conversations, we worked out a scheme that we thought avoided the mistakes he saw himself making and from that generated the idea of joint tactics, techniques, and procedures. The idea died aborning; my successor, not being a doctrinal sort, did not see the urgency of the program, and despite the Chairman's strong endorsement and entreaties, things ground slowly to a halt.

Now, you will find in the new legislation some words about the Joint Staff role with regard to joint doctrine, tactics, techniques, and procedures. At least implicitly, the JCS have had that authority all along. While it is commendable that J5 has recently organized a group to cope with this responsibility, the quality, relevance, and value added of that effort remain to be judged. So, while I am prepared to be hopeful, I am also by nature and instinct prepared for the worst.

How does all that affect your study? To your attention, may I commend study of several critical functions that beg for approved joint doctrine and for which there is none sufficiently relevant and in sufficient detail to be useful to a theater-level joint force commander. I've already mentioned rules of engagement. Go look at those for the Sixth Fleet and those for PACOM, where those two commands come together or into contention, depending on your point of view, in the Middle East. Look also at air defense, close air support, and interdiction in the Middle East and in other theaters, where no command-control infrastructure and so no established procedures are in place. Then, tell me if you think everything is okay. Looking carefully, you'll find it is not okay, and I would argue that it never will be unless some responsible unified command or joint task force commander takes it upon himself to legislate how it is going to be done. The Joint Staff is quite likely not competent to do that and will ever be beset by the parochial bickering between the services so characteristic of our whole experience with that organization since its formal inception in 1949. While new impetus has been given at the Joint Staff level, it will ever be the joint commander's responsibility to tend to his knitting at theater level.

Of the significant challenges inherent in joint forces operations, in the AirLand Battle at the theater level of command, none is more significant than that of command-control.

. . .

For the ugly truth is that, with burgeoning technology sprawling at our feet, we are less capable today of providing the kind of command control devised by Patton, Montgomery, Patch, and Clark, Rommel, Guderian, than were those distinguished soldiers more than 40 years ago. It is almost as if our capability to command-control has proceeded inversely as the square of our ability to apply new technology to communications systems that provide command control for the AirLand Battle.

As CINCRED, I seized our J-Comm with this problem. After much work, he dashed in mid-exercise with a marvelous message. Two hundred fifty thousand words long, it had been transmitted in seconds over a 9600 baud comm system. I asked, "To whom does it go?" "Well, to all recce squadrons, fighter wings, brigades, and regiments," was the answer. "How is it to get there," I asked. "On the teletype," was the response. "Over what net?" "Well, this is important. It has to go on the command net," I was told. Interesting. The command net teletypes in the Army operate at 60 words a minute, in the Air Force at 100 words per minute.

So, here is this marvel of telecomm technology that has moved along the higher echelons at 9600 bauds but that finally comes down to the crunch crunch of the 60- or 100-word-a-minute teletype. So to the J-Comm folks I said, "Not going to tie up command net for 24 hours or more just for your message. I'll take it in my helicopter and on my travels around I'll distribute it direct to the commanders for you." Great! Well, we did that; what I did not say was that I would make another circuit some hours later to see what they had done with those golden 250,000 words. The messages were pretty much where we had left them, on the field tables or wherever. One brigade commander said, "Sir, I really don't have time to read 250,000-word messages."

The situation is bad enough in established theaters where there are established infrastructures; in theaters where there is no communications infrastructure at all, the situation verges on the ludicrous. In one Middle East exercise, we were totally dependent on a single satellite for comm back to the US, and there was some evidence that the Soviets knew that and were at least testing their capability to jam the link. So look to your command-control means. They must be interoperable! They require the commander's personal attention!

Finally, the theater/joint force commander is ever in a dilemma between requirements and capabilities. It is important for you to remember this, for a theater-level commander is likely never to ever have enough to do what he has been told to do. Realizing that, he is then in an ethical dilemma; he either makes believe it can be done with what is available, or reports what can be done with what has been made available, or what it would really take to do the whole job. All too often, you'll find responsible commanders rationalizing their experience, training, and better judgment, driven by a "can do" attitude that holds that it will be done despite the shortfalls. And that is why our political masters all too frequently make less-than-wise decisions. Some overzealous military fellow has reported that he can do it, even though all too frequently faced with sure knowledge that he hasn't enough to do the job.

Some time ago, there was much flurry over deploying a task force to the Middle East to counter a Soviet invasion of Iran. The force under consideration was about three and a third divisions, two and a third Army and one Marine. The threat was substantial, about 20 Soviet motorized rifle divisions stationed in the Trans-Caucasus. Shortly after World War II, following our successful efforts to persuade the Soviets to get out of Iran, we had drawn up a similar plan.

The threat in the 1950s was postulated at 22 Soviet infantry divisions. To counter that threat, we felt we needed to deploy 13 US divisions. Now, those two outlooks on the force requirements for that situation simply cannot both be correct. In the earlier instance, there was considerable concern that the force was not large enough. In the second instance, reputations, promotions, and prestige between the services were all at stake, and the joint force commander produced a plan that was clearly not operably feasible.

So, we are full cycle on the questions I tried to draw out for you in the beginning: Why was it that the theater commander was reluctant to make it clear to his political masters what it would really cost to do what they said they had to have done? What is the obligation—professional, ethical, moral, whatever you call it, of the military actor on that treacherous stage?

This is not a course in ethics and has not been styled as a course in leadership. But it is in so many ways. You should find yourselves face against many, if not most, of those very tough questions time and again if you are honest—and you need have the courage to be honest.

American Military Thought: A Perspective

TRADOC Historians Conference
Fort Monroe, Virginia
17 January 1989

You are most kind to invite one of the ghosts of Fort Monroe to speak on this occasion. Considering what to say to this distinguished body was no easy task. What I thought might be useful is a summary—an audit trail, if you will—of the underlying historical underpinnings of AirLand Battle, the current battle-fighting doctrine for operations at tactical and operational levels of war. For, while much has been written about Active Defense circa 1976 and AirLand Battle circa 1982, there has yet to be set forth a record of the intellectual underpinnings of that doctrine from the point of view of one of its authors. So, while this will be brief and of necessity summary in form, it nonetheless is a beginning—something on which the historical cuspids can go to work as time begins to shape our perspective on the matter.

"Those who cannot remember the past are condemned to repeat it." So opined George Santayana. Rephrasing that just a bit, perhaps it's fair to say that what we most often learn from history is that we haven't learned much from history. Now is that the fault of the historians? Heaven forbid! Surely it must be due to the muddled intellectual apparatus of their students! No more frightening spectacle comes to mind than the meeting of those two old warlords Montgomery and Eisenhower who, standing on the edge of the Gettysburg battlefield at the Eisenhower farm, issued the pronouncement that had they been in charge they surely would have relieved all the principal commanders on both sides in that most famous of Civil War battles. While one would hope that was jest, one fears it was not.

The purpose of history is to inform our judgments of the future to constitute an informed vision of whither we should be trending and how best to get from where we are (and have been) to where we believe we must be.

Implicit is the notion that change—evolution (perhaps minor revolution)—is both necessary and possible. I believe that—strongly. We have a notorious record in this country of summing up our military adventures and misadventures by preparing to do the whole thing over again, only better. In the contemporary world that is likely not sufficient. Things change too quickly. Technology contributes new capabilities; these in turn demand careful consideration as to how best they might be employed at tactical, operational, and strategic levels in our national, political, economic, social, and military endeavors. So a first step is to formulate what we are trying to do—a vision, for it will then be necessary to assign responsibilities and develop operational concepts and doctrinal systems, equipment systems, organizations, and training systems, and allocate resources in order to implement the vision.

It should not be necessary to remark that all these things take time: for equipment alone at least 10 years normally elapse between the time the user describes an operational capability he needs and the time he can expect to see a reasonable development of some kind that purports to match the capability. Five years or more pass before significant quantities of systems are available for field use, and before doctrinal training and organizational rearrangements can be made to accommodate the new equipment. The problem is that no one in charge in the system lasts that long, so the system is always zigzagging as each new actor takes his place on the TRADOC stage, in the Chief of Staff's chair, or in key positions on Congressional committees, all of which makes more imperative some broad consensus about the nature of what Russell Weigley has called "the American way of war," and what it means to those responsible for developing a vision for the future. So what follows is an attempt to pique your imagination. You need not agree—I would hope that some number of you will agree, but that's not totally necessary. This is not an alchemist's formulation of transformation chemistry; rather it is an attempt to suggest what might best inform our vision. Details are sketchy, as they can and should be for this audience.

American military thought reflects no more than perhaps three fundamental influences: Napoleon, the industrial revolution, and modern technology.

After meager beginnings in the American Revolution we disbanded our Army. We came to the War of 1812 with neither a strategy nor any approved doctrine for actions at operational or tactical levels of war. Our performance in that war spoke eloquently of our ineptitude in these matters.

From the War of 1812 onward to the Civil War the United States Army was intellectually dominated by one man—General Winfield Scott. It was an era of westward expansion and growth, of national concentration on affairs other than military—for which there was little need in a country so isolated.

There was of course a brief interlude in 1846 in which we mounted an expeditionary foray into Mexico. Organized and led by Winfield Scott, it was remarkably successful given the state of the expeditionary force art in those years.

The War with Mexico had a longer-term effect on the United States Army, for in its ranks at the lieutenant, captain, and major levels were officers who would command at brigade, division, and corps level on both sides in the American Civil War 15 years later. What these officers knew of land warfare was largely derivative of operational concepts—tactical, theater, and strategic, of Scott, of the 18th century, the wars of Napoleon in Europe, the War of 1812 in North America, severely moderated by the realities of the scarcity of resources in the Army in which these men came of age. However these officers—the professionals, were also, for the most part, graduates of West Point, where Jomini on Napoleon was the standard historical treatise on the military art and engineering [was] the dominant academic bias. So it is appropriate to focus on what and how Jomini's Napoleon was applied by his practitioners in the American Civil War, for these ideas took deep root; in one form or another they persist to this day.

That the proper object of military operations on land should be the enemy's army was a dictum from Jomini that led, in North America, to battles whose fundamental goal was defeat of the opposing army. Once it became apparent that the "Confederates" in the Civil War fought from conviction more than any other motive, destroying the enemy army became a necessary translation of the term "defeat." The idea was perhaps more clearly perceived by General Ulysses Grant than by any of his peers, and so it was Grant who pressed on through the Wilderness and eventually destroyed the Confederate Army and ended the war. Jomini was also well understood by General Robert E. Lee; but, in possession of ever scarce resources, growing more scarce by the year, Lee was forced to embrace a strategy of offensive defense that failed against overwhelming odds from the North. Grant not only realized the preeminent need to destroy the enemy army but also the need to destroy the enemy support infrastructure—the material means to support the fight. Hence Sherman's march to Atlanta and northward to the sea and the 1864 campaign in the Shenandoah Valley.

In their perception of the need to destroy the enemy means for supporting war, as well as the means for waging war, Grant and Sherman were perhaps more Clausewitzian than were they disciples of Jomini. In any event, adoption of the battle of annihilation strategy in the Civil War was destined to cause the deaths of more soldiers on both sides than in all the rest of our wars combined. In conscript armies—the Grand Armee and its successors across the world in time, there is always a tendency to look on manpower as a virtually free resource, one which, if provided in sufficient numbers and properly supplied, is the surest way to win. It is an idea that dominated US military thought at least until the advent of nuclear weapons, but which yet dies hard.

The second important influence on US military thought was the industrial revolution. Armed with our Napoleonic heritage, expanded upon by our Civil War experience, we found industrial revolution concepts of mass production of enormous numbers of things in a short time at relatively lower costs to be made to order for military operations in which the overpowering influence of mass—people and equipment—was paramount.

The convenient, serendipitous relationship between these two intellectual currents was made even more so by two other factors. They were, first, the miserable state and performance of our logistical system and deployment means demonstrated in the 1898 war against Spain in Cuba, and secondly a growing body of military thought about strategy, planning, preparing for, and supporting forces afield that started after the War of 1812, aggravated by equally wasteful and inefficient performance in the Civil War—especially in the North.

In the years after the Civil War, while the active Army pursued the wily American Indian with Napoleonic zeal in the North American hinterlands, thoughtful military men began to set forth proposals about how to fix what was quite obviously broken. High on all lists of proposed reforms were the need for an institutionalized education system for officers and the need for a properly organized and functioning staff system to develop strategy, operational concepts, and plans. With the example at hand of the introduction of military staff systems in Europe, and the demonstrated efficiency of the Prussian mobilization system in the Franco-Prussian War, early 20th century military reforms in our country brought into being a general staff system and an officer education system. Both went to work on the strategy/planning lacuna that had persisted since the War of 1812. Among first staff tasks was development of a mobilization system—programs and schedules to bring together masses of manpower, train them up quickly, and provide them with masses of equipment, supported by megatons of supplies, and send them to war, where they would prevail by force of arms. The mobilization system was a direct translation of industrial revolution thinking—training factories turned out thousands of at least partially trained soldiers; airplane, tank, truck, cannon, and other factories similarly produced volumes of essential commodities of war. Along the great production line people and equipment trained up and together went off to war. It was the military equivalent of Henry Ford's River Rouge—the total and complete factory.

So well did this system work for us in World War I that, after that war, we set about to make the system more efficient. And in 1940 it was that reworked system, activated by Franklin Roosevelt, that commenced our mobilization for and eventually supported our participation in World War II. In both World Wars ocean distance isolation provided time for mobilization to take effect and burgeoning industrialization provided shipbuilding and shipping capabilities to ensure deployment of our mobilized forces. In the First World War the big problem was movement of troops, most of the materiel means being provided by Allies. By the Second War, however, the need for troops and materiel was global and, with some notable but few exceptions, the great industrial system rose to and met the challenge. Impressed by our own success, postwar planners set about developing ways to do it over again, only better.

So it is fair to say that, with very, very few very, very striking exceptions, US military thought to 1945 produced a military system designed to overwhelm by mass in a battle of military and national annihilation using the production techniques and tools of the industrial revolution.

Following World War II, our doctrinal revisions were drawn largely from our own experiences; from preliminary evaluations of operations of Soviet forces; and more particularly from perceptions of the Wehrmacht in the war just finished.

The offense was the dominant form of combat; massing at the decisive point was the operative technique. Together these principles would serve to defeat enemy air-land forces and win the battle, so the war. Inferred was the notion that we would be able to mass sufficiently when and wherever we desired.

As this doctrinal evolution unfolded, it quickly became apparent that the growing conventional strength of the Soviet Union was a most vexing problem. For, inherent in the changing force balance was the unpleasant truth that no longer could we guarantee numerical superiority, even with the help of allies. So our whole concept of mass conscript armies and mass industrial means to support battles of military and national annihilation came to be at risk. In Europe it was General Dwight Eisenhower who as SACEUR set forth the need for 96 divisions and 9,000 tactical fighters to defend NATO. Rejected by member countries unable or unwilling to provide the staggering resources necessary, President Eisenhower finally settled for 26 divisions—12 of them to be West German, 1,400 fighter aircraft and 15,000 theater and tactical nuclear weapons. Of the latter, about 7,000 were deployed and some 4,000 remain. So we come to the third, and perhaps most complex, factor in US military thought—modern technology, in this case nuclear weapons.

In the beginning, and for some time after, the United Sates controlled an overwhelming strategic nuclear capability. Since the mid-1960s, Soviet nuclear capabilities have grown dramatically in all dimensions; now it is quite likely that the United States no longer has the capability to reduce a Soviet counterstrike to reasonably tolerable levels. It was early perception of this developing reality that led the Kennedy administration in the United States to abandon massive retaliation and embrace the flexible response strategy in its stead.

Then, as we struggled to find a suitable vision for ourselves following the Vietnam War, the Egyptians and the Syrians attacked Israel in October of 1973. Soon laid out for all to see were the lessons of the Yom Kippur War: the staggering density of the battlefield at critical points; the presence on both sides of large numbers of modern weapon systems; the increased criticality of command and control; the increased likelihood of interrupted command and control due to large numbers of sophisticated electronic warfare means; the inability of any single weapon system to prevail, reaffirming the essentiality of all arms combat; the outcome of battle reflecting, more often than not, factors other than numbers. In addition, it was starkly obvious that large-scale destruction in a short time was a most likely outcome of first battles in modern war. Time to mobilize behind ocean barriers and deploy to relieve beleaguered forces—both characteristics of our World War I and II experience, was likely at risk as well. This final challenge to the mobilization system spelled for many an urgent need to rethink mass conscript armies, nations in arms, mass industrial mobilization, and battles of military and national annihilation as relevant concepts for the use of military force as an instrument of national policy in our time.

These great historic vectors confronted those of us who set about to put the US Army back together again after Vietnam. First we sought to cope with the problem of numbers. Thus the circa 1976 Active Defense sought to use the technical capabilities of the antitank guided missile systems that we had developed and deployed in considerable numbers in the early 1970s to increase the defender's range and kill ratio advantage at the forward line of troops and thus make the defense a far more formidable undertaking. It did, however, belie to many observers our traditional fixation on offense as the decisive form of combat. So in Active Defense we reopened the classic historic debate surrounding offense versus defense. In general the "defensists" note that one can defend successfully at a ratio of about three to one, but that it takes about six to one to mount a successful offense. Therefore defense is the most economical way of operating. True enough. But about that argument may I cite the following example:

> It is 7 October 1973. The commander of the Israeli Defense Force reserve division charged with defending the Golan Heights arrives at Northern Command headquarters about noon. Here he finds that the 188th Armored Brigade, the regular force defending the Golan, has lost virtually all its officers in the first day of battle, and that the Golan is a nearly open avenue into the Jordan Valley and the Heart of Israel. Initially instructed to prepare a defense as his division arrives from the south by road, the division commander argues that an attack is the better plan. The division arrives during the night and the counterattack begins early morning 8 October. Attacking with 300 tanks, about 75 percent of the force, the division in about 2½ hours reaches the prewar border, threatening to cut off the attacking Syrians who have, by this time, lost about 600 tanks. The attacking division closes on its objectives with 55 tanks of its original 300 still operative.

The lesson is clear: if a battalion can defend four or five kilometers of front against three attacking battalions, could it successfully defend 10 times that frontage against an attacking company? Likely not, so when the area is very large it is advisable for the smaller force to attack at a point, or on a very narrow front, rather than to try and defend everywhere at once and wind up defending nowhere successfully. While I do not pretend to have the last word on the defense/offense argument, I would note that, as we regard ourselves realistically as the outnumbered force, it becomes necessary to look again at some of our long-held convictions drawn from a time when we were the outnumbering force.

It was on this basis that Active Defense evolved into what is now called AirLand Battle. AirLand Battle seeks to moderate the force ratios at the FLOT by merging active defense and deep attack of follow-on echelons into one battle; it embraces the need to seize and hold the initiative through maneuver of forces and fires. Its basic intellectual underpinning is the conviction of the need to fight and win at the tactical and operational levels of war without the need to revert to nuclear weapons.

In a very broad sense, AirLand Battle is a grand offensive defense, one on the style fought so successfully by Washington in the Revolution, with great skill and initial success by Lee and Jackson in the Civil War, with striking success by MacArthur in World War II. That is why, realize it or not, the concept is dramatically at odds with our historic perceptions of what is necessary to win drawn from our Napoleonic heritage, our obsession with the mass concepts of the industrial revolution, and our fixation on substituting technology for numbers.

How to sum up this somewhat intense journey through two centuries of military thought in our country? Having either been party to or authored several such assessments in the past years, for my own part I come back to several fundamental observations.

First, we long have been, and are yet, children of our Napoleonic heritage—some from Jomini, some from Clausewitz, some homegrown—based on our own unique military experiences. This has led to a military system and concepts embracing a mass conscript military force in time of emergency, a force whose primary operational modus has been the destruction of enemy armed forces largely by overwhelming numbers. With few notable exceptions, our principal military names have been men steeped in that tradition.

Second, the systems and processes of the industrial revolution have fit conveniently, indeed reinforced, our convictions about large conscript masses, the nation in arms, and provided the material means for the ultimate battle of annihilation—of unconditional surrender.

Third, the post-World War II development of Soviet conventional military power has created circumstances in which it appears highly unlikely that we will ever again prevail by mass force of arms alone in a confrontation with the Soviets. More recently this circumstance has extended to other theaters, especially to Northeast Asia and the Middle East.

Fourth, the early attempt to substitute technology, in the form of nuclear weapons, for numbers was likely ill-advised at the outset and certainly foredoomed to bankruptcy once the Soviets had realized nuclear parity/superiority. We ignored this reality far too long.

Fifth, the character of modern armored battle unveiled by the Arab-Israeli Wars clearly suggests that winning early and succeeding battles while containing losses to reasonable levels is essential; deployment means and times do not exist to support deployments of sufficient rapidity and strength to overcome potential losses in early battles. Particularly is this the case if surprise is an operative element in initial battles.

Sixth, all this means that the mobilization system—in place since the convenient marriage of Napoleon's warfighting concepts and the systems of the industrial revolution early in this century, is quite likely an anachronism as we move toward the turn of the century.

Seventh, against the threat of nuclear weapons employment as a backdrop and with growing potential for massive destruction by quantitatively and qualitatively more powerful conventional forces, the use of military force as an instrument of national policy has become less and less attractive and feasible as between nuclear armed or supported foes.

Eighth, about technology—men and equipment, it is likely necessary to make one last enduring observation. The more intense and difficult the battle becomes because of numbers, weapons ranges, all the lessons cited a moment ago in regard to the Yom Kippur War, the more men—the soldiers and their leaders, become the critical and deciding element. Which is the better tank? The one with the better crew, of course! And so it is and ever shall be that battles are won by the courage of the soldiers, the excellence of the leaders, and the effectiveness of the training in their units before battle begins. Technical developments in new weapons systems increase range, lethality, and precision. Increased also is the intensity of battle, the effects of fatigue, the destruction of battle, the presence and pervasiveness of fear. No amount of technology can be invoked to solve these problems. And so we're back to the soldiers, their leaders, and the excellence of the training of the units in which they fight.

Finally, the attempt to move US military thought, operational concepts, strategy, tactics, and techniques from a mass destruction—annihilation—brute force—industrial revolution-based set of convictions into a more sophisticated, intellectually demanding, deft and precise set of operating parameters as we move ahead is yet aborning. It is clearly a necessary development. Your role in it—the historians' role, was not at all trivial; indeed it formed the basis, perhaps the bias, for much of the undertaking; how well it will fare, history alone may judge.

The American Way of War

Higher Command and Staff Course
British Army Staff College
Camberley, England
9 February 1989

Military thought in the United States in regard to land warfare, the "American way of war," or air-land warfare, reflects no more than perhaps three fundamental influences: Napoleon, the Industrial Revolution, and modern technology.

After meager beginnings in the American Revolution, we disbanded our Army. For a time, it consisted of one understrength battery of artillerists assigned to guard ammunition stores at West Point. Whatever military thought from you, from the French, or from Von Steuben may have been put in place by the Revolution, it went away quickly. We came on to the War of 1812 with no strategy and with no approved doctrine for actions at the operational or tactical levels of war. Our performance in that war spoke eloquently of our ineptitude in these matters, most eloquently in a battle in Canada at a place called Queenston. I commend it to your study of how not to do things.

From the War of 1812 onward to the 1861–65 war between the North and South, the United States Army was intellectually dominated by one man, General Winfield Scott. An 18th-century warrior, he commanded a brigade in the War of 1812, acquitted himself with distinction, and became an important postwar influence on reforms after that war. Eventually appointed General Commanding in Chief, he held that post for two decades, retiring because of age and physical infirmity shortly after the onset of the Civil War in 1861. For nearly four and a half decades, his was the dominant influence. It was an era of westward expansion and growth and of national concentration on affairs other than military—for which there was little need in a country so isolated.

There was a brief interlude in 1846 in which we mounted an expeditionary foray into Mexico. Organized and led by Winfield Scott, it was remarkably successful, given the state of the expeditionary force art in those years.

The war with Mexico had a longer term effect on the United States Army, for in its ranks at the lieutenant, captain, and major levels were officers who would command at brigade, division and corps levels on both sides in the American Civil War 15 years later. What these officers knew of land warfare was largely derivative of operational concepts—tactical, theater, and strategic, of Scott, of the 18th century, the wars of Napoleon in Europe, the War of 1812 in North America, severely moderated by the realities of the scarcity of resources in the Army in which these men came of age. However, these officers, the professionals, were also, for the most part, graduates of West Point, where Jomini on Napoleon was the standard doctrinal treatise and engineering the dominant academic bias. So, it is appropriate to focus on what and how Jomini's Napoleon was applied by his practitioners in the American Civil War, for these ideas took deep root. In one form or another, they persist to this day.

That the proper object of the military operations on land should be the enemy's army was a dictum from Jomini that led, in North America, to battles whose fundamental goal was defeat of the opposing army. Once it became apparent that the Confederates in the Civil War fought from conviction more than any other motive, destroying the enemy army became a necessary translation of the term "defeat." The idea was perhaps more clearly perceived by General Ulysses Grant than by any of his peers, and so it was Grant who pressed on through the Wilderness and eventually destroyed the Confederate Army and ended the war. Jomini on Napoleon also was quite clearly understood by General Robert E. Lee and his strong right arm, General Thomas Jonathan Jackson, but in possession of ever scarce resources, growing more scarce by the year, Lee was forced to embrace a strategy of offensive defense. Brilliantly conceived and even more brilliantly executed, it failed against overwhelming odds from the North. Grant not only realized the preeminent need to destroy the enemy army but also the need to destroy the enemy support infrastructure—the material means to support the fight. Grant and his principal executor in this task, General William Tecumseh Sherman, conceived of a great devastating march through the enemy homeland—Sherman's march to Atlanta and northward to the sea. Grant's convictions in this regard also led to the 1864 campaign in the Shenandoah Valley, where lay the considerable logistics infrastructure supporting Lee's Army of Northern Virginia.

In their perception of the need to destroy the enemy means for supporting war, as well as the means for waging war, Grant and Sherman were perhaps more Clausewitzian than were they disciples of Jomini. It is not known whether either had studied Clausewitz; had they done so, it would have been in German, for *Vom Kriege* was not translated into English until 1873. In any event, adoption of the battle of annihilation strategy in the Civil War was destined to cause the deaths of more soldiers on both sides than in all the rest of our wars combined. In conscript armies, the *Grande Armee* and its successors across the world in time, there is always a tendency to look on manpower as a virtually free resource, one which, if provided in sufficient numbers, properly supplied, is the surest way to win. It is an idea that dominated US military thought at least until the advent of nuclear weapons, but which yet dies hard.

The second important influence on US military thought was the Industrial Revolution. For, armed with our Napoleonic heritage, expanded upon by our Civil War experience, we found the Industrial Revolution concepts of mass production of enormous numbers of things in a short time at relatively lower costs to be made to order for military operations in which the overpowering influence of mass—people and equipment—was paramount.

This convenient marriage, the almost serendipitous relationship between these two intellectual currents, was made even more so by two other factors. First, the miserable state and performance of our logistical system and deployment means were demonstrated in the 1898 war against Spain in Cuba. Apparently, we had lost whatever magic enabled us to successfully mount an expeditionary campaign into Mexico 52 years earlier, for in Cuba everything went wrong as expeditionary operations put the ultimate strain on a system that probably did not work all that well anyway.

Second, there was a growing body of military thought about strategy, planning, preparing for, and supporting forces afield that started after the War of 1812, aggravated by an equally wasteful and inefficient performance in the Civil War—especially in the North. In the years after the Civil War, while the Active Army pursued the wily American Indian with Napoleonic zeal in the North American hinterlands, thoughtful military men began to set forth proposals about how to fix what was quite obviously broken. High on all lists of proposed reforms were the need for an institutionalized education system for officers and the need for a properly organized and functioning staff system to develop strategy, operational concepts, and plans.

With the example at hand of the introduction of military staff systems in Europe and the demonstrated efficiency of the Prussian mobilization system in the Franco-Prussian War, early 20th-century military reforms in our country brought into being a general staff system and an officer education system. Both went to work on the strategy/planning lacuna that had persisted since the War of 1812.

Among first staff tasks was development of a mobilization system—programs and schedules to bring together masses of manpower; train them up quickly; provide them with masses of equipment, supported by megatons of supplies; and send them to war where they would prevail by force of arms. The mobilization system was a direct translation of Industrial Revolution thinking—training factories turned out thousands of at least partially trained soldiers; airplane, tank, truck, cannon, and other factories similarly produced volumes of essential commodities of war. Along the great production line, people and equipment trained up and together went off to war. It was the military equivalent of Henry Ford's River Rouge—the total and complete factory.

So well did this system work for us in World War I that, after that war, we set about to make the system more efficient. And in 1940 it was that reworked system, activated by Franklin Roosevelt, that commenced our mobilization for, and eventually supported our participation in, World War II. In both World Wars, ocean distance isolation provided time for mobilization to take effect, while burgeoning industrialization provided shipbuilding and shipping capabilities to ensure deployment of our mobilized forces. In the first war, the big problem was movement of troops, most of the materiel means being provided by Allies. By the second war, however, the need for troops and materiel was global, and with some few but notable exceptions, the great industrial system rose to and met the challenge. Impressed by our own success, postwar planners set about developing ways to do it over again, only better.

Operationally, the Second World War once again reflected the happy marriage of our Napoleonic mindset and the Industrial Revolution. However, even so, there was a footnote of significance for, as in the Civil War, there appeared two strikingly different modes of making war. In Europe, with the notable exception of General George Patton, it was Eisenhower—disciple of Jomini and Clausewitz, son of the Industrial Revolution, who led the Allies in a campaign to bring massive forces—men and equipment, to bear to simply overwhelm the foe.

Meantime, on the other side of the world, a quite different course of events was in train. There, confronted by the Allied "Europe First" strategy, that most imperial of all American Caesars, Douglas MacArthur, employing an offensive defense strategy, seized more enemy-dominated area, captured more foe, and lost fewer of his own soldiers in the process than any of his peers anywhere else in the world. He was clearly in the mold of George Washington in the American Revolution, of Lee and Jackson in the Civil War, and of George S. Patton Jr. in the campaigns of the Seventh and Third US Armies in World War II.

So it is fair to say that with very, very few very, very striking exceptions, US military thought to 1945 produced a military system designed to overwhelm by mass in a battle of military and national annihilation using the production techniques and tools of the Industrial Revolution.

Following World War II, our doctrinal revisions were drawn largely from our own experiences, from preliminary evaluations of operations of Soviet forces, and more particularly from perceptions about the Wehrmacht in the war just finished.

Defensive operations were to feature a mobile defense. While this term had various interpretations, there were several fairly common precepts. First was a reconnaissance and security force deployed well forward to detect and report enemy movements and, if possible, to locate the enemy main attack. The US Army called this the covering force; it was almost always composed of armored cavalry units reinforced as dictated by terrain and threat.

Second was a force deployed along the forward edge of the battle area. The general idea was that this force was to be fairly light but capable of fighting sufficiently well to canalize the enemy onto selected avenues of approach. It was almost always composed of mechanized infantry reinforced by tanks.

Third was a large mobile reserve or striking force, mostly tanks, that was to attack and destroy reentrants developing in the forward battle area as a result of enemy thrusts into and through the forces deployed forward. At first, US doctrine called for mobile defense tactics at every echelon from combat command or battle group up to corps and field army. Later, as tactical nuclear weapons became available, particularly cannon artillery-delivered weapons, it was considered that the division was the lowest level of command with the capability of successfully employing both a large maneuver striking force and nuclear firepower and that the combination of the two could with certainty provide the means to destroy any enemy penetration into forward defenses.

While the mobile defense was soundly based, it ignored several realities of the post-War II military circumstances in Western Europe that likely made it, virtually from the beginning, less than an ideal concept for the NATO Alliance. First was the matter of distances necessary to allow a penetration to develop to a size sufficient to make it an attractive target for counterattack by a large force, particularly a force supported by nuclear fires. Transposed onto a map of Western Europe, the deep thrusts of attacking and defending forces in western Russia during War II, for example, quickly ran out of maneuver ground in the west of Europe. Second was the matter of political boundaries and the natural imperative not to surrender any of that precious natural soil or the population, facilities, industry, and infrastructure set down thereon to an invader—particularly to a Soviet invader. This latter consideration urged a defense well forward.

In offensive operations, doctrine echoed our Napoleonic past. The offense was the dominant form of combat; massing at the decisive point was the operative technique. Together, these principles would serve to defeat enemy air and land forces, win the battle, and so the war. Inferred was the notion that we would be able to mass sufficiently when and wherever we desired. In retrospect, it is fair to say this offensive doctrine was also probably ill-suited to the post-War II era, for it virtually ignored the substantial disparity in numbers between the respective sides—especially in Europe and Korea where we faced our most contentious threats.

As this doctrinal evolution unfolded, it quickly became apparent that the growing conventional strength of the Soviet Union was a most vexing problem, for inherent in the changing force balance was the unpleasant truth that no longer could we guarantee numerical superiority, even with the help of allies. So our whole concept of mass conscript armies and mass industrial means to support battles of military and national annihilation came to be at risk. In Europe, it was General Dwight Eisenhower who as SACEUR set forth the need for 96 divisions and 9,000 tactical fighters to defend NATO. Rejected by member countries unable or unwilling to provide the staggering resources necessary, President Eisenhower finally settled for 26 divisions (12 of them to be West German), 1,400 fighter aircraft, and 15,000 theater and tactical nuclear weapons. Of the latter, about 7,000 were eventually deployed, and some 4,000 remain. So we come to the third, and perhaps most complex, factor in US military thought—modern technology, in this case in the form of nuclear weapons.

In the beginning, and for some time after, the United States controlled an overwhelming strategic nuclear capability. At the time, NATO, under the US nuclear umbrella, was likely to prevail in a nuclear war. Certainly was this the case in a first-strike situation; it was true for an even longer time for a strike-second situation. This was so, of course, because the United States had the capability to destroy Soviet strategic forces and/or reduce a Soviet counterstrike to manageable proportions. Since the mid-1960s, Soviet nuclear capabilities have grown dramatically in all dimensions. Now, and for the foreseeable future, it is quite likely that the United States no longer has the capability to reduce a Soviet counterstrike to reasonably tolerable levels. It was early perception of this developing reality that led the Kennedy Administration in the United States to abandon massive retaliation and embrace the flexible response doctrine in its stead.

With these developments unfolding, the United States became deeply embroiled in the Vietnam War. Unwilling to mobilize and pay the attendant political consequences, we expanded our force structure by creating new units and, in the case of the US Army, used the entire force, to include the force deployed in Europe, as a rotation base for Vietnam. Thus it was that, for nearly 10 years, in the late 1960s and into the 1970s, the capabilities of US units deployed to NATO were at an all-time low. Our Alliance partners, whether they realized it or not, simply stood ground for us under the shadow of US and their own nuclear umbrellas, all of which were quickly growing more and more tattered.

It therefore seemed to those responsible for putting the US Army back together after Vietnam that the time had come for serious review of operational concepts and of the force structure, equipment, and training necessary for their implementation, especially against what remained our most contentious threat—the Soviet/Pact confrontation in Central Europe.

While we were gearing up to do that, the Egyptians and the Syrians attacked Israel in October of 1973. Soon laid out for all to see were the lessons of the Yom Kippur War:

- The staggering density of the battlefield at critical points.
- The presence on both sides of large numbers of modern weapons systems.
- The presence in the air over the battle of large numbers of sophisticated air and air defense weapon system.
- The increased criticality of command-control.
- The increased likelihood of interrupted command-control due to large numbers of sophisticated electronic warfare means.
- The inability of any single weapon system to prevail—reaffirming the essentiality of all-arms combat.
- The outcome of battle reflecting, more often than not, factors other than numbers.

In addition, it was starkly obvious that large-scale destruction in a short time was a most likely outcome of first battles in modern war. Time to mobilize behind ocean barriers and deploy to relieve beleaguered forces—both characteristics of our World War I and II experience—were likely at risk as well. This final challenge to the mobilization system spelled for many an urgent need to rethink mass conscript armies, nations in arms, mass industrial mobilization, and battles of military and national annihilation as relevant concepts for the use of military force as instruments of national policy in our time.

Several critical problems confronted those of us who set about in 1973 to put poor Humpty together again. First was the internal trauma in the Army itself—loss of morale, motivation, and esprit that grew out of Vietnam. Second was the nuclear situation just described. Third was the growing qualitative and quantitative disparity between our own and Soviet conventional forces—in their favor. Last was the nature of the modern battlefield characterized so dramatically by the 1973 Yom Kippur War.

We attacked the numbers problem first, for it was and is paramount. One can fight and win outnumbered. However numbers, it turns out, are important. But within a reasonable range of force ratios, say 1 attacking 6 to 6 attacking 1, history instructs that, more often than not, the outcome of battle turns on factors other than the force ratios extant at battle's onset. What are those factors? Primary among them, in the history of nearly a thousand battles, is that the winner was, in every case, the side that seized and held the initiative, whether attacking or defending, whether outnumbered or outnumbering, at the outset. Seizing and holding the initiative means maneuver; the ability to maneuver means that force ratios between the forces in contact must be held to reasonable levels in order for one side or the other to take the initiative and maneuver. So we concluded that, with nuclear strategy tenuous at best, it was necessary to contend with Soviet-style operations concepts—mass, momentum, and continuous land combat, all in a potential chemical-nuclear environment, by dealing with two important matters at one time. First is the fight against the forward assault echelons. Second is the attack against follow-on echelons in order to disrupt, delay, and destroy sufficient of their velocity and mass to prevent them from building the momentum necessary to success in Soviet-style operations. If the forward echelons can fight successfully, if the follow-on echelons and the OMGs can be prevented from loading up on the front-end battle, then the commander fighting the forward battle has half a chance to seize the initiative, maneuver to the attack, and win.

At the front part of the battle then there is a need for a stout defense, a mobile defense, but not of the style brought to mind by that phrase.

Let's call that the active defense. It is designed to defend in depth against leading echelons of a Soviet-style attack. It invokes the gods of technology in quite a different way—in this case, the vast array of antitank guided missiles being deployed in the early 1970s—TOW, HOT, Milan, Sagger. The active defense is sited well forward, using deep belts of ATGMs. Battalions take up battle positions well forward, but organize other battle positions in depth to which they or other battalions move and take up the fight. Reserves are primarily uncommitted or least committed forces. Extensive use of terrain to reinforce the defense was emphasized. In V Corps when I commanded, there were identified nearly 150 battle positions, with more than 25 occupied at any time, but all prepared and planned carefully to make maximum use of terrain and the long-range capabilities of the newly deployed ATGM. Brigades fought in battle areas of many battle positions, divisions in battle areas of several brigade areas. There is a covering force whose mission is to fight, not to fire and move back. Reserves are small until mobilization can provide reserve divisions from the US.

Active defense, in sum, prepares to fight fiercely and, from the first, well forward, believing that we must survive and win the first battle, else there will likely be no chance to try again.

Active defense circa 1976 essentially deals with enemy assault echelons. No sooner had we the concepts in hand than it became apparent that our concept was not complete. So we came to understand that, as the active defense begins against first-echelon forces, a deeper campaign to delay, disrupt, and destroy follow-on forces and OMGs must be brought to bear. Reconnaissance, surveillance, and target acquisition means are directed so as to detect, follow, and target follow-on forces attempting to close on the forward battle. In Central Europe, as an example, allied corps commanders should, as a general rule, be cognizant of, and following, enemy forces that are within about 72 hours of being able to intrude into the corps-level battle. On Central European terrain, that time equates to as much as 150 kilometers. Time is, however, the important feature, for of prime concern is the time of closure, the time within which the enemy is able to intrude himself into the corps battle, requiring the marshaling of forces and fires or posing a threat to corps operations designed to seize the initiative through maneuver.

The missions at that depth are to delay, disrupt, and destroy; attacks are made against chokepoints, command control nodes, logistical support, routes of movement, means of deployment, and troop concentrations. It is essentially an air and long-range missile battle. As follow-on echelons close to within about 24 hours of the forward battle, perhaps as much as 75 to 100 kilometers away depending on the terrain, the divisional commanders begin to follow their progress and to attack as appropriate. Here the mission is disrupt, delay, and destroy as possible; real-time or near-real-time target acquisition is most desirable; the attacking force is beginning to have fewer maneuver alternatives. The battle is an AirLand Battle with close air support, battlefield air interdiction, and ground maneuver forces, as appropriate, being applied against the follow-on forces.

It is against this 72- to 24-hour deep enemy battle array that theater nuclear weapons should be used, if indeed they are to be used at all. For, with the delay times inherent in the decisionmaking process, it must be clearly established at the outset at what point in the battle it is necessary, prudent, and possible to use theater nuclear weapons.

Finally, follow-on forces closing to within 12 hours of the forward battle are the responsibility of forward-deployed brigade commanders. The mission is to destroy and disrupt those forces, defeat the echelon in contact, and go in against follow-on echelons as appropriate to destroy or render the force ineffective before the next follow-on echelon can appear. It is without question a land forces battle with a full spectrum of all-arms capabilities brought to bear to defeat the enemy as quickly as possible. It is the central and critical duel of the AirLand Battle.

That is the essence of AirLand Battle. Its precepts are severalfold:

- Deep attack and the close-in battle are inseparable.
- Seizing and holding the initiative through maneuver of forces and fires is essential to success.
- The objective of the battle is to win, not just to avert defeat.
- Successfully conducted, the AirLand Battle can, with conventional means alone, considerably postpone the time at which the defender must consider the first use of nuclear weapons, thus raising dramatically the nuclear threshold.

In a very broad sense, AirLand Battle is a grand offensive defense, one on the style fought so successfully by Washington in the Revolution, with great skill and initial success by Lee and Jackson in the Civil War, with striking success by MacArthur and Patton in World War II. That is why, realize it or not, the concept is dramatically at odds with our historic perceptions of what is necessary to win, drawn from our Napoleonic heritage and our obsession with the mass concepts of the Industrial Revolution.

It is likely fair to say that active defense (circa 1976) was perhaps more of a defensive battle of attrition and that in AirLand Battle we sought to move the remaining steps to a full-blown offensive defense. AirLand Battle then is the basic generic doctrine for US forces: one finds it embedded in theater war plans for Europe, Korea, and elsewhere in the world where we believe it necessary to fight modern wars outnumbered and win. However, the need for evolution in doctrine continues. Several recent developments in threat and technology make that even more of an imperative than when AirLand Battle was first postulated seven years ago. Those developments are primarily in new technology and an upgraded threat, perhaps most dramatically portrayed in the armor-antiarmor equation. Their story goes like this.

First, the threat: since about 1966, when the Soviets believed they had achieved so-called nuclear parity, they have embraced the notion that they could and should try to win at theater level, avoiding the use of nuclear weapons if at all possible. At first, they filled gaps in their conventional capabilities with nuclear weapons until they could develop the numerical and technical superiority necessary to win quickly at the theater level of war without invoking the nuclear specter in the process. Their conventional weapons developments in the 5-year plans have for 25 years been clearly focused on that goal. Perhaps that focus is clearer in hindsight than it was in the threat projections. They are now completing the most recent conventional development that will make their nonnuclear theater-level winning capability a reality, as they view it.

Over the span of the relevant 5-year plans, we have seen the development and deployment of impressive capabilities in artillery, air defense, radio-electronic combat, combat helicopters, ballistic missiles, fixed-wing airpower, and force structure changes such as OMGs and Spetsnaz. The missing link has been in the armor-antiarmor equation, for with deployment and proliferation of antitank guided missiles in the West in the early 1970s, Soviet ability to break through NATO's antitank defenses without risking substantial loss and possible defeat was uncertain at best, doubtful at worst. So they set about to solve the technical riddle of the shaped charge warheads used on all the world's ATGMs—TOW, HOT, Milan, even their own Sagger. Not only were those warheads roughly twice as effective for armor penetration as counterpart high-velocity kinetic energy penetrators from tank guns, but the long-range accuracy of ATGM systems nearly doubled effective gun engagement ranges and, in so doing, quadrupled the close-in battlespace that would be dominated by well-laid-down defenses built around integrated combinations of artillery, tank guns, and ATGMs.

By the mid-1970s, the Soviets had begun to field tanks in the T64B series featuring their initial response to the problem of how to defend against the ATGM shaped charge warheads. For the T64B and subsequent models—T72, T80, and now what the West calls FST 1 (Future Soviet Tank 1)—contained lighter weight glass composite or ceramic laminate armors that were more than a match for the early model ATGM warheads. While tank fleets with this advanced level protection were growing apace, with production running 3,000 to 4,000 units per year, technology provided the means for even better protection in the form of what is popularly called reactive armor. Reactive armor in the form of appliqué boxes—fixtures about 10 by 12 by 4 inches in size containing specially designed explosive arrays that detonated on impact, destroying the integrity of the incoming ATGM shaped-charge jet—was first fielded by the Israeli Defense Force in the late seventies and early eighties. It went to war in the 1982 Israeli operation for peace in Galilee mounted on 26 IDF M60s and other more lightly protected armored vehicles. Some of these boxes fell into Syrian, and so into Soviet, hands. Soon improved models were available in Israel, and within the last 2 years, similar improved boxes have begun to appear on tank fleets in the Group of Soviet Forces, Germany. In fact, by some estimates, the Soviets have equipped more than half the Soviet tank fleet in GSFG—T64Bs, T72s, and T80s—with reactive boxes and can likely complete the fielding in the next couple of years.

The unhappy fact is that tanks with reactive armor on top of built-in glass composite or ceramic laminate armor will quite likely defeat the warheads of all deployed ATGMs in the world today—TOW, HOT, Milan, even Sagger. While the technology involved is not all that exotic, reactive armor in this form is a very attractive solution to what, for the Soviets, since the early seventies has been a most vexing problem—how to quickly get through the deepening belts of antitank guided missile systems deployed forward in NATO without suffering inordinate losses in Soviet tanks. Here then is a relatively uncomplicated, fairly inexpensive solution, one that can be achieved without building a new, or rebuilding an older, tank fleet. Further, it is a fairly rapid way to complete development of their conventional capability to break through quickly, strike deep to disrupt the integrity of NATO's defenses, and do so quickly enough to preempt a nuclear decision in the West. So now, not only are they in possession of all the capabilities long considered essential for quick victory in Europe without risking a nuclear exchange, but this all comes to pass at a most opportune time. It allows them to offer to the West the opportunity to apparently defuse the theater-level nuclear confrontation by taking out the theater-level nuclear systems that the Soviets have concluded they don't need any more, but that we, given the conventional force imbalance and the technology situation in regard to armor, need even more perhaps than we did before. Thus, we have a technology advance at the tactical level of war that has important implications for theater- and national-level strategies and capabilities as well.

The Soviets obviously believe that they now have the means to execute the operational concept they have embraced for so long—achieve quick, decisive victory at the theater level of war and do so with such thoroughness and dispatch that the nuclear decision in the West is yet in abeyance even as they consolidate their theater-level gains.

Were it not for the strategic situation, the INF Treaty agreement, and the growing clamor not only for additional nuclear arms reductions but for conventional arms drawdowns as well, it might be possible to look on this as just another of those technological flip-flops that occur now and then, one that in due course can and will be overtaken, perhaps reversed. In this case, there's much more to it than that. From the Soviet perspective, eliminating the INF in Europe, coupled with their vastly improved antiarmor capability, significantly improves their capability to execute a long and consistently held operational concept. From our standpoint, we are without a theater-level nuclear deterrent, the close battlespace in our forward defensive areas has been shrunk to its pre-1970s levels, and we are apparently without a carefully thought-out strategy defining the linkages, if any, between the conventional imbalances at theater level and our remaining nuclear capabilities. Therein lies our dilemma and the challenge ahead.

How to sum up this somewhat intense journey through two centuries of military thought in our country? Having either been party to or authored several such assessments in the past 15 or 20 years, for my own part I come back to several fundamental observations.

First, we long have been, and are yet, children of our Napoleonic heritage—some a la Jomini, some derived from Clausewitz, some homegrown—based on our own unique military experiences. This has led to a military system and concepts embracing a mass conscript military force, especially in time of emergency, a force whose primary operational modus has been the destruction of enemy armed forces largely by overwhelming numbers. With few notable exceptions, already cited, our principal military names have been men steeped in that tradition.

Second, the systems and processes of the Industrial Revolution have fit conveniently, indeed reinforced, our convictions about large conscript masses, the nation in arms, and provided the material means for the ultimate battle of annihilation—of unconditional surrender.

Third, the post-World War II development of Soviet conventional military power has created circumstances in which it appears highly unlikely that we will ever again prevail by mass force of arms alone in a confrontation with the Soviets. More recently, this circumstance has extended to other theaters, especially to Northeast Asia and the Middle East.

Fourth, the early attempt to substitute technology, in the form of nuclear weapons, for numbers was likely ill-advised at the outset and certainly foredoomed to bankruptcy once the Soviets had realized nuclear parity/superiority. We ignored this reality far too long.

Fifth, the character of modern armored battle unveiled by the Arab-Israeli Wars clearly suggests that winning early and succeeding battles while containing losses to reasonable levels is essential. Deployment means and times do not exist to support deployments of sufficient rapidity and strength to overcome potential losses in early battles. Particularly is this the case if surprise is a determining element in initial battles.

Sixth, all this means that the mobilization system—in place since the convenient marriage of Napoleon's warfighting concepts and the systems of the Industrial Revolution early in this century, is quite likely an anachronism as we move toward the turn of the century.

Seventh, against the threat of nuclear weapons employment as a backdrop and with growing potential for massive destruction by quantitatively and qualitatively more powerful conventional forces, the use of military force as an instrument of national policy has become less and less attractive and feasible as between nuclear armed or supported foes. Deterrence is the safest goal, for failure of deterrence opens onto a totally uncharted landscape.

Finally, the attempt to move US military thought, strategy, operational concepts, tactics, and techniques from a mass destruction, annihilation, brute force, Industrial Revolution-based set of convictions into a more sophisticated, intellectually demanding, deft, and precise set of operating parameters is yet aborning. It is clearly a necessary development. How well it will fare, history alone may judge.

AirLand Battle Determinants

Letter to Cadet Marty D'Amato
10 April 1998

...

Soviet doctrine for warfare at the operational (theater) level had a great effect on development of AirLand Battle doctrine. We were (1973 and on) faced with a Soviet threat to NATO Europe that had grown in numbers, in quality of fielded equipment, and in operational doctrine during the years when we were away at the Vietnam War.

The question was how to fight outnumbered and win at the operational level of war without serious risk of having to resort to the use of nuclear weapons.

The Soviets had embraced a doctrine of Mass-Momentum-Continuous Land Combat for operational-level warfare. Mass meant sheer numbers; momentum meant setting the numbers in motion, then keeping them in motion in continuous land combat, echelon after echelon, to achieve overwhelming combat power at places where they hoped to achieve victory.

There were four Soviet echelons deployed in the space between the borders of European Russia and the inter-German border. Against the eight NATO corps deployed the Soviets were capable of launching four simultaneous breakthrough attacks, hoping thereby to break the cohesion of the NATO defenses, then defeat NATO forces in detail. Initially they apparently believed they would probably have to resort to the use of nuclear weapons to do this. However, as they grew in strength and modern conventional weapons in the years while we were absorbed by Vietnam, they came to believe they could also succeed with conventional forces alone.

They saw their operational and tactical-level nuclear weapons as a means to victory at the operational level. We, on the other hand, were never able to decouple the use of nuclear weapons at operational and tactical levels without the very real threat of invoking thereby the use of nuclear weapons on intercontinental ballistic missiles to achieve mutual nuclear Armageddon, a really unacceptable circumstance.

Tactically, AirLand Battle sought to take advantage of growing numbers of longer range and increasingly lethal antitank guided missile systems (TOW, HOT, MILAN) to defeat Soviet assault echelons, grasping the initiative in the battle at the forward line of troops. In the 1976 version of FM 100-5 we called this Active Defense. While that doctrine gathered sharp criticism for other reasons, it is also true that the doctrine was basically incomplete, for it ignored the problem of how to cope with follow-on echelons.

While many people contend that AirLand Battle as set forth in the 1982 and subsequent editions of FM 100-5 was a revolutionary change, in my mind at least it was a more robust and complete development of what had gone before, for it added deep attack to the improved ability to take the initiative at the tactical level and so represented a fully developed operational concept.

The primary driver in AirLand Battle development, as with any doctrine, was the threat, in this case the new, improved, more numerous, stronger-than-ever Soviet threat to NATO Europe.

Again, as with any doctrine, technological improvements to existing weapons, and new technology fielded in new weapons—all in the hands of the Soviets and in Soviet client states, especially in the Middle East—demanded action on our part. Available to us were technology advances for our own weapons systems which would enable us to fight and win, outnumbered, at the operational level of war, without invoking the serious threat of having to resort to operational or tactical nukes—and by inference the threat of a nuclear ICBM exchange.

Tactically the technology was antitank guided missiles; operationally it was synthetic aperture radar in what became the JSTARS system, and precision-guided submunitions on longer range and more accurate surface-to-surface missile systems—ATACMS.

Finally, the moribund state of the post-Vietnam US Army demanded we do something drastic to restore confidence in the institution, its soldiers, and leaders.

Doctrine changes as threat and technology (on either or both sides) change. Doctrine changes could also be required by a new or revised national military strategy. However, if that strategy is correctly drawn, it too will begin with evaluations of threat and technology in whatever new environments [we] have on the screen.

At the time, at least, a Carter foreign policy shift was not heralded as such. Whatever the President and his cohorts thought they were doing policy-wise, the stark truth was that we were yet a NATO partner, indeed leader of that alliance, and it was an alliance confronted with more serious problems than ever before. So something like what we were about [in formulating new doctrine] had to be done in any event. To answer your question, if there was a visible [national] policy shift it was—at TRADOC, anyway—"noted."

German and British viewpoints during AirLand Battle development were fully aired and considerably discussed. After all, it was to be a doctrine brought to battle in our common endeavor. They had to be partners, and they were. AirLand Battle is a concept with deep roots in German concepts: the importance of the initiative, of maneuver, of coordination of close and deep battle, of control of battle space, of *Fuhrung im Gefecht*—command in battle. While the staff talks facilitated consensus building, most of the hard thinking came from officers like von Senger (CINCCENT), Wenner (HERRESAMT), Lemm (HERRESAMT), Burandt (DEPINSPEKTUR), and politically from Helmut Kohl. As Minister Praesident of Rhineland Pfalz when I commanded V Corps, he was a full-scale participant in what was then called the Corps Battle, but which would become, several evolutions later, AirLand Battle. As Chancellor he understood better than anyone the need to defend Western Europe in Eastern Europe, even though [it was] politically not feasible to announce that.

On the British side, again the staff talks were aids to consensus building, but the conceptual brilliance of people like General Sir Richard Worsley (I BR CORPS/QMG), General Sir John Stanier (VICECGS), and General Dwin Brammel (CGS), and their help with their political masters, made British acceptance possible.

Possibly the greatest contribution of all came from several thousand miles away. The Israeli influence on AirLand Battle can scarcely be ignored. For it was their open and generous sharing of IDF lessons from the Yom Kippur War against Soviet-style echeloned formations that vividly demonstrated what we had to do. My problem was to find the means and build a consensus in our country.

Defense forward in Europe has been much debated. Most of that debate has been without substance. We had no choice but to defend forward. Take a map, look at the maneuver room on the Eastern front in WWII, at the depth of maneuver with armored forces on both sides. Then compare that with maneuver room west of the inter-German boundary in Cold War times and the dilemma becomes apparent.

Some [argued] that we could not talk about deep attack when that obviously meant attack across the IGB. Not relevant. In large measure deterrence is the uncertainty created in the mind of the enemy when you in effect indicate to him that if he should elect to attack, then the issue will be decided consistent with ground rules of which he has not yet been advised—but on his side of the border.

Recently I met the officer who commanded the Carpathian Front—fourth echelon back. He told me that when he read the article called "Extending the Battlefield" in *Military Review*, long before AirLand Battle became a field manual, he reported to his superiors that if we could do what we said had to be done, then he could not accomplish his mission.

. . .

Sound conceptual thinking is evolutionary. Nine years elapsed between the time I first became seized with the problem of changing Army doctrine and the publication of AirLand Battle in FM 100-5 circa 1982. During that time the concept went through several iterations: "Modern Armor Battle," c. 1973–1974; "The Corps Battle," c. 1976–1977; "The Central Duel," c. 1977; "The Central Battle," c. 1978; [and] finally "Extending the Battlefield," c. 1979–1980. Compare the last with the 1982 edition of FM 100-5 and you'll find they are substantively the same. Many people contributed, as already noted.

For many of those nine years there was nothing but a stack of VuGraph slides. The Exec claims that I gave that speech a thousand times. He may be right. But in the process several things happened. Many people heard it, thought it made sense. Many heard it more than once and it was never the same. The ideas matured and changed, largely reflecting questions or suggestions from one or more audiences. So they came to believe not just that it was a good idea, but that it was their idea in the first place. Armed with those two things, you can change the world, but not overnight or on short notice, [and] not without hard thinking.

Welcome to the Counterinsurgency Century

The Armor and Cavalry Journal
September–October 2008

As this is written, we are but a few steps into the 21st century. It does now appear that it may one day be characterized as the "counterinsurgency century." What might that portend?

The 21st century, even in its infancy, is obviously quite complex; perhaps even far more complex than the worlds of the 19th and 20th centuries, both of which were characterized by warfare, largely between nation states, in conflicts resulting in frightening losses in human resources, as well as other national treasure; indeed the loss of entire nation states, as well as the catastrophic devastation of others—even those said to have "won" the war.

To illustrate the complexity thesis, consider the French experience post-1939–1945, as Japanese forces withdrew and the French attempted to reestablish control over their territorial holdings in what was once called "French Indochina." It was here that the French army was confronted by a considerable and well-developed communist underground who aimed to spread communist governance into Indochina, thus beginning counterinsurgent warfare against the Viet Minh.

French army forces deployed to Indochina were far too few and not adequately equipped to accomplish their assigned mission. Recognition of those inadequacies caused French army commanders on the ground to petition the home government for more units, weapons capabilities, and support to match. Their petitions were largely ignored or outright denied. The best and most relevant histories of this period are set forth in Bernard Fall's books, *Street Without Joy* and *Hell in a Very Small Place*. Both have been extensively read by those attempting to characterize counterinsurgency warfare in Vietnam, as they represent preludes to what took place after the Geneva Accords were signed in 1954 and, at the time, at least token U.S. involvements in Vietnam began.

Surrendering at Dien Bien Phu, the French army leadership considered the rug pulled from beneath them by their political masters who, from the soldiers' viewpoint, had neither tried to understand the situation nor respond to the entreaties of on-site commanders for help. The army literally withdrew into seclusion in army schools and colleges to begin the construct of a relevant counterinsurgency doctrine—at strategic, operational, and tactical levels—in an attempt to determine what they should have done, strategically, operationally, and tactically; what had gone wrong; and how they might have done better.

Over the next few difficult years, they fashioned an operational concept titled *La Guerre Revolutionnaire*, which included concepts for strategy, campaign, and tactical operations. With its new operational concept, the French army went to war once again in a French colonial holding where there was a mounting insurgent movement. It was, however, an involvement quite different from that in Indochina. Algeria had in fact been a French colonial holding; however, it was to most French people part of the homeland—metropolitan France. It was acceptable to give up some colonial involvements, but never the metropole. General Paul Aussaresses, in *The Battle of the Casbah*, provides a striking account of what happened as *La Guerre* doctrine went to counterinsurgency war.

Reprinted with permission from *Armor* Magazine, publisher of *The Armor and Cavalry Journal*.

The campaign ended in 1962 when a French government under General Charles de Gaulle signed an agreement with the National Liberation Front (FLN) granting Algeria independence from France. France thereby gave up a vast colonial holding in North Africa—nearly one million French citizens were forced to abandon their possessions and flee, there was admission to the deaths of nearly 30,000 French citizens, and perhaps as many as half a million Algerians died. Once again, French military leaders considered the rug pulled from beneath them by political masters, the senior of whom was this time one of their own. History had been provided a counterinsurgency situation considerably more complex than had been prepared for, despite the fact that French military doctrine in support of national goals had been drawn from the French army's own bitter experience in Indochina.

It is not at all difficult to transfer from the French experience in Indochina to that of U.S. forces in Vietnam. Once the November 1968 U.S. elections made clear that there would be a Republican in the White House in 1969, it was also clear that there would soon be a move made to redeploy U.S. forces from Vietnam. Further, it was anticipated in Saigon that by some official means redeployment would be ordered soon after the 1969 installation of the new government. This particular directive arrived in the form of National Security Study Memorandum 36 in April 1969.

The commander, U.S. Military Assistance Command, Vietnam (COMUSMACV), General Creighton Abrams, had already assembled a very small group of officers and enlisted and begun planning for the inevitable. The redeployment was called "Vietnamization." There were public pronouncements that U.S. forces would turn over conduct of the war to the Republic of Vietnam Armed Forces (RVNAF). Further, it was announced that funds would be made available to provide RVNAF with capabilities which were insufficiently robust in its existing forces to support its force structure. In the main that meant fire support means—artillery and air, and logistics support of all kinds. Funds were appropriated by the U.S. Congress, earmarked for that support. General Abrams' instructions were quite clear, "We have been directed to do this. There is considerable pressure from Washington to just cut and run. We must therefore very carefully examine the situation—the enemy's and our own, and propose redeployments that do not jeopardize the Vietnamese army's ability to continue successful combat operations against regular NVA forces attempting to infiltrate into South Vietnam, and infiltrations to support the remaining Viet Cong infrastructure in the south."

The first redeployment increment of 25,000 troops departed Vietnam in the summer of 1969. Subsequent increments for redeployment were planned beginning in late 1969, all pursuant to General Abrams' guidance. However, two significant obstacles were thrown into the works by directives from Washington. First, General Abrams and his planners had developed a plan to redeploy by unit rather than by individual.

Despite brisk exchanges of traffic on the matter, General Westmoreland, U.S. Army Chief of Staff, overrode General Abrams and redeployment was to be done by individual. General Westmoreland's decision meant that, once redeployment began, there would be a constant readjustment in Vietnam to fill the ranks of units, still in country and fighting, and replace the long-tenure people in those units who had been redeployed as individuals. The inevitable result was an on-station Army in Vietnam considerably less combat ready than it had been and needed to be. Secondly, as redeployment progressed, the U.S. Congress reneged and withdrew appropriations programmed to provide adequate fire support, transportation, and logistics support to the RVNAF once U.S forces were redeployed. Many military members, and others, serving in Vietnam when this happened were, and remain, convinced that had the United States lived up to its commitment, the RVNAF could quite likely have won the fight against the NVA intrusion from the north. It was that close. A better description is to be found in Lewis Sorley's excellent book about General Abrams, *A Better War*.

One recurring conclusion from the examples cited above, along with many others, is that military forces can perhaps no longer cope with more than part of war. Many counterinsurgeny requirements stem from political, social, demographic, religious, and other situations not directly resolvable by military operations. At the outset, then, there should be serious consideration of precisely what is being attempted, what capabilities are required (what are we trying to do), and how might the tota l capabilities of the nation be assembled to achieve whatever desired outcome has been decided on. However, if one then looks to departments of a federal government for help and finds employees who refuse to serve in an expeditionary environment, then what?

General Eisenhower, during his time as President, created an undertaking titled *Project Solarium*. It was an attempt to focus the U.S. Government executive branch's resources on a select agenda of likely situations with which the President could be confronted and postulate coordinated solutions to those situations. If, however, it is considered that the military forces of the nation are the only resource available for deployment—in a counterinsurgency or any other situation demanding action on the part of the United States—then there must be a defining statement in the National Security Strategy that stipulates this fact. It is only out of defining statements that force structure, manpower, and equipment capability requirements statements, prescribing the size and shape and equipping of the nation's armed forces, can materialize.

The examples cited above also represent involvement of officials in national political infrastructures in the conduct of military operations in the field, which those political entities had directed be undertaken at the outset. Some who have suffered the effects of those intrusions would call it "meddling." And so it is; unfortunately, it may continue to be. Indeed, the increasing complexity of counterinsurgency operations quite likely invites that type of intervention. In the United States, the tendency to attempt to direct operations of a deployed military force in the field from Washington offices has been a serious problem since the Spanish-American War. The problem has been aggravated by the growing ability to almost instantly move information, in considerable volume, from places far distant from one another to far more people than truly have a "need to know."

Advances in information technology have created an information glut that defies description, as well as inhibits intelligent decisions based on analysis of available information. There is more information available than can be digested in a reasonable amount of time, enabling a decision that is relevant to the situation. In other words, there is not time to sort out and think about what all that information conveys. Further, the media—print as well as video—now has a parallel information glut to that in "official" channels. There is "investigative reporting" by people who are neither qualified "investigators" nor good reporters.

A hand goes up in the back of the room! "Is the peacekeeping function considered a mission for counterinsurgency forces? If so, is doctrine for such operations to be found in an appropriate field manual, or elsewhere?"

Several fairly recent events prompt such a question. Most dramatic, although now a matter of tragic but nearly "ancient" history, is the United Nations Assistance Mission that deployed to

Rwanda in 1993 and 1994 to referee the confrontation between the Tutsi and Hutu. The force commander was Canadian forces Lieutenant General Romeo Dallaire, a brilliant, brave, and concerned soldier with an impossible mission. In a long overdue book, *Shake Hands With the Devil*, General Dallaire recounts his experiences, his reports to United Nations Secretary General Kofi Annan, his requests for more forces (all denied), the tragic deaths of 15 of his soldiers (4 officers and 11 enlisted), and the tragic deaths of nearly 800,000 natives in the massacre that ensued. The U.N. failed; humanity failed.

As U.S. forces concluded redeployment from Vietnam, the obvious question became, "What to get ready for next?" Several considerations made answering the question much more difficult than necessary. First was the early decision not to mobilize Reserve Component units for Vietnam. Army Chief of Staff General Harold K. Johnson frequently recounted that he had gone to the White House seeking Presidential approval to mobilize, only to be rebuffed by President Lyndon B. Johnson some five times on the basis that mobilization would threaten the LBJ Great Society program and therefore it was not an acceptable course of action.

The Army then simply created three new divisional structures, then filled them with a combination of draftees and cadre from existing units. Absent mobilization, the authorized end-strength was then considered inadequate to support a one-year tour for those deployed to Vietnam. So the entire Army, CONUS-based units as well as those located in Europe, Korea, and elsewhere, became the rotation base for Vietnam. This resulted in unit turbulence rates well beyond any threshold necessary to achieve and sustain readiness.

Especially hard hit was the noncommissioned officer (NCO) corps—NCOs stationed in Europe could leave families there, deploy to Vietnam, and return after a year, only to find themselves back in Vietnam again in about 18 months. On an average, this occurred three times, then the NCO would retire, divorce, or both. Most unit NCO academies shut down for lack of students as well as cadre. Morale was rock bottom; military jails were full to overflowing; and equipment readiness rates were seldom above the 50-percent level due to lack of parts, mechanics, and trained crews. Units deployed to NATO Europe did not believe themselves capable of successfully defending against an attack by the Group of Soviet Forces, Germany (GSFG), let alone capable of "winning" against such an attack.

On the other side of the inter-German border, it was apparent that the Soviets understood what was happening in U.S. Army Europe and elected to take advantage of the situation. In the roughly 10 years that we concentrated almost solely on Vietnam, GSFG fielded new operational-level doctrine. The new doctrine, "mass, momentum, and continuous land combat," featured reorganization of heavy units, fielding of two and a half generations of new tanks, seven new field artillery systems (six of them nuclear capable), other technically improved equipment, and shorter timelines for follow-on echelons to move forward to reinforce the first echelon fight. It was a new force; it obviously cost them dearly. GSFG exercise data revealed that they intended to concentrate on the northernmost three of NATO's deployed corps. Two of those corps were not deployed; one was only partially deployed. It appeared that they hoped to bring down those corps before the 16 NATO nations could reach a nuclear decision, and do so with conventional weapons. But if NATO did give a "yes" to nuclear employment, GSFG was ready to go nuclear at the tactical and operational levels of war. It was quite clear that the threat from GSFG was much more urgent than anyone could remember, making resuscitation of U.S. forces, especially Army forces in Europe, a first-order requirement.

On the other side of the coin was the U.S. Army's traditional practice after every war of getting ready to fight it over again, only better. This line of reasoning led to a need to determine what we had learned in Vietnam and develop revised doctrine, new force structure and manpower requirements, and new equipment requirements, all for fighting the counterinsurgency war, as well as the war against NVA regulars like those we had just left behind in Vietnam.

One of General Abrams' first challenges as Chief of Staff, having redeployed from Vietnam early in 1972 and been confirmed as Army Chief of Staff later that year, was to resolve the issue of "back to Europe first" versus the pressing need for counterinsurgency doctrine. The best advice was, while we did know a lot about counterinsurgency, we had not yet digested what we knew to the point from which we were ready to write doctrine and spell out equipment requirements, organizations, and related requirements; hence, the decision to fix the U.S. Army in Europe first. Reflecting that decision, the Army returned to its pre-Vietnam 16-division structure, but with a manpower base more than 200,000 smaller than the pre-Vietnam 16-division Army. Manpower of course is money, and the best advice seemed to be to take what could be had and ask for more as time and circumstances allowed. So it is that the 2008 Army does need greater end-strength and that need is a holdover from the post-Vietnam decision to return to 16 divisions, but without trying to settle the end-strength problem at the same time. Relative to that was the decision not to seek renewal of the draft law, which expired the end of July 1973. We knew we would be short end-strength, but we had no experience as to how many volunteers we could recruit. Today's Army lives in the shadow of those long-ago decisions.

In this special issue of *Armor*, the reader will find a potpourri of writings relevant to the question of counterinsurgency doctrine—tactics, techniques, and procedures; equipment; and force structure—all considered relevant to operations in an insurgency environment. The U.S. Army and U.S. Marine Corps Counterinsurgency Center at Fort Leavenworth, Kansas, has set forth a new doctrinal statement on the subject. This new Counterinsurgency Center product is, of course, not the first attempt at providing relevant counterinsurgency doctrine. Indeed, in his Preface to this issue of *Armor*, Colonel Dan Roper, Director, USA/USMC Counterinsurgency Center, notes the emphasis on tactical doctrine rather than strategic- and/or operational-level doctrine, which was the focus of earlier counterinsurgency writings.

In that regard, it is necessary to remember that, as the Army redeployed from Vietnam, while there were many problems, two demanded immediate resolution. One was the rather dismal condition of U.S. Army units deployed and on station in NATO Europe, as described earlier. Second was the advent of a volunteer Army reflecting the decision not to seek extension of the draft law, which expired in July 1973. Given the decision to reconstitute a credible U.S. Army in NATO Europe, that requirement became the focus of doctrine, equipment, force structure, organization development, and fielding for nearly 17 years from 1973 to 1990.

For the Army that went to war during Operation Desert Shield/Storm, and performed so very well, was the product of two doctrinal evolutions that characterized those busy years—Active Defense (c. 1976) and AirLand Battle (c. 1982). Desert Shield cum Desert Storm were together the field test of all elements of that doctrinal evolution. And while not all of it worked precisely as its authors had intended, whatever shortcomings there may have been were overcome by the synergy of sound tactics, well-trained soldiers, and well-led units. As a general rule, really good work is not done overnight.

Finally, some relevant observations about mechanized (armored) forces in counterinsurgency operations are appropriate; Vietnam, for example, where both French and U.S. forces employed a varied assortment of armor(ed) equipment and units. The story commences with armor in Vietnam in the years immediately following the 1939–1945 war. The French, attempting to reestablish their pre-war colonial hold in French Indochina from 1945 to 1954, when French forces surrendered at Dien Bien Phu, experienced a generally unsatisfactory experience with mechanized forces, all equipped with 1939–1945 war vintage equipment.

Observing the French experience, U.S. Army planners in Washington were convinced that armored forces could not operate successfully in Vietnam. There was considerable misunderstanding concerning monsoon climate, jungle, mountains, rice paddies, weather and the Mekong Delta, not to mention the enemy in all those venues. As a result, when U.S. forces, primarily infantry, deployed to Vietnam in the early 1960s, infantry units deployed without their organic tank or armored cavalry battalions or squadrons; once there, they realized that they needed their mechanized components and sent back to have them deployed after the fact.

At the same time, however, considerable investment was underway to create an armored command for the RVNAF, including necessary equipment and a cadre of U.S. advisors. On balance, it was a quite successful effort. Forthcoming from the Naval Institute Press is a scheduled publication of a full-up history of the RVNAF armor command, titled *Steel and Blood.* Written by Colonel Ha Mai Viet, a distinguished member of that command, it is a well-written, authoritative account of RVNAF armor command operations against insurgents, as well as regular NVA forces.

However, it was not until 1967 that the report of the Mechanized and Armor Combat Operations, Vietnam (MACOV) study group, led by Major General Arthur L. West Jr., chartered by General Abrams, then the serving Vice Chief of Staff of the Army, reported after several months of in-theater evaluation that armor units were very effective in a counterinsurgency environment. Further, that the most cost-effective force in the field during all kinds of operations in Vietnam was armored cavalry, best represented by the 11th Armored Cavalry Regiment (Blackhorse). Thus, after eight years of fighting over terrain considered impassable to tanks and other armored vehicles, where climate and weather were said to severely inhibit armored vehicle movement, where fighting an elusive enemy whose tactics put armored forces at considerable disadvantage, the mechanized force—especially armored cavalry—stood front and center not only in close combat, but in pacification and security as well. In 1969, that evidence led General Abrams' redeployment plan to hold off redeployment of armor and mechanized units until the very last.

The remnants of war most often leave behind invaluable lessons to be deciphered and applied in an effort not to repeat the same mistakes. In the case of the aforementioned examples, two undeniable lessons were at least taught: in all categories of operations required of U.S. forces in Vietnam, armored units represented, more than any other force and by wide measure, more firepower and mobility for the least manpower exposure; and especially evident in the Cambodian incursion of 1970, when NVA regular units faced U.S. armor units—especially the Blackhorse—the mobility, firepower, and combined arms capability of the attacking armor force inevitably caused NVA commanders to order their troops to break and run. Herein lies the very important question: Were those lessons well learned, or were they not?

Postscript:
1. A salute to Colonel Dan Roper and his cohort at the Counterinsurgency Center for a masterful job of doctrine writing—an undertaking with which I, amongst many others, have extensive and frustrating experience. Well Done!!!
2. A salute to Armor for undertaking the imposing task of assembling the ongoing reasoning about this important subject.
3. A proud salute to Soldiers, Marines, Sailors, and Airmen of our magnificent Armed Forces.

General Donn A. Starry

US Army Photo

Photo Collection

Starry with his father, Lieutenant Don A. Starry, Kansas National Guard, during a summer camp at Fort Riley in 1933. *Starry Family Collection*

An accomplished horseman from an early age, Starry is shown here at the Fort Riley National Guard Camp where he accompanied his father during several summers. *Starry Family Collection*

Private Starry was inducted into the US Army at Fort Leavenworth on 13 August 1943 and assigned to Company D, 3319th Service Unit (USMA Preparatory), at Lafayette College. *Starry Family Collection*

Cadet Starry at West Point early in his plebe year, September 1944. He graduated in June 1948 as part of the first class to return to the four-year curriculum following World War II. *Starry Family Collection*

While a cadet Starry, appropriately nicknamed "Fish," coached the plebe swimming team to two undefeated seasons. *Starry Family Collection*

Newly commissioned Lieutenant Starry with fiancée Leatrice (Letty) Gibbs at West Point on graduation day, 8 June 1948. A week later they were married in Kansas City. *Starry Family Collection*

Lieutenant Starry, 3d Platoon leader in Company C, 63d Heavy Tank Battalion, in West Germany, with his sergeants. *Starry Family Collection*

Willard Lucas, Starry's first platoon sergeant, a great soldier cited by Starry ever after as an exemplar of dedicated service and a wonderful trainer of his platoon leader. *Starry Family Collection*

Lieutenant Starry as a platoon leader checking communications near Nuremberg, West Germany, in 1951. *Starry Family Collection*

Starry (back row right) with fellow officers in the 63d Heavy Tank Battalion, 1st Infantry Division, in Mannheim, West Germany, in 1950. Next to Starry is Lieutenant George Patton. *Starry Family Collection*

Lieutenant Colonels Starry and Al Haig near Lai Khe in 1966. Starry was escorting General Creighton Abrams, then Army Vice Chief of Staff, on a tour of units in Vietnam, including the 1st Infantry Division where Haig was serving. *Starry Family Collection*

On a second tour of duty in Vietnam in 1969, Starry served as close-hold planner for reployment of successive increments of US forces, reporting directly to General Abrams. *Starry Family Collection*

COMUSMACV General Creighton Abrams passes the 11th Armored Cavalry colors to Colonel Starry as Starry assumes command of the regiment at Quan Loi, Republic of Vietnam, 7 December 1969. *Starry Family Collection*

Starry, with General Creighton Abrams, celebrates his assumption of command as the 41st Colonel of the 11th Armored Cavalry (Blackhorse) Regiment, Quan Loi, Republic of Vietnam, 7 December 1969. *Starry Family Collection*

Starry, while commanding the 11th Armored Cavalry, with his Regimental Command Sergeant Major, Don Horn, south of Quan Loi, Republic of Vietnam, in February 1970. *Starry Family Collection*

During the Vietnam War Starry commanded the 11th Armored Cavalry in the Cambodian incursion. Early on the first day his vehicle threw a track, so he (center, with cigar) caught a ride with some troops who were going his way, 1 May 1970. *Starry Family Collection*

While commanding the US Army Armor Center at Fort Knox, Starry visited the French Armor School at Saumur and training areas in the south of France, where in 1974 he drove and fired French tanks and other combat vehicles. *Starry Family Collection*

While commanding Fort Knox Starry gained some notoriety with this controversial but effective recruiting poster (in which the soldier pictured is the father of the young lady). *Starry Family Collection*

8. Force Structure

Page

Redesign of the Light Division 460
Force Structure 461
Force Structure Strategies 466
TRADOC Regimental System Concept 467
Light Forces 468
Balanced Forces 469
Force Structure Issues 470

Redesign of the Light Division

Message to Lieutenant General Glenn Otis Deputy Chief of Staff for Operations 15 January 1980

Have your 141340: Redesign of the Light Division. Bill Richardson and I, and a cast of thousands, have just finished a review of work done to date on the light division. It's essentially a disaster. We are falling back behind the nearest terrain feature to regroup. In a week or two we will have some concepts on the basis of which we might draw up more definitive guidance to the workers. Meantime the task forces have taken an enforced time out. Fred was here and party to this—he will debrief you. We have agreed with him that we will support and provide what input we can to his work for the Chief on the accelerated schedule. I still plan to meet my commitment to Shy to present something on or about 1 April in order to get into the POM window, but we've a lot of work to do, and we've made a false start. I've told the Chief this. Let's just link arms and press on.

Force Structure

Air University Airpower Symposium
Maxwell Air Force Base, Alabama
5 March 1980

Frankly, any Army officer is a little reluctant to get up before an Air Force audience to give a discourse in force structure. We're having enough problems of our own trying to describe the US Army in the 1980s without giving advice to our sister services.

It appears, however, that, in the AirLand Battle we all foresee, a description of how the Army plans to fight is necessary for your deliberations. Obviously, if we are serious about an AirLand Battle, and I believe we are, then each of us has to be aware of how the other plans to fight. The Army can't win it alone and the Air Force can't win it alone; that's no secret.

In the Army we insist that all the generic aspects of the battle—tactics, equipment, training, and force structure—have a basis in a concept. So let me quickly review the operational concepts of our potential enemy and our own so we have a common ground to examine the force structure contributions to the AirLand Battle.

Our potential enemy, and his surrogate forces, depend on three clear operational notions—mass, momentum, and continuous land combat. These, in turn, are translated operationally into numbers, concentration, speed, depth, and echelonment of forces, at night, using smoke, and in bad weather.

When we studied these concepts, we determined that the basic threat strategy is maneuver-based force disruption. Now there's a lot of firepower and mass connected with this, but it is all intended to assist the maneuver of large forces. These are designed to disrupt the defender before he gets set.

In their practice of these concepts when they are on the offense, they require that the balance of force ratios be in their favor. This leads to some force ratios that they set down some time ago—in the 1930s. For units in a movement to contact, they are three to one; for a concentrated breakthrough they are six to one.

Now our problem is how to counteract or take advantage of these operational concepts and so win the battle. The basic strategy of the US Army is maneuver and firepower-based force destruction and disruption. Maneuver-based because that's what conserves forces and disrupts the enemy maneuver. Firepower-based because, with the numbers that the enemy forces have, we must destroy some of their forces to allow our maneuver. I don't mean an attrition battle, but I do mean we must even the odds somewhat.

Translating that operational concept onto the battlefield, it means that we must *see deep*—into the enemy second echelons—and establish a picture of what the enemy is doing. We must *move fast* to concentrate firepower to oppose his maneuver and position maneuver forces to disrupt his operational scheme. Those firepower and maneuver forces must *strike quickly* to destroy the first echelon and finish rapidly to get set for the second echelon. All the time we must use terrain to the best advantage.

When you lay this out on the ground, whether in offense or defense, some key concepts become apparent. First, we believe that brigades must fight the first-echelon regiments; divisions must fight the first-echelon divisions, including their second-echelon regiments and see the second-echelon divisions; corps must fight the first-echelon army, including their second-echelon divisions, and must see the second-echelon army.

So that this can be more meaningful, we noted that the brigade area of interest is out to 12 hours, the division to 24 hours, the corps to 72 hours. That says that forces that can threaten each of those levels within that timeframe are what's important. The terrain dictates the extent of that timeline, but it's the time-space interval that's important, not distance.

All that's easily said, but how do you handle the enemy echelonment of forces? Let's look at a typical enemy combined arms army attacking. The echelonment is quite clear. The army deploys to a depth of about 72 hours—about 120 kilometers. The first-echelon regiments are from 0 to10 hours away; the second-echelon regiments from 10 to 15 hours; second-echelon divisions 16 to 48 hours; and the elements of the second-echelon armies about 72 hours.

The way this works is that an element's first echelon, no matter what the level—regiment, division, army—attains that element's immediate objective. The element's second echelon attains the subsequent objective. Once it starts to roll, this can be very awesome.

When you look at this second echelon from an analytical standpoint, a certain calculus or formula becomes apparent. Momentum is the equivalent of mass times velocity. The problem then becomes how we can alter that formula. The answer is that mass can be affected by degradation or destruction and velocity by disruption or delay. The questions are what we attack, when and where we do it, and how.

Initially there is no problem, even when the follow-on battalions close. But when the second-echelon regiments come in, the enemy overwhelms the friendly target-servicing capability. That succinctly is why we need to—must—attack the second echelon, and do it deep before it closes.

If you examine the second echelons in terms of potential engagement payoff, close to the FEBA [forward edge of the battle area] destroying combat systems—tanks, BMPs, artillery—has a high payoff. But as you move back—into the second echelon—combat support and combat service support [CSS] have a much better engagement payoff.

Why? No mystery. Tanks near the FEBA are shooting; they are an immediate threat, but away from it they are only a potential threat. But CSS is much more lucrative because, farther back, there's more of it, it's more vulnerable, and its loss to the enemy is much more disruptive.

When you look at the enemy CSS requirements for an army or division for one day, you find those needs are pretty demanding, even to just move. And those requirements must be transported, so there are plenty of targets, all very critical. So it's not just a simple case of finding and killing tanks. There are quantities of targets—command posts, radar installations, logistics vehicles, ammunition and fuel depots—that are lucrative and hard to replace. We need all the help we can get fighting tanks, BMPs, and artillery, but kill off the fuel tankers and a whole lot of tanks won't move. Knock out a command post and a brigade—90 tanks—is immobilized.

So far we've talked about the necessity to see and attack deep and to be discriminate in our target selection, especially in the second-echelon battle. There is one more piece of the AirLand Battle that we must think about. The next battlefield will be an integrated one—with nuclear and chemical weapons. We believe this may occur early in the battle, just as our potential enemies have been saying, writing, and training for the past 20 years. They have the capability; they train with it; they even deploy it into fellow Communist countries—on invitation, of course. We must believe their words and actions. The Catch-22 is that, if we don't plan and train for an integrated battlefield, we make it more likely to occur. Our potential enemies, equipped and trained as they are, are watching, and our only protection is in being prepared.

That means we've got to look at the integrated battle a little differently. Our operational concepts have to be used within a plan that embraces all the echelons of the enemy. If you look at that enemy battle array stylistically, and then plan our actions alongside, we fight and destroy the echelons in contact, either by fire or maneuver or both. We suppress and destroy the enemy supporting artillery. Then in the next echelons we shape those units and cause them to commit according to our plan, not theirs. With the deep echelons we start their destruction, disruption, and delay very early, thus preventing their interfering with our tactical operations.

From all this flows a simple truism that the shaping of the battle requires an increase in our delivered firepower to offset numerical superiority. That means integrated fire support—*deep*. That also means it's an air/land problem, because the Air Force can accomplish the deep mission.

Here are some of the implications that flow from the integrated battlefield.

First and foremost, if we're going to fight the second-echelon fight on an integrated battlefield, we've got to be able to find the enemy and find him deep, before he's able to engage our units. We, on the ground, can't do that very well, and we must depend on the Air Force for information. We need intelligence and surveillance systems that will find targets that have a high engagement payoff. That means an ability to discriminate targets and to search out the enemy command, control, and communications, logistics, and other soft targets I mentioned earlier.

That see deep capability has to be in real time, and it must be accurate enough for targeting purposes. We don't need more 8x10, three-day-old glossies—"Army Happy Pix," as the recon guys call them. This system could be sensors, radar, TV, or a combination, but it must be redundant so all our eggs aren't in one basket. On that note, one word of caution. Every time I mention this, everyone says AWACS [Airborne Warning and Control System]. Great, it's a super system, we all need it, but I don't believe it can do everything unless you plan to put wings on the "Super Chief." We need more than one system, and one with a lower level of mission and direction.

As a corollary to seeing deep, we have another problem. We expect to have an awesome amount of information flowing back to the commander, so much that he is in danger of going into overload. General MacArthur, among others, suffered from this in Manila at the time of Pearl Harbor. It was costly. We can't afford it on the next battlefield. We need a system or an arrangement that will get the critical information needs to the commander, but only the critical ones. We need some sort of preprogrammed discriminating system tied into all our information-gathering devices, both friendly and enemy. Our studies show us that this information engineering, the sorting of information, is a great effectiveness multiplier. TRADOC has a group involved in this, and the US Air Force should be too. The requirement is there.

We need to be able to plan an integrated battle, both air and land. This is an urgent and compelling need. At one time, in the 1950s, we led the world in this. I daresay we may have forgotten more than many have yet learned. We must relearn it quickly.

We need to be able to attack the enemy on the integrated battlefield. We need air/ground forces that can do this. We need more quantity and more versatility in our nuclear and chemical delivery means. It doesn't matter what kind of a suit the person wears who delivers the weapons—blue or green—we need more forces of both. We've let ourselves fall behind in tactical nuclear weapons. If we really need some ideas, I suggest we all go back and read our manuals and professional magazines of the 1950s—*Infantry*, *Armor*, *Artillery*, *Army R&D*, *Air Force Aviation* and *Space Age Weekly*. Many of the ideas on those pages died at birth, overtaken by the events of the 1960s and Vietnam.

When we do attack the enemy, as I mentioned earlier, we've got to be able to get at those high-payoff targets. Now I know that Air Force pilots love to paint tank silhouettes on the sides of their cockpits. It looks neat. But we need some new templates, ones that have the outlines of command posts, ammo dumps, fuel trucks, logistics installations. We need to strip away the combat service support from the tanks and the tanks will stop. It happened to us in World War II in Third Army. It happened to the Japanese in the Pacific in that same war.

We also need to be realistic in what is a lucrative target. If it takes all the artillery of an entire corps to fire to suppress air defenses so a plane can come in and drop two 500-pound iron bombs on a target, we've got a problem. We can't afford that kind of a payoff; we just don't have the forces.

To lash all these forces together and really get a combat multiplier, we need responsive and secure command-control—C2. Frankly, we can't afford, either in the budget or on the battlefield, any more super command-control systems. We need to wire together what we have today in such a way that it works and that it will survive on an integrated battlefield. That means mobile, simple, small systems, something that works with reasonable use of the state of the art and no need to wait for a technological breakthrough.

Another look at what we have on the battlefield now would be in order. The air control centers we send to support the Army corps headquarters are growing by leaps and bounds. All those vehicles and vans are invitingly vulnerable. I know we need the support—desperately, but not at the expense of its own destruction. And if they are damaged or destroyed, what redundancy have we built into the system to allow it to continue to work? If it's computerized, have we considered—heaven forbid—manual backup? The integrated battlefield is going to play havoc with electronic systems.

Finally, in deepening the battlefield we must discipline our planners and our fighters. We can't afford to allow expensive machines to cruise around the battlefield looking for targets. The enemy will make that tactic too expensive. Our force structure—air and land—must be predicated on and used in conjunction with an event-structured, high-target-payoff plan. We can't afford just to buy weapons systems that have caught the public or media fancy. We need to explain and insist on weapons systems and force structure that we need to win. That is not just the job of the generals or the service secretaries, it's the responsibility of every soldier and airman, no matter what grade. I might add that it's the job of every concerned and informed citizen.

We must explain to all our soldiers and airmen what we need, and why, so they can be articulate. We haven't done this well in the past. I suspect that, once we start, we may define our needs even more clearly and the interchange will generate more ideas. There is no one who knows what is important on the runway better than our airmen. Things that don't have a very high payoff only add to their burden. You can't fool them with nice-to-have eyewash, and they're quick to spot things that are organized or designed wrong.

It's no good if the force structure looks good on paper or in the staff. It's got to work for them in a less-than-ideal environment—where it's wet and cold and one of the crew is on detail and another just reported in and a third one never saw this equipment before. It's an environment where the system's manual has pages missing and the stress level is very high because the risk is great and the only law is Murphy's. They don't need sophisticated equipment with sophisticated problems. They don't have time for them and I don't blame them.

Let me end this on one last point. Our force structures—Army and Air Force—must be designed to be complementary and supplementary, not competitive. It's not a roles and missions fight, it's a fight for survival against an enemy who has a significant edge over us. We need everything that each service has to offer the other, and much more of it. *And we need it now*. Time is desperately short. We cannot afford to delay any longer.

Force Structure Strategies

Message to General E. C. Meyer
Army Chief of Staff
15 October 1980

I firmly believe that, when addressing a force structure transition strategy, the best course to follow is to transition to new organizational designs without waiting for new equipment . . . because we derive the earliest benefit of design change, minimize time of doctrinal mix, and speed total force conversion time. There will, however, be strong opposition to that course of action. We cannot modernize incrementally as equipment becomes available. If we do, we'll never get it done. This issue is the one that needs resolution soon. If we resolve nothing else, its resolution will be a significant step forward—or backward, depending on what we decide to do.

TRADOC Regimental System Concept

Message to General E. C. Meyer
Army Chief of Staff
17 February 1981

. . .

2. Following our 22 January presentation of concepts for moving the Army into a regimental system, the Chief asked that we "hang a few regiments on Concept Alpha . . . see if it's feasible."

3. The bottom line is: yes, Alpha, with some important changes, appears to be feasible. It is a natural evolution from where our present Combat Arms Regimental System stopped in 1957 and could be further developed in the future. . . . Alpha will definitely build unit cohesion, family stability, and move our Army toward a more professionally based cadre of soldiers and noncommissioned officers.

4. The first modification made to Alpha was to retain all regimental designations now represented in our force structure. Although Alpha would have limited the number of regiments on our active rolls to 51, we have found that we need not bear the burden that would be caused by a large-scale renumbering of our combat battalions.

5. We have developed a rotation scenario that will allow the Army to routinely deploy combat arms battalions during peacetime and return them to a regimental home base. The ability of the Army to deploy and redeploy soldiers as unit members rather than as individuals will determine the degree of identification and cohesion we are able to build within our units. The two are inseparable. We will also illustrate modifications that can be made to the present replacement system to stabilize soldiers and promote cohesion, albeit with a price: the trade-off is loss of flexibility within the highly disciplined replacement system needed to replenish unavoidable individual losses from deployed battalions. This will require a modified individual replacement system understood and supported by commanders at all levels and of all grades. The other indirect cost is a deemphasis of our present efforts to assure absolute overseas tour equity between individual soldiers. For example, in our proposal some within the same MOS will serve outside the US more often than others. Other major benefits of our regimental-battalion rotation scenario are: experience in routine deployment and exercising the Air Force and Navy lift capability. These are in addition to the increased reenlistments possible by stabilizing soldiers within units and actively promoting family satisfaction.

Light Forces

Message to Major General John B. Blount
Fort Monroe, Virginia
4 June 1981

No certain knowledge about what is meant by lightening up the force. My understanding is that the light infantry guys are having their day at the expense of the mechanization so badly needed. Exactly the wrong way to go. That's trading one scarce resource, equipment, for an even more scarce resource, people. Will tackle this when the time comes. Meantime see if you can improve the quality of our spy network. Call Glenn or talk with him somehow and find out what's going on.

Balanced Forces

Letter to General Henry A. Miley Jr.
American Defense Preparedness Association
13 October 1981

We must continue to try to fulfill our common goal of an efficient, cost effective, well-thought-out national defense. It was particularly gratifying to be asked to speak on the same program as Secretary Marsh. Our views balance each other and, I believe, reflect the essence of a rational military strategy. Ours is a statement of need for light and heavy forces, for a practical balancing of the two, for they are surely complementary, given the scope and diversity of the threat facing us.

A balanced joint combined arms concept is mandatory if we are to respond to situations and contingencies requiring quick deployment of some forces, backed up by heavy forces, or if we must face a Soviet or other heavy force threat. This is our challenge, and I urge that together we explore, define and develop the requirements flowing logically from it. Since we can't afford everything, it's not a question of either/or. It is rather a question of how well we can build flexibility, redundancy, diversity, and multipurpose capability at reasonable and affordable costs.

Force Structure Issues

Conference on Conventional Forces and Advanced Technology
Austin, Texas
30 November–2 December 1989

Infrastructure—that is, the size and shape of the armed services—begins with the interrelationship between Defense Guidance, the Five-Year Defense Plan (FYDP), the service and joint Program Objective Memorandums (POMs), and the budget finally enacted by the Congress.

Infrastructure of the armed forces is generally defined as the combination of force structure—the numbers of divisions, wings, battle groups, and supporting arms and services—and personnel strength—end strength, both military and civilian, that describe both the size and strength of the armed forces—in this case, the Army.

In-place infrastructure reflects the realities of what can be bought for the resources finally made available. Realities of the budget almost never provide enough strength and structure to satisfy requirements set forth in response to Defense Guidance.

The fundamental infrastructure issues, the force structure issues of the day, lie in the imbalances between requirements as set forth in the FYDP and POMs and resources provided by Budget Authority (BA), and also in the structure/strength imbalance in the Army reflected by inadequate end strength to match the force structure. While this is a historic situation, it has been aggravated in recent years by the so-called "Reagan buildup"; it will be further aggravated should the US undertake to move to a smaller armed force as part of a response to force reduction proposals being considered (for Europe) in the CFE negotiations now underway.

While there are many ways of describing the requirements/resource imbroglio, it is useful to consider some gross aggregations of defense budget averages beginning in the early 1950 decade. Using DOD budget estimates for 1989–1990, it is possible to construct averages of defense budgets at FY 1989 economics as follows:

	AV $B BA
1954–1960 Eisenhower slump	202
1961–1965 McNamara buildup	224
1966–1972 Vietnam high	260
1973–1980 post-Vietnam slump	206
1981–1990 Reagan high	292

From this we note that average defense expenditures during the McNamara buildup were a modest 11 percent over the average during the very, very lean Eisenhower slump. From the baseline McNamara buildup, the average Vietnam high budget increase rose but 16 percent, that increase, and more besides, going into operating instead of investment accounts. Then, without either acknowledging or trying to recoup the investment shortages of the Vietnam high, in the post-Vietnam slump average budgets declined by about 21 percent, an amount more than the Vietnam high increase over the McNamara buildup baseline average.

Now consider the Reagan buildup: It was a striking 42 percent over the post-Vietnam slump. However, it was no more than 12 percent over the Vietnam-era high, and 30 percent over the McNamara buildup. It should also be remarked here that the McNamara buildup was, in fact, constrained from the beginning by the notion that nuclear weapons would bear more than a lion's share of any ensuing conflict; therefore conventional force modernization was less urgent a matter than it needed to be and than it might have been otherwise. Since the Vietnam high did not provide modernization, and since the Vietnam slump not only bought no modernization but ate into some of the modest McNamara conventional force modernization as well, we came to the Reagan high years with a long hiatus in modernization behind us. For reasons not at all clear, it was apparently presumed by those planning the Reagan high that Defense Guidance could set forth requirements relatively unconstrained by any backlog in modernization. Thus it was that DOD requirements, reflected in the FYDP, have grown apace, all out of proportion to growth in BA—reflecting the fact that the Weinberger-era Defense Guidance tried, in the main, to solve all problems at once, never acknowledging either the urgent need for discipline in the process or the need for real hard-drawn priorities for modernization. In the end, in the waning Weinberger budgets, the FYDP-to-BA gap came to be on the order of a third of a trillion dollars. Even in the McNamara years, when the FYDP system began, there was an FYDP-to-BA gap. In those long ago times, however, the gap was modest; further, it was considered that, should crisis ensue, mobilization would follow, and with it there would be an outpouring of BA dollars that would quickly act to close the gap.

While hardly anyone foresaw growth of the FYDP-to-BA gap to the dimensions of the "Weinberger gap," it was patently obvious years ago that sooner or later the gap would likely reach a level that taxed the credulity of even the most outspoken defense advocates. And that indeed is what happened.

The problem with which we are now seized has characterized every recent "slump" in defense spending. In lean times, the services tend to do everything—buy things, do things, at levels just a bit less than before. Nothing is eliminated; everything just goes a little short. This "salami slicing" procedure is justified on the basis that there will soon be another "high," and there must be maintained a "hot base" from which to expand. That is precisely what happened in the post-Vietnam slump; the 1981–1990 glut in requirements simply reflects the Babes-in-Toyland reaction to the Reagan cornucopia. Realistically it was, as has been suggested, no cornucopia at all. Had we recognized that fact in context, it might have caused us to discipline ourselves more effectively in the 1981–1990 Defense Guidance/budget process. Given the history of such matters, however, that might have been too much to hope for. It is, however, quite clear that some mechanism must be found to discipline the requirements-setting process. None of the defense reforms set in train by the Goldwater-Nichols Act address themselves to this problem directly; it therefore remains a matter of urgent national priority.

The structure of the Army—the numbers of divisions, separate brigades, and supporting units, while basically determined by the Defense Guidance, is determined in the main by a series of force planning documents that describe how many units of what type there are to be and how many soldiers it takes to field and support the force. As the Vietnam War ended, the Force Planning Guide, Europe, called for divisions with initial and sustaining support increments that totaled, in manpower strength, about 56,000. Already constrained by a variety of factors, budget planning focused on a division force equivalent of about 45,000, with 15,000 each in the division, initial support increment of combat support troops, and sustaining support increment of combat service support troops. Later this number—45,000—would be further reduced, as we shall see.

As the Army redeployed from Vietnam in 1970–1973, there was a temporary hiatus in our ability to describe the objective force structure of the future—how many divisions, separate brigades, and so on. Reflecting this situation, force planners simply dealt daily with disposing of units (mostly flags only) redeploying, and the personnel management system dealt with discharging soldiers from the Vietnam-era Army of 19 divisions and supporting units. The end result of all this was that, at a point in 1973, the Army stood at roughly 11 2/3 divisions, with a strength of about 765,000 soldiers. In an attempt to stem the tide—there was conversation about a 600,000-man Army of 9 divisions in 3 corps, the Chief of Staff, General Creighton Abrams, negotiated with Defense Secretary Schlesinger to stabilize the Army at 16 divisions. This agreed, the question of end strength was raised; the result was the beginning of the Army's present dilemma. Secretary Schlesinger agreed to a minimum number of about the 765,000 level with the proviso that, if the Army could justify more end strength, that justification would be the basis for further negotiation—budget negotiation. By way of perspective, the last 16-division Army—the Army of 1968, had in it 986,000 soldiers. Without claiming that latter number to be precisely correct, the gap between 986,000 and more than 200,000 fewer is ample evidence that the correct answer may well be somewhere in between, but certainly not at the lower level. Thus was created the overstructured and understrength Army—called a "hollow" Army by at least one of its Chiefs of Staff. But it was an Army added to in a subsequent spasm of "light" force additions that saw more structure but not substantially more manpower. And so an already difficult problem was simply aggravated.

At least two fundamental problems characterize an overstructured and understrength Army. One is the tendency to shift more and more of the active force burden into the Reserve Components, particularly to migrate active combat units into the National Guard, to the end that there are not 16 divisions in the active force at all—the number is more like 13, with "Roundout" units of the National Guard designated and equipped to mobilize quickly and round out the fighting units of divisions scheduled for deployment. This practice has the added effect of linking the mobilized and deploying forces to a Presidential decree announcing a state of emergency, for the National Guard cannot be federalized without such a decree. The Congress meantime began to post increasingly stringent limits on how long the President might mobilize forces under a state of emergency, further restricting executive power and flexibility.

The second effect of the overstructure-understrength combination is the increase in turbulence in units brought about by personnel shortages in the first place, aggravated by the need to rebalance in critical positions to keep fighting crews together. Army studies of the 1970s clearly established that a turbulence rate of more than 20 percent—new faces in positions—per quarter led to a situation in which no meaningful training could be accomplished. Presumably, also, that means no meaningful fighting can be done. At the time of the study, nominal turbulence rates were on the order of 15 to 20 percent in most skills—it was a marginally trainable (fightable?) Army. Additional structure simply adds to the demand for critical skills and increases the rate at which people are moved around in an attempt to achieve balance.

All factors considered, an Army numbering about 765,000 can support a structure of about 12 divisions and 6 or 7 separate brigades—not more. And yet we stand today at 18 divisions, most bobtailed somehow, all organized at authorized levels of organization (ALO) below full strength in an attempt to make too few soldiers fit into too much structure. It is the companion issue to the requirements/resources confrontation already described and is in the view of many the single most difficult issue we face today. For, confronted with the likelihood of force reductions in Europe, how now are we to structure the Army for whatever missions might be considered relevant?

9. Intelligence

Page

Intelligence Assessment 476

Intelligence Dissemination 477

Intelligence Assessment

Letter to Major General E. R. Thompson
Assistant Chief of Staff for Intelligence
29 November 1978

Thanks for . . . sending me a copy of the DIA report entitled "An Assessment of Soviet Non-Nuclear Breakthrough Operations." . . . While you are quite correct in suggesting that this evaluation of Soviet tactics might be useful to us in developing or revising our own tactics, it is also true that such information is virtually useless to us in classified form. I know only too well that this argument has been going on since the beginning of TRADOC, and that in the eyes of those who classify things we stepped over the line with our descriptions of how the enemy fights which we published unclassified in 100-5, 71-1, and 71-2. But I still believe, as I did then, that we must put that kind of detailed enemy information in the hands of the soldiers in usable form—unclassified, or we won't get to first base in having our Army understand what it is up against. As a matter of fact, most of the information in the DIA report is provided in unclassified form in a DNA report done for us by Joe Braddock of BDM. It is a super piece, and will appear virtually verbatim in 71-100, *The Corps*, and other pubs to follow. While I applaud the quality of the DIA work and congratulate the author(s), we can't use much of anything that has a SECRET stamp on it.

Intelligence Dissemination

Letter to Major General Albert N. Stubblebine Jr.
US Army Intelligence and Security Command
12 May 1981

We must do more . . . to be effective in intelligence dissemination. We are still using the technology of the 1950s, while planning for intelligence dissemination using the technology of the 1970s. I contrast our efforts in applying technology with those of the Wright Brothers. In their time, they were using the latest technology to produce the first iteration of a succession of very sophisticated machines. The Army today has not yet begun to exploit technology, which is already becoming obsolete, to do something as simple as communicate better. This is not a criticism specifically of the intelligence and communications communities. It is just an observation to point out that, to achieve the end of timely intelligence dissemination, we must take advantage of smaller and cheaper computers, or "word processors" which can be used as computers, of high speed and secure communications capabilities. We must take advantage of these in the near to mid-term to disseminate intelligence by miniature terminals at the users' locations and by teleconferences between intelligence agencies and the users. It is not inconceivable that the Army and the intelligence community could reprogram resources in order to provide a communications satellite system dedicated to the intelligence support of major Army commands and tactical units worldwide by the end of 1983, using off-the-shelf equipment.

10. Leadership

Page

Chief of Staff Criteria 480

Leadership and Branch Bias 482

Command Qualifications 483

Duties of the Command Sergeant Major 484

Sergeants' Business 486

Officer Leadership 496

Leader Skills 497

Command Sergeant Major Qualifications 502

Leadership Instruction 503

Branch Qualification of Command Sergeants Major 504

Leader Development 505

Ethics and Leadership 507

NCO Criticality 508

What's Next 509

Command In Europe 510

Chief of Staff Criteria

Letter to the Honorable Robert F. Froehlke
Secretary of the Army
11 January 1972

This letter responds to yours of 3 January soliciting thoughts on criteria and considerations you should use in arriving at your ultimate recommendation with regard to selection of General Westmoreland's successor.

To begin with may I suggest that, first, and somewhat apart from the main issue, should be consideration of the makeup of the Chief-Vice Chief team. I do not know General Palmer's plans, but I assume that, in view of his long tenure, he will soon be moving on, so it is important to think about this aspect of our leadership now. It is true that the Chief is the visible man; visible to press, Congress, Secretariats, Joint Chiefs, and, most importantly, visible to the Army. Each of these facets of visibility will be important to you in considering qualifications, capabilities, and personalities of candidates. However, there is possibly nothing more important than matching the qualifications, capabilities, and personalities of the two top men of the Army. Let me be specific.

We require at the top one man who moves primarily in the operations channel. He should be the strong presence in the JCS, with the field commanders, and with you and Mr. Laird regarding roles, missions, strategy, requirements, and the more operational Army matters. This is the Chief's role. His teammate, the Vice Chief, must be the manager—capable of making, and empowered to make, the hard detailed decisions about manpower, equipment, force, and resource programs. The Chief should probably devote about half his time to Joint matters and a quarter of his time to purely Army matters of an operational nature; the remainder will no doubt be consumed by the demands of official appearances at functions to which he should lend a presence. The Vice Chief, on the other hand, except when standing in for the Chief in the Joint arena, should devote all his time to the running of the Army—especially should he manage the detailed day to day decisions that have to be made about resources, their procurement, distribution, organization and employment. And there must be a rather explicit understanding between the two men about what is to be done by whom.

So in selecting your Chief of Staff nominee, you should also consider who his Vice Chief will be, even though the change in Vice Chiefs might not occur simultaneously with General Westmoreland's departure. For most of us it is normally not an easy task to be another man's second in command. It goes against the grain. And so it is an enormously difficult task you face in matching people for this team; but at the same time it is utterly vital to the Army that they be matched.

Now for the person of the Chief himself. Of the many criteria I know you must consider I would cite three as most important. They are: credibility, toughness, and empathy.

Whomever you select must be credible—as a professional, as a leader, and as a manager. Professionally he must have qualified himself as a commander in Vietnam, but elsewhere as well; qualified so as to be accepted by the Army he will lead, the Congress he will confront, and the DOD management which he must serve. He need not be a charismatic leader, but the Army must have confidence in his image and ability as a leader. He must enjoy the respect of his contemporaries, the respect and admiration of his juniors—especially among the field grade and junior general officers, and the confidence, respect and admiration of the NCO corps and enlisted ranks. He must be credible as a leader for our day and time. He must be a man who understands how to manage an Army with dwindling resources. And again, especially in the field and junior general officer grades, there must be a feeling that he knows what he is doing and that what he is doing is best for the Army under the circumstances. He must be credible as a manager in the contemporary framework of Defense management in our country.

We need a tough man at the Chief's desk. Especially must he be tough in the areas of discipline for the Army, innovation within the Army, and taking a stand for the Army in the JCS-OSD and Congressional in-fighting. There's just no way to have an Army without discipline. We've spent too much time trying to make ourselves popular, and too little time trying to find for each man a useful task that he can identify with a mission that appears important to him, and that has high standards of performance associated with it. At the soldier level discipline saves men's lives; ultimately, I'm afraid, discipline means national survival. The right kind of discipline must be brought back to the Army. It will require a tough man to do it.

The Army needs to innovate—find new ways of doing old things, ways that are not only more effective, but hopefully more attractive to the soldier. They can be found; it takes imagination to innovate, but it takes a tough man to manage the risks of innovation. But that's what we must have. Finally, the Chief must be a consummate poker player—tough, resilient, persistent, master of bluff and ploy, for these are the tools he needs to do business in the Joint arena and, to a certain extent, with OSD and the Congress. He must be tough *for* the Army at the right time and about the right issues.

Finally, and possibly most important of all, is empathy—the ability to see and understand the other person's point of view. In this case your candidate must have empathy for his young soldiers and officers, his NCOs, his field grade and junior general officers, and for the Army as a whole in terms of how it will understand and judge what he does for it. Empathy in this sense is not relaxing standards of performance so that the training and discipline associated therewith will be less onerous. It is, however, understanding how other ranks see the same problem and intelligently integrating that understanding into decisions. Correlative to empathy in this sense is the ability to communicate goals and programs meaningfully to each of the important groups who make up the Army. Too much of what we have done in the recent past has been misconstrued in the field because we could or did not communicate clearly what it was we were trying to do. One man can't solve that problem, but he can provide the leadership requisite to its solution.

Reading this over, I see that your nominee should be a demi-God, if not a deity with full credentials. Not so. However the standard should be high. You asked that names not be named, so I will simply say that on the 1 December 1971 roster of Army generals there were two four-star and five three-star generals whom I would judge to meet the criteria I have set forth above. And you're really only looking for two, under the matching team concept.

Leadership and Branch Bias

Letter to General Bruce C. Clarke
Palmyra, Virginia
11 December 1975

Since I've been in the Army it has been most difficult for me to understand how we could have a predominantly mechanized force run by a bunch of airborne, straight leg and now airmobile generals. We seem to be light at the helm in an Army designed to have most of the hard work done by armored folks. However, in both Korea and Vietnam I noted the preponderance of Armor officers in staff assignments—the only place they could serve due to the dearth of Armor units in the theater and the infantry-airborne syndrome rampant at the top. This just reinforced infantry's conviction that they are destined to run everything—that, in spite of an Armor-heavy force structure, when the fighting begins it's the infantry that really is used in the end.

Historically, I suppose, this is all part of our Pershing, Marshall, Eisenhower heritage. In fact one could almost make a case for an antiarmor conspiracy. For, discouraged at the prospect of service in other than armored infantry, our truly great armored infantry leaders of War II transferred to Armor rather than risk a career dominated by the airborne-leg syndrome. And in this movement they were vigorously encouraged by Infantry branch. In addition, looking at Knox itself, I have to conclude that, except for your tenure here, we haven't been very selective in our choice of leadership, nor have we ensured sufficient tenure to get anything done. Since Adna Chaffee, the average tenure of my distinguished predecessors has been 19 months. And for a year or more at one point there was no Center Commander—only an acting fellow. And it was one of my least distinguished predecessors who gave away proponency for mechanized infantry to Benning rather than fight an infantry-dominated CONARC over the issue.

Conspiracy is a pretty strong charge, and I doubt that one existed, at least in the street sense of the proper definition of the word. Nonetheless many in my generation saw Abe's appointment as Chief as an opportunity to at least begin breaking up parochialism at the top. But it was not to be. And so we have a Secretary who leaps from airplanes in the company of a corps commander who typifies the infantry syndrome; and a Chief who once recently said that the only reason for the tanks anyway was to support the infantry. Daily almost I am called from Washington by my Armor 05 and 06 network decrying the fact that no Armor general was present when an important decision about tanks, or whatever, was made. Where was I—hell, I wasn't invited, and probably would not have been welcome had I been there. Bill Maddox testifies before Congress about airplanes, but the Commander at Knox does not testify about tanks—nor does any other Armor general!

Despite all that, we are moving many things in the right directions—all because of Bill DePuy—an infantryman, definitely not of the traditional mold. A brilliant man by any measure. Armor today has a voice, and a strong one, because of him, and in some measure at least because of my relationship with him. He's the best friend we have. I do fear that when the two of us leave these jobs things may revert to type. But for now we're getting something done. If we can but institutionalize what we're doing before we leave, then we'll leave a large footprint. We hope to be able to do enough of that to make it last.

Command Qualifications

Letter to General Bruce C. Clarke
Palmyra, Virginia
7 January 1976

The general problem here is that we have always assumed that everyone has to command, and in due course assigned everyone to command. Now the fact is that not all of us are good commanders. In fact my experience is that about 40–50 percent of us aren't—even though by all indications the personnel managers can show that by their criteria we can and should command. Therefore we must be willing to weed out those who can't, and do it early in their tour before they screw up the troops too badly. It's a command responsibility—it requires a certain ruthlessness. More than that it requires guts, honesty, integrity, and some empathy on the part of the commander who does the hiring and firing. You can't let your subordinates fear to make mistakes, worse yet they cannot fear to let their subordinates make mistakes lest they themselves as first parties get fired.

It's a fine line. The only guy I've known well who could do it well was General Abrams. However he said to me many times that the first place he looked when a subordinate didn't do well was at himself. Once he had satisfied himself that his instructions were clear, then he could proceed to act without hesitation to get rid of the marginal performer. It's a good rule.

Duties of the Command Sergeant Major
Memorandum to Command Sergeant Major Wayne F. Schildauer
V Corps
24 May 1976

1. The following instructions set forth duties and responsibilities of the Corps Command Sergeant Major.

2. Your duties and matters pertaining thereto are considered to be governed by the general provisions of AR 600-20, *Army Command Policy and Procedure*, especially those parts of Chapters 2 and 4 of that regulation relating to noncommissioned officer channels and the position of Command Sergeant Major. To further amplify the provisions of AR 600-20, the following are included within the scope of your responsibilities. They are matters to which you should devote the bulk of your time, on which you should report to me, and which you should emphasize through noncommissioned officer channels in the Corps.

a. The discipline and responsiveness to command of the noncommissioned officer and other enlisted ranks of the Corps.

b. The performance of soldiers of the Corps in soldierly matters such as the operating efficiency and proper care of weapons and equipment—organizational and individual, wearing of the uniform, personal appearance, observing of military courtesy, and related activities.

c. The general proficiency of the soldiers of the Corps in their military occupational specialties, and their opportunities for advancement, career development, and progression under current provisions of the Enlisted Personnel Management System.

d. The adequacy of provisions that have been made for the welfare of the soldiers of the Corps to include barracks, dependent housing accommodations, recreation facilities and services, other community activities and services, and off-duty activities, services, and facilities.

e. The adequacy of provisions made to ensure correct, fair, and equal treatment of all ethnic groups and both sexes with regard to assignment, advancement, promotion, grievances, complaints, and related matters.

f. The ability of the noncommissioned officer corps to advise and assist the junior officers of the Corps, especially newly assigned platoon leaders.

g. The adequacy of advice and counsel provided the officer chain of command in the Corps by noncommissioned advisors at all levels.

h. The functioning in the Corps of noncommissioned officer channels as described in paragraph 2-2 of AR 600-20.

3. As the Corps Command Sergeant Major, you should maintain your own circuit of visits, inspections, and observations. You should not accompany me except when our respective schedules make it convenient for us to travel together to the same location or area, or when the ceremonial nature of the occasion makes it appropriate for both of us to be present for whatever events are scheduled. When we do travel together in the interests of sharing transportation, you should pursue your own itinerary, looking at those things which are within your responsibility, just as I will be looking into matters which are my responsibility. Nothing is worse than to find

all the Command Sergeants Major gathered together, gawking, at the rear of the Commander's entourage. It is a waste of your time and mine.

4. Periodically you should report to me on matters, circumstances, or incidents bearing on your responsibilities as set forth in paragraph 2 above, or any other matter not included therein which in your judgment needs to come to my attention. Your reports should be oral rather than written, informal rather than formal, unscheduled rather than scheduled. In other words I expect you to tell me what you've been doing, what you have seen, what problems you may have discovered, what you have observed—good or bad, that in your judgment deserve my personal attention.

5. You must keep the Deputy Corps Commander, Corps Chief of Staff, and the Corps staff informed of observations you may make that pertain to their respective areas of responsibility. In the case of the Corps staff you may do this with the staff section head or through his noncommissioned officer channel.

6. In all matters the advice and counsel you provide me must be based on fact, completely honest, candid, and straightforward. You must tell me what is, in your best professional judgment, the absolute truth. If the truth is unpleasant, which all too frequently it is, then that is all the more reason for telling it to me just as you see it.

Sergeants' Business

US Army Sergeants Major Academy
Fort Bliss, Texas
3 November 1977

With the onrush of Task Lists, Skill Qualification Tests, and Army Readiness Training and Evaluation Programs, the responsibility for individual training of our soldiers has been changed. Unfortunately, we weren't clever enough to emphasize this change when it occurred, so that's what I'd like to talk to you about, what I call Sergeants' Business.

Now not everybody understands what Sergeants' Business is. Some of the sergeants don't understand it and most of the officers don't understand it, and therein lies the problem. So what I'd like to do is try to explain to you what I think it is and why it's important to us. In order to do that I'm going to go back in history a little bit. Some of you may recognize "Old Bill," because he represents the old noncommissioned officers of the Army. I don't know how many of you have ever read Frederick Remington's explanation of why he painted that picture. That's a real person; it isn't just a stylized cavalryman. His name was John Lennon and he was a sergeant in the 3d Cavalry whom the artist met in Florida during staging for the war in Cuba. Mr. Remington thought so highly of him in the sense that Lennon, in Remington's eyes at least, epitomized the old line sergeant of cavalry that he painted that picture of him. I'd like to talk about that fellow, not Sergeant Lennon, but the image he represents of the highly professional noncommissioned officer, and what's happened to him in our Army since I joined it as a second lieutenant in 1948.

The first platoon sergeant I ever owned, or to whom I belonged, as a second lieutenant, was an old gent named Leonard Lucas. He was a crusty old fellow, and he said to me as I walked in and reported to him, "Well, I see I have another lieutenant to train." He issued me a tool box, made me inventory the tools and identify them, made me sign for it, and then he said, "Okay, lieutenant, I know you've been to West Point. I know you've been to Fort Riley to school. I know you've been to Fort Knox to school, and all that was very useful, but now what we are going to do is make that practical. I'm going to do that, and the maintenance instruction begins in the motor park tonight at 1930. Please be on time. Bring your tool box."

So I thought about that and said to myself, "Well, who is in charge of this mess?" But I figured maybe I'd better keep my mouth shut and listen for awhile and see what happens. I went to the motor park, and it was an unfortunate thing. He made me so proficient as a mechanic that I had to later do a tour as a battalion motor officer, which is not always the finest job in the world. He did a good job all around. The first time we went to the field and pulled into a bivouac area, Sergeant Lucas came up to my tank and said, "Now if the lieutenant would be so kind as to go over and sit down under that tree, I will put the platoon in position. I will report back to the Lieutenant when we are ready for inspection." I thought, "Well now, I really should, as the platoon leader, be doing something besides sitting under that bloody tree."

But I did as he said. So there I was, sitting under a tree reading my Soldier's Manual for lieutenants, trying to decide what I ought to be learning out of that. I also watched him as he went around. He made them go into position; he made them put up camouflage; he made them make out range cards. He had a kind of a checklist of crew duties. He had a little inspection and he checked to see if they had been doing their maintenance. Then he came over to me and reported, "Sir, the platoon is ready for inspection. If the Lieutenant would please accompany me, we will inspect the platoon. Here's what I want you to look for in tank 31, 32, 33." Then he told me some things, because he knew the sergeants and the crews better than I did and he knew where they were weak. He knew the things he wanted to emphasize with them, and we did just exactly that. When we got all through, he said, "That's fine, Sir. Thank you very much. You did what I asked you to do, but you weren't tough enough. Now that may be for one of two reasons: one is you are inclined to be too easy on the soldiers, and we can correct that. I'll tell you how to do that. The second reason is you don't know enough about the skills, the things that we're checking on, the little tasks that I want them to do. You don't know enough about that yourself in order to ask questions, and you may be a little nervous about that, so we'll fix that. We're going to improve that, because we are going to have classes for you and I'm the instructor."

Now, he was always very respectful about all this, but it was clear to me that his job as he saw it was to train that platoon leader, and he was not alone. We had a battalion of those sergeants. This was 1949. They had all fought World War II, every one of them. Some of them had been officers in World War II and reverted to enlisted rank, or gotten out and come back in, a whole combination of things. They had gone through some pretty rigorous training getting ready for that war. They had all fought the war, and most of them had done very, very well at it. Every tank commander in the first platoon that I commanded had been a tank platoon sergeant in that war. They were super sergeants. Every platoon sergeant in that battalion took the attitude that it was his job to make sure that his lieutenant was the best platoon leader in the battalion. There was a competition among the platoon sergeants as to who had the most proficient lieutenant.

Now, as to individual training of the soldiers in the platoon, Sergeant Lucas was very clever about that. For instance, we'd go to the range and come back in and clean weapons. The first time I went down there to watch this, he had the soldiers all gathered around in a circle. They had a couple of bins of solvent and cleaning materials, and they were cleaning the weapons. This particular day in question it was machineguns. But they weren't just cleaning the weapons. He was conducting a class in assembly, disassembly, nomenclature, functioning, care, and cleaning as they cleaned the weapons.

I really didn't realize what I was looking at, but as I look back on it, it was the most effective kind of thing I've ever seen. It's the kind of thing that I call Sergeants' Business, and it's the kind of thing that we've got to get the sergeants in our Army back into the habit of doing. The other problem is, we've got to get the officers in the Army back in the habit of letting the sergeants do that business, because that's the Sergeants' Business. That old man never missed an opportunity to take something that would just be normal maintenance or normal care and cleaning or normal duty and to make a little training session out of it.

He was always checking his people. He had a little notebook in which he kept notes on me, on the men in the platoon, and probably on the battalion commander for all I know. But he had more knowledge in that notebook than all the manuals in the Army. Now today we call that a Job Book. But, back then, that was Lucas's job book, and he had everything in it. One of the first things he made me do was to get one of those things myself. I had everything in there from shoe sizes and hat sizes for the men who had big feet and big heads or little feet and little heads to how many times they had been married and how many kids they had. It was all in my platoon leader's job book.

As a matter of fact our battalion commander at that time was named Lieutenant Colonel Creighton Abrams, and he also had a job book on everybody in that battalion. If you couldn't compare notes equitably with him, and your job book didn't match his, you were in trouble. Now what happened to all that? I don't think we have that in our Army today and, as I look back at the last few years trying to reconstruct what happened to all that, I think it goes something like this.

We had a corps of noncommissioned officers in World War II, most of whom had been trained in the years before the war, many of whom became officers, or at least exercised positions of responsibility as platoon sergeants or acting platoon leaders. If they didn't actually get a commission, they were in fact the leaders in their units during the war. That whole system, in my mind at least, pretty much survived the Korean War. We lost a lot of noncommissioned officers in the Korean War, but not all that many. Not many of them were armor soldiers, because we didn't have much armor in that war. So in my own experience, in the armored force, we survived the Korean War with that body of noncommissioned officers pretty well intact. For the most part they had the kind of attitude that I just described.

When I went back to command a battalion in the early 1960s, we still had a lot of that. Not quite as much as we had before, but it was not so far gone or so far away from us that we couldn't reconstruct it with a little training for the NCOs. That's what we did, and that was a pretty satisfactory time as I recollect it. The Vietnam War wiped the system out. People came and went so fast that we couldn't even remember their names. We wiped out our noncommissioned officer corps. I think we did it unconsciously in the rotation policies that we had. A lot of guys just got disgusted and left, retired with 20 years.

So we find ourselves today, at least in my summary, in the situation of having to rebuild our noncommissioned officer corps. Particularly is this so when looking at not so much the sergeants major, but at the E5, E6, and the E7 levels especially. We'll never get back to the days of Sergeant Lucas and his contemporaries, because we just will never have that much experience again unless we fight another war. We've got to somehow get back enough of the kind of thing that they were doing, for their officers as well as for their soldiers, to restore the high professional quality of the noncommissioned officer corps.

The result of all this change has been, at least from what they tell me, a perception by NCOs that there has been a loss of NCO authority. Perhaps you would say that NCOs have lost a lot of authority and a lot of ability to do things, but have not lost the responsibility. In fact, one of the things that NCOs tell me frequently is that someone has taken away all NCO authority to act, but they have not taken away any of the responsibilities charged to the NCOs. So the sergeants still are supposed to be doing the same things, and even more things than ever before, yet they have not been given the authority to do them.

Now the problem is that we've had a loss of not only authority, or perceived sensing of the loss of authority, but we've actually had a situation where we in the volunteer Army relaxed.

Here is an interesting case. A young soldier's outfit had just completed a successful ARTEP, and he was relaxing after the strenuous effort. The sergeants in his outfit were all terribly busy. The officers were all extremely busy, but that soldier was bored, bored stiff. I woke him up after I took a picture of him and asked him what was the matter. He said, "Well, you know, we had a good ARTEP. Now I don't know what we're going to do next. The sarge said we were going to go back and clean the stuff up. Then we were going to start over again with first aid and CBR training and that stuff. I'm tired of that. I'll take a test. I think I can pass it, and if I pass it, how about letting me go do something else?"

Now that's our fault; that's my fault; that's your fault; that's our fault collectively for that problem. So the question we have to ask ourselves, I think, is where did we go wrong? Some of the answers I guess are in the history that I tried to review for you, but it seems to me that there is more than that. So in the last year or so we conducted a survey. As part of that survey we looked at time utilization on the part of E1s through E4s, idle time during the day. If you took the total population, they said that well, out of every day this many of them spent a half of that day or more just doing nothing, at least as they perceived it. See the problem is not what you or I think they are doing, the problem is what they think they are doing. The problem is their perception, not ours. The officers are busy, the sergeants are busy, but the soldier perceives that he is not busy doing useful things. What we found in the survey was that many of them said that they were wasting half or more of the day. Almost half said that they had about one to four hours out of every day in which they really weren't doing things that they perceived as being useful, *that they perceived.* I've forgotten the exact number there, but it was between 45 and 50 percent.

Now that's awful, and that isn't the soldier's fault, that's our fault. That's their perception of what you and I are doing with their time.

What was the one thing that they all agreed upon in the questionnaire as the single thing the Army could do to improve their perception of the little world they lived in? It was a question that involved meaningful training, and we examined the responses for the enlisted men and the officers. What we learned was that 75 percent of everybody in the Army, from grade E1 through grade O6, said that meaningful training for the soldier was our biggest problem—meaningful training for the soldier as the soldier perceives it, not as you and I might perceive it. It's his perception that's the most important part of it.

Now, what can we do about that? Well, what do we want to train the Army for in the first place? We've all got to agree on what that is. We talk in terms of the modern battlefield, where we say that this is all of the material that all of you should have seen as you memorized Field Manual 100-5, which I presume you have all done. We're talking about quality weapons. There has been a gradual, but very, very steady, increase in the quality of the weapons systems in the last 30 years to the point that, today, they are dramatically more effective than they ever were before. The problem is, we don't recognize that very well, because it has happened to us in a sort of evolutionary way.

The fact of the matter is that tanks today are something like four or five times more effective than they were in World War II. Artillery today is something like 10 times as effective as it was in World War II. All that has happened gradually, in an evolutionary way, over a period of years. Some of that quality has brought with it a certain amount of sophistication, and so it takes more on your part and mine to train soldiers to use the equipment. It makes the leader/trainer challenge much greater. It requires us to do some things better, better than we've ever done them before, because the things that we are doing them with are simply not as simple.

We are talking about a battlefield with the Soviets or the Warsaw Pact or an enemy that has a lot of weapons and uses Soviet tactics. We are talking about large quantities of weapons in the main part of the battle, quantities of weapons systems the likes of which we've never seen in one place on the battlefield before. It's a very, very busy place.

In the Syrian attack on the Golan Heights in the Yom Kippur War, there is a place northeast of Kuneitra where Syrians came down from Damascus in a column that was about 3,000 meters wide and 10 kilometers long. You can see it all from the hill outside of Kuneitra, the whole thing. A 3,000-meter-wide strip of tanks, BMPs, BRDMs, and ZSU-23/4s that stretched back toward Damascus for a total distance of 10 kilometers. That's frightening, particularly if you're standing out there with a handful of men trying to defend that thing. You feel like General Custer at the Little Big Horn. Where did all the Indians come from?

The other thing that is inherent in that quantity of weapons is that there is a great variety, such a great variety that it's quite apparent to us that no single weapon system is going to be able to cope with the problem. So we come back to the old combined arms team. It's going to take a little bit of everything, and I won't belabor them because they are in the books. You've read them, I'm sure.

The high levels of electronic warfare that are present on the battlefield are going to make command and control difficult. They are going to make the leader's job harder, because it's going to be more difficult for him to communicate with those whom he has to talk to in order to get on with the battle. The air defenses on the battlefield are going to make the air over the battlefield a much more dangerous place in which to operate. The battlefield is going to be dense with obstacles of one kind or another. It may be just blown-up equipment, or antitank ditches, or mines, and so on.

All that is a very difficult environment, and it poses some training problems that we never had before. In the end, however, I still believe that wars are won by three things. I have no question in my military mind about the courage of the soldiers, it's there. The problem is that you and I have got to get it out in a usable fashion and put it to work on the battlefield. That's a leadership problem, so the problem we run into is the quality of leadership and the excellence of training. These both are training problems, because we ought to be able to train the leaders to train the trainers and train the organizations to do the job, to use the courage of the soldiers effectively in battle. I believe the courage of the soldiers derives from their confidence in themselves, confidence in their small unit, their tank, their squad, their section, and their confidence in their leaders. They derive confidence in their leaders as a result of their recognition of the ability of the leader to do the job as well or better than they can. The excellence of the training derives in my view from the ability of the leaders to think logically about difficult problems.

Let me tell you a little story about leadership, and let me use the Israeli Army as an example. They have a draft. Boys are drafted for three years, and girls for two years, at age 18. In the Armored Corps in the Israeli Army, the soldier goes through 18 weeks of what you and I would call OSUT, One Station Unit Training. It's a combination of BCT and AIT. Eighteen weeks; we give thirteen. They give him a test beforehand to determine whether or not he should be a gunner/tank commander or a driver/loader. If he becomes a driver/loader he goes to driver/loader training, then goes to a division where he serves the rest of his three years. Then he goes into a reserve division, which is to be mobilized in time of war.

If he is selected by tests beforehand to be a gunner/tank commander, he goes through gunner training as a crewman gunner and those that pass that test at a successful level go on immediately to tank commander training. It is 21 more weeks long to learn to be a tank commander, or a total of 39 weeks after he joins the army. In 39 weeks he graduates from a school that has made him a tank commander. He goes to a unit and spends six months as a tank commander. If he passes certain tests above a certain level, performance tests as well as written tests, he goes to platoon leader school, and that is 26 weeks long. After about a year as a platoon leader, certain individuals are selected to be company commanders and go to a company commander's course, which is 20 weeks long. That's not the exact number, but it's about that.

Now there are some things about that that I'd like to emphasize. The point is that the best soldiers by test are the leaders. The best tankers are tank commanders; the best tank commanders are platoon leaders; the best platoon leaders are company commanders; the best company commanders are battalion commanders. The soldiers have great confidence in themselves, but also in their leaders, because they know the leaders are best by test. Their battalion commanders are 26 to 28 years old; their brigade commanders are 30 to 32 years old, no older than 33 or 34. Their division commanders are 34 to 36 years old. They retire at age 40, because they claim war is a young man's game. At every step in the chain, the best man is the leader by test, and he has the complete confidence, admiration, respect, and obedience of all the other soldiers in his outfit because they know he's the best by test.

There is another side to that, and let me tell you about it. The other side is that they don't have any noncommissioned officer corps, and they envy us. If my friend Major General Musa Peled were standing here, the commander of the Israeli Armored Corps, he would be very envious to look out and count the stripes in the audience, because they don't have anybody like our NCOs in their army. They envy us and our ability to gather a group of NCOs together in a room and have a conversation. They envy us the fact that we've got NCOs in our Army, because they don't have anybody like that. They make up for it by this very, very rigorous training and selection process that I just described for you.

There are, however, two sides to the coin, and I don't want you to go out saying that I am proposing that we adopt the Israeli model. It has a lot of things arguing for it. The thing that argues most strongly, in my mind, is the excellence of the leadership and the quality of the training and therefore the confidence of the soldiers and the units in themselves and their ability to perform in battle. As far as the noncommissioned officer corps or lack thereof goes, we in our Army have got to capitalize on the fact that we have NCOs and they don't, because in the end it may make the difference.

Some of the things that they do in their training we simply cannot do. The age brackets, for example, are just completely outside the bounds of possibility in our service. Somehow we've got to make up for the things that we cannot do that make them so good by using some other advantages that we have. In my view, the NCO corps is part of a singular big advantage that we've got over them.

Now how do we do that, and specifically where do our NCOs fit into our training system? Well, generals assign missions and resources and give broad outlines and so on. The colonels and lieutenant colonels allocate time, set priorities, provide support, train and coach the trainers, and do some other things. The sergeants, the NCO chain of command, do these things: prepare training, conduct training, and evaluate training.

Let's talk about the NCOs and training, because that is at the heart of what I'd like to leave with you. Individual training, in my view, is the total and primary responsibility of the noncommissioned officer corps. If that's true—and I must say not everybody agrees with that, not everybody agrees with me, not all the sergeants major in the Army agree with me, not all the officers in the Army believe that. But I believe it, because the NCOs all represent an asset we have in our Army that the Israelis, for example, don't have. Many other armies in the world don't either. The Russians don't have a corps of noncommissioned officers like we do, and I'm sure they envy us our corps of noncommissioned officers, although I doubt that they would admit it.

What am I talking about here? I'm talking about the job of the first-line supervisor, the job of the next level supervisor, the platoon sergeant, the job of the first sergeant and the job of the sergeant major. Let me tell you what I think, in my view, each one of those jobs amounts to. The first-line supervisor's job is to make sure that the soldiers that work for him—in his crew, in his section, in his squad, in his fire team, in his little office, if you will—are proficient in the tasks that they are responsible for maintaining proficiency in, according to the Soldier's Manual definition of tasks at their level.

What I'm saying to you is that, if you are an 11 Bravo and you are in an infantry unit, what the first line supervisor holds as his responsibility is the proficiency of his soldiers at Skill Levels 1 and 2 as specified in the task list in the 11 Bravo Soldier's Manual. It's that simple. It's even simpler with the Job Book that we've put out, since that is just a list of those tasks, even though a lot of people think that it is some kind of manual that they've got to memorize. All that is is a notebook. That's Sergeant Lucas's notebook. That's where we got the idea. We called it a Job Book because it sounded exotic, I guess, but that's what it is.

Now, how do you maintain proficiency in the soldiers you supervise? Well, you can go to the Soldier's Manual, and the Task List is fairly easy to come by. You can read that; you can copy it down; you can number the tasks; if you use them often enough, you can memorize them. You eventually get to know what they are. You know what the standards of performance are. You know the conditions under which the soldier is going to be tested, and whether those tasks are included on an SQT or if they're included as individual tests in an ARTEP. You can get all that out of a Soldier's Manual. Then the first-line supervisor has got to look for opportunities to check and see if his men are proficient in their tasks and make sure that they do them often enough so that they maintain proficiency. That says something else. That says that the first-line supervisor has to be proficient enough in those skills himself to do two things: one, teach the soldiers those tasks as they ought to be taught, and two, evaluate them properly. The truth of the matter is that not every first-line supervisor is proficient enough at Skill Level 1 and 2 tasks to do those things that I just said, train the soldiers and evaluate them, and that's the problem.

How do you get at that problem? That's the job of the platoon sergeant and of the first sergeant and of the sergeant major, and my contention is that the command sergeant major has got to get that organized. It's the responsibility of the command sergeant major in the battalion, the first sergeant in the company, and the platoon sergeants all together to understand how much each one of those first-line supervisors knows about each one of those skills that their soldiers are supposed to know. Can they train them? Can they evaluate them? There is a third job for the platoon sergeant, because he has to know whether or not his E6s are proficient at Skill Level 3. That means that he, the platoon sergeant, has got to be proficient enough to make sure that he can check the E6s and that he can teach them the job if they don't know it that well.

What does the first sergeant do? He checks the whole platoon, but he is responsible primarily for the proficiency of the platoon sergeants at Skill Level 4. Do they know it well enough to teach the soldiers? Do they know it well enough to check the soldiers to see if the soldiers know it, and do they themselves maintain proficiency at Skill Level 4? The same is true for the sergeant major with the first sergeants. Now that's the way the thing is supposed to work. That's the way this whole thing was designed, against the little scheme that I just described for you. That means that you are going to have to do two things: one, you're going to have to have some instruction for the first-line supervisors, and probably for the platoon sergeants, and two, the first sergeants themselves are going to have to have some instruction.

Let me tell you first how not to do it. When I was the commander of V Corps, I went around to every battalion in V Corps, all 71 of them, and I asked the battalion commander to tell me how he was going to program training. By programming training I mean taking those task lists out of the Soldier's Manuals and the task lists out of the ARTEPs and putting them together against resources—time, money, range areas, gallons, and bullets—and coming out with a training program in the end. The other thing I required was that his command sergeant major tell me what he was doing in his battalion to do the things that I just described with regard to the first-line supervisors, the platoon sergeants and the first sergeants.

One command sergeant major said, "I've got a course of instruction going for these guys," and I thought, "Boy, he's really with it." He said, "Here's the POI, General," and the first thing on the POI was methods of instruction. I asked for the lesson outline for that, and it had to do with the podium and the pointer and the screen and the slides, and I said, "Wait a minute, Sergeant Major, that's right out of an NCO Academy POI, isn't it?" "Yes, sir, that's where I got it," out of the something or other school he'd gone to. Well, that's fine. It was super for an NCO academy. But I said, "Sergeant Major, you are not running an NCO academy. You are trying to teach people how to teach other people things in the motor pool, at breaks during training, on the range, during concurrent training, in the barracks, out behind the motor pool. When there is nothing else going on, you are trying to eat up the time of all of those bored soldiers that I showed you a moment ago, and doing something that the soldier sees as useful. If it relates to tasks in the Soldier's Manual, that the soldier knows he's likely to be tested on when SQT time comes along, that's a motivator in itself. That's what you're trying to do."

Don't, therefore, have a class and have a POI that includes a class on methods of instruction: hold the pointer in the hand closest to the board, point vigorously, and put the pointer back down. All of that stuff we used to do so well in places like Fort Benning and Fort Knox, where the instructor marches on and goes like this, you know, like a damn wind-up toy; that's dumb, that's dumb. What you've got to do is teach them when it's raining, when it's night, when people are cold and tired and scared. What you're trying to build in them is enough confidence in their ability to do those tasks under all those circumstances so that they'll do it under pressure when they're frightened. That's what wins wars. So that's the problem, and that's the way the system works, at least as I see it.

I've given you the concept of the system, but let's dig a little deeper into it and touch specifically on how the sergeants accomplish this training. I'll use the battalion command sergeant major as an example, and you can modify that example as needed for the other levels or positions in the NCO chain.

Now the command sergeant major is responsible in the battalion for the overall NCO individual training system. In addition, he has an immediate and direct charge for the training and job skill proficiency of the first sergeants in the battalion. To do this he can use a myriad of methods: training bulletins, letters, information sheets. All these deal with printed media that you and I know are often unread and put in file 13. The most effective training the sergeant major can do is eyeball-to-eyeball, cheek-to-jowl.

He can use the Sergeant Major's Meeting to cover such things as methods of leadership for AWOL prevention, individual skills to solve administrative glitches, supervisory role of appointments for the units, and a host of day-to-day duties with which first sergeants must contend. All of these can be skill or job-related and discussed in plain, everyday language. Another method might be in individual counseling sessions with the first sergeants, one on one. He can get down to brass tacks and discuss the strengths and weaknesses of the first sergeant and point out areas or ways to improve. We've been giving lip service to counseling for years, but here's a way to make it pay off in SQT results. He might suggest nonresident skill courses for the first sergeant in training or training management for the company. Any area is open, but the key is informal, open discussion. That doesn't sound like training. Some might say it's a bull session, but if it's done right it's more valuable than a dozen formal classes.

Finally, the sergeant major could and might conduct or cause to be held formal or informal individual training. He can teach these himself or use the battalion staff experts, both officer and NCO. No lesson plans, no methods of instruction approach, but again good, honest communication. It could be jotted down on a note pad and the battalion command sergeant major could give it. Just make it worthwhile and as short as necessary.

Now, all these methods are not the complete list. They are examples, but they can be used; don't be limited, use your imagination. What are each of the levels of the NCO training system trying to get across? At the first sergeant, platoon sergeant, level it's much more troop-leading oriented than job skill, but as you move down the chain the balance begins to shift more to skill-level training.

At the squad leader and fire team level the effort is almost solely on skill level training. This is the first line supervisor. He's the one who must pass on the legend and lore of the Army. The whole system is built upon him, but it's organized and supervised by the command sergeant major of the unit. He's the one who gets the NCOs back to Sergeants' Business.

I had a sergeant major report for duty in a battalion in V Corps. He was a very, very fine sergeant major with a very, very fine record, and I'm sure he was a super noncommissioned officer. He said, "Hey, Sir, I'm an administrator," and I said, "What do you mean?" He said, "I've been a first sergeant or a sergeant major for eight years, and all that anybody has ever asked me to do was be an administrator. When I was a first sergeant I ran the orderly room, kept the clerks straight, kept the company commander's administration going, and that was all that was required of me. When I became a command sergeant major, the battalion commander never told me what to do, but he seemed to be satisfied if I went around and kept the battalion headquarters running and I worked for the S-1. I've been an administrator for eight years. I couldn't pass an SQT at Skill Level 2." "Well," I said, "what are you going to do about it?" He said, "I've got a lot of work to do and I'm working on it, but it's going to take me some time. What you're talking about for sergeants like me is a complete change in the way we look at our job and in the way that our bosses look at our job."

Now, I don't think the whole Army is like that, but so much of it is like that that it's frightening to me. One of the problems is that, after that sergeant got to be a command sergeant major, no commander had ever sat down with him, eyeball-to-eyeball, and said, "Sergeant Major, here is your part of our job and here is my part of our job. Here's what you're going to do and here's what I'm going to do. We're going to talk about once a week, or once a day, or as necessary. Let's get on with it." That's our fault. That's the fault of the officer corps and the fault of the senior noncommissioned officers. If your boss doesn't communicate clearly with you, or if you don't have some kind of an understanding with him about what you want, what you expect of him, fairly explicitly, then you have no one to blame but yourself when he doesn't do it and when the thing comes unglued.

I have been in a circumstance where I had to write my command sergeant major a letter of instruction telling him exactly what I wanted him to do and holding him responsible for it. I've only had to do that once, and I've had some super, super command sergeants major working for me over the years. I have never once failed to sit down with that sergeant and reach an agreement with him about what he thought was his part of our outfit, our battalion, our regiment, our installation, our corps, and what was my part. That has to be done, and we have to do that at the top. We've got to do that.

Now, let's talk a little bit about SQTs. The SQT is simply a way of checking whether or not we've done our job with regard to the Soldier's Manual, and it operates with a feedback loop. I want to make a point with you, and the point is this. If we get to the place where the Skill Qualification Test is something that we stop and stand the whole unit down once a year, or once every two years, then we have defeated the whole purpose of the Soldier's Manual and the SQT. It becomes nothing more or less than the old MOS test. One of the reasons we got rid of the MOS test is that we were standing down to give the men time to study for the MOS test. We started this whole thing by saying that's not the right way to do it. But, everywhere I go, I find units that are standing down or doing something like that. Some of them don't admit that they're standing down, but they are. I don't object, when the SQT task list comes out, to a little brushup. I think it would be dumb not to do that. What I'm talking about is saying for the next month or six weeks we're going to get ready for the SQT in old Bravo Company or in the old xxx Mechanized Battalion or Artillery Battalion or whatever; we can't do that.

How do you avoid that? You avoid it by integrating the tasks in the Soldier's Manual into the everyday training. It also means that the command sergeant major and the first sergeants together are going to have to draw out of the NCO training system, the individual training system in their unit, enough evidence for whether or not they need time on the training schedule to train. Then they must go to the company, battery, and troop commanders, and to the battalion and squadron commanders, and say, "Boss, we have got to have some time. We need some practice in this skill or that skill or this group of skills. To do that we need to go to a range, and in order to go to a range we've got to have some time and some bullets and some money and so on. Can we do that?" They have to convince the commanders to say, "Yes, we can do that. Here's how we're going to do it."

You have an obligation to bring that up to the top as you try to keep the soldiers proficient enough in the Soldier's Manual skills so that we don't have to stand down for SQTs. If we do our job right, we will in fact be able to win the first battle or battles, if you will, of the next war. But the problem goes back to the courage of the soldiers and the quality of the leadership and the excellence of the training. What I've been trying to tell you is that the excellence of training is the glue that ties the whole thing together. It's the cement. It's the cement that fastens this thing together in a way that puts an effective fighting team on the battlefield, where people survive to do their job and come home and wear all their ribbons and talk about it, brag about it, so that the battles get better every time they're refought.

Officer Leadership

Letter to Captain Louis P. Cantolupo
Fort Sam Houston, Texas
13 January 1978

. . . Let's see if I can describe for you what I think officers are to do, and why I believe they must do it just that way.

First, may I say that the image of the crisp young subaltern, crop at the carry, striding along midst whizzing bullets and crashing shells, is not an American image. It never has been, never will be. It is even more incongruous in the contextual setting of leaders and led in our society today than ever before. No doubt it had its utility to the British—in the army and at the time from which it is derived. I doubt that even in their army today is it much more than a romantic remnant of a time long gone.

The reason for that is quite clear—warfare has become a terribly complex operation, so much so that everyone intimately involved in it must be very, very technically competent. The day of the instant hero is gone—charisma will help, but being a good shot is better. Therefore it follows that the officer's contribution to battle today is quite different than his contribution needed to be years ago—in olden times, perhaps.

On the other hand it is my contention, after considerable study of the matter, that officer leadership of the kind I believe most appropriate in modern battle has always been the best way—even though from time to time the image may have been different. So what do officers do? Officers lead—they lead best by example, they lead best by doing well all the things their soldiers are supposed to do and more besides. The soldiers learn to respect the officer's prowess at soldierly skills, and they also learn that the officers know some things—officer skills, call them, the soldiers don't know. All that builds soldier confidence in the leadership. Once the soldiers have confidence in themselves and in their leaders, then confidence in their unit begins to build. And so builds confidence in the Army, the country, the "cause," and so on.

Now it's also true that the skills demanded of officers differ from grade to grade. The platoon leader is pretty much a "follow me and do as I do" fellow. That's why we call him a platoon "leader" instead of a platoon "commander," platoon "director," or something else. Quite clearly his bag is high technical proficiency in weapons, tactics, maintenance, commo, and how to get things done with his platoon. Surely he's not an armorer, not a mechanic, not an electronics repair person. But to the extent that he's better at each of those skills than any of his soldiers—to that extent his stock as a leader goes up in their eyes.

The company commander is also a technician—he retains his skills learned as a platoon leader and adds new company commander skills—how to get platoons to do things together, train the company, maintain the equipment, counsel the soldiers, cope with the admin. Quite clearly his skill base is broadening—and his SQT, or whatever we call it, will reflect that.

Leader Skills

Pre-Command Course
Fort Leavenworth, Kansas
22 January 1979

In the next few weeks, particularly in the week devoted to command development, you're going to wrestle with the real, but often hidden, problems of the preparation for war, those that relate to the humans who fight it.

You all have some idea by now of what the modern battlefield will look like. Our potential enemy on that battlefield will be strong in numbers, well-equipped, and trained. We hope to fight and win on that modern battlefield, but have to do something to overcome the enemy's advantage in numbers, organization, and force structure. We don't know the answer yet, but we're starting to look. You'll discuss it yourselves later today, so I won't belabor it. But let's explore the question and some possible perspectives to it.

We have plotted [displaying chart] on the ordinate a probability of victory and on the abscissa an attacker versus defender ratio that runs from one to five to five to one. In other words, over here, we have one attacking five, and over here, we have five attacking one. If you believe the mathematics that come from comparing numbers, whether you use a Lanchester Inverse Squares Law or whether you use a stochastic model that simply relates the firepower scores, you get a model that says if you were one attacking five, there is really not much hope of winning. It also says that if you were five attacking one, you always win—almost always. Now, if you believe that you're going to have to fight a war outnumbered, that is not very encouraging if you're the defender. Even if you're the attacker, with inferior numbers trying to attack the guy with more numbers, you can't win.

What that says to you is that, if you're outnumbered, you can't win if you attack and you can't win if you defend. Well, that's great. Now what? Well, if you look at the history of a thousand tank battles, going all the way back to World War I, you get a curve that says you can win and that there is more to winning in battle than numbers. It also says to you that if you are outnumbered perhaps you can figure out how to win outnumbered—attacking and defending.

What makes that happen? There are a lot of ways to analyze it. Today, in TRADOC, we are involved in trying to do some things to make that happen. We are involved in trying to apply technology to weapons systems, and even more relevant, we're trying to improve the ability of the individual soldier to use the weapons systems that we've given him. Our training plans reflect that.

What we're trying to strike at here is how you put soldiers—however well trained they may be—and weapons—that we are trying to improve by applying technology—together into some kind of an organization, a system if you will, in which we've got the right kind of controls, the right kind of people, and the right kind of internal relationships established that produce the synergism that comes from those relationships.

Let me tell you something about how that might happen in a unit. We know, at certain times of the year, tank crews fire better than they do at other times. The reason is that they get ready for annual gunnery qualification. TOW crews get ready for it, rifle squads get ready for it, machinegun crews get ready for it, helicopter crews get ready for it. Everybody gets ready for annual qualification.

Now, what happens in this process? Well, if we plot the probability of a hit of a given system, whether it's a weapon fired from the shoulder of a soldier or an advanced attack helicopter firing some kind of missile, there are certain times of the year when over range you get a pretty high performance curve. Some systems like TOW can go on out to maximum range with little change in performance. A typical curve of a tank crew at 1,500 meters begins to get about 50/50.

We also know that, some time after that maximum performance is achieved—in some units, it's a matter of hours, in others, a matter of days, in some, a matter of weeks—the ability to perform in that fashion falls off rather dramatically. You thus get an area that reflects the difference between the high-performing crews and the low-performing crews. This difference can only be changed by training.

If you apply low band performance curves to target servicing situations like the ones we face in Europe, what you achieve is a verification of Lanchester mathematics. In effect, you start outnumbered, and you end up still heavily outnumbered in the ratio. You aren't going to win that kind of battle. You can't go out on the battlefield with crews performing at low performance levels and hope to change the orange curve into that red curve. It won't work.

If you go to high band crews, you find that you survive at a slightly higher rate and that you kill enough more enemy so that the ending force ratio is about half of that at the start. You may win, but it's not sure. The one point that is sure is that only with high band crews do we begin to get ourselves in the ballpark of being able to win the battle.

History and our battle situations in Europe show us if we put those high band crews in well-trained units where there is stability on the part of the followers, stability on the part of the leaders, and there's a climate of good leadership in the unit, then we get a real improvement. I'm not talking about individual leadership. I'm talking about the kind of synergism that accrues to units where there is real cohesion.

If we do put those well-trained crews in well-trained units and again apply the performance curves to the calculus of battle, a significant change occurs. We find that the ratio at the end has shifted in our favor. It means now that we could attack. Even more important, it means that there is an extra plus—a synergism, if you will—about the combination of well-trained crews in well-trained units. That synergism, organizational performance, is what the command development week is all about. It wins battles, and it's the real difference between Lanchester mathematics and the historical results of battles.

Let's recap this and see what the real potential is. The advantage provided by technology may gain something like 3 to 5 percent, taking full credit for the sophistication of guided missile guidance systems, tank fire control systems, whether we have a laser designator or thermal sight and they don't, and so on. Crew performance, the difference between the performance of low crews and high crews, adds about 15 percent to battle outcome. Organizational performance, the difference between high performance crews and high performance crews in a fully trained unit, adds about 25 percent to the battle outcome. There are some studies that say much more, up to 300 percent, but 25 percent seems more reasonable and conservative for now.

Now, if anything like this is true—these numbers are not precisely correct, but they are about right—then we have no choice but to try to get well-trained crews and put them in units in which we've done something to try to improve the effectiveness of the organization and the context of the organization as a system—the mechanisms of control established in the organization—and we're going to have to do that through people.

What we're trying to tell you is that, in some way, to some extent, careerism has kind of overtaken us. Entrepreneurship of the wrong kind has kind of overtaken us. We are more concerned with *my* efficiency report and *my* outfit and *my* this and *my* that than we are in *us*, than we are in the results of the calculus that I've tried to describe for you. You have got to change that. Your leadership has got to build synergism in your units so that something like that red line happens instead of the very high level of very low efficiency we have today.

It's got to start with you. You've seen this term—US Army—almost one-half of your life. Take a close look at it—only six letters. What has happened is that, instead of concentrating on the first two letters as a word—US—we've fallen into the habit of concentrating on the last two—MY. We do it at all levels. Maybe it's a sign of our times. "What's in it for me?" "Do your own thing!" "Individual life styles." These are all symptomatic of the problem in the Army and in our society.

In the December 1978 issue of *Military Review* there is an article titled "The Meaning of Freedom" by Herman Wouk. You may remember him as the author of *The Caine Mutiny*. In his article, Mr. Wouk develops the theme that there is a contradiction, a dichotomy, between individual freedom and national freedom. He further submits that only the US has been able to successfully nourish both these freedoms simultaneously. He then goes on to point out that there is a price to pay for this achievement. He suggests that "the national freedom of America rests on the willingness of an able few to give up personal freedom in the necessary discipline of a military arm." That, in essence, is what the Army is all about. We must concentrate on the *US* at the expense of the *MY*. You and your fellow soldiers at all ranks are asked to make that sacrifice. You must put aside *my* career, *my* efficiency report, for the good of the whole—the team. When you do that, your soldiers will know and follow your example.

Our Army—any successful army—is not an individualistic entity; it's a team. It is interdependent, not independent. It needs a climate of leadership that nourishes the development of teams and units and produces the synergistic effect that wins battles despite the odds. It begins with you, the leaders, and you have got to understand that equation.

There's a great tendency to go charging off in all kinds of programs that address themselves to individual problems that have surfaced in individual systems. In the soldier world, there are all kinds of programs: soldier welfare, recreation facilities, etc., and all those are very important. All of them sort of pale in the face of the analysis we've done here. I maintain that, if you do this sort of thing, if you put the soldier in the unit where there is that cohesiveness and the climate of leadership, many, many of those other problems will take care of themselves.

Now, it isn't going to be easy. There are all kinds of distracters. So many requirements on time—yours and your unit's—that the environment can best described as downright hostile. You and your unit will be battered from all sides by important things—fund drives, extra duty requirements, inspections, parades, on-duty education, deadlines, schools, crises, crises, crises. You can command from day-to-day if you choose, stamping out the fires as they pop up. You won't be moving toward any goal and you may even survive, but you won't create the type unit that wins battles.

There is another way to go about your task. You've got to look at the problem of command, both objectively and systematically. It doesn't mean to replace the art of command with management, but it does mean organizing for the job systematically using some selected management techniques. The best one I've ever seen for this is the one taught here at Fort Leavenworth—the Estimate of the Situation. If you use it right, it'll help you build an input-output model of your unit. Once you do that, you can compare the unit to the model and see where improvement is needed.

For instance, let's take a look at a possible battalion model. First of all, what's the mission? That's easy; for almost all units it's some aspect of combat readiness. There are a few units with operational missions right now, but they also have a task of being combat ready. Now, how do you get that? Well, I don't know any miracle workers who can wave their hand and achieve it, so I fall back on the old, sure way—training. Training is the process that achieves combat readiness. How often have you heard people, leaders and followers, say, "If we didn't have to train so much, we'd be combat ready"? That makes me cringe. I'd love to ask, "How then do you get combat ready?"

Anyway, we have a goal (an output), a process (training), and we need some inputs. What are they? First are soldiers, then trainers (leaders) and resources—money, equipment, POL, time, training areas, ammunition. Now, you put them all together in the training process—a plan with scheduling. Then place them in the environment of your unit. If you look honestly at that environment, you'll find it's very hostile, all those distracters I talked about, but they can be coped with and some of them even solved.

Now, we need one more thing, a feedback loop that gives good communication on how all this is working. That feedback loop is where you can apply some of the techniques of systems management and organizational effectiveness.

Seems easy, doesn't it? However, your work has just begun, because you have to analyze each of those inputs to determine how much you have and whether it's enough. Because they're all interdependent/interrelated, you also have to analyze how a change in one will affect the others. If you increase the numbers of troops in training, you need more of almost everything else. If you decrease the time, you'll need more instructors and training area so you can get the training done.

You can tinker with a model like this as you get to understand it before trying it out in the unit. You can troubleshoot the process. For instance, all things, all systems, living or nonorganic, have certain common system characteristics. You might call these warning signs. System designers would call these warning signs discrimination processes. Let me give you an example: If your unit is working overtime, that's a form of cueing and it tells you something isn't up to the system load. If your unit isn't getting information through the chain of command, both up and down, you've got filtering or, maybe worse, omission.

Historically, in any system, and particularly Army systems, the answer has been multiple channels—informal and formal, deliberate or accidental. But remember, these warning signs are just symptoms, and it does no good to cure symptoms; your job as a leader is to find and fix the root cause, the bottleneck in the message center or whatever it is. With a little model or plan of your unit, you can examine it in some detail.

This, then, is what the model looks like when it's all in place. Every unit won't have exactly the same type model. If you don't like the word model, call it an estimate of the situation or a plan. Whatever you call it, it's something you'll need to do if you expect to have an effective unit with effective organizational leadership. I submit to you that you'll have to do something of this nature or you'll always end up in an action-reaction cycle—crisis management. You'll always have warning signs flashing—cueing, error, filtering, etc. Your unit won't achieve the necessary organizational cohesion, and you'll never get the benefit of the added plus—25 percent and more—the synergism we need to win.

Now, to some of you, this may sound like thinking that stems from bearded, long-haired, sandal-clad wonders who have sold the Army a bill of goods. Well, I came here personally to tell you it's not. I don't have long hair, in fact just the opposite. What I am trying to prepare you for is the fact that we've done a lot of introspective studying in the Army in the last few years. Some was surely generated by the bitterness of Vietnam, other by the turmoil of the early 1970s that burst upon the Army. Whatever the motive, it hasn't been wasted time. We learned some things in depth for the first time. We found out that there was more to leadership than 7 or 9 or 11 principles and a lot of charisma.

Charisma is wonderful. If you have it, flaunt it. But for most of us who don't have it, it's impossible to fake it. More important, charisma is part of the problem. It causes us to focus on leadership as an individual thing, more of the *my* syndrome I talked about earlier. We found that really successful leadership that results in an outstanding organizational performance focuses on the *us*. Leaders really do lead organizations—platoons, companies, battalions. These units are usually only mildly responsive to individual leadership, less so the higher up you go, and not at all if it's phony.

Units need organizational leadership, the kind that draws on the strength of the organization itself. It comes from planning ahead; from the confidence of the leaders themselves and the confidence of soldiers in those leaders; from motivation; and from training, training, training. It doesn't just happen, and if you wait around for it like a knighthood, it won't. It takes hard work and a leadership climate that only you, as the commander, can create.

Let me congratulate all of you for having been selected for command. There is no finer, more rewarding, and more important job in the Army and for the Army. You only get one chance at it; I know you'll give it your best shot. Just make sure that the shot you give it is from the perspective of *us*, not *my*.

Command Sergeant Major Qualifications

Letter to Lieutenant General Robert B. Yerks
Deputy Chief of Staff for Personnel
21 April 1980

This letter expresses my deep concern over the change in policy governing the selection and assignment of Command Sergeants Major as announced recently in Army Personnel Letter No. 2-80 dated February 1980.

You will recall the long and arduous fight I had with Bernie [General Bernard Rogers, Army Chief of Staff] and others to revoke the policy of nonbranch assignment of CSMs. Now we are retrenching on that hard-won victory. While the needs of MILPERCEN personnel managers in filling vacancies are appreciated, the more important issue is that we meet the needs of the units being served.

Today, more than ever, units, especially at battalion and squadron levels, require a branch-qualified CSM, one with the technical and tactical knowledge and credibility developed by experience doing the same tasks that their soldiers perform.

We made the right decision when we ruled that the best branch (MOS) qualified noncommissioned officers be selected as CSM and assigned to battalions and squadrons where their technical and tactical expertise can be put to best use. Any other system is, in my judgment, counter to the real needs of units and the Army.

Leadership Instruction

Letter to Colonel Rollin T. Steinmetz
Kent, Washington
22 May 1980

During the past few years the Armor School has moved away from teaching the qualities of leaders and the principles of leadership as separate subjects. Instead, the focus is on the tasks leaders perform. Students receive instruction in such subjects as motivation, communication, counseling, decision making, planning, professionalism, and ethics. Traits and principles are included in this instruction. Leadership responsibilities are also emphasized in tactical and weapons training to the extent feasible. In summary, the thrust at the Armor School is to integrate leadership training into as many subjects as possible in the program of instruction.

Branch Qualification of Command Sergeants Major

Message to Multiple Addressees
28 May 1980

1. Several years ago some of us were members of a cabal that eventually persuaded Bernie Rogers of the wisdom of insisting on branch qualification of command sergeants major, especially at the battalion/squadron level. You recall at the time we had CSMs with medical background in artillery units, CSMs with 71L background in cavalry units, and so on. While the sergeants themselves struggled mightily to meet the requirements of their new environment, many just gave up, others couldn't make the transition, and none were happy with the system.

2. Recently, in a DCSPER newsletter, Bob Yerks announced a return to the system which we went to great pains to convince Bernie we should change. I have written Bob about this. His response sets forth all the MILPERCEN arguments about best qualified and one-cut versus two-cut board proceedings. In all, it sums up the views of those personnel managers who believe that equal opportunity for everyone in everything is in the best interests of the Army, and it reflects the normal personnel manager's bias toward making the system of selection convenient for them as opposed to making it functionally right for the Army.

3. If we believe that the command sergeants major of the Army are essential links in the NCO chain and that that chain's primary business is in the training of the soldiers, then it follows that nothing could be more important to us than that we have branch qualified command sergeants major, especially those at battalion and squadron level.

4. Therefore I intend to make an issue of this with Shy, and to try once again to get us back on track. It would help a great deal if I had your support, and that's the purpose of this message—to see if I'm a Contubernia of one, or if the Legion is with me. What say you?

Leader Development

Message to Lieutenant General Glenn Otis
Deputy Chief of Staff for Operations
14 July 1980

1. In my absence Bruce Blount responded to your requirements for MACOM comments on the leader development plan; my perception is that response was not strong enough to express my own judgment of the idea.

2. The leader development system of the Army lies in its schools, training centers, unit schools, and training activities. Indeed the whole Army is in the leader development business; the formal and visible part of that process is in TRADOC and unit schools and other training activities directed specifically at leader development.

3. That the Army at the top should have some pretty well-thought-out leader training strategy I do not question. Indeed we spent a couple of days last week trying to lay out one for NCOs and for officers. However I question that, beyond establishing amongst ourselves the thrust of that strategy and perhaps expressing that in a general way in some appropriate document, we need do more. What we do not need is a plan establishing a series of detailed matters to be monitored, reported on, and otherwise bureaucratically looked after.

4. The plan itself as proposed has serious flaws in its underlying assumptions. Especially is this true of the notions that all soldiers, or even officers, enter the Army with a burning desire to become leaders. The second is that Monographs 7 and 8 provide a doctrinal basis for leader development. Neither of those assertions is correct.

5. If Army and unit school systems periodically train leaders, I fail to understand the need for or advantages of an LDP. Are not commanders in the field giving subordinates an opportunity to develop leader skills with appropriate training, teaching, coaching, and counseling? If not, then we need to buck up leadership training in the TRADOC school system and in the Army in the field.

6. The LDP asserts that "Army personnel management systems have at least as much potential for positive impact on leader development as do schools and command practices." I disagree. MILPERCEN actions may facilitate the development of leader skills through appropriate and timely assignments, but the magnitude of the personnel managers' impact on the production of desirable leadership traits in our soldiers and leaders is not comparable to that brought about by critical school and command practices. Indeed one can make a strong case for keeping the personnel managers out of the leadership business. I would argue that training managers have a stronger claim on leadership proponency than personnel managers. Therefore I take issue with the LDP notion that the DCSPER have a DA Staff responsibility that is clearly the business of the DCSOPS, who is responsible for individual and unit training, the areas wherein leadership is practiced.

7. The idea that TRADOC is to "ensure that leader development initiatives are integrated into and aligned with all other related initiatives, e.g., Army 86, RETO NCO Development Plan, etc." sounds nice, but what does that mean? That's far too sweeping a requirement. What does Army 86 have to do with this?

8. While there may be some useful notions in Monographs 7 and 8 as a conceptual framework for developing the Army's doctrinal leadership material, the concepts contained in those documents are in no way ready to be lifted whole and placed in our training system without major modifications and a healthy dose of "does it work in the field" analysis. No one has blessed those two monographs; now they are to be the Bible. They might work fine for Sears Roebuck or IBM, but they are far too "organizational" and not reflective of military leadership requirements. It is TRADOC's responsibility to decide what the doctrine for leadership is. The rest of the "players" should step back and let us get on with it.

9. Finally, it is true that we've a lacuna in the matter of leadership proponency, and we have for some time been trying to close that gap. At the moment we believe CAC should have that role. We will, however, decide soon, and you will be informed of the outcome of our deliberations.

Ethics and Leadership

Message to Combined Arms Center
24 September 1980

1. This message reestablishes proponent responsibilities for research in ethical behavior and its assimilation into leadership doctrine.

2. For some time now we have been discussing who should be proponent for the study of ethics, especially as it relates to leadership. Having digested all parts of this discussion, it appears the following division of responsibilities is in order.

3. CAC, as the proponent for leadership, must also be the proponent for ethics, since ethics is the underpinning upon which good leadership is based. For this to be done effectively CAC will have to draw upon various other TRADOC sources. Just as CAC is now proponent for combined arms doctrine and must draw from Knox, Benning, Sill, and others for specific doctrinal inputs, so will it be necessary for CAC to write leadership and ethics doctrine based on input from the SSC, the OE Center/School and others. Only in this way can we have a fully coordinated, logically supported developmental process of leadership doctrine which uses all possible sources of information.

4. CAC must assume responsibility for research which identifies, justifies, and catalogues the attitudes and values which are the foundation of our profession and contribute to success in military units. For almost a year now these have been referred to in terms of the four Cs—competence, commitment, candor, and courage. Though this is probably a good start, we need to continually study this subject in greater detail, since the cohesion which makes good military units, like all living things, is constantly changing. Finally, CAC as the leadership proponent must study, define, and articulate the techniques, qualities, and attributes which leaders can use to modify, or otherwise transform, the soldier's entry level attitudes and values into those which produce cohesive and effective units.

5. SSC must continue to contribute to this effort in several ways. Among them is the conduct of research to identify, evaluate, and catalogue contemporary society's attitudes, values, and standards. The product of this effort should be a description of the entry level soldier's social and behavioral patterns and mores.

NCO Criticality

Letter to Brigadier General Kenneth A. Jolemore
19th Support Command
24 March 1981

Your concerns about Logistics Readiness are shared by all of us. We do need to better understand and utilize our people in the current logistical system, rather than reorganize or develop new systems. The Army's greatest potential for increasing combat effectiveness is through well-trained people and teams in well-trained organizations. Your example of the lack of integration of first and second line supervisors into the NCO ranks and the resultant impact on the performance of the activities for which they are responsible is indicative of our failure to realize the full potential of our personnel. Even worse is that the problem is not restricted to the logistical system. The care and training of the mid-grade NCO is our most serious and persistent problem overall.

TO:

C/S _____ DCSCD _____ DCST _____ DCSRM _____

DCSDOC _____ DCSPAL _____ DCSROTC _____ DCSENGR _____

SJA _____ IG _____ PAO _____ POST CDR _____

SURG _____ CHAP _____ C-E _____ FLD ELM _____

OTHER _____

ACTION OFFICER

ONE COPY TO EACH AO

HQ TRADOC
ATSC

INSTRUCTIONS/GUIDANCE: As I hand over TRADOC to General Glenn Otis, I must thank each of you for your loyal support and hard work in our very important business. I must also ask of each of you that you give General Otis the same kind of good help you've given me. These last four years have not been easy ones. The next four will likely be equally tough, albeit in somewhat different ways. Just keep up the good

MEMO FOR RECORD: work. We've done some good things for our Army -- of that I'm sure; there are equally good things waiting to be done -- of that I'm equally certain. That's your challenge -- take it -- Press on -- Good luck -- God Bless!

gen

Action Officer's Signature: Phone No.: Date:

ACTION

SUBJECT: WHAT'S NEXT

BASIC DOCUMENT:

HQ: TRADOC	DATE ASSIGNED: 31 July 1981
FILE: CDR	ACTION NO.: XXX
DATE: 31 July 1981	SUSPENSE: ASAP

ATCS FORM 119 Jul 80 Previous editions are obsolete

Command in Europe

Letter to General Frederick J. Kroesen
Commander in Chief, US Army, Europe and Seventh Army
29 March 1983

I couldn't let the occasion of your passing out of command in Europe go by without just thanking you for what you've done for our Army in Europe during your time there.

As you know, a considerable part of my adult life has been spent wandering around that landscape, to the end that professionally and emotionally as well, I've a considerable interest in and bias toward armies there, their mission, training and well-being. For years, even in the heyday of the early sixties when funds were relatively abundant, we were beset by lousy management at the top. Some of those old warlords bore distinguished names as combat leaders. Good leaders they may have been; good managers and programmers they were not—even the best of them.

The Vietnam War period was, for USAREUR, an unmitigated disaster. Everyone there from top to bottom just gave up, refraining from doing anything because everything that needed doing looked too tough. For years they programmed four quarters of activity on three quarters, sometimes two quarters, of money. As a result, there was never a decent range and training facility improvement program—Hugh P. Harris was the last Seventh Army commander who even gave it a try. BEMAR or its equivalent was always years and gigabucks behind. Even the barracks renovation program could not bail us out, for from the outset we failed to recognize the need for recurring barracks renovation—assuming that once around would carry those old facilities for another fifty years, or perhaps more.

While I know you've had a lot of favorable things going for you on this side of the Atlantic, it is nonetheless true that you've been willing to insist on and hold to a set of priorities that made the necessary improvements. As a result, my perception of USAREUR today is that, while there's work yet to do, the place is in so much better shape in so many ways than I can remember seeing it in nearly thirty-five years. And in very, very large measure the credit for that goes to you and you alone. So, folks who have spent much of their careers trying to do many of the things you've done, just have to stand up and salute your accomplishments. The generations serving there now should grow up with a far better standard set for them than those of us who had to make do with too little for so very long.

11. Logistics

Page

Readiness Rates 512
Maintenance Organization 513
Heavy Equipment Transporter Requirements 514
120mm Gun and Ammunition Program Management 515
Tank Replacement 516
Logistics Planning Factors 519
Ammunition 520
Transportation Planning 521
Recovery of Battlefield Losses 522
Maintenance for Battalions 523
Concept-Based Requirements Strategy 528

Readiness Rates

Letter to Brigadier General Clay T. Buckingham
1st Armored Division
18 March 1974

It's always a little discouraging to note that we seem to have to start all over again with every new generation of subordinate commanders.

When I got here last summer I found the 194th Armor Brigade had not had an OR rate over *50* percent for over 3½ years! In addition to most of the techniques you used—I just reinstituted what amounts to a CMMI. We give a grade—people pass and fail—commanders who don't get with the program get moved on. Also I had to put about $1 million in my GS maintenance facility to get it moving—they had drawn down on it in the two 5 percent civilian manpower reductions to the point that they simply couldn't do the work. We actually had tanks deadlined for 6–8 months and everyone thought that was OK. As of end February the 194th is pushing 70 percent OR and going up. We'll get there, but God what a struggle!

Maintenance Organization

Letter to Colonel Grail L. Brookshire
3d Armored Cavalry Regiment
12 November 1976

Supply, less automotive, yes—that can be consolidated at battalion level, and probably should be. Maintenance and the supply of spare parts cannot and should not be run by the S4. The battalion motor officer has to be a principal right-hand staff officer for the battalion commander—there isn't any other way. I'll continue to resist that to the bitter end.

Heavy Equipment Transporter Requirements

Message to Multiple Addressees
3 March 1978

1. LOGC is reviewing Army requirements for HET. I have been over that review with them, and have subsequently reviewed inputs provided by Knox and Leavenworth. . . . I find the Leavenworth and Knox position statements well short of the mark, analytically as well as conceptually.

2. The Army requirement for HETs must be viewed in the context of total requirements to (1) move disabled vehicles to and fro as they are repaired and replaced in the fighting-fixing part of the battle, (2) to move replacement vehicles in substantial numbers forward into corps and division areas, and (3) to move divisions from one corps area to another in Europe, or from a lodgment area to a forward location in a contingency corps operation. If one simply adds up modest estimates of what this might require, the stated positions from Knox and Leavenworth are almost ludicrous.

3. LOGC has what I believe to be a rational case for a considerably increased HET requirement. While it may be short of the total requirement by some margin, it does make a rational case for the to and fro problem, somewhat for the replacement problem, and recognizes USAREUR's stated requirement for the means to move at least the tanks of one armored division from one corps area to another by HET. I support the outcome of the LOGC study.

120mm Gun and Ammunition Program Management

Message to General Walter T. Kerwin Jr.
Army Vice Chief of Staff
25 April 1978

. . .

2. . . . I'm convinced that separation of the gun and ammunition development effort could add considerable risk to an already risky effort. I'm remembering Sheridan now—we had the weapon system ready before we had acceptable ammunition available. Then the developer tried to con the Army into deploying the thing—a 300K machinegun platform, to avoid the embarrassing admission that we hadn't a complete system. In the end we accepted when we should have refused to take the system. Never again!

Tank Replacement

Letter to General Frederick J. Kroesen
Army Vice Chief of Staff
6 November 1978

The Tank Force Management Group . . . recommended development and testing of personnel and tank replacement systems which would be more responsive to requirements of modern battle. Obviously the problem is one encountered not only with tank systems, but with all other weapon systems as well. . . .

The present method for replacing tanks and all other crewed weapons systems is not systematic, efficient, or feasible under conditions envisioned on the modern battlefield. Personnel are managed, requisitioned, and supplied separately and without coordination with the equipment they are to man. In the case of the main battle tank, the ancillary equipment such as radios, machineguns, and fire control instruments essential to operation of the tank must be ordered separately. Consequently, a tank weapon system is not normally ready to fight until the battalion commander assembles the components in his field trains area. The same is true of all other fighting vehicles and crewed weapons.

The Current Personnel Replacement System

- Each battalion reports its personnel status via the Personnel Daily Summary (PDS) and the Periodic Personnel Summary (PPS). These reports are transmitted from the battalion adjutant to the division G-1. The G-l forwards these reports to the Adjutant General (AG) for use as an assignment guide. . . .
- Upon mobilization, personnel replacements for the first 60 days of war are based on a "shelf requisition" recommended by the theater army commander and prepared by the Department of the Army Military Personnel Center. . . . Replacements will be transported from CONUS to the air port of debarkation in the corps rear. A detachment located near the port coordinates and processes all replacements and arranges their transportation to the division replacement detachment, where the AG assigns personnel to units.

The Current Equipment Replacement System

- With the onset of hostilities, fighting vehicle losses are reported by the battalion S-4 via the Battle Loss Report (BLR) and verified by the Logistics Status (LOGSTAT) Report. The BLR is transmitted on the command radio net sequentially from battalion to brigade to division. The LOGSTAT Report is forwarded through logistics channels. The reports are cross-checked at the Division Materiel Management Center, which then prepares a requisition for tanks or components and transmits them to the corps. If the corps cannot fill the requisition, it will pass it to the theater army, which will either fill the requisition or pass it to the National Inventory Control Point (NICP) in CONUS.
- Pre-Positioned War Reserve Stocks (PWRS) are located in the Theater Army Area. A portion of PWRS may be located forward in the corps area. In peace, PWRS are under theater army control. Upon mobilization, PWRS in corps area may be transferred to corps control.
- PWRS must be deprocessed by a Heavy Materiel Supply Company or its equivalent. Because of current manpower and structure austerities, Heavy Materiel Supply Companies are not deployed. Unique peacetime procedures have been established in which local civilian

labor under US cadre deprocess tanks and other vehicles. The Civilian Equipment Groups, Europe (CEGE) in effect replace the Heavy Equipment Supply Companies in the active force structure. The Heavy Equipment Supply Companies are assigned to the Reserve Component for deployment after mobilization.

- Replacement systems come from either CONUS or PWRS in theater army or corps. They are deprocessed and made ready to issue. By present definition, for example, a "Ready for Issue" tank has the main gun mounted and it is removed from preservation. It is transported by rail or Heavy Equipment Transportation directly to the battalion field trains in the brigade support area. The tank is not "Ready to Fight," however. First the fire control instruments must be mounted by division direct support ordnance teams; then radio mounts, radios and antenna mounts must be installed by the battalion communication section. The crew must then stow the basic issue items (BII), mount machineguns, stow ammunition, and fuel the tank. Then the crew is ready to bore-sight and zero the weapons, finally making the tank "Ready to Fight." The same routines apply to other systems.
- Again considering the tank system, if a tank is damaged in combat, the extent of damage is assessed by maintenance personnel from the forward support company. If there are surviving crew members, they normally remain with the tank if it is repaired within the division by direct support maintenance. If the tank must be evacuated to general support maintenance in corps it becomes a loss to the unit. Once the tank is repaired at GS level it becomes an asset in the supply system and the corps Materiel Management Center assigns the tank as a replacement.

Shortcomings of the Present Replacement System

- There is no weapon system management above battalion. The separate management of personnel and equipment components seriously limits the corps and division commanders' ability to set priorities, concentrate forces, and maintain a current view of command capabilities.
- Final processing, loading, and link-up of weapon system components in the battalion field trains, located in the brigade support area, are time-consuming and delicate operations. This area will probably be under enemy indirect fire and will be displaced frequently.
- Crew members arrive in the battalion field trains as individuals and are not linked up with their weapon until it is in the brigade support area.
- Direct fire weapon systems are therefore not capable of engaging the enemy or defending themselves until they are forward in the zone of combat and vulnerable to enemy fires.

The enclosed operational concept describes Weapon System Replacement Operations for the tank. The concept is obviously applicable to other critical weapon systems with appropriate modification.

We believe we need to agree on this concept for test, then test it. We further recommend that the Weapon System Replacement Operation concept be field tested in USAREUR. Field tests should particularly evaluate the most effective location for loading main gun ammunition and other necessary equipment onboard a replacement tank. Options are:

- In COSCOM by labor service personnel.
- At Corps Ammunition Supply Point by ammunition handlers or labor service personnel.
- In Division Support Area by crew. This would require establishing an ammunition point in DSA.

- In Brigade Support Area by crew. This option precludes tank being ready to fight when it leaves the DSA.

I believe the fixing of our Weapon System Replacement Operations is of the utmost urgency to the Army. First, I'd like to solicit your agreement to the concept, then agreement to test it in USAREUR. While I know full well the drawbacks to testing in an operational theater, I am also persuaded that's the only place we can give a system a good wringing out. Therefore, I come down on the side of doing the test in Europe. I solicit your support.

Logistics Planning Factors

Letter to Major General Homer D. Smith
US Army Logistics Center
10 January 1979

This responds to your letter . . . regarding days of supply as a means of expressing logistics requirements to support the battle. While I tend to agree with your rationale for retaining the concept of days of supply as a logistics planning factor, I am also concerned that it does not adequately describe two very important requirements which must be met if we're to succeed in battle. These requirements are: first, the demands of the first few critical days of the battle when numbers and positioning of stocks will be all-important to survival; and second, the surge rates that can be expected as second-echelon forces arrive to participate in the first-echelon battle. There is a tendency to apply a day of supply factor to every day of the battle, and as you know that just won't work. Nor does it help us much in evaluating whether or not we've the right stockage levels, safety levels, delivery rates, emergency procedures, and transportation to support surge requirements generated by periods of increased intensity.

While you probably will need to solicit USAREUR's help sooner or later, I would prefer that we try to bracket the problem first, then go to the field and validate our hypotheses. To do this I'd like for you to use the work being done at Leavenworth in response to my question about what we must do to the Soviet second echelons in order to enable friendly forces fighting the first echelon to attack. What we need is a summing up of what will be required to support the defense against the first-echelon divisions—the losses derived from my Corps Battle presentation should suffice for that. Second, we need a summing of what it will take to support the offensive action we see as following on the defense, to include the logistics requirement generated by forces deploying from CONUS should there be time for that. Third, we need a summing up of what it would take to support the battle against second-echelon divisions and the second-echelon army when they arrive, or when we attack them. Fully developed, these summaries would provide us the distribution function of losses, replacement requirements, repair requirements, and evacuation requirements—materiel and medical, and from that we could develop our structure, equipment, and organizational needs to satisfy the distribution function. In any event, what you've got to do is figure out what it takes to support the battle we believe we're going to fight, then assess whether or not we can do it, identify the shortages, and from those state the requirements. Your analysis of the V Corps battle and structural requirements for logistics forces in the corps slice is also a good place from which to start.

Ammunition

Letter to Major General William R. Richardson
Office of the Deputy Chief of Staff for Operations and Plans
25 April 1979

I hope we can get the ammo thing lined up soon in a way that will allow us to make a more coherent argument for how much we need. As you know, the whole ammo account has long been used as a sort of balancing account in the budget process—some would say it's a big slush fund. While I wouldn't go quite that far, it is still a big dollar item and has a long-term get well time if it's sick. Therefore it is even more important that we be more precise and definitive about what we really do need to do the job. We'll keep at it.

Transportation Planning

Message to Lieutenant General John R. Thurman
Fort Leavenworth, Kansas
3 May 1979

. . .

4. Transportation Sizing and Basic Load Derivation. Okay to publish as outlined. This will result in a lot more trucks than we have now. Therefore proponents should look to capacity—weight versus cube, numbers of drivers, and so on. I shouldn't have to say that, but just this week I've found one bunch computing their lift capability at 24 hours a day with one driver per truck, and another group with 5-ton trucks on which you can put 2½ tons of missiles versus 10-ton trucks on which you can carry 8-plus tons, and so on. Remember Starry's third rule of combat developments—if you have to do something dumb, please fix it before I find out about it.

Recovery of Battlefield Losses

Message to Major General Fred Mahaffey
Fort Leavenworth, Kansas
29 May 1979

. . .

2. I haven't seen the studies referred to, but sense from your comments [message 252105Z May 1979] that they are askew. Accordingly I'd like to do the following:

a. CAC assume responsibility for setting forth operational concepts for recovery of battlefield losses of major end items of equipment. CAC then develop, based on the operational concept, what kind of equipment is required to meet the needs of the concept, how many, how organized, and at what echelon it should be assigned.

b. The brigade is not a logistical echelon, nor is it operative in the recovery and evacuation net. Therefore the brigade is not an appropriate echelon at which to centralize recovery and evacuation assets or responsibilities.

c. There is sufficient data on the unsuitability of M578 for support of armor units that we need not have another test. My Vietnam monograph and my after action report as CO, 11th Cavalry should be sufficient evidence, and there is more besides. The M578 is probably marginal for support of SP artillery units for which it was designed in the first place. We should have a single series recovery vehicle for armor units, and that should be the M88.

d. Part of the recovery and evacuation issue has to do with HETs as part of the recovery and evacuation team. The questions again are those of how many, at what echelon, how organized, and how controlled. No study of this subject can be complete without taking cognizance of this part of the problem.

3. CAC, with LOGCEN and others as required, draw up the operational concept and report that to me by late June. We'll decide where to proceed from there.

Maintenance for Battalions

Message to Multiple Addressees
15 October 1979

1. This message responds to Tom Lynch's KNX 0459: Consolidated Maintenance Company—Division 86. It sets forth an operational concept for maintenance operations at battalion without reorganizing, and without adding the spaces required by the maintenance company concept. I propose that we use it in describing battalion-level maintenance operations in Division 86 units. In it you will see that I've used a mech battalion as an example, and I've used a USAREUR mech battalion MTOE. It will now be necessary to work that back into a TOE, and work out of that into other battalions to which the concept is to be applied. I would like for you all to start work on that course of action, setting aside maintenance companies and other alternatives for the moment.

2. Background. The Army's repair parts and supply systems are designed to be responsive "wholesaler to retailer" operations. Today's maintenance system of direct support, general support, and depot/rebuild has been in use through two wars and 30 years of peacetime experience with an increasingly mechanized Army.

Nevertheless, despite our considerable experience with the system, we know that the maintenance status of our equipment is marginal at best. All over the Army, METs, AGIs, roadside spot checks, MAITs, Army Audit Agency reports, and random visits to any motor pool indicate that we don't do well in maintaining our equipment. For example, in one division over a two-year period, inspections indicated that only about 1/3 of the vehicles declared operationally ready (green or amber) were in fact ready. Two-thirds were actually red—red for basic reasons, not for things that required a depot-level expert to find.

The apparent anomaly of well-designed maintenance and supply systems, but poor maintenance status, suggests the need to examine carefully the organizational maintenance structure. The following discussion describes why this is so and, in the concept which follows, proposes an operational concept that could be adopted without an increase in resources.

3. Discussion. The organizational maintenance system consists of two principal parts: (1) TOE structure at battalion and company and (2) the operator and his first-line supervisor, the actual users of the equipment.

In order to understand the hypothesis that our maintenance problem lies fundamentally at this level, let us consider, for the sake of developing the hypothesis, a USAREUR-based mechanized infantry battalion.

. . .

The data show 184 vehicles are to be maintained by 76 maintenance people, with the help of 184 drivers and 184 supervisors. The procedures established are designed to maintain retail parts and a basic repair capability at company level; these are backed up by increased technical capability at battalion.

Now it is important to understand two things: preventive maintenance and basic fault detection depend upon the driver—the least trained and the least technically qualified link in the maintenance chain. In addition, fewer than half the TOE maintenance people actually perform maintenance. Let us expand that idea.

Let's look first at the driver and his first-line supervisor. How do they perform maintenance? The driver is charged with before, during, and after operation checks. He must know how to drive on roads and across country, with lights and blacked out, buttoned-up or not. He must be able to select a position on the move, handle the intercom, swim the vehicle, and keep his own gear maintained. He must also know how to fill out a 2404, perform "operator maintenance" on his vehicle, and record appropriate information in the vehicle log book. Dozens of other tasks are assigned to drivers, but already it is apparent that the lowest paid, lowest grade individual is charged with the largest part of the total organizational maintenance system. Moreover, the driver's supervisor is an NCO who surely has several other tasks. His supervision of the driver is often perfunctory—limited to making sure the vehicle has fuel, oil—that it runs.

"Command attention" to driver maintenance comes and goes with changes of command. Motor stables, predispatch TI, roadside spot check, are in and out. Why is it that maintenance is still poor? The answer is that it's wrong at the foundation. The driver does not, and probably cannot, cope with all the tasks required of him.

Now let's consider the people listed as maintenance personnel. There are 5 PLL clerks, 5 TAMMS clerks, 2 dispatchers, 9 recovery specialists, 6 drivers, a maintenance warrant officer, a battalion maintenance officer, and 6 senior NCOs—among the maintenance personnel—all these are either paper handlers or supervisory personnel, not maintenance doers. This leaves 41 people divided amongst 5 companies and the battalion maintenance section to actually work on 184 vehicles. If supervision were nearly perfect, the job would still be almost impossible. We also know that it's a rare battalion in which all 5 company and 1 battalion motor sergeants are in fact well qualified and present for duty. Supervision is therefore another weak link in the maintenance chain. Less than 54 percent of the maintenance strength is actually performing maintenance.

The problem is balance in the organizational maintenance system. The driver must indeed be in the scheme, but in his proper place. There must, however, be more hands-on work from full-time maintenance people. This can be done by scaling down the paperwork and upgrading inspections and repair service. And that can be done within the current resource allocation (people and equipment) and without upsetting the DS-GS-depot maintenance system.

4. Limitations. The concept in this paper can be applied to mechanized infantry and tank battalions, cavalry squadrons, self-propelled artillery battalions, and other equipment-heavy battalions.

5. Organizational Maintenance Concept—Garrison. The concept is based on five functional operations, all conducted by the battalion maintenance organization, assisted by the driver:

a. Post-operation inspection.

b. Qualified maintenance service.

c. Quality control inspection.

d. Driver assistance.

e. User care—mechanic service.

(1) Post-operation inspection (post-op). When a vehicle completes a dispatch, the driver with vehicle reports to the postoperation section. At that time his vehicle must be clean and refueled. The post-op NCO makes a technical inspection of the vehicle and debriefs the

driver concerning maintenance symptoms or problems. The post-op inspector executes a 2404 on which he records all deficiencies. He checks vehicle 2404-14 (uncorrected fault record) to verify deficiencies previously recorded and to verify parts on requisition. He takes the 2404 to the maintenance service section NCOIC, who then assumes responsibility for the vehicle and for fault correction. The driver is present, with his dispatch ticket, ready to assist the service section during the next step.

(2) Maintenance Service (MS). The service section NCOIC checks the 2404; he estimates the time it will take and the number of mechanics required to repair the vehicle. He schedules mechanic(s) to do the job and directs the driver to assist. When the mechanic reports completion, the NCOIC checks, properly annotates the 2404, and takes the 2404 to the quality control section. The driver still has his dispatch ticket and is with his vehicle.

(3) Quality Control Operation (QC). The NCOIC of quality control accepts the 2404 and 2404-14 from the repair section NCOIC and assigns an inspector to the job. The inspector checks the entire vehicle, with emphasis on the corrected faults. If all work has been done properly, and all uncorrected faults are properly recorded with appropriate requisition action noted, the quality control NCO collects and releases the trip ticket, authenticates the 2404, and certifies that the vehicle is prepared for the next required dispatch.

This simple three-step process can be accomplished within current personnel assets. However, several details need clarification. We'll start again with post-op. When the driver returns from dispatch, he must get a release from his dispatch. To do so, he has to report to post-op with a clean, refueled vehicle. If he returns after duty hours, he reports first thing in the morning. The important thing is that he cannot turn in his trip ticket until he gets through post-op, MS, and QC. If post-op inspection finds no deficiencies, or deficiencies that are already covered by 2404-14, then postop can skip MS and take 2404, 2404-14, and driver directly to QC.

MS is the second step, and it is here that unit pride and individual pride of workmanship can make a real difference. The payoff to MS is measured by: (a) number of vehicles on deadline and (b) number of vehicles rejected by QC. MS is also the connection with the repair parts system. MS puts the demand on the retail parts supply systems. They initiate DX requirements. The NCOIC of MS will be a high pressure, talent-demanding job, for the incumbent must know how to estimate labor requirements and repair time. He must be an organizer and a manager. He must schedule vehicles in and out of shop. His section is the largest of the maintenance group. The OR status of the battalion will to a very large degree reflect his ability.

QC must be an independent section. Its NCOIC reports to the top—the battalion maintenance officer. He must be the honest broker of the organizational maintenance system. QC should be a small section, say one NCO and four or five inspectors. Each man in the section, to include the NCOIC, must be a hands-on person with the courage to turn back sloppy work.

. . .

6. Operational Concept—Field Operations or War. Will the system described above work in field operations or war? The answer is in two parts. When the battalion is fighting "pure," without reinforcement or without losing a company, organizational maintenance remains at battalion level; company commanders are responsible for recovering and evacuating vehicles needing more than user maintenance to an area where a battalion contact team with parts and

tools will make battlefield repairs, or will direct further evacuation to the direct support repair area—normally the brigade trains area.

When the battalion is cross-reinforced—that is, when a company-sized unit is cross-attached to another battalion, a contact team with repair parts and organizational maintenance tools goes with the company. This contact team supports the company under the control of the gaining battalion's maintenance section, operating exactly as before. The company retains its recovery equipment and has full responsibility for user maintenance of its assigned vehicles. The battalion has backup recovery and evacuation capability.

7. Notes on the Concept. It can be argued that this system takes the company commander out of maintenance, that his motivation to field and to keep operational all of his vehicles will be weakened by this emphasis on maintenance at the battalion level. There is overwhelming evidence to show such arguments to be wrong. For example, all communications equipment and all weapons are required at battalion or higher level, not at company level. The proposed system simply extends that notion to automotive maintenance. The system actually provides for a more efficient structuring of the total maintenance effort. Each command echelon is assigned specific responsibilities and given assets to do the job, to wit:

User Servicing	Co Cdr (driver/vehicle cdr).
Organizational Maintenance	Bn Cdr (BMO).
Direct Support	Bde Cdr (maint bn/fwd spt).
General Support	Div Cdr (maint bn/HQ & A Co).
Depot-Rebuild	Corps Cdr/Theater Cdr.

So the company commander is still responsible for his vehicles—for oil, water, fuel, for reporting system failures, for proper use of the vehicles, for ensuring drivers are properly supervised and available to the organizational maintenance echelon as required.

Meanwhile the technical supervisor and qualified repairmen are concentrated at battalion. Clearly the number of mechanics, compared to the total vehicle density of the battalion, shows that with only average supervision the mechanics can easily keep each vehicle in a high state of readiness. There are 107 track vehicles and 77 wheeled vehicles in the battalion. With 26 mechanics in the MS section, this means that each mechanic needs to handle on the average only seven vehicles throughout the year. Obviously such a ratio allows for highly professional care for each vehicle. The good battalion commander will find ways to associate maintenance crews with vehicle crews to develop team pride and unit esprit in the OR status of the battalion.

Nothing in the proposed system is designed to downgrade or eliminate the requirement for recordkeeping and for a disciplined entry to the established repair parts system. In fact, the only change proposed is to reduce the number of people doing duplicative jobs. For example, currently there are five PLL clerks maintaining five separate PLLs. The workload requires only two, but the necessity for each company to have a separate PLL generates the requirement for five PLL clerks. In the proposed system there are only two PLL clerks. The number of lines of PLL and the volume of requisition actions require only two men. Obviously the parts can be stored on five separate trailers in order to deploy with company-sized units when and if necessary.

8. Conclusions. Today's maintenance indicators show clearly that reported OR status reports do not accurately reflect actual vehicle conditions in today's Army. In addition we use more spare parts, major components, and assemblies, and have less mileage traveled, then we expect in battle. Part of the reason for all this lies in the diffusion of maintenance resources down to company level, with the additional overhead that such decentralization automatically demands. User maintenance is a key part of the total system. The driver, first-line supervisor, and company commander are crucial to the user echelon of maintenance. No changes are proposed in this echelon. Organizational maintenance is fragmented today between company and battalion. The new system concentrates responsibility for organizational maintenance at battalion level. It provides for an independent trouble-shooting team to find maintenance faults (post-op), a technically qualified and supervised repair team (MS) to correct those faults, and a separate "honest broker" team to validate the maintenance work and to ensure a finished product (QM). It focuses on each vehicle as a separate and necessary system. It gives the battalion commander a clearly identified activity to keep his vehicle rolling. It acknowledges that the driver should and must keep his vehicle clean, keep the BII accounted for and serviceable, and service the vehicle with fuel, water, and care for batteries and tires. But it also admits that, beyond such user functions, real maintenance requires a technically qualified service person and an organized supervisory system.

Concept-Based Requirements Strategy

Message to Multiple Addressees

30 January 1981

1. At the fall 1980 TRADOC Commander's Conference we introduced to you a new materiel acquisition strategy that looked at the conceptual focus and resources for mid-term (83-early 90s) and long-term (mid-late 90s).

2. To address the mid-term resource problem we received approval at the Army Commander's Conference for a joint DARCOM/TRADOC effort at selective modernization. That effort, as you know, is ongoing and will be briefed to DA in early February. For the long-term problem, we proposed at the Commander's Conference a concept-based strategy that would drive our long-term needs and thereby provide a sharper focus for investments by the DARCOM tech base. While the strategy emphasized how the user will communicate materiel needs, it is also applicable in stating future needs in force structure, doctrine, and training. Both the DCSOPS and the DCSRDA gave presentations amplifying the importance of these long-range concepts to the long-range RDA planning process. That concept-based strategy was endorsed by CSA. The purpose of this message is to sketch out how TRADOC will implement that concept-based acquisition strategy.

3. The idea is relatively simple. The keystone to this approach will be the concept for the Air/Land Battle of the 90s that you have recently staffed. We now need to use that broad operational concept as a basis for writing eight functional operational concepts. These operational concepts will describe the TRADOC mission areas. Following development of these eight concepts we will be able to begin MAAs that will allow us to describe needs for the mid-90s out to about 2000.

4. A little more explanation of the eight mission area concepts is in order. What we have done is to move from the battlefield functional concepts used in BDP I and II and Army 86 studies to the emerging mission area breakdowns that are more closely aligned to the OSD and DA mission areas, yet still retain all the critical battlefield tasks that were described by the old BDP functions. What we have then are eight functional mission areas: close combat, fire support, air defense, communications, command and control, intelligence and EW, combat service support and combat support, and engineer and mine warfare. Aviation and NBC are inherent across most all these functional areas and will be integrated as appropriate into the eight concepts. MAA related to these mission areas, as well as MAA for aviation, NBC, and battlefield nuclear warfare, will then be performed based upon these functional operational concepts.

5. Our experience in developing the Army 86 organizations reinforces the benefit of forcing those TRADOC elements involved in a particular function to collectively address a problem. It therefore appears prudent to capitalize on our Army 86 experience and designate general officer leadership for the development of these functional concepts, and also to designate the support teams assisting in that effort. Complete methodology for concept development and task force organization was presented to your action officers during the January concepts and doctrine conference at TRADOC.

6. Army 86 brought about concepts and force structure to best use the equipment already under development. If we do it right, the concept-based acquisition strategy will guide investments in today's tech base that will result in materiel that fits a concept of how our Army should fight in the 90s.

12. Management

Page

The American Military Profession Today 530

Managerial Skills 538

Consolidated Administration 539

Battalion-Level Consolidation 540

Consolidation at Battalion Level 541

Centralized Management 542

Living Systems 543

The Over-Under Army 544

Centralization 549

Pressing On 550

Extended Command Tours 551

Standardization and Evaluation 553

Strategies for the 1980s 556

Effectiveness of Army Units 558

Improving Cohesion 560

OSD Characterized 564

Current Army Culture 565

Completed Staff Work 568

Technology Advocates 571

New Readiness Command Functions 572

Readiness Command Status 573

Joint Staff Issues 577

The American Military Profession Today

US Army War College Student Research Paper
Class of 1966

What of the faith and fire within us
Men who march away
Ere the barn-cocks say
Night is growing gray
Leaving all that here can win us;
What of the faith and fire within us
Men who march away
Thomas Hardy, *Song of the Soldiers*

It was Alexander Hamilton who once observed, "Safety from external danger is the most powerful dictator of national conduct." He might well have added that, as external danger appears more threatening, the national conduct more and more reflects universal concern for the safety of society. Thus it is in our time that war, once the special domain of the military man, has under the nuclear shadow suddenly become everybody's business. And the American military profession, long isolated from society, abstemious of political affairs, reluctant to cast judgments other than purely military, suddenly finds itself the object of scrutiny from every quarter. Civilian strategists challenge the traditional role of the military in strategy and tactics. Sociologists examine the composition and attitudes of the membership of the profession. Economists apply their art to comparing the relative worth to society of various expensive and complex modern weapons. In Samuel P. Huntington's *The Soldier and the State*, and in Morris Janowitz's *The Professional Soldier*, the academic community began a searching evaluation of civil-military relations and attitudes of the military profession. In the pages of their professional journals the military themselves have sought to reappraise the role of their profession.

Precise descriptions of the challenges confronting the American military profession are as varied as the professional terms of reference of authors who have addressed the subject. However several common features emerge from the welter of debate about war, weapons, and the military. The discussion which follows summarizes what seem to be the most significant challenges facing the American military profession today.

First, the nuclear weapon has brought to many the feeling that war is now too important a matter to be left to generals. The civilian academic community has entered the field of military strategy with a zestful profusion of books, articles and studies. Civilian "think factories" analyze, debate, and test new ideas about strategy. The military, preempted in their traditional role as designers of strategy, generally view this activity with alarm. One military author has remarked that virtually all the important military policy decisions of the Kennedy administration were made by, or on the advice of, civilian strategists. Another military writer has complained of the declining prestige of military opinion offered in the councils of government.

Second, science and technology have contributed to a management revolution as profound as the General Staff reforms which shook the Army at the beginning of this century. Modern communications have made possible direct contact from the White House to the destroyer bridge and battalion command post. Electronic computers offer management innovations undreamed of a decade ago. Advanced theories of economic analysis have provided new decisionmaking techniques. A vigorous Secretary of Defense has brought to his department new ideas and young men to combine these tools into a new pattern of defense management. The young civilian defense manager, understandably impatient with the bureaucratic inertia and complex procedure of military staff systems, has struck out on fresh avenues to solve the complex problems of defense. These innovators describe the military mind as narrow, dogmatic, prone to oversimplification; and they look upon senior military officers, men of another generation, as antiquated—in education, idea, and action. The military complain of excessive supervision in the name of civilian control, of the burgeoning growth of civilian hierarchy in the Office of the Secretary of Defense, of the ineffectiveness of the joint staff system, of interference by the Whiz Kids in purely military matters. Traditional prerogatives of the field commander, long hallmarks of the American military way, have fallen—they say—to the civilian penchant for oversupervision from the top. Outside the civilian and military population of the defense establishment, liberals complain that the defense civilians have become militarized, their military counterparts have become civilianized, and together the group represents a threat as dangerous to liberal democracy as the military man on horseback once was all by himself.

Finally, the changing nature of international relations brought on by fear of mutual nuclear annihilation and the demise of colonialism has thrust the military profession into a new role in policy deliberations at home and action programs abroad. Samuel Huntington has written of the changing nature of war—from intergovernmental to intragovernmental conflict, from the use of force in total conflict outside the borders of most participants to total strife contained within the boundaries of a state. Lacking the ability to effectively coordinate the economic, social, and political strength of the nation in effective foreign policy programs, the United States has increasingly employed its military establishment as the primary instrument of foreign policy. As seen by the nation's leaders, the new involvement of the military in foreign policy and changing conflict patterns demands a new outlook and broadened skills on the part of the military profession. Perhaps the most perceptive spokesman to define this challenge was President Kennedy, who remarked in an address to the graduating class at Annapolis in 1961:

> You must know something about strategy and tactics and logic . . . economics and politics and diplomacy and history. You must know everything you can about military power, and you must also understand the limits of military power. You must understand that few of the important problems of our time have, in the final analysis, been solved by military power alone. . . . You must be more than servants of national policy. You must be prepared to play a constructive role in the development of national policy....

Although acceding to demands for an awakened awareness of the consequences of military action beyond traditional "military considerations," the military profession as a whole expresses frustration and some confusion about the precise goals of its increased participation in economic, social, political, and other programs with a distinctly nonmilitary flavor. And the American liberal, true to his colors, complains of military intrusion into the nonmilitary affairs of the democracy.

If the general statements just presented fairly summarize the challenges faced by the American military profession today, what should be the profession's reaction? Here again the debate is extensive, the spread of opinion wide.

Many writers have addressed the problem of response to changing patterns of civil-military relations. Almost all recommendations about this complex problem include proposals for some revision of the Pentagon staff system, both in the service departments and in the Joint Staff. There is widespread feeling of a need to clean up staff functional and administrative processes to enable military opinion to be developed more rapidly and efficiently and to be offered in a more timely, forceful fashion. One military writer has questioned the utility of the Joint Staff as it presently operates, expressing the view that the Joint Chiefs of Staff must begin to use the Joint Staff as a decisionmaking agency rather than as a device for compromise and conciliation between service viewpoints. Others point up the need for the military to develop staff methods adapted to the new black art of systems analysis and modern computerized management techniques.

Several authors have opined that the military must revise its education in terms of the changing nature of war and new patterns of management. Assistant Secretary of Defense Edward L. Katzenbach Jr. . . . has charged the service War Colleges with declining professionalism in curriculum content and in their product—the professional officer. Air Force Colonel Robert N. Ginsburgh has written of the need for military education to keep abreast of new developments, proposing the system be revamped to meet the need for education in a broader range of skills. There is rather sharply divided opinion about what turn service education reforms should take. The military, in response to the demand for a better grasp of socio-political-economic affairs, tends to favor, with Colonel Ginsburgh, emphasis on producing broadly trained generalists in senior service colleges. Others contend, with Dr. Katzenbach, that what the War Colleges need is a set of courses in economic theory and operations analysis, military strategy, and force development to improve professional performance in activities of primary professional concern over which the profession has lost sway as a result of declining professionalism.

In the opinion of many, the military profession must assess its own position, indulge in self-criticism, and restore professionalism by "taking our own Clausewitzes and Mahans out of the dust and making use of their intellectual talents."

Finally, several military authors set forth the idea that bright young men must be brought to the top of the profession early in their careers to put into the decisionmaking apparatus young, intelligent military men who can successfully cope with their civilian contemporaries in the Office of the Secretary of Defense.

Both challenge and suggested response have been well and objectively described by spokesmen for many points of view. However, virtually no one has explored the deeper rent in the fabric of the professional ethos. For, if the profession has failed to adjust to changing times, what must be done is to isolate elements of the professional ethic that have failed to grow and from that base suggest how to adjust to the needs of the years ahead.

Perhaps we should begin by describing what the military profession is that other professions are not. Samuel Huntington has broadly categorized professional characteristics in three parts: (1) expertise in a significant field of human endeavor; (2) responsibility to use that expertise for the good of a larger group—society; and (3) a sense of belonging to a corporate body apart from the layman. Writing of his own part of the military profession, Navy Admiral James Calvert has given a clearer view of characteristics peculiar to the naval profession. For the naval officer Admiral Calvert lists: (1) knowledge of the sea, both technical and folklore; (2) technical competence with the hardware of the profession; (3) the ability, judgment, and imagination to command naval forces; and (4) the skill to function effectively as a member of a military staff. If we broaden these categories to include other military services, a picture of skills and abilities characteristic of the military professional emerges. Its key elements are:

(1) Knowledge of the employment of military power—seapower, landpower, airpower; of the history, traditions, theories, extent, and limits of American military power.

(2) Competent technical skill with equipment peculiar to the military profession—ships, weapons, armor, aircraft; and in the employment of that equipment in military operations.

(3) The ability to command military forces; to provide the leadership and management necessary to marshal men and equipment into effective military formations and with them conduct successful military operations.

(4) The ability to participate in planning, management, and direction of military operations as a member of a military staff.

If the profession is to respond effectively to the challenges set forth in the opening paragraphs, it must do so in a way that will preserve and strengthen the professional character of the military. It is well, for example, to say the military education system must adjust; it is quite another problem to describe how change should be effected, preserving at the same time the requisite skills, knowledge, and ability identified as essential characteristics of the profession. Similarly, overhaul of the Joint Staff must penetrate more deeply into the military staff process itself. For, if the military staff system is to be reformed, it must retain its essentially military character and the peculiarly military validity of its product.

There are at least four primary areas in which old ideas recast in new molds may accomplish most suggested reforms and at the same time keep faith with the professional character. They are: military personnel management, the military education system, the military staff system, and military leadership. Enumeration of the details of suggested reforms is beyond the scope of this brief recital. It will serve the purposes of this discussion to set forth some essential tasks to be accomplished in the four functional areas already outlined.

Military Personnel Management. Office career programs must provide progressive development and advancement opportunity for the officer corps. Compared to other professions, an extremely small percentage of each year group entering the officer corps has any chance at flag and general officer selection. There must therefore be intermediate goals attractive enough to persuade quality personnel to stay in service.

Noncommissioned officer career programs are essential to a sound military system. Today only the Navy provides reasonably effective enlisted career management. Army career noncommissioned officers must be in the right place at the right time with the right qualifications to meet local promotion criteria, which have no common denominator and vary widely from command to command. If it is reason for concern that so many career noncommissioned officers retire at 20 years of service, we have only to contemplate the ineffective system devised to manage their careers.

Fringe benefit incentives for all service members are frequently addressed as a means of increasing officer and enlisted retention rates. We should recognize that no amount of direct or indirect benefit incentive can recompense for family separation, the hazards of combat, and the tremendous responsibilities shouldered by military leaders. What could the nation possibly pay the man who commands the lives and fortunes of hundreds of men and millions of dollars worth of equipment? Incentive programs can hope to offer only reasonable fringe and monetary benefits. The real incentives must stem from pride of service, devotion to cause, the feeling of serving a purpose essential to society, and the other hard military virtues that have historically infused the professional spirit.

Billions have been spent to advance technical progress at an exponential rate, yet linear progression of the human engineering curve testifies to neglectful inattention. A senior military figure recently remarked to the effect that we have the finest equipment science can devise, but the people who use it inevitably abuse it through ignorance and misuse. Have we so succumbed to the lure of the technical revolution that we can forget the primacy of the human equation?

Finally, the trend toward centralized management has taken away more and more of the ship, squadron, and battalion commander's freedom in what should be called field personnel management. Computerized central control of some personnel matters may well achieve efficiencies. There is yet no substitute for the sound, considered judgment of the unit or ship commander or skipper in the management of his personnel resources. Critical aspects of this important function must be restored as a command prerogative at battalion, squadron, or ship level, for there is simply no way to transform the human bit into a computer bit.

The Military Education System. The military education system must provide progressive education for the career noncommissioned officer. Again the Navy seems to lead other services in this field. There is no progressive education program for Army noncommissioned officers, with the possible exception of a requirement to attend a noncommissioned officer school for promotion to upper enlisted grades. Of the hundreds of courses offered noncommissioned officers at Army schools, few are mandatory and only a handful mention leadership as a requisite for attendance or as a goal of the course. Yet the noncommissioned officer is the key leader and instructor in the entire military training system. On his ability to communicate essential skills to other men rests the combat effectiveness of the entire military establishment.

Field education and training programs for troop units beyond basic and advanced unit training need the help of human research and engineering. Army training programs standardize training for units progressing from basic training through unit training and test prior to combat. But there is no aid, save his own inventiveness, for the unit commander trying to maintain an acceptable level of unit readiness. Most military training is accomplished in situations not compatible with the standardized training program, yet the least time has been devoted to providing guides and aids to such training. Basic human engineering data on which the field commander can design intelligent training programs is badly needed for units whose primary concern is combat readiness.

Officer education must treat force design and development, force deployment and employment in an increasingly complex variety of missions, and design of training programs to provide forces with skills to meet the broadening mission patterns characteristic of today's military operations.

Command and General Staff and War College courses must treat force development, deployment, and employment for larger forces of all services and military mobilization systems for all forces. Curriculum content must reflect the joint service character of modern military operations—seeking to equip each officer with a thoroughgoing knowledge and understanding of the other services' methods of operations and tactical and strategic rationale. Students at this level must study new developments in management, decisionmaking, information theory, and organization, learning to apply these to the solution of military problems. The military has balked at systems analysis, introduced as a management tool into the Defense Department by the Whiz Kids. In 1909 military staffs began to use a form of system analysis in the commander's estimate, long the basis of military decisionmaking. Because the estimate process was not modernized to incorporate new and more objective methods, it appeared a less satisfactory method than system analysis. A good housecleaning would have disclosed the basic tools in our own closet.

Above all the military education system must teach principles and develop in its students the ability to reason from principles, objectively, to sound conclusions. Rote thought patterns, ideas set in the concrete of Pentagon clichés, are simply not adequate tools with which to compete in today's dynamic decisionmaking forum.

The Military Staff System. In purpose and function the military staff has changed significantly from Secretary of War Elihu Root's vision of the General Staff as the "Brain of an Army." From a small group of Army planners whose task it was to provide plans for formation, expansion, employment, and support of military forces, the military staff has developed into a many-tentacled giant. It is at every step in every service deeply involved in operations as well as planning. Yet, since the very beginning, no military staff has produced so monumental a contribution to the American military system as the mobilization plans drawn up by the Army General Staff in the years before World War I. Influenced by the work of Emory Upton 40 years before, the General Staff in its infancy rejected the traditional militia system and created the mobilization system—the hallmark of American military management in every crisis until the Vietnamese War. If the military staff is to serve its traditional function, it must produce for the nation a new military system for the last decades of the Twentieth Century.

Together with military schools, the military staff system must adapt new organizational methods, decisionmaking tools, academic theories, and the products of advancing technology to the solution of military problems. While, at first blush, administrative reforms, reduction of redundancy, and continuing struggle with the effects of Parkinson's Law may seem of paramount importance, these do not strike at the heart of the staff problem. It is a far more significant goal for the profession to set to restore the proper function of its staffs, and at the same time learn to use new tools in the solution of peculiarly military problems.

Military Leadership. "I waited 20 years for this command, only to find I didn't command at all—everyone up the line commands for me." This all too frequent complaint reflects two phenomena. As one sociologist has indicated, generations educated in the last two decades are more resentful of oversupervision and interference with their prerogatives than were generations before them. At the same time, a certain inflexibility characteristic of the military way of thinking makes elders less willing to accept innovative solutions by younger subordinates. The result is a combination of oversupervision from the top and resentfulness from below.

"The trouble with the profession today is the ensigns and lieutenants aren't what they used to be." To be sure they are not, and well it is that they are not. Indeed, the ensign and lieutenant today are probably much better prepared educationally, socially, and psychologically to take their place in the profession than was the author of the opening statement when he joined 20 or 30 years ago.

Morris Janowitz has written of contrasting leadership requirements for the military profession today in terms of a required balance between heroic leader and military manager-technologistspecialist. Indeed, as he advances in rank, today's officer is required to move alternately from command to staff, from field to research project. In one assignment he is primarily a manager; in the next he must quickly recast himself in the role of the heroic military leader.

There is no doubt a growing need for management skills, in addition to leadership ability, on the part of the professional military man. There is no doubt a growing gap between generations educated in the 1930s, who comprise most of the flag and general officer corps, and those schooled 20 or 30 years later. Military management and leadership skills simply cannot be mutually exclusive. If they are, the profession must separate into leaders on the one hand and managers on the other, at the inevitable expense of cohesiveness and professional character. The profession must close the gap between generations with an intelligent effort by elders to come abreast of the times and patient persuasion by juniors to sell the excellence of their innovative genius. The martial spirit of the heroic leader cannot be cast aside—it continues to give the profession its fundamental character. The ultimate requirement for leadership in combat must inevitably stamp the entire professional outlook, both in management and leadership. The fighter spirit must be translated to oncoming generations. And, although the civilian management may control strategy, policies, even weaponry with which the profession fights, they cannot supply, nor can they dispense with, heroic leadership. The challenge of getting things done through people, on the assembly line or the battlefield, is one to which the military can and must respond with vigor, intelligence, imagination, and dedication.

The New Professionalism. At the outset we defined challenges the military profession faces today: the nuclear weapon—changing the nature of war; the management revolution—changing civil-military relations; and the contribution of destructive nuclear energy to the changing rationale of international politics.

In response to these challenges we reviewed reforms suggested: staff reforms, both to clean up administration and to strengthen the position and functions of the military staff in the Defense Department; management reforms to restore the prestige of military advice in councils of government; education reforms to put new tools of professional management into the hands of the military.

There are some who say professional reform can only be accomplished from outside the profession, that in fact no professional group is capable of significant internal reform. Others hold the view that only by reform from within can the military profession hope to adjust to change and still preserve its essential military character.

Reform from without is upon us—an accomplished fact. The history of this century will undoubtedly record that the two great reformers of the American military in our century were Elihu Root, civilian Secretary of War, and Robert McNamara, civilian Secretary of Defense. For 12 years after enactment of the General Staff Act in 1903, entrenched military bureaucrats in the Army fought among themselves over the internal reforms required to match changes imposed by the civilian Secretary of War over their strong protest. At the same time a few military men of inventive genius set about quietly to rewrite the entire military system of the nation. History will not allow us the luxury of 12 years of bickering at mid-century and beyond. Men of genius must set about the task of internal reform—now.

The beginning of the century saw the American military profession emerge from years of isolation in the Indian Wars and in far-sailing naval squadrons to become a truly professional body. Paradoxically three significant reforms characterized these changes and in many ways served to give the American military its distinctive character. They were: Army General Staff reforms, which brought the professional military staff idea to this country; reforms in the military education system which saw creation of the Army War College, following the Navy's lead of some 18 years before; and reforms in personnel management, especially in the Navy. It is no accident or happenstance that we have proposed reform in these self-same areas as an answer to the modern professional dilemma. Reform can be accomplished from within—indeed it must be. For only by internal reform can we advance that knowledge of the employment of military power, that competent technical skill with tools of the military trade, that peculiar ability to command military forces, and that competence to organize and train military forces and plan for their employment which are the essential ingredients of our professional character.

The American military profession is a product of the society it serves. Flowing from the liberal social system of the democracy, it changes in attitude as society changes. The nation's changing education system, as well as professional education and training, shape and change its intellectual presence. Challenges to the profession are as vital and varied as those to the whole society, but the professional response must bear the particular and peculiar stamp characteristic of the military profession.

As military men, we often like to cast ourselves in the Hamiltonian heritage of devotion to nation, strong sense of duty, and brave and purposeful action for society and state. Traditionally we tend to regret the decline of these cherished values. But, if they have declined, we cannot restore them, we cannot revive them, we cannot win society back to their call. For, if we try, then we have grown out of our role of service to the state and it is effectively the end of the profession and the state.

Nonetheless the professional military character which we have proposed to revitalize is a necessary ingredient in American society. It speaks with the voice of caution, offering the reminder that—if democracy is ultimately to succeed—there must be stronger and more persuasive instruments of foreign policy than the continued application of military power; that beyond national security there are pervasive limits to the use of military force in the pursuit of democracy's goals.

Above all, in this time of change, the American military profession must strive to understand itself, to understand others, and to speak so as to be understood in the councils of government. We must again and again consider what we are that others are not, what it is that makes us distinctively military, what is the source and strength of the "faith and fire within us." For only by so doing can we come to a deeper appreciation of the unique contribution the profession we serve can make to America and ultimately to all mankind.

Managerial Skills

Letter to General Bruce C. Clarke
Palmyra, Virginia
6 September 1974

I was continually appalled during my force structure and budget days by the singular lack of ability on the part of senior officials with whom I came in contact. I must say they were not all in mufti. I hold the view that while better in many ways than their average civilian peer, Army officers are not very good managers. For example, I'm the mayor of the 3d or 4th largest city in Kentucky, with an annual operating budget of over $100 million. Nothing in my background, except my three years in ACSFOR, equipped me adequately to hold this job. And I'm trying to straighten out a lot of pretty bad situations left me by some great guys who preceded me but who, like me, really hadn't been trained for the job.

Consolidated Administration

Letter to Colonel Grail L. Brookshire
3d Armored Cavalry Regiment
12 November 1976

Your concern about consolidation of administration at battalion is one I share. While I can't argue with the principle, I do have to say it isn't practical unless we first take away the paper that created the problem. What we should have done is take away all the reports and administrative burden, then look around to see if by chance we had done enough so that we could now take away the clerks and typewriters as well. We backed into it by taking the clerks and their machines away first, without ever addressing ourselves to the mountain of paper we have generated. Unless we can stop the generals from putting out instructions to companies, we ain't never going to stop this. We have a task force in session here to try and get rid of the paper, while at the same time we press on with an organized program to move the clerks to battalion level. So far it is working well, but only because we have been very careful and almost deliberate about it. We won't be done until mid-year 1977 sometime. By that time we will have reduced the paper burden at company level from some 300 reports and other requirements nine months ago to what I hope to be less than 25 or 30. We are already below the 100 mark, so there is some promise. Much of this is of course created by DA, and USAREUR over here, and so they have to get in the act too. In the end I'm sure I'll just have to tell them that here's a list of things you have sent out that we aren't gonna do anymore—so there, too.

Battalion-Level Consolidation

Message to General George S. Blanchard
US Army, Europe
19 September 1977

. . .

2. As you may recall, CABL in today's multifaceted version grew out of a rather straightforward desire to get some of the paperwork out of the orderly room. From that modest beginning, it has grown into a monster which is about to devour us. Now included are proposals to consolidate personnel administration (PAC), maintenance (MAC), mess (DFAC), supply (SAC), and training (TMA) at battalion. It is now my considered judgment that of those concepts only personnel administration and mess have real utility, especially in a TOE environment. Unfortunately, I find the zealots among us have made some exaggerated claims for the success and the future potential of many of these programs without realistically documenting some very real shortcomings. It's time to "fess up." We have so badly watered down most of the original proposals that they are not now worthy of their original title. Therefore I believe we should just stand up, admit that, and be big enough to call a halt to those programs that appear to offer no real payoff.

3. My proposal is that we cancel SAC, MAC, and TMA, with the proviso that TRADOC draw up training circulars wherever appropriate, incorporating useful data from tests into "how to do it if circumstances permit" kinds of descriptions. It is true that, in some types of units under some garrisoning circumstances, many things can be consolidated. We can draw up guidelines describing that and issue them in TC format. That should suffice.

4. PAC is too far along to turn back at this time. Therefore I propose that we proceed, but that we take the necessary steps to ensure we have fully, clearly, and honestly documented manpower, equipment, transport, and other requirements with regard to integrating PAC into the structure.

5. The mess problem is somewhat a different matter. Since battalion-level consolidation is so facility dependent, especially in USAREUR, I believe we must work on your TOE unit problem and the combat messing problem apart from the broader issues of large-scale messing at Stateside posts or in some of your high density community or headquarters areas.

6. I've not done anything about all this yet; wanted to try it on you for size first. We're trying to haul in a lot of things that appear to have proliferated beyond a useful stage and put our chips where the payoffs appear greatest. I would very much appreciate your reaction. . . .

Consolidation at Battalion Level

Message to Lieutenant General John R. Thurman and Others
17 October 1977

1. Last week at the DA commander's conference the subject of CABL was raised. Without reviewing the gory details for you, I proposed at the end that we do the following:

a. Proceed with the personnel administration consolidation, but take necessary steps to ensure that we have fully, clearly, and honestly documented manpower, equipment, transport, and other requirements needed to integrate PAC into the structure. We must not lose sight of the fact that, when we had battalion-level personnel sections several years ago, we also had clerks in company orderly rooms. To the extent that the PAC needs augmentation based on that fact, we must provide it.

b. Cancel supply, maintenance, and training consolidation efforts and eliminate those acronyms from our lexicon. TRADOC should then draw up training circulars and type manning documents wherever appropriate, incorporating useful data from tests into "how to do it if circumstances permit" documentation. In considering new organizations, it may be possible to consider once again consolidation of maintenance and consolidation of supply at battalion, just as we are presently doing in DRS. But we cannot and should not try to superimpose those concepts on the existing organizations of our Army.

c. With regard to consolidation of mess activities, TRADOC should publish training circulars and type manning documents to be used as guidelines when physical circumstances permit commanders to consolidate mess functions at battalion or higher level. Then TRADOC should set out to decide how the Army in the field should eat and take necessary steps to get rations, equipment, personnel, and other matters resulting from that decision lined up properly.

2. I also agreed to get the proposal together in a briefing for the Vice [Chief of Staff] and lay the issue to rest one way or the other. Accordingly I'd like for the three integrating center commanders to get the brief together under Roy's [Lieutenant General John R. Thurman's] direction. Let's shoot for trying to get back to General Kerwin no later than mid-November. Meantime let's hold in abeyance work in progress to change further any TOE or other doctrinal literature, except for that related to PAC.

Centralized Management

Letter to John B. Bellinger Jr.
Office of the Secretary of Defense
4 June 1979

Thanks for sending me the Defense Agency Review Report. I read it—between two and four one morning as you suggested. It is interesting. I would however observe that the Defense Agencies, to some significant degree, represent the notion that somehow centralization of anything has to be a better way of doing business than to let everyone do things for himself. While this may be true in some instances, I would hasten to say it is not universally true and that no one has ever conceptually established the metes and bounds of when it is and is not true. Until that is done, these agencies will continue to proliferate. And in so many instances they represent unwarranted, unnecessary, and even inhibiting intrusions into the business of the services.

The second observation I would make is that, so long as you task the USD PAE and the ASD MRAL to do some more work on matters pertaining to some of your discoveries, you simply invite additional proposals to be made with regard to further centralization of something—God only knows what, but they'll find something. It didn't all start with Robert Strange [McNamara], but don't forget that many of the key players in today's Farce Along the Potomac came down from the Harvard Business School, or from under some related flat rock, with the "management gang." The difference is that today we are looking at the third string in McNamara's stable. God knows the first string was incompetent enough—I won't comment on the competence of the backup players.

Living Systems

Message to Dr. Walter LaBerge
Under Secretary of the Army
6 June 1979

1. For some time our Task Force Delta has been searching out answers to questions about how we get the most out of military units. How do we run good outfits? More importantly, how do we teach the officers of the Army to run good outfits?

2. Needed to do this in a systematic way is a general notion of how organizations work. That has been provided by the work of Dr. Jim Miller, now the president of the University of Louisville, whose work of 30 years, "A General Theory of Living Systems," has attracted considerable notice since its publication a year or so ago. Jim has been kind enough to help us explore our problem and we have several people working at Louisville with his staff on military applications of living systems theory.

3. Recently reviewing what's been done, I am persuaded that we are about to be able to describe for ourselves and others what makes organizations work, and how to make them work, and do it in terms of a systematic theory which can be "painted green." It is true that we are some ways from having that done, but it can be done.

4. There is a need, however, for a longer term view of this problem, a requirement for continuing research, data collection, analysis, evaluation and feedback into some activity that can collate, analyze, and draw useful conclusions from work in progress. This is so in several fields of endeavor. Jim Miller is working with the National Science Foundation, agencies in the medical field, and many others in an attempt to organize some larger effort with regard to the application of living systems theory to social structures and organizational parts thereof on a wide scale. We'd like to be part of that, albeit the green part.

5. I know you would be interested in the theory itself, as well as in its application to our problem. We tried once to expose you to some of this, but did not pursue it vigorously enough. It does, however, strike me that both the Army Science Board and the Defense Science Board could well consider broadening their horizons to take in living systems—organizations and how they work, as well as the more traditional and inanimate systems to which so much of their effort is currently dedicated. I promise you it's technical enough for the most discriminating technocrat.

6. The Army Science Board is to conduct its summer study session at Charlottesville starting this month. It's probably too late to add something like this to their agenda, but that course of action might be considered. Unfortunately we are not aware of the Science Board program far enough in advance for many of us who are invited to various parts of their work to be present. Such is the case this year, when I shall not be able to be present for either the beginning or the outbrief. Neither will my principal staff. Last year this happened, and I had to send my CD science advisor, Art Christman, as a surrogate for all the rest of us who couldn't be there. I would like very much for some of us to participate in the work of the Science Board. If we don't it will remain directed on hardware technology and conducted pretty much in isolation from any of the "green gang." While there may be some who believe that's an okay state of affairs, I am not one of those.

7. I'm prepared to have some people come talk this over with you if you are at all interested. Just let me know.

The Over-Under Army

Message to Major General Walter Ulmer
Director of Human Resources Development
10 July 1979

1. I have reflected long over your memo of 3 April, "Implications of an Overstructured Contemporary Army." Herewith some observations which might be useful as a basis for further discussion.

2. We really have an over-under Army. We are over in quite a few things; under in others. Probably not all of the overs are bad, but not all are good; the same can be said for the unders. Taken in their entirety the overs and unders produce the situation you and I have talked of so often.

"The Army is overstructured." We haven't been able to recruit enough manpower to fill up the structure, yet we are unwilling to give up structure; for, if we do, the budgeteers will take away the manpower at a rate of 48,000 per division. The result would be a smaller Army, but one that would still be overstructured.

"The Army is overtasked." The Army has been assigned, and has accepted, too many missions. Unwilling to admit there's something we can't do, we accept everything. The result is nothing gets done quite as well as it should, or quite as well as it might were we not to accept some missions. To make matters worse, the Army overtasks itself. This is done with programs developed in the several staffs, to be implemented in the several units—in addition to their other work. The result is an environment in which these highly visible specified priority programs are completed first, then, if time remains, some training may be undertaken. Some of us perceive that to be just backwards from what it should be.

"The Army is overmanned." The Army may be overstructured, but the type units in the structure are not those which optimize firepower and mobility in the interests of saving manpower. Five-sixteenths of the Army is manpower-intensive at a time when the equipment-intensive eleven-sixteenths is begging for more manpower to fill out its ranks and to man new equipment. Five light divisions in an overstructured, manpower-short Army appear to be almost a luxury.

"The Army is overoptimistic." A "can do" attitude is commendable. We all like to see that comment on our own OERs, and like to be able to write it on the OERs of others. But that attitude has led us to the notion that it is somehow less than honorable, certainly less than professional, if we question whether or not something can and should be done. Two things drive this—individual careerism and unit careerism—the desire of unit commanders to have themselves and their outfits known as guys who can do anything and everything. There are two outcomes of this—not very much gets done very well at all, and the same good outfits are tasked to do everything and quickly burn out from the overload they take on in their exuberant desire to be "the guys who can do everything." I do not suggest pessimism as a counterfoil to overoptimism, but some realism would be a help.

"The Army is oversupervised." Technology allows platoons of the Army, individual ships of the Navy, and separate aircraft of the Air Force to be commanded from the White House. Bombing targets deep in the enemy homeland can be selected quite accurately across a luncheon table in Washington and missions conveyed to the operating commands by advanced telecommunications. The fact that this can be done does not argue that it should be done. But it is done, and has been since the early 1960s. The perceived need to retain close control of nuclear weapons employment contributed, as did the "new management" philosophies brought to Defense about the same time. Following that example, we too oversupervise—layers of command helicopters in the air over a platoon firefight in Vietnam and the corps commander in Europe routinely taking personal action with regard to readiness reports of the companies of his corps are but two examples.

"The Army is overcentralized." Overcentralized is most often associated with new programs designed to solve some shortcoming identified as critical. The prevailing notion is that centralization of anything is better by definition. It is a deeply rooted fundamental concept held by most civilian defense "managers" today. Uniformed "managers" hasten to ape our civilian masters. So it is, for example, that an organizational effectiveness program is stovepiped on the model of the equal opportunity race relations program.

"The Army is overloaded." As in any system, output declines as overload occurs. Most organizations of the Army today display many or all of the symptoms identified with overload—work stacks up, things don't get done well, if at all. In fact most of the overload is on the leaders. Officers, and to some extent noncommissioned officers, are terribly busy all of the time. Some even take pride in how busy they are, how early they are at their appointed places and how late home therefrom. And so the units are all busy, while many soldiers are bored to death with tasks that to them don't seem to contribute much to anything that seems important.

"Finally the Army is oversexed" in a couple of ways. Too many of the too many women we have are pregnant, a situation which would not exist had we not taken on girls in such numbers in all those jobs in the first place, or if we'd been willing to take on realistic and much-needed sex education with our new coed outlook. Oversex in the Army is but one manifestation of our societal overreaction to the equality movement; like prohibition, we have once again embraced with exuberance concepts with which we were but ill-prepared to cope practically.

"The Army is understrength." We have been unsuccessful in finding an equation which will attract volunteers to all elements of the Army in sufficient numbers to fill the structure. Aggravating declining military manpower, we have been required to draw down the civilian workforce. The result is an Army whose baseops workforce is operating at 75 percent or less of its needs, and whose mission accounts are short of sufficient people, in uniform and out, to accomplish the tasks demanded by an increasing workload as the Army attempts to bring on and make ready a new generation of fighting gear, tactics, and organizations. The last time the Army had 16 active divisions its military strength was nearly a million. Today we are trying to do as much or more as we were 14 years ago with a quarter of a million fewer men in uniform. This understrength is a contributing factor to the most serious problem facing the Army in the field—turbulence. The turnover of 30 to 40 percent of a unit every quarter prevents sustenance of any reasonable level of training and readiness.

"The Army is undertrained." Tasking the Army with a multitude of first-priority special programs has resulted in a situation in which the unit commander who tries to comply with all he's been tasked to do is really able to train his soldiers almost as an afterthought—after all the other important demands have been satisfied. Every other army in the world spends from four to seven times the amount of time in individual training of their soldiers and noncommissioned officers as does the US Army. The German Bundeswehr, for example, is equipped with the US Hawk air defense missile system. The Bundeswehr trains its individual soldiers in specific system-related skill training seven times as long as does the US Army on the identical system. On the side of collective training, the absence of a long-term range and training area and development and improvement program, and the advent of many new long-range weapons, has left the Army with but marginal training areas for its units.

"The Army is underattractive or undersold." How can we attract young people to want to serve? Virtually since the onset of the volunteer force we have treated recruitment as a marketing exercise and the Army as another marketable commodity—a job. Bonuses, promises of job transferability, and "normal" hours have been but marginally successful in filling the ranks of support forces; they have failed to recruit enough fighters for the combat arms, the hardened edge of the Army. Even well-intentioned promises lead to trouble, as postenlistment depression develops when soldiers learn the Army is not the "job" they had been led to expect. The result is a leadership problem that aggravates the hostile training environment. The Army must decide why it wants young men and women to join. The Army can never join them—they have to join the Army.

"The Army is underintelligent and underliterate." There is considerable discussion about intelligence levels and functional illiteracy among today's soldiers, and the effect of those factors on soldier trainability. Based on scores used to indicate intelligence levels, there has been an overall decline in intelligence levels. It's no secret that 60 percent of 1978's new male soldiers had ASVAB scores in mental category IIIB or IV. However this parallels a general decline in scores on intelligence tests given in high schools, colleges, and universities in recent years. We know they are trainable; we also know training them takes more time—both because of declining intelligence levels and the increased sophistication of greater numbers of systems which soldiers must be able to operate successfully. Further we know that, in that hostile training environment, time is the most precious resource. Trainability is a variable most easily dealt with in terms of time. Every other army in the world gives its soldiers more than four times the training given in the same skills by the US Army. That fact suggests that either we know something about training they don't or their soldiers are much less trainable than ours—neither of which is true.

"The Army is undermotivated." Soldier motivation is a root problem in the hostile training environment. Many of today's young soldiers bring with them from our society a strong attitude of social alienation. Produced by isolation, cynicism, and a sense of meaninglessness, social alienation is completely at odds with what it takes to make an Army.

Isolation—social, political, and emotional—produces introspection and self-induced separation from others, the antithesis of teamwork cohesion which is the backbone of an Army. Cynicism denies the virtues of honesty, integrity, and patriotism. In its grip, soldiers cannot find a moral code to which they can subscribe, or leaders in whom they can trust and believe. Meaninglessness is believing it senseless to risk your life for your country because nothing—even the country—is worth preserving at that price. Soldier motivation today is a formidable leadership challenge. To be effective, an army requires cohesiveness—a sense of community. Liberalizing an army does not help the soldiers of the Army. Armies—our own more than most—need a unique set of values to be effective. These include discipline, obedience, integrity, a high order of technical excellence in military skills, and dedication to a well-defined purpose—defense of the country. Even if those values be somewhat different from those popular in the society at large, we must state clearly what values our community demands and make that value system an integral part of the training of our soldiers.

"The Army is underled—the other side of overmanaged." The hostile training environment abounds in challenges to leadership, both noncommissioned and commissioned officer leadership. It is also true that both NCO and officer leadership have suffered through some trials.

The Vietnam War virtually destroyed the NCO corps of our Army. Only in the last few years has our NCO education and training system begun to bear fruit in new generations of competent NCO leaders. The business of sergeants is a very important matter, for it is sergeants who train soldiers and teams; it is sergeants who lead small units when officers are killed or wounded. In large measure our Army's expansion from a 1939 pauper force of less than 200,000 to a fairly effective force of several million in a few short months was made possible by the exceptional professional competence of our NCO corps. Highly trained sergeants are essential to a successful Army.

The Vietnam War had a severe effect on the officer corps of our Army. The most serious effect reflected two things: the decision to expand without mobilization, and the entrepreneurial mentality made vogue by defense managers of the Vietnam years. Thus a young inexperienced officer corps became terribly confused by the conflict between a traditional military goal—winning the war, and a plethora of imprecise and ever-changing managerial objectives.

What happened to the Army in the last months of Vietnam was not that the ethical value system of the officer and NCO corps collapsed, as some have alleged. Rather it was that, in redeployment from Vietnam, the centralized individual replacement system demanded redeployment of individuals, not units. Those who remained were reassigned to remaining units. As the pace of redeployment quickened, this constant shuffling ensured lack of cohesion in the residual force—in the leadership and amongst the soldiers. Careerism there may have been, and may still be, but the root problem was that the sense of community was destroyed—there simply was no cohesion. In that hostile environment the leadership was overloaded and it behaved accordingly. In many ways today's hostile training environment is very much kin to that of the last days of Vietnam. If leadership is to be effective, then we must do everything in our power to reduce the obstructions which clutter the environment in which leadership must do its work.

Twice in the brief time of my own service it has been my good fortune to serve in exceptionally good outfits in peacetime. In the early 1950s I served in a battalion in USAREUR in which personnel were stabilized for nearly two years—a result of the crisis in Europe which accompanied onset of the Korean War. Again, in the early 1960s, the Berlin crisis resulted in long-term personnel stability in the battalion in USAREUR in which I was serving. Both were superb organizations. Here are some reasons why:

- There was a clear and urgent focus on the mission. Crisis was upon us; we worked hard at the essential business of soldiering because we were certain we might have to demonstrate our soldierly skill very soon.
- There were enough soldiers assigned to man the equipment and allow the organization to function as a unit. Motivation was high. In both units the personnel were largely volunteers. In the former, they were remnants of the pre-Korea volunteer army. In the latter, about 93

percent of the unit was volunteer. Trainability may have been a problem, but motivation was not. While we can't recapture why they enlisted, it clearly was not for the "job opportunity." Soldiers in those times were not paid that well, and both periods were times of crisis—crises which clearly could ultimately call for the clash of military forces.

- There was stability of soldiers and leaders. Many left those battalions having served with the same team or crew, under the same sergeant, for three years or more.
- There was cohesion. Soldiers, NCOs, and officers were pressed together by crisis, shared a common danger, concentrated on a mission all saw clearly—the immediate need to be able to fight and win.
- There was well-trained leadership—technically competent, able, dedicated, and trained officer and NCO leadership which had trained together long enough to become effective as a team.
- There was a shared system of values and priorities. We told the truth about important things. We did important soldier business well, and we believed in ourselves and our outfit. We knew our lives might depend on those things, and we were certain the success of our unit in its mission surely depended on them. Therefore we put training for the mission ahead of all the priorities that did not contribute to winning.

3. Now what shall we do about the over-under Army? First we should stop studying it. I believe we know enough about what's wrong. Second, let's do something creative and positive about it, even at the risk of making a lot of people unhappy. There's enough meat in the above to argue long over, but the answers in my mind are getting more clear and simpler every day. Finally we should just stop doing dumb things that aggravate already aggravating situations—like, for instance, opening up recruiting to 17-year-old non-high school grads.

4. Well, that should give us something to talk over on the 18th. It probably adds to your overload. But your thoughtful memo deserved a response. This may be the only one you get, and it helped unload my overload. Cheers.

Centralization

Letter to General Edward C. Meyer
Army Chief of Staff
15 August 1979

We have reviewed the draft "Implementation Plan for Centralized Execution of the CONUS RPMA Mission." Detailed comments concerning this draft plan have been forwarded to the Office, Chief of Engineers. This plan is based on the assumption that centralization is the panacea for the resource constraints that face us today in RPMA. I question this approach.

Over the past years we have attempted to find easy solutions to difficult problems through higher and higher levels of centralization, only to learn that the real solution is attainable only by the person who has the responsibility—the commander. This plan offers no solutions and should not be approved or even tested; for, to do so, would have a detrimental effect on the entire Army. The Corps of Engineers should not be built up at the expense of installation management and the destruction of an ongoing viable facilities engineer organization.

The plan assembles a number of ideas, none particularly new and most previously nonconcurred in by the MACOM. It does not discuss the overall impact of so large an undertaking. The benefit of the plan fixes the cost of the CITA review and analysis program at a definite $17M, but does not address the price tag of space requirements to form new COE military districts and divisions or increased overhead, design and construction supervision costs chargeable to the installations. The plan splits responsibilities for management of RPMA, usurps or ignores the installation commander's management prerogatives, and would cost more in both manpower and dollar resources.

The plan is primarily justified on the basis that the Chief of Engineers has experience in and can manage RPMA better than the installation commanders. I strongly question this. The installation commander and his engineer are the real experts in RPMA management. Their record in this area is outstanding, especially in light of the constraint on personnel and dollar resources. If better solutions are to be found, they with the help of the MACOM will find them. It is critical that our commanders retain the flexibility to control and manage people and dollars to meet overall installation requirements.

In summary, I strongly oppose this concept and can see no benefit to the Army in its adoption.

Pressing On

Message to Arthur Woods
Fort Hood, Texas
28 August 1979

Have your 19 August status report on TOTEC [TRADOC Operational Test and Experimentation Command]. . . . On the day of retribution we will stomp on the midgets who swarm about knee deep. Press on!

Extended Command Tours

Message to General E. C. Meyer
Army Chief of Staff
4 March 1980

1. Since we announced the extension of tours of command I've been to Carlisle, Newport, Norfolk to speak to the classes. There, and in several other places I've been, command tour extension is the topic of most concern amongst the majors and lieutenant colonels. Of immediate concern is the fate of those already posted to command who may be delayed due to extension of the incumbent. Equally urgent, but less immediate, is the feeling of all that all too many who want and should command will not get a chance to do so. Finally there's a lot of concern that Army leadership isn't too clear in its own mind why this is necessary. I'll talk about those in reverse order.

2. Apparently several senior people have talked at the schools and used as a rationale for extending command tours the fact that it will reduce turbulence. Some who urged tour extension have used this argument from the beginning. Turbulence has its most debilitating effects among the ranks. While it might help some to stabilize commanders, it surely does not of itself get at the root of the problem. We shouldn't use that argument. It doesn't ring true.

3. While most of the people who talked with me might agree that we should develop a corps of commanders, they question that we are really serious about it. They do not believe the selection system can pick the right guys to command all the time. They question that the senior commanders of the Army have the guts to replace quickly those who do not perform to standard. They don't believe they won't be penalized in the long run because they weren't selected for command.

4. We probably made a mistake by making tour extension retroactive. Several real screwups are in the offing in which the incoming commander will have to wait in some limbo for up to a year while the incumbent gets extended. Some of these we aren't handling too well.

5. The response out there is very, very negative.

6. I've told them all, as I have you, that if we're to extend the command tours then three things must happen.

a. MILPERCEN has to be reoriented out of the "everyone must command" mentality.

b. Selection boards must demonstrate by their actions that there is no less chance for promotion or other career advancement for not having commanded. Instructions to boards must somehow ensure this.

c. The general officer leadership of the Army must have the guts to replace quickly those who obviously are not fit to command. The problem is that many of us do not know who is and isn't fit to command—we don't observe them firsthand that much—we shy away from quantitative scoring of their performance—they really have to do something wrong and have it come to command attention before they get fired. Unless we can change that, there are a lot of units out there which will never get much better than they are until the incumbent leaves—many, many months hence.

7. I've been ambivalent about this from the beginning. While I believe in the idea of a corps of commanders, I've never agreed with George Blanchard and Bob Shoemaker that extending

command tours would help at all in overcoming the effects of turbulence. I don't know whether or not we've done the right thing. I do know that we'd better be candid about why we did it, and that in instructions to the MILPERCEN, in the outcomes of boards, and in the actions we take to remove the incumbent we'd better demonstrate to the officer corps that we know what we're doing and are serious about it.

Standardization and Evaluation

Message to General E. C. Meyer
Army Chief of Staff
13 May 1980

1. Some time ago you mentioned that you wanted to make the school commandants/center commanders titular heads of their branches. You recall I supported that notion, but counseled that unless we give them the authority to do something in their newly acquired status it will not be any easier for them to gain access to the major commands of the Army than it is now. We've made giant strides in the last year in being able to get in and do our evaluation work; we are progressing well in developing a consensus about standardization. But it all goes very slowly, and we are always in the mode of persuading everyone else to go along. Indeed, as I've told you before, the only reason they are all for standardization now is that you are. Your successor may not be—your predecessor surely wasn't; I couldn't get to first base with most of this until you came up on the net as being in favor. However, if you are willing to consider branch chiefing again, let me suggest a way of doing what you and I have been talking about in a manner that should not get everyone too exercised over turf intrusions.

a. First: By regulation, establish TRADOC center commanders as titular heads of the branches.

b. Second: Among their responsibilities list evaluation and standardization. In this system we would simply extend to other branches the responsibility and authority now held by the commander of the Aviation Center for standardization and evaluation of aviators, flight crews, and airplane operations worldwide. Commander of the artillery center functions in a similar manner with regard to the missile system evaluation group.

c. Third: Delegate to Commander TRADOC the responsibility for establishing and operating, as the executive agent of the DCSOPS DA, the Army Standardization and Evaluation System worldwide. Commander TRADOC should be charged with seeking a consensus among major commanders as to what should be standardized and how it should be done. Commander TRADOC should be charged with promulgating instructions to be issued by CSA with regard to standardization and evaluation worldwide.

d. Fourth: To accomplish the program, TRADOC would use the evaluation directorates of the schools; these would now become evaluation and standardization directorates. They would work in conjunction with such other teams as may be operating for the major commanders of the Army to conduct evaluation and standardization of Army units worldwide. This would be similar to the system I now have with Bob Shoemaker, but it would be formalized for the whole Army, and it would extend the system Army-wide, not just between ourselves and FORSCOM.

2. Standardization of Organization. Having established as above the basis for the Army Standardization and Evaluation System, a second step would be to standardize TOE for like units of the combat and combat support arms; perhaps even for the combat service support arms. Branch chiefs and TRADOC would be responsible for the evaluation and standardization of these TOE. Such is in fact the case now, but there is now no discipline in the system. We send fully developed TOE to DCSOPS, who approves them. However it is then necessary for the major commands to MTOE those TOE in order to implement them in the field. The MTOE

process is the first step in creating turbulence in units of the major commands, for the MTOE system is the modern day successor to the old general order line deletion of spaces in order to generate manpower authorizations for those baseops functions we can't otherwise afford—the communities in Europe, for example. What I'm suggesting is that we should adopt a system that does not force the major commander to rob from Peter to pay Paul. We should make those TDA resources available on the order of 12–15 percent over the TOE authorizations on a major command basis. Then we should standardize all the TOE and hold to those standards unless the standardization and evaluation systems prove to us that they should be changed. This would act to reduce to near zero the MTOE perturbations that are to be required as the transition to Army 86 ensues. It would also ensure that the TOE which we develop at such cost in study and other effort do in fact get implemented—now they become part of some PARR process, frequently with different priorities in different commands. I have never understood how the department could support an Army in this fashion; the system virtually guarantees that nothing is ever quite the same anywhere. I believe we should seek to have it as much alike as we possibly can—everywhere.

3. Standardization of Job Specific Training for Leaders. Branch chiefs/TRADOC should be made responsible for job specific training of leaders in their respective branches. This would require a subtle but important change in NCOES—not at all unmanageable, I believe, for it would mean that a sergeant would now attend a squad leader's course, a section or crew chief's course; he would do this regardless of his grade. In other words, when a man becomes a tank commander he goes to tank commander's school. If he's an E4 or an E5 and will be in the job for more than a very short time—if we expect him to have to fight in that position, he must be trained for it. Officer basic courses become platoon leader's courses; officer advanced courses become company commander's courses, and there are specialist courses in addition—along the lines we are already pursuing. I've become convinced that we must be more specific and direct about the training we give our leaders—NCOs and younger officers. This is in my judgment the way to do it.

4. Establish a Regimental System for the Branches. Separately I'm sending you a study done by my British and Canadian liaison officers. Several months ago I tasked the two of them to develop for me a regimental system for the US Army. Patterned after the Canadian and British regimental systems, it would nonetheless have to adapt to the American way of doing things. They have just finished this work and have in my judgment done a superb job. It could provide the blueprint for a long-term scheme to provide us with a great deal of stabilization and a bit of the unit association which is always so important to a military body.

5. Establish a Standard Training Strategy for Branches. A study I've just had done reveals that in TRADOC we train somewhere between 14 and 100 percent of the initial entry skills needed for Skill Level 1 proficiency, depending on the MOS. For example, we train 14 percent of the Hercules missile crewman Skill Level 1 tasks and 100 percent of the laundry and bath specialist Skill Level 1 tasks. In reviewing the whole system, there appears to be no coherent strategy. I'm trying to develop one now, but that should be an institutionalized responsibility of the branch chiefs under the aegis of TRADOC guidance concerning what's within the resource windows and so on.

6. Standardize Rotation Policies for Leaders and Key Specialists. In my long personal study of the branch chief proposal I've concluded that, while it might be possible for the branch

chief to become involved in personnel management, assignment, reassignments, and so on, it is probably better to charge the branch chief with only the responsibility for ensuring that there are enough trained leaders and key specialists in the units of the Army. Branch chiefs under such a scheme would have responsibility for ensuring that squads, crews, sections, platoons, companies/batteries/troops, and battalions/squadrons of the Army are manned with qualified leaders and specialists trained in their specific jobs.

7. Footnote. Nothing we do will enable us to get very much better unless we can solve the numbers problem. I'm not talking about the numbers of soldiers; I'm talking about the right numbers of leaders in the right grades, or nearly so, who are qualified by training to lead the unit they head. The perennial shortage of armor NCOs has for years been the single greatest inhibitor to the effectiveness of our tank force—tank commanders and gunners. Therefore we must do two things. First, we have to ensure that we have enough NCOs in the force; secondly we must figure out how to accommodate to those administrative and baseops drains on the force which always seem to be tearing the NCOs away from their units. Finally we must adopt rotation policies that permit people to stay in units of their choice for however long they want to stay and the unit wants to keep them. It should be possible for us to assign officers and sergeants to units for five or six years. Others can come and go, but the leaders and key specialists must remain as long as possible.

8. Someone will charge that this is another power grab on the part of old TRADOC. Not guilty. The platter is full now. I just believe that we have to do some of these things if ever we are to get much better than we are now. I said this in the TRADOC *Green Book* article last fall; I've been saying it for years. We just have to get some discipline back into the overall system without risking the dangers of overcentralization. It seems to me that what I've laid out above represents a reasonable compromise, and yet is not too rich for our blood.

Strategies for the 1980s

Message to General E. C. Meyer
Army Chief of Staff
4 June 1980

1. I recall your telling me that this month, or was it last, is to be your month for deciding what the derms and pachyderms are to consider when next they assemble in October 1980. Also recall promising to send you suggestions. They follow.

2. Last year we set ourselves to the task of setting up a "direction for the Army for the next four years"—your words. We tried to focus on policies and issues, to chart the course, to ratify or amend what we may have already started on. We began our deliberations with your chart showing division, corps, and EAC in the middle of a box; in the four quadrants surrounding the box we considered training issues, modernization issues, manning issues, and mobilization and deployment issues.

3. As I review the tasker(s) that flowed from that meeting and visible progress toward task accomplishment I find that, while some progress has been made, most has been in areas, and with regard to matters, which are not really at the heart of the issues to which we addressed ourselves. In all too many instances we are indeed further behind, or certainly no further ahead, than we were a year ago. This is particularly true of TRADOC and DARCOM. Jack Guthrie can speak specifically to his unique problems, but the growing burden of the modernization program in the face of increasing constrained resources in both commands has combined to produce a situation in which we both just get further and further behind. Jack gets behind in material things like backlog of rebuild and unsatisfied workload requirements; TRADOC gets further behind in the scheduled delivery of products to the field, and in our inability to train the load as Max Thurman's revitalized recruiting initiatives begin to pay off. So, while I suppose it might be useful simply to recount what's transpired as a result of our hopeful pronouncements of last fall, it seems to me that is not enough.

4. The format we used last year is useful, if for no other reason than to permit us to sum up what's been done since last time, where we've fallen short, and perhaps what needs done to shore things up. Those four issue subsets are indeed the matters of continuing interest and concern to us; they still have at their core the structure of our Army; implicitly the structure must reflect our conceptual notions of what the Army is supposed to do and how it is to get done. Therefore I suggest you consider keeping that framework for our deliberations this year.

5. Last year, implicit to most of what we considered was our underlying strategy of how the war ought to be fought, and indeed of what war we were considering fighting. We didn't talk very much about that specifically, nor did we consider the issues in terms of the longer-term strategies that might be pursued with regard to each. This was well in a sense; your first conference as Chief and, as you yourself stated, you were interested in charting out where the Army should go on your watch. While we need to continue to work on what happens on your watch, we must also begin to consider what happens after that, recognizing that even you may not be immortal. Some new problems have risen to beset the country, and some new programs in defense have been commenced in response thereto. The RDF is the best example. In addition, despite some brave pronouncements concerning increased expenditures for defense, the balanced budget mentality still seems to be at the heart of administration policies with regard to all budgetary considerations. In the Army, at least, it appears that there will be

no real increase in TOA; in many cases we will indeed show a decrease in monies available to execute programs already short-funded or behind. Modernization is being mortgaged in the interests of operational accounts. This says that we should look again at our modernization strategy. In addition to what happens to the pace of the current modernization program, which is obviously going to be slowed if this administration continues in office, we have the longer-term consideration of how to begin modernization programs for the nineties. For, if we are to have new fleets, new organizations, new anything in 1995, we have to begin now.

6. We should not try to extend ourselves too far out—that always gets us into the ethereal and out of the real. Might we consider as an overall title "Army Strategies for the Eighties," those strategies being:

a. The Army strategy for meeting our commitment to NATO and to the contingency areas—specifically the Middle East and northeast Asia. In this I would suggest we force ourselves to flesh out in some more detail the possible goals of our involvement in those areas, some outline scenarios of how such involvement might begin and proceed, our deployment and employment strategies to prosecute the missions we perceive for ourselves, and how the campaigns might be conducted and the battles fought.

b. The force structure strategies which we now employ, assessment of how well these meet our needs, and alternatives which we might adopt to meet the demands of the military strategy.

c. The Army training strategies which we are pursuing now, an assessment of how well that meets our needs, and alternative strategies which we might employ to better satisfy the requirements.

d. The Army materiel acquisition strategy for modernization which we are now following, an assessment of how well that strategy meets our needs, and alternative strategies which might be considered to better meet our needs.

e. The Army force manning policies which we are pursuing now, an assessment of how well these meet our needs, and consideration of alternative strategies we might adopt to better meet our requirements.

f. The Army force mobilization and deployment strategies we are now pursuing, an assessment of how well they meet our requirements, and consideration of alternative strategies we might adopt to meet our requirements.

7. Folded into that framework we could include all or most of the actions we have underway, or know we need to consider, in order to meet the demands of what last year we called the issues. One of our persistent problems is that the strategy we are pursuing with regard to most of these matters has been determined more as a result of resource constraints which salami slice away at what we laid out in the beginning than by any conscious reevaluation of strategies. Eventually we reach the point at which the original strategy can no longer be pursued, but we seldom recognize when we're there because we never step back and examine all that's happened in light of what we started out to do in the first place. We now need to take a step backwards and decide if we are embarked on the right path and, at the next turn, in which direction do we go.

8. If the notion I've sketched in above has any appeal at all let me know; I've been sketchy in the interests of relieving your tired old eyes from having to read more.

Effectiveness of Army Units

Message to General E. C. Meyer
Army Chief of Staff
9 June 1980

1. Our research program into the effectiveness of Army units which we've had in train for more than two years now is nearing completion of its current phase. We attacked this problem on two fronts: Mike Malone and Task Force Delta here in the headquarters, and a program with Dr. Jim Miller and his systems science staff at University of Louisville. Headed by Army Major Jim Cary, the Louisville Task Force will soon have been into the entrails of half a hundred Army battalions trying to figure out what makes the good ones good and the not so good ones the way they are, and to describe those differences systematically. Jim Cary will soon have done all of Walt Ulmer's battalions, and that will complete this phase of the research. We have throughout both efforts tried to describe what we know about all this in terms of the general theory of living systems, Jim Miller's creation, and the subject of a very large book which I believe I sent you some time ago.

2. One purpose of all this has been to try and pin down what we mean by organizational effectiveness; to get the subject out of the hands of the behaviorists who understand but a small part or it—and can't paint that part green, and to decide what the Army needs to do about building effective units. We are on the verge of being able to do that. Malone and company will carry over at Carlisle what they started down here. Meanwhile, when Cary's research is done we believe we will have enough evidence to draw some frames of reference on which we can begin to build courses of instruction, modules for existing courses of instruction, and further research goals to pursue whatever unanswered questions we may believe we have remaining.

3. It is becoming more and more clear to me that we need some system of continual research into how our units are doing in the field—a feedback system that will bring us diagnostics about the unit, its performance, the unit leadership situation, and other circumstances on which we might take action as appropriate. The branch chief/school commandant scheme, consisting of the branch school commandants and their evaluation and standardization efforts for the units of the Army—about which I sent you a recent signal—was an outgrowth of our work in this area. Jim Miller and I have talked at length about how we might team up to take advantage of what we've learned, and to conduct further research. Not only does the Army need something like this, but so do other military organizations, industry, business, government—perhaps especially government. Several weeks ago we met to block out the framework of a Systems Science Institute. It would be an organization sponsored by and located at a large university; it would include interdisciplinary capabilities to analyze organizations of all kinds, and to describe and prescribe measures for improving their effectiveness using the methodology embedded in the general theory of living systems. Jim Miller himself is resigning as president of Louisville effective next year; he wants to move back into the research business for the next few years, and he wants to be part of the initiating mechanism for this new endeavor. He has support from the University of California, and somewhere in that university system we could probably get started. If we are serious about wanting to improve the units of our Army, and I know we are, we have to get in on this from the beginning.

4. I tell you all this now because you mentioned to Sin Melner the other day that you wanted to leave the OETC alone for a while in reordering ADMINCEN. With that notion I agree—OETC

has served its purpose, however that purpose might have been viewed by its various supporters, advocates, and critics. We need now to take the best of what we've learned in that effort and decide what we should do next. That's what I've been trying to do, and the synopsis of the work is the subject of the preceding paragraph. When Jim Cary is done with his work in the 3d Armored, I want to ask you to listen to what we've done, and then ask you to meet with and hear out Jim Miller and me on our concepts for a systems science institute—its organization, research, and application to Army problems. This is perhaps a couple months away; I'll send a subsequent signal. Sin's debrief of his conversation with you prompted me to write now to tell you that I've a scheme for what to do next in the organizational effectiveness business, and so we should go along as we are for the time being.

Improving Cohesion

Message to Multiple Addressees
30 June 1980

1. The Chief's read-ahead package should have in it a message in which I summarized for him the results of an extensive investigation of the armor force—its cohesion, stability, and related problems. Some 12 investigations over the past three years are summarized therein. From my own perspective the thrust of that report is not unlike the outcome of related evaluations as early as 1973. From that total experience are drawn the following suggestions related to your 11–12 July meeting.

2. Numbers. Relationships between the Army's end strength and its structure are important. The last time the Army had 16 divisions its military end strength was larger than today by nearly a quarter of a million. For this reason every dysfunctional circumstance that befalls us today is aggravated far more than before. It is not the 16,000 that we fell short of our recruiting objective last year—it is that that took place at a time when we were at least a quarter of a million short of having enough manpower to provide the resiliency we had in that earlier 16-division force. The only reasonable way to cope with this problem is to reduce structure. If we consider we should not reduce from 16 division flags, and that I believe is the judgment of most of us, then other structural alternatives must be undertaken.

In that earlier Army of 16 divisions many, if not most, of today's problems were present. The division in USAREUR in which I served during the nearly 1960s enjoyed about 115 percent strength for nearly two years; however at no time during that period was there a battalion in the division which had assigned more than 94 percent of its authorized personnel. There were, in addition, NCO and junior officer shortages very much like those about which we talk so much today.

All its difficulties notwithstanding, the armor force today is at about 101 percent of its authorized strength. The problem is in the distribution of those numbers across the grade structure—in all grades save E1–E3 there are significant shortages. The same phenomenon is observed in other bulk MOS forces, although it happens at the moment to be more acute in armor than elsewhere. About this problem two principal things can be done—reduce structure to make available NCO and officer faces to fill the spaces in units; adjust promotion and training systems to provide requisite numbers of trained leaders.

Therefore our problem is not one of numbers alone—it is rather of grade imbalances which create shortages in NCO and officer faces to fill the spaces, and in the structure of the force as a whole, which denies us any chance of the requisite measure of resiliency to provide a decent assigned strength in units of the force.

3. Turbulence. In today's combat and combat support battalions, turbulence averages about 40 percent a quarter. Of that amount, half is created by circumstances over which the battalion commander has no control. The remaining half he creates himself trying to cross-level to get the job done. Above 20 percent we conclude it is difficult to conduct much meaningful training. Given a three-year rotation policy, we ensure a turbulence rate of about 12 percent counting rotation, early discharges, and other individual personnel actions. Inefficiencies in the assignment system and other causes quickly bring that number up over the 20 percent mark. We cannot get much better unless we are willing to change this system. Sole reliance

on an individual replacement system in peacetime may be an attractive management practice. However, it is also true that in time of war we will require unit as well as individual replacements. We have, however, neither a peacetime system for providing unit replacements nor a system for providing experience in fielding unit replacements in time of war.

Contributing to turbulence is the percentage of the force serving overseas. In some branches, this number is nearly 50 percent. Time in CONUS is therefore short for the career soldier. Not only does this aggravate turbulence in CONUS-based units, but it is a root cause of low reenlistment rates in some branches. For the NCO whose wife must work to supplement their inadequate income, whose children are in school, and who is struggling with shortages in manpower and a host of mission requirements in his unit, going back overseas after 20 to 24 months in the United States is not at all an attractive prospect. Family pressures drive them out faster than we can persuade them to stay. Short of restationing the whole force to place a larger part of it in the CONUS, and even if that could be done, some combination of structural alternatives must be developed to try and solve this problem.

Priority arrangements which fence, or give preferential treatment to, parts of the force add to turbulence—especially when the parts of the force to which priorities are applied are other than those experiencing significant turbulence problems. Removing fencings or priorities may or may not solve this problem, for if the structure will not support the demands we've placed on it, virtually no system of priorities can ease the problem. Indeed, close study of this problem suggests that we can probably afford no more than to give priority to our deployed forces, using the CONUS base in its entirety as a replacement pool of personnel and units to support the force deployed, and perhaps in addition to satisfy some, but not all, of the requirements we have traditionally levied on our forces stationed in the United States.

Therefore our problem is not necessarily one of how we might wish stability into being. Rather it is one of finding the right combination of individual and unit replacements, rotation policies, and structure in the CONUS base to reduce turbulence to manageable levels in at least part of the force—that part being the force deployed overseas and, in that part of the force, amongst key small unit leaders.

4. What to do. If we believe we must live with a 16-division force that is at least a quarter of a million men short of its reasonable requirements, several things must be done. What follows is a proposed strategy:

Maintain the strength of the deployed force with priority on key leaders and technicians—NCOs and officers necessary to man, train, and command that force.

Maintain a very small contingency force—no more than two divisions in the CONUS. These would be maintained at strength and with proper numbers of NCOs and officers to man, train, and command that force.

Use the remainder of the CONUS-based force as a basis for individual and unit replacements for the deployed force. Lower grade individual replacements would come from IET, wherever it might be conducted. Higher graded individual replacements would be drawn from the remainder of CONUS-based forces. In addition there would be a unit replacement system which would rotate battalions between the CONUS base and overseas stations.

Structure for the MOSs embedded in the military baseops force overseas, and in the US, should be additive to the force structure as a whole. The MILPERCEN models which define grade structure requirements must be made representative of the total forces and produce the required numbers of small unit leaders and their key assistants to the TOE force and the TDA force.

IET in the CONUS-based force should be increased to provide a trained overseas replacement proficient in the minimum number of skills essential to this branch and skill level requirements. This would mean more AIT training in units in the CONUS. It would also mean more BT in TRADOC.

One important mission of the CONUS base would be to provide NCO and officer job specific training to NCOs and junior officers. BNCOC and ANCOC, all conducted under TRADOC, would be job specific. Officer basic courses would become platoon leaders courses. Advanced courses would be company/battery/troop commanders courses. In CONUS, unit schools would provide further job specific training for officers and NCOs involved in training individual and unit replacements for the deployed force.

Structurally all brigades in the Army should be divisional brigades. The special mission account should be eliminated. TDAs should be drawn down according to a set of formulas which provide for only the minimum numbers of any branch for baseops missions. For example, of the 40,000 MPs in the force structure, the formula suggested would leave abut 15,000. The structure related to the other 25,000 should be converted to the purposes described above.

Standardization and evaluation of all training under such a system would be an important matter. To provide that the following scheme would be appropriate:

- First: By regulation, establish branch center commanders as titular heads of the branches.
- Second: Amongst their responsibilities list evaluation and standardization. In this system we would simply extend to their branches the responsibility and authority now held by the commander of the Aviation Center for standardization and evaluation of aviators, flight crews, and airplane operations worldwide. Commander of the Artillery Center functions in a similar manner with regard to the missile system evaluation group.
- Third: Delegate to Commander TRADOC the responsibility for establishing and operating, as the executive agent of the DCSOPS DA, the Army Standardization and Evaluation System worldwide. Commander TRADOC should be charged with: seeking a consensus among major commanders as to what should be standardized, how it should be done, [and] promulgating for DCSOPS instructions to be issued by CSA with regard to standardization and evaluation worldwide.
- Fourth: To accomplish the program, TRADOC would use the evaluation directorates of the schools; these would now become evaluation and standardization directorates. They would work in conjunction with such other teams as may be operating for major commanders to conduct evaluation and standardization of Army units worldwide. This would be similar to the system we now have with FORSCOM and TRADOC, but it would be formalized for the whole Army and it would extend the system Armywide, not just between TRADOC and FORSCOM.

Standardize Organizations. Standardize TOE for like units of the combat and combat support arms, and perhaps even for the combat service support arms. Branch chiefs would be responsible for the evaluation and standardization of these TOE. Such is in fact the case now, but there is now no discipline in the system. We send fully developed TOE to DCSOPS, who approves them. However, it is then necessary for the major commands to MTOE those TOE in order to implement them in the field. The MTOE process is the first step in creating turbulence in units of the major commands, for the MTOE system is the modern day successor to the old general order line deletion of spaces in order to generate manpower authorizations for those baseops functions we can't otherwise afford—the communities in Europe, for example. What I'm suggesting is that we should adopt a system that does not force the major commander to rob from Peter to pay Paul. We should make those TDA resources available on the order of 12–15 percent over the TOE authorizations on a major command basis. Then we should standardize all the TOE and hold to those standards unless the standardization and evaluation systems prove to us that they should be changed. This would act to reduce to near zero the MTOE perturbations that are to be required as the transition to Army 86 ensues. It would also ensure that the TOEs which we develop at such cost in study and other effort do in fact get implemented. Now they become part of some PARR process, frequently with different priorities in different commands. Today the system virtually guarantees that nothing is ever quite the same anywhere; we should seek to have it as much alike as we possibly can—everywhere.

Establish special rotation policies for small unit leaders and key specialists. Branch chiefs should be made responsible for ensuring there are enough trained leaders and key specialists in squads, crews, sections, platoons, companies/batteries/troops, and battalions/squadrons of the Army.

Training assets of the USAR. Its divisions and schools must be integrated into the Army's total training capability. Correctly done, this could require some reorganization and relocation, perhaps traumatic, albeit necessary.

A pay raise is clearly needed. Additionally, an incentive package, largely based on a revitalized education benefit program, must be provided to attract and retain the soldiers we want to keep as small unit leaders—NCOs.

Finally, establish a regimental system for the US Army. Separately I'm sending each of you a study done by my British and Canadian liaison officers. Several months ago I tasked the two of them to develop a regimental system for the US Army. Patterned after the Canadian and British regimental systems, it would nonetheless have to adapt to the American way of doing things. They have just finished this work. It is sound enough to provide the blueprint for a long-term scheme to provide us with a great deal of stabilization and a bit of the unit association which is always so important to a military body.

5. Long message, for which length I apologize. No time to write a shorter one. Not to preempt ARCOST, which did a super job in the time available; simply to provide the perspective which comes from having studied this whole matter longer than perhaps anyone else, and longer than it should have taken to get something going to solve the problems.

OSD Characterized

Letter to Honorable Walter B. LaBerge
Aptos, California
11 March 1981

You have been most kind not only to ask for our views in the "debate," but to listen patiently and even heed when we seemed to make sense, what a refreshing difference from the normal attitudinal mindset of secretariats—especially OSD. Having looked at the zoo from a service as well as an OSD vantage point, I conclude that sooner or later our inability to march to a common drum will redound to our disadvantage, perhaps in some disaster, but quite probably in a more insidious but less dramatic way. The power to say nay or aye without a charge to bear the responsibility for the outcomes is power ill given and ill used. OSD is full of that power, as are the staffs on the Hill, in OMB, and even in service secretariats.

Current Army Culture

Message to General E. C. Meyer
Army Chief of Staff
20 March 1981

1. Someone said the Chief of the Staff of the whole Army is pushing training, favors maintaining equipment, is against turbulence, is for stability. Just looking around and listening the last few months, here's "a" perspective on how all that's going.

2. A recently-in-command mech battalion commander reported that, as he took command, he noted his unit hadn't done an ARTEP in more than eighteen months. So he drew up an ARTEP; went to see the brigade commander to show what he wanted to do. The brigade commander thought that was such a good idea that he said he would consider telling all his other battalions to have an ARTEP, since no one had done one recently. On their installation, the most important daily matters are the blotter, post police, and prompt and full manning of post details. In fact the battalion commander gets a phone call from the post chief of staff almost as soon as one of the battalion soldiers, or dependents, gets a traffic ticket. The brigade commander reports he's very concerned about his unit's image, and so he's paying attention to the "image things." Nearly as I could tell, no one from the brigade staff has visited battalion-level training in more than six months. That includes the brigade commander. The battalion commander says, "I know the Chief is pushing training, but there are a whole lot of folks between me and him—and they obviously have other priorities."

3. A tank battalion commander reported his battalion hadn't been to the field as a unit in over a year. Reason—the chief of staff, the major command commander, and the division commander are pushing "maintenance" and "training." If they take the tanks out and run them around, they break down and the OR rate goes down. That's bad—so they leave the vehicles in the motor park and go to classes most of the time. Until recently they were not allowed to put maintenance on the training schedule—the assistant division commander doesn't believe in it. So they maintain on weekends and at night. Haven't had a quarterly inspection on battalion vehicles in nearly a year—division says they can't afford the time.

4. A mech brigade commander reported he couldn't keep his maintenance up because he wasn't getting the proper DS and GS maintenance support. On his training schedules there is "no" maintenance—he doesn't allow it. When asked about this, he said that he really doesn't understand what the problem is. In his last unit they never had maintenance on the schedule, and they never had this much trouble with maintenance. Asked what his last unit was, his response—"the 82d Airborne."

5. Listening to a group of company commanders in a brigade in an armored division, their principal complaints were that they couldn't get any training done for all the interruptions and distractions, and that they couldn't keep their equipment running. Quarterly services were not being performed; their deadline rates were much too high for their activity levels; they had four hours weekly scheduled maintenance. They reported that they had got together, realized they had to schedule some maintenance, and agreed that the four hours was about all they could get away with, and all scheduled it so they'd all be together. They showed me a list of all the things division, brigade, and their battalions had from time to time listed as number one priority; the list contained twenty-seven items; none had to do with training or maintenance. The division commander disclaimed all this, but on my invitation went to see; later he admitted to me that he had confirmed my observations.

6. The MILPERCEN is apparently sending a team around to talk with middle grade NCOs about all the things we're trying to do to improve their lot. In one brigade the team explained about pay raises, GI Bill, and so on, then asked for questions. A sergeant asked about turnaround time on CONUS tours—the MILPERSON said that E5s, E6s, and E7s in the predominant MOSs in that brigade could expect to get orders about fourteen months after returning to the US, and that situation could be expected to continue for several years. They hooted the MILPERSON. The PACs in that unit had an immediate surge of NCOs wanting to apply for recruiting duty, ROTC duty, Readiness Region duty—anything to get a tour with a promise of more than fourteen months on station.

7. A tank battalion commander reported that, in the division his battalion is in, two major concerns dominate daily affairs—the blotter and the deadline report. The division expects by 0800 daily a report about what action is being taken by the battalion against all those appearing on last night's blotter. Also by 0800 daily, a report is required as to what action is being taken by the battalion, vehicle by vehicle, to get deadlined vehicles operationally ready. In order that this division may appear to be on top of maintenance matters, the division has set the OR rate five points above the DA average. As a consequence, they leave the equipment in the motor park a great deal rather than risk the wrath of the division chief of staff, who descends on them when they run the equipment and it breaks down.

8. Several company commanders in one brigade commented that, while everyone is pushing training, no one seems to know what they are supposed to be training for. One captain said that he understood at first that the purpose of all the training emphasis was so that the unit would have high SQT scores, but that he wasn't too sure about that now because battalion was pushing ARTEPs at the moment and the heat on SQTs was off. They were somewhat less than coherent when asked how their training related to their mission; no one could describe that linkage very well, if at all. It was as if training is a matter to be considered in isolation, apart from any consideration of how it might relate to what they're to do if they have to go to war.

9. An infantry battalion commander, after about twenty months in command, was notified, one, that he'd been selected for the War College, and second that he would be extended to thirty months in command so that he could go direct from command to school. Rather than go another ten months in command, he has elected to turn down the War College and retire. Apparently he is a good battalion commander, at least so say his brigade and divisional commanders. The officer and his wife both report they're exhausted, fed up with rushing from crisis to crisis, tired of constantly having to start over again with a new group of officers who come and go much more often than do the battalion commanders.

10. A brigade commander newly in command gathered his battalion commanders together and announced that the way for them all to succeed was to "stay ahead of the averages." Don't do anything spectacular one way or the other—good or bad, but just "stay ahead of the averages." He was sure they all wanted to be colonels, and he for sure was going to be a general. The way he'd got ahead so far was to just "stay ahead of the averages." Things fell into place that way.

11. A Senate subcommittee staff group visited three posts looking into your allegations and those of others that we've a baseops support-level problem. On two posts they were met by the commanders, who stayed with them for their brief stay and laid out to their satisfaction what the staffers described as a pretty objective accounting of the baseops resource levels, requirements, and related matters. At the third post they were met by a lieutenant from protocol who showed them around. The commander popped in briefly from his busy schedule and apparently had no idea what his baseops resourcing problems were, or if he had any. Finally he took the group to

see a WWII wooden barracks and said he guessed those buildings should be replaced sooner or later. The good done by the first two was largely undone by the last one.

12. Just so you know it isn't all bad, in one division they've laid out a tactical mission for the division and are trying to relate everything else to that. Maps on the wall, sand tables on which they work tactical exercises, field training in which they simulate as best they can the area in which and circumstances under which they might have to fight. In an old building they've some simple sand tables on which they lay out situations taken from S.L.A. Marshall's books, from the Benning reprint of *Infantry in Battle*, and others. Their junior leaders go through these situations and, at the end, the exercise leader recounts what actually happened in the real situation. Super training—related to a mission; they all seem to understand what the emphasis on training is about and what they're training for.

13. It's still pretty spotty, not much better overall than when last you and I talked about it. The units that were pretty good then are still pretty good. The same can be said for the not so good ones. As I commented to you once before, when we talked about this, we're probably never going to get much better until we institute widespread standardization and put someone in charge of checking to see that people do what we say and that they don't overload their units with nonessentials. Turbulence is not diminishing—in some grades and MOSs it's worse than when Rick Brown looked at it three years ago. Extending commanders has had no positive effect that I can discern, and in too many instances of which I am aware it has been counterproductive. Even the pretty good divisional and brigade commanders have trouble understanding the world of the company-level commander and are relatively ineffective at keeping their intermediate commanders and staffs at all levels from cluttering up the company-level world with an overload of trivia. In the past six months I've found only one commander at any level who remembered that, some time ago, the Chief of Staff of the Army sent all his commanders a letter about training, standardization, and what has to be done if ever we're to have a much better Army.

Commander's Notes...

no. 6

14 December 1981

COMPLETED STAFF WORK

1. Staff members are essential to the process of continually updating information. Staff work -- studies, projects, action papers, briefings -- is the lifeblood of any command information flow. The research, coordination, fact-finding, study and presentations by staff officers directly affect and fuel the decision-making process at every level. Unfortunately, as important as staff products are, they are only useful if the staff completes the work.

2. Completed staff work is the only reason for being for any staff. Staff work must be completed in such a way that the correctness of the solution or action is easily recognizable. Completed staff work requires only the commander's or staff chief's approval or very minor modifications prior to approval. Disapproval is most frequently the result of failure on the part of the staff officer to thoroughly recognize the problem, and to analyze all its details and facets -- complete the work.

3. Occasionally, if you are an inexperienced staff officer, you will arrive at a point where you feel there is no logical solution. You become certain, then, that your only answer is to ask the commander or staff chief what to do. This occurs more often when the problem is difficult. This kind of thinking leads only to frustration for both the staff officer and the commander or staff chief. As a staff officer you succumb to it only when you don't know your job. It is your job to advise the commander or staff chief what to do -- not ask him what to do. The commander or staff chief needs answers, not questions. The staff officer's job is to study, write, restudy and rewrite until he has evolved a single proposed action -- the best one. The commander or staff chief then merely approves or disapproves.

4. In USREDCOM/JDA, besides exhortations, what is completed staff work? First of all, it is a thoroughly thought-out, logical answer to a problem. Almost always, in the case of written staff documents, it results in an action paper. Writing an action paper to the commander or staff chief is not completed staff work; writing an action paper for the commander or staff chief to send to someone else is. A bundle or book of recommendations or evaluations is not completed staff work, but it is when that bundle or book is followed with a series of action papers to correct deficiences. Your views

1

COMMANDER'S NOTES NUMBER 6
COMPLETED STAFF WORK 14 December 1981

should be in such a finished form and your work so thorough that the commander or staff chief can and will make them his views and actions simply by signing his name.

5. In briefings, the same rules of thoroughness and clarity apply, but in a briefing, there are some other points to remember. In most cases, it is not necessary to start off with "the first book of Genesis" to bring the commander or staff chief "up to speed." Too much time is wasted. Completed staff work requires that the briefer determine the background and the relative knowledge of the person he is briefing. How do you do that? Ask aides, executive officers, fellow staff officers who have the knowledge. Much time is wasted by briefings that are too basic. The staff officer's objective is to save the commander's time.

6. Briefings should also be just that -- brief. Lay out the facts and get to the point. Rehearsals quickly expose briefings that run like marathons. It's possible that, as you rehearse a briefing and eliminate the unnecessary, the briefing will become so brief, it can be reduced to a written document. Fact sheets are always better, more quickly prepared and more quickly read and understood than briefings.

7. A final reminder on briefings, while "eyewash" is a waste of everyone's time, courtesy and professionalism demand that briefings be organized, training aids be readable and that the atmosphere be conducive to information transfer and decision-making. It's not necessary that multitudes of observers or straphangers be present to hear everything that's said. Control of the briefing room and the audience is a responsibility of the briefer and just as much a part of the briefing as the substance. Rehearsals will aid in eliminating attendees at the main briefing and in insuring that slides, charts, graphs are correctly and professionally handled. The USREDCOM Eagle or JDA Crest on each slide is not a substitute for professionalism.

8. Two other points are pertinent and applicable to completed staff work. First, the job is not done when the paper is signed, the action passed or the telephone call made. It must be followed up. It does no good for a staff officer to give the excuse, "I gave the message/paper/action to so and so," when the work is never completed. The staff officer must follow through. At least a third of good staff work should be dedicated to following-up, checking upon work that was done previously. Too many so-called "completed" staff actions stop dead because no one follows up on the implementation. Even after the action paper is signed and sent,

COMMANDER'S NOTES NUMBER 6
COMPLETED STAFF WORK 14 December 1981

completed staff work demands that it be checked for results. If there are none, the staff work isn't completed.

9. The last point to remember about completed staff work is that a staff directorate or division or even an action does not exist in a vacuum. Staff sections are not stovepipes. Coordination, cross-fertilization and discussion between staff sections on an action can only make the solution to a problem better. It is vital that staff directorates talk to each other at all levels, action officers up to directors. This is a team effort and one we don't do very well. USREDCOM/JDA completed staff work requires that all directorates that are interested in an action or can help with it do so. If not asked, they must insist that they be part of the action. Briefings should be handled the same way. All those who have a stake in the briefing should be present and prepared to answer questions and participate -- observers are not needed.

10. In the final analysis, the staff officer must be a specialist in his field. The end product of a staff officer's labors must be as nearly perfect as humanly possible. It must be of such a caliber that he is willing to stake his own professional reputation on the completeness, accuracy, logic, and workability of his solution. This results in the commander being protected from half-baked ideas, voluminous memoranda and immature oral presentations. It also allows the staff officer, who has a real idea to sell, to more readily find an audience. The final test of completed staff work is this:

a. If you were the commander, would you be willing to sign the paper you have prepared and stake your professional reputation on its being right?

b. If not, take it back and rework it -- it is not completed staff work.

DONN A. STARRY
General, United States Army
Commander in Chief

1 Incl
Annex A -Completed Staff Work

3

Technology Advocates

Letter to Professor Leslie G. Callahan
Georgia Institute of Technology
29 June 1982

. . . in some areas, things have gotten worse, not better, since 1964! The answer, if indeed there be one, is surely elusive. At the top, as between the OSD and the services, we are ever at the whim of an ever-changing group of technocrats in USDRE (DDRE) (ARPA) (whatever). Each comes as administrations change (some more frequently even). Each believes that technology will solve all, but not necessarily the last guy's technology. Each knows that if he does well enough by his part of the industry during his tenure, he's sure to go back to a better job than that from whence he came. Given how slowly the government scientific cycle moves these days, all that means is that we're in an ever-changing priorities routine every few years. Further, the people who are forever reordering things never have to bear the eventual responsibility for what they do whilst in office.

With that sort of system, nothing much good is possible!

New Readiness Command Functions

Letter to General John W. Vessey Jr.
Chairman, Joint Chiefs of Staff
17 September 1982

This is the last of a series of letters I promised you concerning JCS functions that could possibly be assigned to USREDCOM. My staff has completed a review of the JCS Exercise Program and it appears that USREDCOM can assist JCS in the management of that program.

I am again proposing that USREDCOM be named as the JCS Executive Agent with responsibilities in the areas of: development of specific prioritized joint/combined exercise objectives in coordination with other CINCs (these objectives will focus on joint doctrinal issues currently under study and will incorporate the needs of the other operating commands); participation in the JCS analysis of the entire JRX program; and development of procedures for improved control and monitoring of the expenditures of all JRX funds.

Our participation in the management of the JRX program will further enhance JCS control and provide more needed detailed planning guidance to all CINCs. It will cut down on the redundancy in joint/combined exercise objectives/duplicative evaluation efforts and help resolve the overcommitment of forces/resources. Basically, we will assist JCS in their efforts to ensure that we get the maximum benefit from every exercise dollar spent by aligning objectives-priorities-resources to produce an optimally effective exercise program.

This concept is the last link in the chain that ties together the elements of test and evaluation, simulation/wargaming, and development of joint doctrine—all interactive and part of the process of developing joint service interface and connectivity.

Readiness Command Status

Report to the Joint Chiefs of Staff
27 October 1982

For the next few minutes I'll relate for you the directions in which Readiness Command and the Joint Deployment Agency are headed. First, our missions: support to the deployed CINCs by deploying to them and sustaining the Army and TACAIR forces stationed in the United States; land defense of the CONUS and Alaska; military aid to civil authority; military support for civil defense. Essentially, therefore, the planning process is our business; plans are our product. How is all that going? Our joint planning requirements are derived from missions assigned in the Unified Command Plan or by the JCS in some other form. We do not maintain a general war plan as such. Our primary task is to support *other* unified commands.

To that end, we currently maintain plans in support of over 40 contingency plans of the other unified or specified commands. We maintain two plans in direct support of the JCS and four plans as a supported command. These plans range in scope from minor contingencies to general war. We have entered into agreements with seven other commands—EUCOM, SOUTHCOM, Alaskan Air Command, PACOM, LANTCOM, Military Airlift Command, and Aerospace Defense Command—with regard to our respective responsibilities in the development of these joint plans. The planning cycle normally begins with a draft concept plan, a plan for the employment of forces in joint operations according to the current doctrine for joint combat operations.

In building a plan, we attempt to identify several feasible and affordable courses of action. With these defined, simulation, in which war-gaming concentrates on the functioning of forces in joint operations, provides the most effective method of comparing alternatives. Then, having selected a preferred course of action, a plan is produced. This plan, or parts of it, is tested in a joint exercise—CPX or FTX. This is accomplished through a coordinated program of evaluating joint objectives in all JCS- and CINC-sponsored exercises. This evaluation provides feedback from which plans, concepts, and procedures can be validated or revised as necessary.

That is the full planning cycle. The sequence of drafting plans, conducting war games and simulations, exercising and evaluating exercises, is designed to develop and refine contingency plans, war plans, operational concepts and in the long term the employment, deployment, and sustainment issues for any given mission. Identification of what is needed to do the job, the costs and risks involved, and the provision of plans, modernization of joint tactics, techniques and procedures, and joint force training are the products of the planning cycle.

Now to describe some specific programs underway to develop these products. First, JTTP. Bearing in mind that our basic goal is not to direct or change service doctrine, we are nevertheless responsible for the joint interfaces between the services, those interconnections without which we cannot conduct joint operations. It is necessary to tie service concepts together with missions, concepts, and battle-fighting techniques for joint force employment. To do that, we have laid down a framework for joint doctrinal development.

First, we have entered into a Memorandum of Agreement with the Air Force's Tactical Air Command and the Army's Training and Doctrine Command to develop and publish joint force tactics, techniques, and procedures. Beyond this, we're working to bring the maritime forces into this endeavor. This past June, Glenn Otis of TRADOC, Bill Creech of TAC and

I, and representatives of the Navy and Marine Corps, met to discuss how we can all work together. We hope to soon send to the JCS for their consideration the first four-service manual, *Joint Psychological Operations*. Under the aegis of the MOA, 21 joint publications are under development, 12 by REDCOM and 9 others by REDCOM, TAC, and TRADOC; hopefully many of these will also have JCS blessing in due course.

Beyond this, in being is a Joint Tactics, Techniques, and Procedures Review Group made up of representatives of all the unified commands, TAC, FORSCOM, and TRADOC, with observers from CAC and the JCS J-3 and J-5. This group works the face-to-face problems and is a working level conduit to the operating commands.

The second major initiative is a Joint Deployment Agency responsibility. Since our ability to move forces to wartime locations is essential to their employment in any theater, we need to consider the Joint Deployment System as an integral part of the planning process and vital to the successful execution of any mission.

The Army Force List for 4102, the deployment plan to reinforce Europe, consists of 19,000 individual lines. Each is a unit, everything from a division to a two-man tooth-drilling detachment. As you know, it takes about 18 months to fully develop these lists—the infamous TPFDL. Last year, during REFORGER, we planned to deploy 360 of those lines. We published the flow plan five months in advance. In the ensuing five months, there were 1,100 changes to those 360 lines. Obviously we've not got a functioning system. How to fix that?

We have proposed to solve the problem I've just described and a related problem, that of how to develop a sound plan in less than the 18-month deliberate planning cycle. You were briefed on what we now call JDS Prototype last week, so I won't detail it here. It is, you recall, designed to cut the planning cycle from 18 months to a matter of days. Beyond this, throughout the crisis, we can respond quickly to changes. Additionally a proposed revised set of terms of reference will give the JDA directive authority over the transportation agencies during wartime deployments; this should eliminate many bottlenecks we see during our peacetime exercises.

Finally, the JDA's OPLAN refinement process is being streamlined and modernized. This should result in more realistic operation plans. For example, we no longer allocate strategic lift to move nonexistent supplies. We recognize the needs of the supported commands; however, if the supplies are not available they are not available. It does no good to move things on paper if they aren't going to be available when and where they are needed.

The efforts of the JDA tie directly into the next REDCOM initiative, war-gaming and simulation. The very best plan is useless if it is not executable. One of our greatest planning shortcomings has been the lack of an ability to evaluate a plan thoroughly and at reasonable costs. To do this we are developing a theater-level simulation system that will allow us to evaluate alternative courses of action to see what works or, equally important, what doesn't. We are developing this capability in such a way that the system may be used by the entire joint community to conduct simulations of joint operation plans. Our goal is to be able to analyze the feasibility of a plan, providing an objective assessment of risks and implications of different courses of action. We must produce plans that are both force and transportation feasible. The simulation development is being accomplished in concert with the War Colleges, the Naval Postgraduate School at Monterey, Lawrence Livermore Lab, Jet Propulsion Lab, and several civilian agencies. The goal is to develop a theater-level simulation capability that we can use to analyze deployment, employment, and sustainment options in contingency situations; aid in the development and refinement of joint planning procedures and concept development; and support the control requirements of the Joint Readiness Exercise Program.

The simulation system is being designed to analyze force requirements, employment options, deployment schedules, and sustainability requirements through theater level. The Army's McClintic Theater Model is being used as the theater-level model; other models will deal with specific areas of evaluation. For example, TACSIM will provide the intel module; Livermore's JANUS is the AirLand Battle module; NWIS—Naval Warfare Integrated Simulation system, a 2000-platform surface, subsurface, and air model—will model the sea battle; the JDS Prototype will model deployment. We have several logistics model candidates at the moment. Our main task is to create linkages between these models and to bring them together into a viable theater-level simulation system available to all CINCs through secure satellite terminals.

Now the Joint Exercise Program. First, we no longer conduct exercises just to be running exercises. JRXs, particularly field exercises, are just too expensive, both financially and in terms of the few select units that seem to play in most exercises. We have entered into agreements with the other unified commands to join in manning and evaluation of all joint exercises. These agreements provide for the joint development of exercise objectives and for REDCOM involvement in evaluation of joint exercises wherever they are scheduled. We are also investigating other mediums—CPXs or simulations if these will allow us to evaluate a plan or check out our procedures at less cost than with a full-scale exercise.

This should also lead to a better coordinated JRX program. This doesn't mean we are taking over anyone's program; however, as the major provider of combat forces to the overseas commanders, we provide advice, evaluators, and a link between Stateside augmentation forces and the in-place forces. Our evaluators are people with special expertise who look at specific procedures in which we or the sponsoring command have an interest. We have just completed our first action under this agreement, an evaluation of joint suppression of enemy air defenses during Coldfire 82 in EUCOM. The RDJTF and PACOM have asked us to help evaluate Jade Tiger and Team Spirit 83. This is not to say that there won't be REDCOM exercises. In 1983 we will conduct Brim Frost in Alaska and Gallant Knight in an RDJTF scenario. Future adjustments to the exercise program will depend on how well our evaluation system works.

Two other areas are worth mentioning. We are working on communications linkages at the joint operational interfaces. As you are well aware, service problems in C3 are immense, doubly so in the joint arena. We have suggested to the Chairman that we be tasked to oversee the development of these interfaces in several joint programs.

Finally, the ability of REDCOM and the Joint Deployment Agency to continue operations in wartime is marginal. At MacDill, we are well within the range of Cuban-based MiGs; our current alternate locations at Forts Gordon and Rucker are likewise vulnerable. We have been unable to find an untargeted area or one which can provide adequate protection for the staff and equipment.

Given the threat, we are examining the feasibility of relocating to a nonmilitary, no-threat area using a mobile command facility. We're examining all modes of transportation, to include railroads. I mention railroads because they provide increased flexibility over other transportation modes. Basically, by loading road-mobile C2 equipment on the train, we can go by rail or off-load and go overland to a secure area. In any event, we are working directly with FEMA to ensure we tie closely together our co-op plans in a mutual effort.

That is REDCOM, a different sort of unified command. Still a combatant command, we are primarily a command supporting the other CINCs and the JCS. As a result, our major interest is figuring out ways of doing things better *jointly*. Development of modern aids to planning, the JDS Prototype, war-gaming and simulation, exercises and exercise evaluation, development and publication of joint tactics, techniques and procedures—we intend to attack the joint issues so that, no matter where the crisis may be, we can respond rapidly, economically, and with appropriate joint forces.

Joint Staff Issues

Letter to General John W. Vessey Jr.
Chairman, Joint Chiefs of Staff
14 March 1983

You should not have to be bothered with this matter. Neither should I. It is however typical and symptomatic of the difficulties we've had getting done the things that you, I, the Chiefs and others have agreed to do. It is for this reason that I've elected to make an issue of it.

The subject is the proposed joint PSYOP [psychological operations] manual. You'll remember that I sent it to you on the 29th of October last year, pursuant to our agreement that here might be a test case; one we might use to test out our ability to get doctrinal matters worked out in the system out here into the joint staff and perhaps into the form of a JCS Pub.

On 3 March this year the J3 signed out a memo to the J3 REDCOM rejecting the proposed PSYOP manual. May I just make several observations about that memo. Mind you, we've no pride in authorship in this—had we done what truly needs to be done, the proposed Pub would have been far more directive in nature, and much more precise. It is written the way it is because of our desire to get something we thought the Chiefs could agree upon. So there's no air of the jilted suitor in what I'm about to say.

. . .

The J3 memo asserts that there is considerable duplication between the proposed Pub and the JSCAP Annex D. Right on! It is deliberately that way; we wanted to get it in words that would be relatively less difficult to get agreement on. A couple of extracts enclosed also will show that there is some improvement in the wording in the proposed Pub; additional material is included; and the whole is unclassified—vice the SECRET classification of the JSCAP annex, and its limited distribution restrictions. In short, we thought we had provided some attractive features which would help along the idea of making this paper into a Pub.

Finally, a matter sent to you over a CINC's signature has been returned to the CINC's staff by a staff principal on the Joint Staff. Sent to some of the other warlords out here that would have elicited a blast far more vituperative than what you're reading. Obviously, the Joint Staff needs a lesson in protocol.

The cavalier tone of the J3 memo, and the absence of substantive comment on the proposed Pub, are quite typical of the kind of response we get from much of the staff to almost everything we submit. The J5 is the only staff agency with whom it is at all possible to work in a half rational way. The J3 and J4 vie with one another for top honors in obstreperous obstructionism. Until receipt of the J3's memo they were neck and neck. However, with the instant memo the J3 has forged ahead in the race to determine which of the two staff groupings is least effective and most difficult to deal with. We have (as nicely as we know how) asked the J3 for substantive comments on the proposed Pub. If any are forthcoming, we will fix what needs fixed.

Then I believe you and I should talk again about whether or not we want to continue this caper or anything like it. Perhaps we should just give up on trying to get anything done through the Joint Staff. It takes forever to get even this kind of lousy work out of them.

While I regret shoving this back to you this way, I'll not be pushed about or backhanded by some miserable staff poltroon. Especially is this true when we're out here trying in good faith to do something we thought you and I had agreed on. Cheers!

13. Manpower

Page

Recruiting Poster I 580
Recruiting Poster II 582
Personnel Management System 584
Replacement Systems 585
Recruiting and Training 588
Turbulence and Readiness 590
Personnel Doctrine 592
Training and Force Modernization 593
Recruiting 594
Recruiting and Retention 595
Recruiting Strategy 596
Turbulence in Armor Units 598
NCO Reenlistment 601
Managing Manpower Shortages 602
ROTC Program 604
Overstructure and End Strength 606

Recruiting Poster I

Message to General William E. DePuy
Commanding General, US Army Training and Doctrine Command
5 February 1974

1. Yesterday I got a curt little note from General Abrams about our "We Will Guarantee You Almost Anything" recruiting poster. His note was attached to a letter from Dewitt Smith explaining to me why he (Dewitt) is against further release of the poster. The gist of the letter and CSA's note is that the use of a good-looking girl on the poster is a "gimmick," to use the CSA's word. I wanted to warn you of this in case the Chief brings it up with you.

2. So that you may be a little forearmed as well as forewarned, here's a synopsis of what has happened as we know it.

- We released about a thousand posters in early January.
- On 18 January the Associated Press put a picture of the poster and a related story on the AP wire.
- Our responses indicate the AP story has been published in every major and many minor newspapers in the US. It was published in *Army Times* this past week. It has been publicized on many radio and television stations. In most cases it has served as a lead-in on recruiting activities in that area, and from that standpoint is publicity that probably would not have been received by USAREC and certainly not by Fort Knox.
- A total of nine unfavorable responses have been received at Fort Knox. Three of these came from chapters of N.O.W., the National Organization for Women; one from a little old lady who identified herself as "Secretary of War and regular General of the Army"; one was from a high school student; and four [were] copies of the article with comments scribbled across the text. All were from women.
- Favorable responses have been numerous, to include about 50 phone calls, some from recruiters requesting copies for distribution. The television quiz show "Split Second" requested a copy for use; this was cleared with CINFO, Washington, DC, and sent to Hollywood. Three letters requesting copies have been received: one from a recruiter in San Antonio; one from a private citizen; and one from a retired USMC veteran who serves as a VA counselor at Northern Illinois University. Two letters were received requesting assistance: one from a woman in Perris, California, requesting help in getting her son enlisted in the Army, and one from a USMC first lieutenant desiring information on a lateral transfer from the USMC to the US Army. The final response was a call from Fort Benjamin Harrison USAIA, which teaches the Army recruiting course to EM/EW of USAREC, requesting several copies of the poster for use in their course as an innovative example and for inclusion in the Recruiting Museum.
- From May through October 1973, Fort Knox averaged less than 200 acquisitions per month through SOC/UOC programs. Our November input was 430, December 337. In January we had 608, 348 of whom enlisted after national coverage of the poster; at least some of the dramatic increase we know was occasioned by the poster story.

3. In sum, to my knowledge the only people who have taken exception to the poster are the nine women's libbers and Dewitt Smith. Not only do I believe Dewitt is overreacting, but I'm more than somewhat put out at the technique of causing the CSA to bite me without at least giving the Chief the whole story. Except for Bernie Huffman, who had the courtesy to call my

SGS and ask us to hold up distribution, no one has bothered to talk with me or my staff about the whole matter. Needless to say, I'm less than overjoyed by the whole affair.

4. Finally, the Chief's note was attached to some colorful but prosaic posters from the 2d Armored Division, and both he and Dewitt suggested those as more appropriate than ours. It's interesting to note that our per-recruit costs for FY 74 are $42 compared to $228 for Hood. Whatever happened to cost effectiveness?

Recruiting Poster II

Letter to Major General DeWitt C. Smith Jr.
Assistant Deputy Chief of Staff for Personnel
8 February 1974

This responds to your letter of 30 January regarding our "We'll Promise You Almost Anything" recruiting poster.

The poster was prepared to target against one of our most serious problems, a growing perception that many recruiting promises are not fulfilled. We tried it out on RMS Louisville with whose people we work very closely and have a fine relationship, and they agreed it filled the bill.

In early January we released about a thousand posters.

On 18 January the Associated Press put a picture of the poster and a related story on the AP wire.

Our responses indicate the AP story has been published in every major and many minor newspapers in the US. It was published in *Army Times* this past week. It has been publicized on many radio and television stations. In most cases it has served as a lead-in on recruiting activities in that area, and from that standpoint is publicity that probably would not have been received by USAREC and certainly not by Fort Knox.

A total of nine unfavorable responses have been received at Fort Knox. Three of these came from chapters of N.O.W., the National Organization for Women; one from a little old lady who identified herself as "Secretary of War and regular General of the Army;" one was from a high school student; and four copies of the article with comments scribbled across the text. All were from women.

Favorable responses have been numerous, to include about 50 phone calls, some from recruiters requesting copies for distribution. The television quiz show "Split Second" requested a copy for use; this was cleared with CINFO, Washington, DC, and sent to Hollywood. Three letters requesting copies have been received: one from a recruiter in San Antonio; one from a private citizen; and one from a retired USMC veteran who serves as a VA counselor at Northern Illinois University. Two letters were received requesting assistance: one from a woman in Perris, California, requesting help in getting her son enlisted in the Army; and one from a USMC First Lieutenant desiring information on a lateral transfer from the USMC to the US Army. The final response was a call from Fort Benjamin Harrison USAIA, which teaches the Army recruiting course to EM/EW of USAREC, requesting several copies of the poster for use in their course as an innovative example and for inclusion in the Recruiting Museum.

From May through October 1973, Fort Knox averaged less than 200 acquisitions per month through SOC/UOC programs. Our November input was 430, December 337. In January we had 608, 348 of whom enlisted after national coverage of the poster; at least some of the dramatic increase we know was occasioned by the poster story.

My assistant IO gave the *Army Times* an improper summation of the responses, and that was reflected in the *Times* story.

We had seen, evaluated, and set aside the colorful but more prosaic 2d Armored Division posters before we went ahead with "Almost Anything." It is interesting to note that Fort Hood's per capita recruiting costs for FY74 to date are reported at $228; ours are $42.

Finally, it is truly regrettable that the perceptions of how we should approach recruiting differ so dramatically between here and Washington. I did not and do not consider that poster a "gimmick." It was designed for a specific purpose and, especially with the free publicity we've had, it has done a lot to dispel the notion against which it was targeted.

Personnel Management System

Letter to General Bernard W. Rogers
Army Chief of Staff
20 November 1978

This is the status report of the concept and plan of attack for getting on with the top-to-bottom analysis of our personnel management system. . . .

The goal is the development of a system that will place and sustain trained soldiers in cohesive units that are capable of effectively employing our modern weapons systems on the battlefield of the 1980s. The concept is to look at where we are now, where we should be, and determine how to get there. An initial milestone schedule is enclosed. As we develop the program further I'll keep you posted.

In the analysis we are going to consider man-machine interface as it relates to individual weapons systems and the complications associated with employing these systems in units on the lethal, continuous operations battlefield. Our ability to man the systems in units in both a peace and wartime environment will be a product of this effort. We must ensure that our personnel system has the capability of supporting our programmed modernization efforts. To accomplish this, we plan to use a "total systems approach" to prevent suboptimization. Our concern is that, in fixing one part of the system, we may be creating problems in another. I have told Ben to take both a short- and long-range look at the problem. In so doing, we will develop a series of time-phased actions. In the short range we are going to attempt to develop solutions to problems that now adversely impact upon our force modernization and near-term readiness. Principal areas to be examined include the following:

- What can be done to reduce the impact of turbulence on our force readiness.
- Determining the training necessary for S-1, G-l, DPCA and battalion administrative officers.
- Integrating instruction into officer basic and advanced courses on how commanders can best deal with the human component of their organizations.
- Developing a package for commanders to assist them in analyzing the human component of organizations and generating the human element of the commander's estimate of the situation.
- Determining a framework for institutionalizing the combat developments, training developments, and training concepts growing out of the top-to-bottom analysis.
- Establishing an ARI field office at Fort Benjamin Harrison to initiate a long-range development capability.

Replacement Systems

Message to Major General Fred Mahaffey
Fort Leavenworth, Kansas
1 May 1979

. . .

2. First, we need a combination replacement system—one which provides individuals to the Army in the field to fill in combat and other losses, and also provides units to fielded divisions, corps, and armies.

3. Now how might that work in the framework of our present situation?

a. Some individual replacements will be available in the early days of a war from the training base—that is, those who are in training when the crisis begins. The last of those individuals will complete training about 90 days from mobilization day, assuming they entered on that day. Were we to mobilize today, over the next 90 days we would send to the field about 50,000 individual replacements—the number in training today, less estimated TOP and EDP losses.

b. Where would these replacements be assigned? They would go immediately to fill shortages in deployed units, and in early deploying units, starting with the D to D+30 package. There would probably be an imbalance of MOSs which needs to be analyzed.

c. Deployed units in Europe would have some individual replacements assigned from the 10 percent or so of their strength that has been siphoned off for baseops functions in USAREUR. Here again there will be an MOS imbalance which needs to be analyzed.

d. Some replacements might come from the individual ready reserve (IRR), although it should be recognized that no more than a quarter to half of those called can be expected to report, all who do so will need some training, and a considerable skill mismatch can be expected. In sum, at today's strength the IRR should not be considered with confidence an early source of individual manpower replacements on mobilization.

e. Were we to enact selective service on the day of mobilization, it is estimated that at least seven months—210 days—would elapse before the first draftees could appear in units as trained replacements. Considering a minimum of 90 days' training time, and backing that off from 210, draftees would arrive to be trained in the training base at about day 120. If the training base turned out its last on-board volunteer at day 90, it would virtually stand idle for nearly 30 days before the first draftees arrived to be trained. This might not be all bad, for the training base would need some time to gear up for its expanded load. However, in any analysis of the replacement system it is an important consideration.

f. Individual replacements might be made available from military personnel assigned to baseops functions in the CONUS and those assigned to the training base. This is a course of action and is one that has been pursued in partial mobilizations in the years since World War II. It is not a desirable solution, for it means that just at the time baseops and the training base are trying to gear up for expanded workloads and accelerated schedules under mobilization, the very people most needed to support expansion and acceleration are being removed for other assignments. It is therefore a course that should be avoided; however the potential for replacements from this source should be evaluated as part of an overall analysis of the larger problem.

g. Finally, some individual replacements could be made available by drawing down on units low on the TPFDL—units undermanned anyway due to priority fill of early deploying units and the overall shortage of manpower in the Guard and Reserve. The Army uses up its second division structure in a mobilization in about 150 days, therefore the practice suggested above would shrink that time to less than 150 days. Whatever the result—100 days, let's say, it is clear that long before they were available in any numbers individual replacements would be needed to refill this depleted structure. In addition there would be a substantial MOS imbalance, since most of the late-deploying units are CSS units. While the MOS imbalance would mitigate the force structure impacts of the suggested practice, it would at the same time militate against the development of large numbers of usable MOS individual replacements becoming available from this source.

4. The unit replacement part of the equation is equally complex. Here are some notions about that:

a. The Total Army of 24 division force equivalents runs out of structure in a mobilization in less than 150 days—alluded to above. The force packaging methodology used by PAE will result in considerable numbers of units in the Army that are being undermanned upon mobilization. Still others will be underequipped as a result of normal policies with regard to equipment distribution, priority of issue policies concerning new equipment to modernize the force, and the drawdown on CONUS-based equipment to create in-place pre-positioned equipment sets overseas on which units from the CONUS fall in to fight. Any analysis of this problem must make some reasonable estimates about the total effect of all this in terms of what viable force structure might be left and filled at what levels after all these factors take their toll.

b. In any deployment there will be left behind the equipment of units deploying onto pre-positioned stocks. Left behind equipment will not be in unit sets, and even near sets will be incomplete. There will therefore be an enormous task of inventory and cross-leveling that must take place before additional units can deploy. Some estimate must be made of the threshold at which this effect begins to lengthen the deployability of units against their TPFDL stated requirement.

c. It can probably be reasonably anticipated that all the individual replacement generative measures described above will not suffice to fill out the existing structure, and there will be a lacuna during which there will be virtually no trainees in the base and no source at all of individual replacements. Nor will there be unit replacements available, since whatever units remain will be scrambling around trying to put together some unit sets of equipment left from the residue of equipment left behind by deploying forces. We must at least estimate the dimensions of that equation.

5. If, at some point, all that could be brought into some balance, then the Army should have a combination system in which some forces are raised and trained as units and some individuals are provided as individuals. Units could be created by cadre from active command and staff cadres, from combat-depleted units, and draftees from the training base. In World War II this

was a 100-division program. We in fact ran out of manpower before we reached the 100-division level. Fortunately the war ended before we had to cope with the problem of what to do about that, so the issue was never really resolved. We must try to figure out what in today's circumstances a reasonable mix might be of unit replacements and individual replacements and what might be the resulting structure of the training base, both for unit training and for individual training.

6. That just scratches the surface. It is an enormous problem. However you can see that I believe Jack Faith's first-30-day problem to be but a tip of a much larger iceberg. I don't believe, Jack, that I agree with what I believe you said to me the other day—that after 30 days or so we would go back to the individual replacement system and everything would settle down. For all the reasons cited above, and more besides, I can't be persuaded that is a rational assumption. The question, therefore, is how much of this should be tackled by whom, when and against what time lines? It has to be done sooner or later. I had intended to do it, but not for the next year or so simply because we haven't the capability. Perhaps, Fred, you could mull this over and we can discuss it when I come out in late May.

Recruiting and Training

Message to General E. C. Meyer
Army Chief of Staff
10 July 1979

1. I was surprised to learn yesterday that the Army has decided to recruit 17-year-old males, NHSDG, mental category IIIB and above, who score 50 or higher on the military aptitude predictor test. My concerns, which extend beyond my own sensing that we may be buying into a larger early discharge and disciplinary problem than we can reasonably handle, are summarized below.

2. On 25 June I sent John McGiffert a letter laying out the increasingly demanding impact on your training base of a seasonal, boom and bust pattern of recruiting. Didn't want to bother you with this issue during your first days in office, but the possibility that we may be about to exceed our already marginal and overloaded capability prompts this message:

- In order to be able to give training to the 50 percent of the annual program that DA estimated we would have to train during the period June–September, we have taken a number of extraordinary measures: loaded units and barracks to their absolute capacity; accepted the stress and risk of back-to-back training cycles; moved to double and triple-shift training where equipment limitations exist; gotten PCS movement deferred for a number of drill sergeants and other trainers; retained selected AIT graduates to act as assistant instructors; and borrowed heavily from FORSCOM units on TRADOC installations, with Bob Shoemaker's full cooperation and support.
- The total effect of these measures was to permit us to train 105,000 in this four-month period, as compared to the 77,000 we are structured to train. In March, when we planned to receive this tremendous summer surge, we were prepared to train the programmed end strength of the Army at end FY 79, provided that recruiting matched MOS training capability—which it almost never does.
- Subsequently, recruiting shortfalls during the spring have left us in a position where we cannot train about 15,000 of the program, a number which currently matches our recruited shortfall. Since a training vacancy once lost is lost forever, any additional recruiting shortfall automatically reduces training capacity.
- I am told that an optimistic estimate of the effect of the new enlistment criteria is that they may enable USAREC to recruit as many as 5,000 additional, thus reducing the recruiting shortfall from about 15,000 to about 10,000. The bottom line is that, if this happens, we cannot train them without establishing a long queue. In that event, the recent changes in criteria will have exacerbated an already severe problem rather than helped to solve it.

3. Also understand that the lowered recruiting criteria are to remain in effect only until 30 September 1979. From where I sit, that is exactly bass ackwards: if we are going to use lowered standards to maintain the programmed structure of the Army, we need to do so early during the fiscal year, when enlistments sag far below training capacity, then raise standards for summer enlistment, when our capacity is stretched to the limit. This would dampen a sine wave that has become unmanageable.

4. Since what we do or do not do for the remainder of FY 79 will have a direct impact on our program for FY 80—for which our capability is essentially already fixed—strongly recommend that you tackle this relationship between recruiting flow and training capability as an urgent matter. I am ready to help in any way I can. I was neither consulted nor informed on the most recent change, or I would have provided this assessment earlier.

Turbulence and Readiness

Message to General Frederick Kroesen
US Army, Europe
12 October 1979

1. For some time the two of us [General Starry and General Robert Shoemaker] have been talking about the adverse impact of turbulence on the readiness of our forces. We have agreed that an Armywide standardization program offers considerable potential for reducing its effects.

2. A strawman list of the most potentially profitable areas for standardization has been drawn up as a rough start point for such a program. The list comprises four categories: tactical procedures, gunnery/crew drill, support procedures, and training management. The first three focus on the central battle. The fourth, of course, is independent of any scenario.

3. We have reviewed this initial effort and have agreed on the following list, prioritized by categories, for initial consideration:

a. Tactical Procedures. Standard operational drills for combat arms units, ARTEP-oriented, would include such drills as "Move" at company, platoon, and squad level.

b. Gunnery/Crew Drill.

(1) Standard gunnery courses for all applicable weapons systems.

(2) Standardization crew drill—all crew weapons.

(3) Standard weapons positions—includes construction of the position itself and the positioning of necessary equipment and ammo within the position.

c. Support Procedures.

(1) Standardized stowage schemes for combat vehicles.

(2) PLL procedures: for both garrison and field operation establish worldwide standards for processing, cataloging, and storing repair parts.

(3) Standardize basic load configurations.

(4) Trains operations: addresses company trains and battalion field and combat trains procedures for given tactical situations. Emphasizes the who, what, and how.

(5) NBC alert procedures: includes standard alert warning devices and individual through company protective procedures and training therefore.

d. Training Management.

(1) Standardized battalion training management system.

(2) Standardized training schedules: substance and format, lead times, etc.

4. Your requirements in Europe will be a major consideration in worldwide standardization. We hope to discuss standardization at the commander's conference in Washington the end of the month. Consequently we would greatly appreciate your views and comments on the following prior to that time:

a. Standardization in general as a means to combat turbulence.

b. Our short list and, in particular, the priorities we have allotted.

c. How best to promulgate standards to the Army.

d. How we might check to see that units are meeting them.

. . .

6. We are fully aware of the emotional issues surrounding any talk of standardization and recognize that there may be some resistance to such an effort. Nevertheless we feel confident that, if approached properly and implemented sensibly, standardization offers the potential for mitigating the vexing problem of turbulence.

Personnel Doctrine

Message to Lieutenant General William R. Richardson
Fort Leavenworth, Kansas
13 November 1979

1. Our respective staffs have helped draw up a set of definitions, relationships, and guidelines for separating concepts, policy, and doctrine in order to assist ADMINCEN in its task of developing people doctrine for the Army. Reviewing that work, I am impressed that we've got it about right—finally. The purpose of this message is simply to send that out to all of you so that you, as well as your staffs, are aware that the process is going on, and that I believe we've got a useful product. I'm still open to suggestions about what follows; I send it to you as a basis for our further discussion.

2. The following relationship is set forth:

a. Concepts lead to doctrine; compliance with doctrine is directed by policy. Concepts→Doctrine→Policy. These are defined as:

(1) "Personnel concepts" are ideas, thoughts, and general notions about equipment, systems, procedures, organizations, tactics, and training methods necessary to maximize the human potential.

(2) "Personnel Doctrine" describes how people are integrated into Army organizations to produce combat-effective fighting forces.

(3) "Personnel Policy" describes missions, responsibilities, and administrative procedures necessary to ensure uniform compliance with personnel doctrine.

b. Concepts are usually published in informal papers; doctrine is usually published in field manuals; and policy, which must be compatible with doctrine, is promulgated in Army regulations. Informal Papers→Field Manuals→Army Regulations.

(1) The above suggests this relationship:

What	**Who**
Concepts	Originate anywhere
Doctrine	ADMINCEN
Policy	DA DCSPER

(2) Which parallels the logistics community relationships:

What	**Who**
Concepts	Originate anywhere
Doctrine	LOGCEN
Policy	DA DCSLOG

Training and Force Modernization

Letter to Major George L. Doolittle
Albuquerque, New Mexico
14 November 1979

Thanks so much for your letter . . . in which you applaud my *Green Book* article on training and force modernization. I need not tell you that often definition of a problem is a lot easier than knowing what to do about it. Such is the case this time. We can get a fairly wide consensus on what's wrong with the Army, but no one wants to make the tough decisions that will spell out what we do about it. The Army's senior leadership is now struggling with what we have to do to get on with laying out for ourselves the tough courses of action that will solve the problems I defined in that article. Meantime we need all the help we can get from people like you, and from our other civic and business leaders all over the country. If present trends continue, we simply won't have enough soldiers to do the job, no matter how good the programs we devise might be.

It might have escaped you that the last time we had a sixteen-division Army—in the mid-1960s, we had nearly a million men in Army uniforms. Now we have about 750,000 men and we're trying to do about the same thing we were doing then. Just in terms of numbers alone, that's an almost impossible undertaking! Added to that are all the problems related to mental qualifications, unit cohesion, [and] officer and NCO training that I tried to define in that article.

Recruiting

Letter to Lieutenant General Glenn K. Otis
Deputy Chief of Staff for Operations
7 January 1980

Recently one of those ubiquitous General Accounting Office (GAO) reports fell into my hands concerning the "Difficulties in Selected Army Reserves Recruiting Under the All-Volunteer Force" (20 August 1979). In view of the Chief of Staff's guidance concerning the manning of the force, I gave this report more than a quick glance and found some interesting problems.

It appears that we have adopted a unique philosophy for recruiting soldiers into the Reserve Components. We have established wartime and peacetime strength requirements for these components. However, for a variety of reasons, our recruiting objectives are based on expected enlistments rather than on manpower needs. The thought being that, as we reach these expected enlistments, we will raise the recruiting goals and thus eventually reach the established peacetime strengths.

Since fiscal year 1977 the exact opposite phenomenon has occurred. We have been unable to meet our recruiting goals, so we keep reducing the goals and our end strengths. The shortage in our Reserve Components thus grows and is fed by a strategy that I think should be declared bankrupt. As of today, the shortage in the Reserves is greater than our recruiting goals. If we keep this up, we could end up with no one in the Reserves and no requirement for a recruiting goal.

This bankruptcy is enhanced by our own reports to the public and to Congress in which we say that we are achieving over 80 percent of our *recruiting* objectives. When you compare the recruiting objectives to our requirements, we are actually achieving less than 50 percent of our *needs*.

I submit that we can no longer support a strategy that has proven to be self-defeating. In addition, our lack of candor in reporting only adds to the problem. It may be helpful if you read the GAO report, for it delineates other problem areas that bear on this. Our corporate Army leadership must come to grips with this Reserve Component shortage and let everyone know how really serious it is.

Recruiting and Retention

Letter to Herbert J. Budnick
Manhasset, New York
27 March 1980

The solution to recruiting and retention—especially the latter—lies in meaningful training that challenges soldiers and makes them want to increase their professional knowledge. The important corollary benefit is that the soldiers' image of their leaders is enhanced if the leaders present worthwhile training. No amount of money or benefit programs can buy this result.

Moreover . . . the same synergism is created in the Active Components. In the US Army Training and Doctrine [Command], we set training standards, *insist* that young soldiers meet them, and *assist* them to do so. The same theory should apply to all ranks.

Recruiting Strategy

Message to General E. C. Meyer
Army Chief of Staff
2 June 1980

. . .

5. In the course of our briefings we went through the input statistics—ASVAB scores, etc. He [Bill Peacock] rose to the bait and tried to get us to talk quality. As you know I refuse to use the word, and try to have everyone in TRADOC refuse to use it as well. What I told him went something like this:

a. We should stop talking about quality. The statistics we use to describe qualifications can be interpreted in too many ways, therefore everyone who wants to debate quality can make his own case. It is therefore fruitless for the Army to debate the quality issue. I can prove to you that nearly fifty percent of the armored force at the end of World War II was in what we would now classify as Category IV. Our viewpoint is that in any expansion of the Army we'll get about what we've got now, however we measure their mental qualifications, and indeed we might even get less well qualified people mentally just based on our World War II experience. Therefore it's well that we learn how to train people in that aptitude range—we will probably have to train a whole lot of them.

b. It is counterproductive and equally fruitless for us to undertake any study or other effort to revise the ASVAB test battery. The scores tell us something now; we understand what they tell us pretty well; if we change them we must learn a whole new set of norms; other studies will follow to redo the ARTS study and other work according to the new norms. Finally, given the public debate with Pirie and others that brought on the current effort to revise the norms, we could develop the best set of norms in the world and they'd still be suspect and subjected to the same kind of dialogue we're now having about the current norms.

c. The Army should talk instead in terms of motivation. What motivates men and women to enlist, serve, and stay? We should concentrate on the motivators that best serve the Army's needs. It is not appropriate to try to attract young people to enlist solely because of the civilian convertability of the job for which we train them. It is, however, quite appropriate for us to use an educational incentive package or packages to persuade them to enlist. Further, we need to develop in our leaders the notion that only meaningful training and meaningful off-duty opportunities will produce for us the kind of Army we need in this crucial time; that it is the duty and responsibility of the leadership to provide that opportunity or those opportunities. Lay it on the chain of command. The soldiers are bored—bored because we've not trained the leaders to be sensitive enough to the need to keep the soldiers challenged.

d. We should pound this approach home to the members of the Congress who claim they are trying to help us, but inevitably wind up being a pain in the neck. . . .

e. We have to be against the draft for military service in peacetime. I believe I can certify to you that the emotions which blossomed in the late sixties and early seventies over the process of who has to go and serve are still lying just beneath the surface. Surely, in the process of trying to make a military draft equitable in peacetime, they will rise up again. We can't stand that. Therefore we must solve the problem through motivation and other appeals to the young people we want to enlist. The Army just must find better motivators—I am not confident that we have exhausted or even fully explored the ways in which this might be done. Max has made a lot of much-needed change in the Recruiting Command; we are still far from having used all the tricks in the bag.

. . .

Turbulence in Armor Units

Message to General E. C. Meyer
Army Chief of Staff
9 June 1980

1. I don't know what you intend to do next in the stability/turbulence area. We have just compiled the results of some twelve studies and evaluations that have taken place over the last three years, all designed to tell us something about the causes, effects, and possible remedies of and for turbulence in the armor force. The ARI unit at Knox pulled this together, validating or modifying early data so that what we have now is up to date—April 1980 is the cutoff. Unfortunately this was not ready in time for the Olsen task force, hence this report.

2. After evaluating all active duty armor units this past year, we conclude that the major problem is not a shortage of people overall; rather it is the severe shortage in middle-level NCOs. It is that shortage that generates most of the turbulence at crew and platoon level. In April this year CMF 19 stood at 101 percent of authorized, but 11E grade E6 tank commanders were at 74 percent of authorized. Specifically the Army is 759 tank commanders short, and there are shortages in grades E4 through E9. In grades E1–E3 we are grossly overstrength, accounting for the disparity between numbers like 101 percent and 74 percent.

3. What does this do in units in the armor force?

   a. 17 percent of the companies are commanded by lieutenants.

   b. 27 percent of the platoons do not have officer platoon leaders.

   c. 27 percent of the first sergeants are not E8s.

   d. 35 percent of the PSGs are not E7s.

   e. 30 percent of the remaining tank commanders are not E6s.

The result is that leader turbulence is more severe than turbulence as a whole.

4. Study of the MILPERCEN model for CMF 19 indicates that CMF 19 is not theoretically grade feasible at grade E6. Current grade distribution policies will produce only 2,752 E6 tankers against a current authorization of 3,430. The difference of 678 is fairly close to our calculated shortage of 759. This situation was the genesis of the fifth crewman action several years ago. We finished the evaluation of the fifth crewman, but the basic problem is still with us—the system hasn't changed a bit. In addition, there are 457 armor NCOs assigned to Readiness Region, ROTC, and recruiting duty. Additional numbers are drill sergeants—we couldn't determine that number except at Knox. So the fundamental problem is that there aren't enough NCOs to begin with, and the MILPERCEN model isn't designed to produce enough to meet the requirements—TOE and TDA—demanded by the fielded force.

5. CMF 19 soldiers reenlist at about 56 percent, 12 percent below the career force Army average. Why? Two reasons: the training climate in units is frustrating and discouraging due to the constant milling around; repetitive overseas assignments with too short a time in CONUS create so much family discord that they retire or quit rather than go back overseas. Contributing to the dissatisfaction with overseas duty, especially in USAREUR, is the constant hassle over concurrent travel and the lack of responsiveness of the European housing system to the needs of the NCOs—or so they perceive it. So in fact do I, even though we have made

some improvements in it in recent years. Of CMF 19 soldiers, 43.5 percent must be assigned to USAREUR; for infantry that number is 26.5 percent; for artillery it is 37.9 percent. Armor career E6s are spending 24 months in CONUS; infantry E6s 40 months; artillery E6s 27.5 months. If we continue to insist on a three-year overseas rotation policy for individuals, and the percentages overseas stay about the same, the austere rotation base in CONUS will always mean short turnaround times. Olsen's two-thirds/one-third policy may be the way to attack this, but at least we should look at the rotation policies which drive the train.

6. In surveys we conducted among mid-grade armor NCOs, five principal reasons were cited for getting out/not reenlisting. They were:

a. Inadequate pay and eroding benefits.

b. Family separation and disruptions.

c. Short turnaround for Europe.

d. Lack of discipline in the Army and NCOs' lack of authority to enforce discipline.

e. Dissatisfaction with the quality of new soldiers.

For the first time in some years the most persistent problem cited is inadequate pay; it joins the familiar problem of decline of benefits. By a wide margin, that and family trauma associated with constant moving back and forth to Europe are far and away the award winners.

7. Study of crew performance brings these conclusions:

a. Time in position for gunners and tank commanders is positively related to gunnery performance to a high degree. The longer the tank commander and gunner work together the better their gunnery performance.

b. Time in position for loaders and drivers is not necessarily related to good crew performance. As a whole, officers tend to believe that the whole crew needs to be stabilized, while the NCOs believe, to the contrary, that if the gunner and tank commander can be stabilized it is good to rotate the other crew members so that cross training can take place. I side with the NCOs.

c. A tank designed to be operated by four crewmen simply is not combat effective with fewer than that number, and those who fill in for those who are not there need to have at least a modicum of training to enable them to be crewmen, even though that's not their primary job.

d. A well trained tank commander can largely overcome the effects of turbulence in the rest of the crew—less the gunner.

e. For loaders and drivers, recognition-type awards such as gunnery patches, letters of commendation, awards, plaques, and so on, have a high correlation with individual motivation.

8. We asked mid-level armor NCOs to rank order the incentives they would most like to see implemented. Their top five suggestions boiled down to:

a. Tax-free base pay for men assigned to Europe—on the order of $500 a month.

b. Monthly combat arms bonus paid to those who pass their SQT—about $100.

c. Monthly maneuver battalion bonus for personnel assigned to such battalions—about $100.

d. Longer time in CONUS between overseas tours.

e. Guaranteed quarters and concurrent travel when assigned to Europe.

Significant again is their concern for adequate pay and for reducing the family trauma associated with overseas moves.

9. Unfortunately I wasn't smart enough to start this kind of evaluation in some of the other MOSs, but sufficient data exist to suggest that the outcomes wouldn't be too different from the summary above. In the other bulk MOSs, where the shortage of NCOs overall is not so great, we might find somewhat less turbulence in units, but probably not too much. I need not remind you that the captain problem is also a factor in this whole situation.

10. What to do? Before we undertake any large-scale sweeping actions to fix things through unit rotation or some other scheme, we have a lot of basic housecleaning to do. For, were we to go to some other rotation scheme, the inherent ills of the system would only appear in some other way and place. So I'd suggest we do the following:

a. Fix the total numbers and grade distribution problems in the larger career fields, to include fixing the model that MILPERCEN uses to generate promotions and so on. If you want to promote people rapidly to fix the problem, I will be happy to run schools for them so that they are qualified to be crew chiefs, squad leaders, or tank commanders. But we have to fix the grade imbalance, and do it quickly.

b. The authorizations system and the models which implement it must provide sufficient mid-level NCOs for TOE, TDA, and sufficient additional to allow for all those assignments outside TOE units of their branch. Theoretically our manpower accounting system is designed to provide for this; somehow it seems to have gone agley.

c. Review the three-year overseas tour rotation policy; change it to provide for people staying longer when they want to do so.

d. Fix the concurrent travel/housing assignment problem in Europe. It has needed fixing for 30 years, and despite our best efforts it is still unsatisfactory.

e. Until such time as we get the NCO force up to strength in grade, we should organize a program to train and promote people into those positions. It would require some changes in the BNCO course and perhaps in others as well, but I believe we know how to do that.

f. Whether or not we press for some incentive package like the NCOs suggest, or just press on in the general thrust to get pay raised, is a choice you're better able to make than am I. The problems cited are not unique to the armor force—that tells me we need more pay across the board. I believe in the long run that's the way we should go.

g. Standardization of training and drills from theater to theater and unit to unit would help a great deal. It was frequently mentioned in our survey responses; I didn't highlight it because I've already sent you a message about it. My suggestions in that earlier message stand. We have a lot of standardization work under way. My success in getting the force to adopt it really depends on the extent to which someone is designated to be in charge of the whole matter. We can work our way around the turf problems. In due course I will bundle up the study and send it in, but its essence is in this message—the study itself won't add much, and if we get staffs to arguing about its detail we may lose sight of the larger issues, things about which we have to do something soon.

NCO Reenlistment

Message to Lieutenant General Robert G. Yerks
Deputy Chief of Staff for Personnel
24 November 1980

1. Your letter does an excellent job of summarizing ongoing efforts to improve service conditions, although it does not specifically define the problem. I do not agree, however, that the letter should be forwarded directly from the MACOM commander to the NCO corps. To do so would exclude local commanders who are key to our reenlistment efforts. Also it might be perceived as just another canned letter and not accomplish its intended purpose.

2. I intend to address the final product to commanders and add an outline of a general TRADOC plan. Our plan will place greater emphasis on personal involvement by the chain of command in explaining the Army's efforts and its commitment to their welfare. This approach will convey an impression of genuine interest, sincerity, and add greater credibility to the Army's efforts. Commanders will also be tasked to use the OESO in helping to identify and remove factors unique to their organization which hinder reenlistments.

3. We will also include a PAO blitz directed at the whole family. This is an often overlooked area which will pay good dividends. ("If we reenlist the wife, the NCO will follow.") As a capper, we will use a series of handouts of your information and also target them for the whole family.

4. We will need some assistance from the DA Staff for the PAO blitz and in developing the handouts. You might get a head start in this area now by revising your letter. The information is quite voluminous and would be better understood if it were reorganized under broad subject areas, i.e., compensation, personnel policies, quality of life, etc.

5. I am convinced that the key to success is in developing an effective ongoing program rather than resorting to periodic blitzes. I feel certain that this framework will get us on the right track.

Managing Manpower Shortages

Letter to General John W. Vessey Jr.
Army Vice Chief of Staff
4 March 1981

The continued use of the 197th Infantry Brigade to support infantry IET [initial entry training] at Fort Benning causes me the same concern expressed by you and Bob Shoemaker. Rest assured I am committed to terminating that support at the earliest opportunity; however, I am unable to predict when that will become practical.

Your assessment that the problem is a shortage of qualified CMF 11 NCOs is on target. The problem is concentrated in grade E6 NCOs in non-drill sergeant status. Despite priority assistance from MILPERCEN, latest fill projections for 3rd Quarter FY 81 are well under authorized spaces needed to train the infantry load this summer. Given current levels of authorization and other fill priorities, significant additional percentages of fill cannot be anticipated. Although long-range forecasts are favorable for increasing the availability of personnel in CMF 11, they offer no immediate relief at Fort Benning.

I am constrained from further shifting current assets by a 40 percent shortage of CMF 11 E6s in non-drill sergeant positions at remaining TRADOC installations. MILPERCEN has provided good fill in the grades of E5 and E7. However, we have used these personnel to fill the E6 gaps and we are still in a bind. The situation at Fort Benning is unique in that we simply cannot use E4 as lane graders in the Ranger Course or as instructors/jumpmasters in the Airborne Course. It's too risky, and the quality of training will fall significantly. Bob Elton projects Fort Benning's CMF 11 strength (E5–E7) to be at 91 percent by June. We will still require additional NCO support from the 197th. To free the 197th, I need your support in obtaining additional personnel, approximately 75 in CMF 11, 13 in MOS 54E, and 3 in MOS 91B.

The situation at Fort Benning offers a microcosm of the problems generated by manpower shortfalls permeating TRADOC. In FY 81 allocations were 20 percent less than requirements. Actual fill spread the gap to 24 percent. These shortages continue a downward trend begun in 1977 which continues throughout the current PBG cycle. The shortages have been absorbed thrcc ways:

(1) We have curtailed essential projects at the expense of necessary investment in maintaining or improving the Army's long-range capital base.

(2) We have borrowed military manpower for missions which require military labor at the expense of FORSCOM readiness.

(3) We have accommodated overly compressed workloads by expanded platform time, back-to-back training cycles, cadre shifted into mission roles for which they are neither assigned nor trained, and extremely high ratios of trainees to cadre. We accomplish this at the expense of our immediate human resource base and product quality.

I propose, with your assistance, a better way.

We are currently developing a comprehensive manpower strategy to free military spaces in TRADOC for realignment into more critical areas now experiencing authorization and personnel operating strength shortfalls. Many of the programs in this strategy have a longer-range focus, including increased mechanization efforts, and elimination, consolidation or contracting of functions not within the current CITA [commercial/industrial type activities] scope. We will submit detailed proposals on these at a later point. Two programs offer more immediate relief.

Through FY 81, the TRADOC CITA program freed 1,224 military spaces for reapplication to the force structure outside of TRADOC. This number represents 68 percent of the total Army effort for that period. Backfill with civilian or contract labor has permitted the implementation of CITA without degradation to base support operations. However, I have been unable to retain any of these military spaces for application against internal military manpower shortfalls. The TRADOC CITA program is anticipated to free 750 military spaces for reassignment in FY 82 and 1,010 in FY 83. I must be allowed to retain those savings to apply against current shortfalls in critical military skills. Failure to do that perpetuates the dysfunctional cycle of borrowing military manpower to perform essential training tasks that cannot be performed by civilians, such as has been recently demonstrated at Fort Benning.

My staff is currently working on another proposal. If DA can provide up to 6,000 additional civilian spaces, I can move immediately to start filling the major gaps between the TRADOC requirements and authorizations. A comparable number of military positions have already been tentatively identified as eligible for fill by civilian incumbents. Freeing these spaces of military incumbents would permit me to increase specific allocations in training, training developments and combat development activities which require military labor, and are now disastrously understaffed. These spaces are distributed piecemeal throughout TRADOC and could be freed without incurring the detailed study process or political backlash of realigning or eliminating a function or activity. Successful realignment of these spaces, however, will depend on the ability to provide backfill with civilian authorizations. Without your assistance in obtaining the increased civilian allocations necessary to make the trade-offs, TRADOC cannot free the identified military manpower.

ROTC Program

Letter to Lieutenant General Robert G. Yerks
Deputy Chief of Staff for Personnel
13 April 1981

The performance of duty demonstrated by the vast majority of lieutenants commissioned through ROTC is a source of national pride, not concern. Throughout the past fifteen to twenty years at the Officer Basic Courses, ROTC commissionees consistently have received more than their proportionate share of academic honors as represented by the number of officers graduated as Distinguished or Honor Graduate or on the Commandant's List. Further confirmation of the academic and intellectual quality of ROTC commissionees is accentuated by the fact that by the 8th year of service more than 20 percent of the ROTC officers obtain advanced degrees.

The legal and regulatory requirements for entry into the ROTC precommissioning program (contracting) and for receiving a commission have remained relatively constant throughout the past twenty years. . . . As an internal check on cadet quality, average test scores on the Cadet Evaluation Battery (CEB) subtest II were examined for 1971, 1975 and 1978. In 1971, the average was 22, in 1975 this average declined to 17, however by 1978 the average had climbed back to 20. It is important to note that for the two years (1975 and 1978) Scholastic Aptitude Test (SAT) data has been available for all ROTC cadets, the mean ROTC SAT score remained constant at 1,041 while the national mean dropped from 906 to 896. These figures clearly indicate that the average ROTC MS III and MS IV cadet was, and is, a cut above the average college student.

The environment in which the ROTC program has had to operate for the past two decades, both at the national and campus levels, has undergone dramatic changes of mood and interest. This has been particularly true concerning faculty and student attitudes toward ROTC and military service. Certainly the most significant aspect of these changes was the transition from a draft-based peacetime Army, to a draft-based Vietnam War Army, to our current all-volunteer peacetime Army. The aftershocks from these shifts are clearly evident in the areas of ROTC cadet enrollment; the number and type of institutions hosting ROTC; the required versus elective nature of the ROTC program; and many more direct and indirect influences that have impacted on the precommissioning phase of the ROTC program.

Over the past fifteen years Army ROTC has withdrawn, or been asked to terminate, at 25 institutions. During this same period, however, 63 new host institutions have been added. Using *Barron's Profiles of American Colleges* ratings, the current ROTC distribution of quality schools rated "Competitive Plus" or higher exceeds the national distribution, while those rated as less than competitive are below the national norm. Obviously, there is no way to replace a Harvard (while Harvard is not a host institution, there are presently more than 30 Harvard students enrolled in ROTC at MIT on a cross-enrolled basis) or Yale except with Harvard or Yale; however, Army ROTC is still represented in more than its share of high quality colleges and universities across the nation. In addition, this current high level of academic excellence has been carefully retained in our ongoing Expand-the-Base Program when 28 new hosts will be selected for promotion from the most successful extension centers.

The Vietnam War and the end of the draft were not the only factors that contributed to the preceding changes. The universities and colleges that host Army ROTC programs were also changing—their economic status had gone from the booming sixties to the financial crunch of the late seventies; grade point averages have increased while SAT scores have decreased; mandatory courses and distribution requirements have been replaced with electives; the pool of potential students will soon begin declining for the first time in many years; and the relative power of students/faculty/administration continues to change vis-a-vis each other and the university/college at large. During this period, the ROTC program adapted to meet the sometimes competing demands of the Army and host institutions. As a result, many new initiatives were developed that helped us maintain the quality of ROTC and the officer corps.

One of the greatest measures instituted to increase quality and quantity in ROTC was the Scholarship Program. It has permitted us to attract quality students and gain vastly improved retention as compared to nonscholarship cadets. In addition, the ability to advertise and offer educational assistance provides Army ROTC entry into high schools and colleges through guidance counselors and other influencers within the academic community. Scholarships are a proven means of creating awareness, generating interest, and, most importantly, increasing quality and quantity in ROTC.

While we have always been concerned with ROTC graduates who perform poorly at the OBCs, it has been only in the past few years that this problem has become critical. Whereas in the sixties and early seventies the bottom 5 to 10 percent of ROTC graduates were fully able to complete OBC and meet required levels of proficiency, in recent years, due to the decline in national educational standards, this group is often unable to accomplish either. Recent internal studies have led to the development and implementation of the ROTC Quality Assurance System to raise minimum contracting and commissioning standards. These standards will permit us to select only those candidates capable of meeting Army Total Force requirements. This, when combined with the Assessment Program, will ensure that we commission only fully qualified, highly motivated second lieutenants. In addition, the planned transition to a new Military Qualification Standards (MQS) I-based curriculum during the 1981–1982 school year should also favorably impact on the quality of our product. . . .

The preceding information demonstrates that not only has ROTC survived, but it has coped successfully with change, with dilution of quality, and is once again on an upward trend in all areas.

Overstructure and End Strength

Message to Major General John B. Blount
Fort Monroe, Virginia
6 June 1981

. . .

2. We should not get too emotional about the shifting around of RC and Active, nor about adding light divisions to active force—at least for the moment.

3. The two problems of most concern to us should be overstructure and end strength. The Army is already overstructured. To get the correct structure well would require at least 200,000 more end strength. Adding two light divisions adds dramatically to that problem—up front. Adding mech or armor tends to add more to the tail as opposed to up front. In any event, even at 48,000 for a DFE, the two divisions add 100,000 to our ES requirements. That's lots of folks! We must not let our end strength and structure get any more badly out of alignment than they are now. And we should strive to correct today's imbalances, whatever we do. As before, I still believe we are grossly remiss in not making a concerted effort to police ourselves internally before we start asking for more end strength. Proposals we've made at the last three Army Commander's Conferences are still relevant.

4. Related to above is the draft question. We can't have any large share of 300,000 more ES without it. The Navy and USAF also need more ES. The manpower problem must be considered in its entirety. We'll not be asked, I'm sure; whether or not we should speak up remains to be seen. I'm still opposed to drafting for military service only, unless we can figure out how to make it equitable as perceived by those subject to the draft. And I'm still convinced we have too many soldiers who believe their talents are not being productively used to draft any more. The social malaise of the late sixties and early seventies is still a threat—buried, I fear, all too close to the surface. The media and the left will for certain drag it up. It could ruin us—not just the Army but the country.

14. Military History

Page

Military History Research 608
The Principles of War 609
Military History (Multiple Addressees) 615
Lessons of History 616
Military History (Richardson) 617
Military History (Goodpaster) 618
Military History in TRADOC 620
Historic Preservation 621
Principles of War 622
Military History (American Military Institute) 623
Tenth Principle of War 625
Military History (Carey) 626
Military History (Holley) 627
Command Historian 628
Military History (Kleber) 629

Military History Research

Message to Lieutenant General John R. Thurman
Fort Leavenworth, Kansas
26 September 1978

. . .

2. The reason I raised the issue of military history research and instruction at Leavenworth is twofold:

a. First, the study of war is an important aspect of every officer's professional kit. In fact it is essential. We don't do enough to stimulate our officers to study military history, and reductions in course lengths in recent years have all but driven it from most school programs of instruction.

b. Second, detailed studies of battles and campaigns—statistical studies, analytical studies, command studies, all kinds of studies, are essential to further our understanding of military operations, organizations, tactics, equipment, structure, and training. We have never, to my knowledge, undertaken in our Army a program designed to study the history of battle and derive therefrom lessons to be applied in our own preparation for, and conduct of, war.

3. If one ties these two things together, it seems to me an opportunity to let our officers grow in an area where growth is necessary, and at the same time provide the Army with a coherent body of studies which in the aggregate help with devising operational concepts, organization, materiel requirements, and training systems. Such an effort obviously has to be directed—I suggested the technique used by most university professors working with their graduate students. Each has some overall framework into which the research of their students fits over time. Each student is assigned a research effort within his capabilities, and within the time allowed him is able to produce a gradable product. The professor collects, collates, integrates, and massages all this, and one day publishes his book.

4. What is necessary, therefore, is for us to provide the guidance, direction, study topics, and areas of interest, then guide and direct the research effort. Not knowing how many of your inmates would take us up on this, I'm not able to judge the validity of your resource requirements. I would judge that we should try to do it within the total we now have, for all the reasons I cited this morning on the phone. However, it may be that some changes in the skills behind the faces now present would be necessary to get the program moving.

5. Please work up a proposal for doing something like I outlined above. Then let's talk about it. I'd be happy to provide the guidance for the first couple of years' work.

The Principles of War

Army Historians Conference
Washington, DC
2 May 1979

Let me assure you that I come before you as only an amateur military historian. But I do share with most of you a firm belief in George Santayana's succinct statement that "those who cannot remember the past are condemned to repeat it." In time of peace, with no universally accepted threat and no apparent sense of urgency about learning from experience, Santayana's warning appears to carry no penalty for those who fail to heed it. In war, however, the failure to heed could well be a matter of national survival, particularly in this day and age, a fact that is all too frequently realized too late.

There is, of course, another side to the study of military history, one that says we cannot take the lessons of the last war and apply them directly to the next. All too often, we have historically sat ourselves down to get ready to fight better the war we have just left behind. Unfortunately, or fortunately, wars do not nicely repeat themselves. The student of history who can draw the necessarily fine line between these two opposite views, ignoring history and repeating it, will be successful in applying the appropriate lessons to the future battlefield.

The key, of course, is how you do that. I'd like to share with you some perspectives on military history, the principles of war, and military doctrine.

Our fundamental military heritage comes to us from the Napoleonic wars. More precisely, it derives from the writings of two men who reported on Napoleon's campaigns. Each drew inferences that were to have far-reaching influence on the American military system. The two men were Jomini and Clausewitz. We will come back to them, but let's remember for a moment the time when they wrote.

European wars of the 1700s were largely fought for limited objectives. They involved the participation of armies only—sometimes mercenary armies. Their purpose was to maneuver the enemy into some compromising situation from which he deemed it too costly to try and escape, but situations so contrived that he could escape with sufficient grace and face so that he did not view the consequences of surrender, defeat, or termination as disastrous.

When Napoleon Bonaparte appeared on the scene, first as general and later as emperor and general, the object of war began to change. Bound up in the Napoleonic wars was the spirit and idea of the French Revolution, a revolution that was threatening and frightening to the crowned heads of Europe. Faced with the prospect of a victorious foreign army spreading a new ideology and sweeping them from their thrones, the European monarchs set about to break the back of the French Revolution.

The will to resist, fight, and win was so great in Revolutionary France that it created new and revolutionary forms of war. France's stake was high—survival. It is not hard to understand why it was necessary to marshal all the nation's manpower and resources in defense of the Republic. Nor is it difficult to understand why, once the enemy was driven from French soil, he should be attacked in his homeland and brought under French control so that he could not attack France again and also so he could experience the benefits of the new ideology.

Bonaparte found victory by lavishly expending manpower while imposing even greater losses on the enemy, thereby creating an immense morale loss and forcing the enemy literally to abject surrender. Napoleon's idea of battle—annihilation—embellished by Jomini and Clausewitz, was abetted by new developments in artillery and innovations in organization that formed armies into independent divisions, each capable of acting on its own in pursuit of assigned missions.

It was Jomini's interpretation of Napoleon that became the foundation of military strategy taught at West Point. This was so largely because Clausewitz was not translated into English until 1873. Clausewitz influenced Emory Upton and the military intellectuals who brought sweeping reforms in the American Army in the early 1900s. But Halleck, Grant, Sheridan, and other familiar Civil War names all spoke in the context of Jomini on Napoleon. In many ways, Jomini was less well equipped to interpret Napoleon than was his Prussian contemporary Clausewitz, for Jomini's intellectual roots were deep in the 1700s. He was repelled by Napoleon's indiscriminate bloodshed. He abhorred armies that lived off the land and left a wake of destruction, but he seized on the essential Napoleon—the massing of one's forces to bring the greatest possible weight to bear at the decisive point, at a place where the enemy could bring but part of his force to bear. He faithfully recorded Napoleon's conviction that the offensive was the military operation necessary to victory.

While Jomini's concentration on the decisive place tended to put him more in the 18th-century tradition of a quest for geography rather than destruction of enemy armed forces, this subtlety was lost on his American readers, all the more so once Clausewitz's more powerful interpretation of Napoleon was available in English. The early American strategists, Halleck and Dennis Hart Mahan, were Jominian in outlook and espoused limited operations for territorial objectives rather than destruction of the enemy army.

But the next great American war, the Civil War, was on the style of the French Revolution—a war for survival, and so it evoked all the intensity in Americans that Clausewitz correctly reported in Napoleonic France. While the early Union generals were perhaps more Jominian than otherwise, it was an essentially Clausewitzian Grant who bore in on the enemy; forced his army to a standup fight, inflicting on him unbearable losses, albeit at the cost of heavy losses on his own side; and so became the first of a long succession of American strategists in the Napoleonic mold of Clausewitz.

Interestingly enough, it was not until after World War I that the United States Army tried to codify the fundamental essence of war, despite the early influence of Jomini, who sought a set of *timeless principles*. The first principles of war espoused by the United States Army were set forth in War Department Training Regulation 10-5 of 1921. They were objective, offensive, mass, economy of force, movement, surprise, security, simplicity, and cooperation.

Not much has been done to change these principles in the ensuing 50 years. In their modern counterparts, movement became maneuver and cooperation became unity of command. Whether this reflects the ultimate wisdom of their first drafter, intellectual bankruptcy, or some other circumstance would be hard to say.

Other armies of the world have codified their experience into principles. All European armies, including the Soviets, are basically students of the Napoleonic experience, interpreted by Jomini and Clausewitz. Remember, Jomini reported on Napoleon from both sides, French and

Russian, a point not at all lost on modern Soviet strategists. But, perhaps foremost among all European nations, the Soviets are advocates of the classic Napoleonic battle—annihilation. At the moment, they can afford both the manpower and the weapons. Time may change that, as it has done in our own country.

When the United States Army entered the nuclear era, it was immediately confronted with an obvious dilemma—war of annihilation had become so complete as to be virtually unthinkable. But what was the alternative? John Foster Dulles, in the Napoleonic tradition of annihilation, announced massive retaliation. Led by General Maxwell Taylor, military intellectuals and some in the academic community objected on the ground that there simply had to be an acceptable alternative short of massive retaliation. This was the genesis of deterrence. But no one ever said quite what to do if deterrence failed, and the antagonists were really not serious enough to move on to annihilation.

And so it was that we came to Korea where, for the first time, the bankruptcy of Napoleonic strategy was suggested. Somehow, we were not clever or clairvoyant enough to divine the lesson of Korea. We embraced first massive retaliation, then deterrence, and finally backed into Vietnam with an ill-conceived and poorly thought-out theory of gradual application of military force. This deterrent strategy also proved bankrupt. There were, in addition, some ill-devised theories of airpower that proved equally bankrupt.

And so in this, the third decade of the nuclear age, we find ourselves still pretty much disciples of Clausewitz, in the main possessed of a strategic bias, which more and more seems inadequate amid the realities of modern world power balances. But our principles of war are with us. There are some who say that they are shopworn, out-of-date vestiges of an era now long gone with no application to the modern battlefield. Some of these critics even use our Field Manual 100-5 to illustrate that the principles are no longer in vogue.

Now, as one of the ancient sages who had a hand in that book, I find that amazing. I guess we failed when we didn't put a list of the principles on the first page. It wasn't our intent, I assure you, to ignore those principles of the history from which they have been repeatedly distilled or to even give that impression. We deliberately left the principles out, wanting them to be understood—not memorized. Instead of giving a bland, dry listing of principles, we tried to apply them appropriately in modern terms to the battlefield.

Thus, if you read Field Manual 100-5, you'll find discussions on the concentration of combat power, surprise, deception, command and control, the spirit of the offensive, and so on. The US Army believes, as did Sir Archibald Wavell, that the principles are accepted as "common sense and are instinctive to the properly trained soldier." Our doctrine today contains the principles of war treated as common sense.

I don't think I have to tell historians that Army doctrine is rooted in history. What should also come as no big surprise is that some of that history is very recent and didn't directly involve the US Army. I'm referring, of course, to the 1973 Yom Kippur War. That war came at an extremely critical moment in our deliberations about the future of Army tactics. For almost eight years prior to 1973, we had concentrated on the war in Vietnam to the exclusion of modernization and doctrinal and organizational development for other theaters of war.

In the swift period of 19 days, the war riveted our attention on a new battlefield, one for which we would have been ill-prepared, materially and psychologically. Initially, all kinds of conclusions were drawn, and for a short period, a confusion of voices resulted. But cooler heads, more information, and the maturing process of time let us derive some lessons that could be applied to the Central European battlefield. The two antagonists used equipment and doctrines that represented those used by the most technologically advanced societies. The resulting high losses from lethal weaponry and the rapid pace of the battles were eye-opening to many.

In the US Army, we haven't in the past learned very well from our own mistakes, and even less well from those of others. Vietnam was an example of this. But we think we have this time. FM 100-5 and the supporting doctrinal books are the result. We believed in 1973, and we believe now, that we must prepare to win the first battle of the next war and the battles that follow.

Historically, as you know, we have not been prepared to win the first battle. Instead, we have relied for almost 100 years on a system that assumed we would absorb the first blow and then mobilize both militarily and industrially to win the war. Probably the last war in which we won the first battles was that against the forces of General Santa Anna in Mexico in 1845.

In the short period of 15 years from that date, our great war of national fratricide loomed. Misapplying one of the great principles—the offensive—in the opening days of that war, the northern politicians demanded "forward to Richmond—on to Richmond." Mr. Lincoln summoned a council of war. The three-month volunteers had but a few weeks to go. There would be political repercussions if they went home without striking a blow. General McDowell was ordered to attack, to prepare a great victory for the Fourth of July. Regiments were not organized into brigades. There was no cavalry. McDowell was visited daily by a Soviet of elected colonels, all demanding to be made at least generals of divisions. But he did as he was told.

Now, let's pause for a bit of geography. Across Virginia slantwise runs the Blue Ridge, shutting the eastern plain in against the sea. Between it and the crest of the Alleghenies lies the valley of the Shenandoah, its northern end issuing forth behind Washington at Harpers Ferry, its southern end opening into a passage behind Richmond. Midway the barrier is pierced by a gap through which a road and railroad issue to meet the main north-south line from Washington to Richmond at Manassas Junction. Capture the junction and control the gap, then you can dash through to intercept any force operating in the valley.

In the valley was a Confederate force commanded by Joe Johnston. Pierre Beauregard, with 23,000 men, lay along the banks of Bull Run in front of Manassas Junction. McDowell could bring 28,000 against Beauregard, defeat him, dash through the gap, and gobble up Johnston. To keep Johnston from joining with Beauregard before all this could be done, Mr. Lincoln employed the services of General Patterson of the Pennsylvania Militia, whose position had been secured by a healthy stipend into the political campaign coffers. Patterson with 22,000 was to entertain Johnston; McDowell was assured that would be more than adequate.

The operation was planned for the 9th of July; by the 16th, the Army managed to get off its duff, onto its feet, and saunter to Manassas. It was infernally hot. Carrying ammunition was too much trouble, so the men emptied cartridge boxes into ditches alongside the route of march. At every halt, the swarm of Congressmen, following the caravan to see the fight, drew a regiment or two out of the column to demonstrate oratorical prowess. Correspondents came too; on the 17th, the Washington papers carried a full plan of the campaign, together with maps, an account that was well received in Richmond.

On the 18th, McDowell met the pickets of the Army of Northern Virginia along Bull Run. In the Shenandoah, a light cavalry force entertained Patterson while Johnston marched double quick to join Beauregard. As McDowell fumbled with the Confederate Army in the brush around Manassas, Joe Johnston, the Louisiana Tigers leading, fell on the flank of the Union Army. Trains, artillery, wounded, and two Congressmen were left on the field; the rest of the Army never stopped until the Potomac River bridges were in sight. That was a first battle of a war.

The principle of offense had been followed but so, too, on the other side, had been the principles of security, mass, economy of force, maneuver, and surprise. The lesson could almost have been written before the battle. Any leader who adheres inflexibly to one set of commandments is inviting disastrous defeat from a resourceful opponent. That's why we didn't list the principles of war in our book.

The final lessons we derived from that war, and the later altercation with the Spanish in 1898, resulted in the rejection of the old volunteer system and development of its successor—the mobilization system. It better suited our sense of democratic fairness, and it was cast in the mold of industrialization. The 2,000-mile ocean moat between us and our enemy made it possible. We pursued this doctrine through two World Wars and, although we came perilously close in the second one to losing all the battles, including the first ones, we were successful in the end.

Time has moved on, however. Distances have decreased, and weapons have increased in range and lethality, until even a 2,000-mile moat was no protection. With the dawn of the missile age, the luxury of time to mobilize was denied us. It became apparent that the sanctuary of the North American continent was nonexistent, and the Army must prepare to win the first battles in what has been termed a "come-as-you-are" war.

Our doctrine needed rejuvenation too, but this time with a strong eye toward history and a foundation built upon the time-honored principles of war. If you look closely, you'll find those principles are all still viable, still in use. But they are, as Admiral Mahan said, "not so much fetters, or bars, which compel war's movement, as guides which warn us when it is going wrong." In that sense, they are applied in modern doctrine along with the recent history that illustrates their application.

They are part of our heritage. Some say they are immutable. What I have tried to suggest is that our view of them must change as times change. Success in battle rests in some measure, a great measure, on an understanding of the philosophical bias that underlies those nine principles of war. In our lifetime, success may rest on our ability to cope with the lacuna between our Napoleonic heritage and the realities of deterrence in the modern world.

One final word: Beauregard could not have been successful without a McDowell as his opponent, nor could Crazy Horse and Sitting Bull have whipped the 7th Cavalry without a Custer. Bonaparte needed a Ferdinand and a Mack to produce Ulm, and Lincoln finally found Grant who in turn found Richmond and Appomattox. Regardless of what principles are embraced by which side, there is a fine line between victory and defeat that is virtually impossible to define. There is a great deal of failure in every success, a large measure of defeat in each victory.

So, to win, any army must be master of the situation at every turn of events; know all that can be known about its own situation and that of the enemy; prepare itself in advance with all the philosophical speculations about the principles, how and when they might and should be applied. An army must study the actions of others in history who have done the same thing.

In the end, these principles are but tools of the trade. Victory comes to the army that can adapt them to changing times, apply them to new situations, accommodate them to a changing enemy, and win in the end. Only the study of history and the correct application of its lessons can make an army victorious.

The wisdom to learn from experience, without just getting better prepared to relive that experience, is not easily won. We owe it to the thousands of brave Americans who have answered their country's call, helped us learn the lessons, and paid the price of learning. They left us a large legacy—larger perhaps than we deserve.

Military History

Message to Multiple Addressees
17 July 1979

1. A knowledge of military history—the acquisition of a sense of historical mindedness—is a necessary component of an officer's technical competence, whatever his rank or position. This knowledge won't give an officer a precise solution or blueprint for solving current problems—history doesn't work that way, and those who believe so will be disappointed.

2. History does provide examples of the logical approach to solving problems and the reasons for and the results of previous solutions. If an officer knows what was attempted in the past, the conditions under which it was attempted, and the results achieved, he can deal more positively with his immediate problem. The formulation of FM 100-5 and its supporting manuals relied heavily on a knowledge of the historical past. Our current efforts in force modernization have a historical precedent in our efforts in the early 1940s.

3. A recent survey of the military history program at each TRADOC school indicates there are as many approaches to historical study as there are schools. This is as it should be, since each school has differing requirements and resources. As an additional resource for everyone's use, I recently directed the Combat Studies Institute be organized at Fort Leavenworth. Its primary purpose will be the production of historical studies to support doctrine and training.

4. I am concerned, however, that, as a necessary discipline, military history has only our collective judgment as its proponent. I do not want to tell any of you to put military history into your curriculum or to take it out. Rather what is needed are your ideas on how best we can accomplish a sense of "historical mindedness" in the officer corps, particularly lieutenants, captains, and majors.

Lessons of History

Letter to Major General Mike Forrest
Fort Snelling, Minnesota
28 September 1979

Many thanks for your kind note . . . concerning our fledgling Combat Studies Institute at Leavenworth. I truly hope we can somehow bring ourselves into some more active interest in the doctrinal and policy lessons of history. That is the intent of CSI. So far all we can claim as an Army and a nation is that the lesson we most often learn from history is that we don't learn much from history.

Military History

Letter to Lieutenant General William R. Richardson
Deputy Commanding General, US Army Training and Doctrine Command
5 November 1979

During the past summer, I solicited the thoughts of the tribal elders of TRADOC as to the status and future potential of a military history program in the Army. More specifically, they were tasked to provide comments on how to instill a sense of "historical mindedness" into the Army officer corps. The response was overwhelmingly in support of such a program. There were, of course, some realists who mentioned our constrained resources, but even these noted that there were things we could do right now.

My review of all the replies indicates there is a consensus on the direction and purpose of a history study program. Your predecessor's message on this subject and the Combat Studies Institute (CSI) paper on military history reflect the kind of comprehensive approach for which I was looking. I thoroughly agree that CSI is the logical and best choice to develop a total Army program for the study of military history. As an added benefit, it will make their mission and their capability more widely known throughout the Army community.

. . .

As general guidance, we must make sure that the study and use of history relates to today's problems. Ideally, any study of history should be stratified at the level of the student—captains concentrate on captains, lieutenants on lieutenants. This foundation must be built before the "great captains of history" are studied. More importantly, historical study and appreciation must start at the beginning of a military career—the pre-commissioning phase, ROTC/OCS/USMA. Reinforcement of this study, both formally in schools and informally in units and by self-study, must continue for an entire career. Your program should include this total approach.

As always, resources are constrained, so any program must fit within those currently available. Our plan should include a core of actions to be taken as a minimum by every school and some suggested additional steps that would be at the discretion of the commandant. The current US Army Armor School program may provide some ideas in this vein.

In addition, there should be an exportable package to the field, and even to individual officers if appropriate. This package could include reading lists, sources, agencies and materials that are available for historical study. This approach will ensure that responsibility for study becomes part of each officer's professional kit, not just another class or requirement.

Military History

Letter to Lieutenant General Andrew J. Goodpaster
Superintendent, US Military Academy
14 April 1980

As you may remember, in 1970, when BG Hal [Pattison] stepped down as Chief of Military History, he expressed deep concern over the declining attention paid to military history by the officer corps and the Army school system. One result of Hal's concern was the creation of the Department of the Army Ad Hoc Committee on the "Need for the Study of Military History." The central finding of that committee, which was chaired by Colonel Tom Griess, was "that the Army has a valid requirement for studying and using military history." The committee recommended that there should be a centrally directed, integrated, progressive program of instruction and study in military history for the officer from precommissioning through senior service college level. Nearly ten years have passed since that recommendation was made and the Army is . . . little closer to achieving it today than it was when it was made. Many of the committee's recommendations were set aside or overlooked as a result of our consuming focus on Vietnam and the subsequent institutional turmoil of readjusting from the demands of Southeast Asia to those of the Central European battlefield.

During the last three years we in TRADOC have reexamined the issue; I have personally devoted considerable time and energy to building a consensus within the TRADOC school system which would enable the development and introduction of a meaningful military history education program for officers into our present program. The fruits of that effort will soon be realized. The Combat Studies Institute (CSI) at Fort Leavenworth has prepared a comprehensive proposal which is the focus of talks at the annual TRADOC Military History Conference this month. After that conference we intend to refine CSI's proposal and introduce a program beginning next fall. . . .

The more we examined the problem of developing a comprehensive program for officers, the more we realized the importance of establishing a solid foundation in military history at the precommissioning level. As you know, our newly commissioned lieutenants come from a wide variety of sources (USMA, OCS, and 270 separate ROTC detachments). Some standard has to be established. The excellent program of USMA's Department of History immediately came to mind. In February I asked my deputy, Bill Richardson, to explore the desirability and feasibility of exporting part of West Point's military history program to ROTC and perhaps OCS. The result was two days of informal discussions between Bill; CSI; Colonel Roy K. Flint, deputy head of your Department of History; and Colonel Ed Kelly, Assistant DCSROTC, HQ TRADOC. Those frank and informative talks led me to conclude that we must do something to ensure a common level of knowledge of military history on the part of all newly commissioned officers, and that elements of the West Point history program contain the standard that we ought to achieve.

In this regard I am soliciting your help. Specifically, we would like to ask that the Department of History prepare a "suggested course plan" for a three-credit-hour offering in the American Military Experience. . . . DCSROTC will use this "suggested course plan" as the basis for presenting a common military history course in all ROTC detachments. For a number of reasons we have concluded that the most acceptable way to present such an offering would be to solicit civilian instructors from the host institutions to give the instruction, augmented by officer members of the ROTC detachment whenever possible. A suggested course plan prepared by USMA's Department of History would carry the weight needed to introduce such a plan to many of the host institutions, because it would emanate from a nationally recognized college academic department, long established as a leader in the field of military history.

Once we identify members of the civilian academic community who are interested in teaching military history to ROTC cadets, we should do something to assist them in the preparation of their offerings. This brings me to my second request. We would ask that West Point explore the feasibility of having the Department of History offer an expanded (perhaps double the present enrollment) ROTC Military History Workshop in the summer of 1981 and possibly 1982 for the civilian professors and military officers who will teach military history for ROTC.

As indicated above, we do not intend to divorce uniformed ROTC faculty completely from the teaching of military history. While civilian faculty will give our program academic credibility (important to host institutions), uniformed faculty will give it the relevancy so important to the cadets. Thus in years to come we would want to continue the present summer ROTC workshops for uniformed faculty. Our goal should be to keep at least one officer in each detachment sufficiently conversant in military history to assist in the course at all times. To facilitate this goal I have asked Bill Richardson to look into the possibility of having CSI conduct a "Workshop West" at Fort Leavenworth, beginning in 1982. This would allow us simultaneously to increase production and ease the present burden carried solely by USMA. We would like to send someone from CSI to West Point for a week or two this summer to explore that possibility. If we decide to go that route, CSI will need some help in developing a curriculum for a Workshop West; USMA's expertise and experience would be indispensable.

Your help in the areas outlined above will go a long way toward making TRADOC's efforts to upgrade the study of military history by all officers a reality; we hope for a long and productive relationship between our commands in this endeavor.

Military History in TRADOC

Message to Lieutenant General William R. Richardson
Deputy Commanding General, US Army Training and Doctrine Command
2 August 1980

. . .

2. It . . . seems to me that we might be well advised to be more specific about what we want done in order to accomplish our goals. For example, instead of specifying a number of "set aside" hours for military history coverage in various courses, could we not designate what we want covered subject-wise and let the schools figure out how to do it? We might begin this process ourselves with some idea of about how many hours we believe should be spent—but we wouldn't specify it that way. Were we to do it as I suggest, we would specify the battles, campaigns, and actions we want covered to illustrate what we want the students to know about military history. In courses in which we strive to ensure a basic knowledge of the military art, we specify reading the "hard books" and say which ones, so that we are assured the student comes away knowing his Jomini, Clausewitz, Upton, Wilkinson, Fuller, Liddell-Hart, and so on.

3. Such a method on our part would, it seems to me, help us direct our own study and research toward the development of the teaching materials, guides, and databases that we need to support the military history program and to move it ahead toward the horizons we see for it.

Historic Preservation

Message to Major General John B. Blount
Fort Monroe, Virginia
3 December 1980

Forgot to tell you before I left. Subject of fireplace screens in the old quarters. Issue instructions to DFAE that there will be no modifying of old fireplaces, in any way, to accommodate his notion that we need glass screens for energy control. This includes boring holes in tile, in mantelpieces, removing metal framing or grates, or anything that defaces or changes the historic and esthetic value of the old hearths. You'd better do this personally. They've already speculated on all those courses of action for 33 Fenwick. Brutus [dog] might scare them away, but I fear they will go elsewhere. If they screw it up, the crosses may go up before Easter.

Principles of War

Letter to Colonel Harry G. Summers Jr.
US Army War College
5 March 1981

Thanks for sending me your "occasional paper" on the principles of war. . . . Hopefully we're not trying to reinvent the wheel with our current principles of war effort. War is an art, not a science. In art you can do as you damn well please so long as people like it, buy it and critics comment favorably about it. In war you can do as you damn well please so long as it turns out well. The problem with war is we've fuzzed up what is meant by turning out well—what it means to win. Therefore it's probably well to sort out some generalities about what seems to work fairly well most of the time, more often than not, even though we're too often not too clear about what we're trying to do. Maybe that's a principle!

Military History

American Military Institute
Fort Monroe, Virginia
28 March 1981

Brooks Kleber asked that I explain a little of my approach to history and the influence it has on how I perform my duties as the TRADOC commander. Well, on reflection, that's a tall order, simply because this thing called the Training and Doctrine Command is so big that, sometimes, I question whether I really know what all those duties entail, let alone how my appreciation of history might influence them.

I guess when I really think about it, though, I have to agree with General MacArthur, who is reported to have said that if there is one thing he had learned during his long years of experience it was that, if we were to correctly solve the problems of the present and chart a sure course into the future, we must study and weigh and understand the manifold lessons of which history is the great—indeed the only—competent teacher.

I know that for a while many in my profession were allowed to stray from the study of history. We're trying to turn that around now because of the conviction that military history is our best insurance against repeating mistakes of the past. I'm convinced that officers should have a sense of "historical mindedness." This does not mean they should dwell on the past necessarily. Rather, they should be more aware of the past and the evolutionary nature of things as they wrestle with the difficult problems that confront our Army and society today.

I think that a person who is "historically minded" habitually solves problems by searching for the broad themes that trace developments over lengthy periods. He tries to identify cause-and-effect relationships, analyze past events and actions in the context of his own times, avoid retrospective judgments, and consider present circumstances in the light of the past.

Such a historical perspective is important to today's officers in order that they might develop a common understanding and sensitivity to the intellectual and functional values of military history and, in so doing, learn from the past that which may be relevant for today and tomorrow.

I've found the study of the past offers collateral benefits to our officers. Routine reading increases communication skills, the deterioration of which has become a matter of official concern in the Army today. A better understanding of the Army's past can also contribute to a sense of corporateness, esprit, and continuity. Finally, for a profession that cannot practice its craft except in wartime, history offers "laboratory experience."

How do we hope to institute such an appreciation and awareness? Well, it will start in precommissioning training that ROTC cadets receive. Each of them will be required to complete about three semester hours in military history. Some of you who are university professors might be involved in that program, since we're trying to get professors from the institution's history department to teach those courses for us on a "contract" basis.

Then, after commissioning and while he attends his branch school, each officer will receive 6 to 10 hours of instruction centered on the role that historical inquiry plays in professional self-development, the Army historical program and its resources, the history of his branch, some of the evolution of American military professionalism, and finally some battle analysis using the historical method.

The officer advance course will build on this by presenting instruction in the evolution of combined arms warfare and advanced battle analysis and emphasizing the continuing need for professional self-development.

Once we get the new course, the Combined Arms and Services Staff School, underway at Fort Leavenworth, all captains will attend it, and there they will learn about historical development of military staffs and participate in numerous exercises that will include historical case studies.

When an officer attends the Command and General Staff College, he will participate in a survey course of 20th-century US military institutional and operational history and also a number of advanced-level electives.

The Sergeants Major Academy and Warrant Officer Senior Course will develop instruction consistent with the professional needs of their respective student bodies, generally along the lines of the CGSC curriculum.

Finally, in between each of these school periods, officers will be required to participate in a historical reading program as part of fulfilling their requirements for promotion and further schooling.

In all, we think we've got a pretty good comprehensive program, one that I hope will solidify an interest in the study of military history and the development of individual reading and study programs. If we can do all this, I'm sure it will help develop in our officers a mature, informed conception of their role as leaders in the Army.

Tenth Principle of War

Letter to Professor I. B. Holley Jr.
Duke University
24 April 1981

The tenth principle of war will remain but a good idea—I couldn't persuade Shy [Army Chief of Staff General Edward C. Meyer] that this was a propitious moment to make visible to a new administration the idea that they need have their public political act in order before they turn to the use of military force as an instrument of national policy. We'll let that idea germinate a bit before we air it again.

Military History

Letter to Lieutenant General Richard E. Carey
US Marine Corps
25 January 1982

As you no doubt know, budget constraints over the past six or eight years virtually swept away our military history capability in Army service schools. In an effort to reestablish that we tried not to set up military history as an entity unto itself, but rather to imbed the teaching of history in case studies and historical examples closely associated with or interwoven into other instruction. This makes the history part less time consuming, less identifiable as a separate instructional subject and somewhat less vulnerable perhaps to cuts. Finally, that method might just teach our officer corps to look first at the history of any subject before they charge off trying to reinvent the wheel. Part of that whole effort was creation of an institute [the Combat Studies Institute] at Leavenworth. It has the mission to produce historical case studies investigating the background of problems we face today. This was done in the hope that we might derive doctrinal perceptions which would help us toward solution without so much stumbling about oblivious of history related to the subject. I think their first four products are quality work; they've been useful—I hope the system survives.

Military History

Letter to Professor I. B. Holley Jr.
Duke University
24 March 1982

The program [of military history at Fort Leavenworth] is doing exactly what we intended that it do when we started, albeit somewhat more slowly than one might like, as you quite correctly signal. Whether or not we were able to imbed the revised history program in the Army school system sufficiently to have it survive my passing from command of TRADOC, only time will tell. My concern is that it is not all that deeply imbedded yet, and we have a generation or two of officers, in the middle grades especially, who do not understand the usefulness of the study of the military art, and don't know how to study or use it in their own professional lives. That's where the problem lies principally, for it is those fellows who teach at service schools and do the daily work that imbeds historical perspective into succeeding generations.

With emphasis from the top, they, too, can learn as they teach, so the top lions must continue to roar. I hope that will happen; I fear it will not. However, I'll continue to push it as long as possible; sooner or later, it will have to go it alone.

Command Historian

Message to Lieutenant General Wallace H. Nutting
Panama
6 December 1982

1. Recently USREDCOM/JDA's first command historian arrived for duty. It was imperative that the activities of this command be regularly and systemically preserved, recorded, and analyzed for our operational use. Moreover, in virtually every aspect of joint staff work, historical methods, properly utilized, can be of material assistance. The program now underway has as its first objective the placing of military history at the active and direct disposal of the staff; in other words, it is not a passive program, but one aimed at providing a substantial and effective service in the furtherance of this headquarters' mission.

2. Until earlier this year, only USREDCOM and USSOUTHCOM, of all the unified commands, had no permanent command historian. With the increasing significance and wide range of USSOUTHCOM activities, the time may be more appropriate than ever to establish a USSOUTHCOM historian. The importance of effectively documenting and analyzing USSOUTHCOM's involvement in major events in Central and South America has probably never been greater. If you concur in this assessment, we would be happy to assist your efforts in establishing a USSOUTHCOM historian. The USREDCOM/JDA command historian, Dr. Roger J. Spiller, will advise and assist as you desire.

Military History

Letter to Brooks E. Kleber
Center of Military History
11 March 1983

I believe that to fully understand this difficult business in which we're engaged, we just must understand thoroughly from whence we have come and how we arrived in our present circumstance. Of all the things we've done, none perhaps is so rewarding as to see the increased interest in and emphasis on the study of military history. I pray that we've done it right; that it will be a lasting thing; and that we have institutionalized it to the point that it won't fall prey to the budget (or some other) axe once again in some future lean time.

15. Modernization

Page

Who Plays What in the Symphony 633
Survivability Gains 635
Night Operations 636
Communications Innovations 637
Position Location Reporting System 640
Combat Mission Failure Criteria 641
Rationalization, Standardization, and Interoperability 642
Position Location Reporting System 649
Acquisition Cycle Compression 650
Technocrat Mentality 651
Information Explosion 652
Tank Development 656
Perry Initiatives 657
XM1 Tank Issues 658
Operational Testing 659
Computer Systems Needs 661
Telescoping Materiel Acquisition 662
Improved TOW Vehicles 663
Weapons System Development 664
Acquisition Cycle Compression 665
Top Attack Systems 666
Command-Control and Communications 667
Army of the Future 668
Modernization 673
User Influence on Laboratories 677
Modernization Issues 678
Tank Cooperative Development 680

Page

Future Close Combat Vehicle Concepts.......681
Man-Machine Interface.......683
Proposed Test Command.......684
Mobile Protected Gun (22 October 1980).......685
Mobile Protected Gun (8 November 1980).......687
Soldier-Machine Interface.......689
Modernization Timing.......690
Current Imperatives.......691
Joint Tactical Cruise Missile System Rationale.......693

Who Plays What in the Symphony

Message to General William E. DePuy
Commanding General, US Army Training and Doctrine Command
25 September 1975

1. Yesterday I was briefed by some hucksters from Fairchild about the A-10. Have in hand your letter to the Chief supporting A-10. The briefing played heavily on A-10's ability to destroy tanks; gonna do it with GAU-8, with missiles, with rockets, and I guess with pilot's revolver.

2. Last week I reviewed the AMORES study. This study of 78–82 ammo rates done by CACDA showed in a Europe scenario nearly 16K of 20K or 80 percent of attacking tanks destroyed by CLGP. This devastating display of firepower, taken I'm told from the CLGP COEA, left the Reds with but 600 tanks to face the Blues, whose 6,600 (80 percent) surviving tanks and pure hearts promptly dealt the enemy mortal blows.

3. Cited incidents are but two in a growing series of similar developments that I find quite alarming. It is encouraging that everyone wants to kill enemy tanks. There are lots, and we expect tank killers in modern battle will have a busy time. While it is comforting to us to find many people wanting to help us kill enemy tanks, we really aren't outnumbered now in Europe, for example, much more than we were 15 or 20 years ago. So to those of us who grew up outnumbered it's not new, but it's a novel idea to many who can't remember back that far. The problem is that the enemy has lots of infantry, artillery, air defenses, and so on, and he believes in the complementariness of his systems. So should we. We should also believe strongly in the complementariness of our systems. The Fairchild guys explained that they emphasized tank killing with A-10 because that's what everyone seemed to want, and Army guys were always complaining that they couldn't get any USAF close air support—now they could have it close and effective against the main ground target—tanks. While that's all good, I happen to believe that if USAF can perform well their first two TAC missions—control of the air and interdiction, I'd be very, very happy with them. If they have any left over for us in close air support, all the better. For it just may be that the best thing they can possibly do for us is to keep enemy air off our back and get the second-echelon division or even the second-echelon regiment of the lead division. For that's what we can't do; that's the thing they do best; that's their complementary contribution to the armed brawl in progress below them. So it is with the artillery. While we really appreciate their wanting to shoot at tanks, while they're doing it, who is going to shoot at all the infantry, air defenses, and fire counterbatteries? The very, very best thing the artillery can do is get the enemy infantry and artillery off our backs, and suppress air defenses when we want to move air cav and attack helos around. That's what the tanks and panzer grenadiers can't do; that's artillery's bag, their complementary contribution to the armored melee.

4. I fail to comprehend why it is necessary to justify every weapon system, and judge the quality of every analysis, on an ability to kill tanks. In fact, we may just be degrading the primary mission capability of some system to the point that it can no longer do best what it is supposed to do best. In addition the whole process smacks of a typical systems analyst's approach to any problem—one that has got us in trouble before and will again. The fact that a system is complementary does not make it any less valuable to the battle outcome. I respectfully but strongly suggest that we better stop all this intellectual chicanery, lest we soon be unable to tell truth from fiction ourselves.

5. If I were conducting the symphony it would be most disconcerting to come to the part of the 1812 Overture where the guns are sounding, raise my baton to the drums, only to find that a violin salesman passing through had persuaded the equipment manager that he had a bunch of fiddles that did just exactly what the drums did and so sold a bunch. And so instead of a thundering finale all we could manage was a few squeaky wails at the end.

Survivability Gains

Letter to Dr. George H. Heilmeier
Defense Advanced Research Projects Agency
30 March 1977

The STAGS analysis is just super. . . . The analysis to date strongly indicates clearly what we must do next to flesh out the gaps in our knowledge, and to determine if the trends indicated by the test data thus far are indeed continuing trends. So that will tell us what to do when test time comes for the HIMAG vehicle. It will also tell us, indeed it already strongly suggests, that there is a considerable payoff in survivability to be gained from high performance ground vehicles—those that can deliver speed made good over the ground. This alone is significant, for when we started there was strong evidence that, beyond about 30 horsepower per ton, there was little survivability gain. Indications now are that there continues to be considerable gain up to the limit of our current knowledge—about 70 horsepower to the ton. If those trend lines continue, even more survivability is available from higher performances. How much, and where the margin is, we have yet to find out, but we're well on the way to being able to do that. Since that was the ultimate goal at the outset, I see progress in the right direction. I will communicate this sentiment to General DePuy and others, urge that the work continue apace, and that the Army pursue it to a logical conclusion. Therefore, I'm most appreciative of your support for this project and hope that you'll continue to help us with it. It's 30 years overdue, but it does appear that we are closing in.

Night Operations

Message to Major General John W. McEnery
Fort Knox, Kentucky
19 September 1977

Last week I visited Detroit to look at the XM1 mockup. During the visit I asked that they look at ways to rig the tank for night operations. Believe we know enough about what passive night equipment can see of even very low light levels reflected through vision devices, and escaping from improperly sealed vehicles, to require a serious review of how we rig our vehicles for night ops. Armor Center start work with PM XM1 and Chrysler to develop a statement of what we believe we need to rig the XM1 for night, and what are acceptable ways of doing it.

Communications Innovations

TRADOC Army Communications Command Representatives
Fort Monroe, Virginia
1 February 1978

I'd like to share with you some observations on communications and the TRADOC mission. Most of you must realize that TRADOC's mission, to prepare the Army to win the next war, depends on the development of the correct tactics, organizations, and equipment to do so. Cementing these three categories together is TRADOC's interest in training.

On the modern battlefield of the next war, communications, both tactical and long lines, have a large stake in all of these areas. Without the necessary communications, the tactics of attack and the active defense become unmanageable. Communications are their lifeblood. The same is true in the organizational and equipment development fields. The problem is that the modern battlefield that we face is going to be a very busy and complicated place. The high levels of electronic warfare that are present on that battlefield are going to make command and control difficult. They are going to make the leader's job harder, because it's going to be difficult for him to communicate with those with whom he has to talk in order to get on with the battle. In fact, the battle will start in the electronic arena long before the first bullets are fired.

That, in turn, will make the communicators' job infinitely more difficult. Their ulcers will start early, and the solution must come immediately. When you look over the problems I've mentioned and take a quick glance at what we're doing, I'm afraid that you must have a few moments of hesitation.

We have so many complicated and complex systems that seem to be highly dependent on ideal conditions. We seem unnecessarily wedded to the need to hear the voices of those with whom we communicate. We seem to be caught up in the electronic field in a vicious circle of development and instant obsolescence, so much so that we appear to be constantly promising new systems or families of communication devices and fielding very few. To the uninitiated, and even those in the know, we present a picture of developing duplicative and redundant systems that often proceed to the field without attempts to integrate them with the other systems in development or even with those we have in the field now. When you draw up a menu of the data, digital, and voice systems in development or fielded, it reads like a Notre Dame football roster and requires a separate translation index. The communications field may have cornered the market on alphabet soup.

To the soldier and leader in the field, the sense of bewilderment is overwhelming. All he knows is that, with all the systems, with all the expensive equipment, and with all the miracles, he still has problems, serious problems, in communicating. Notwithstanding that some of those problems may exist between the headsets, it doesn't take too much effort to share his bewilderment.

Your reaction to all this may be the question, "Where do we go from here?" After all, TRADOC is supposed to have all the answers. I don't want to discourage you, but we don't. What we do have is an idea of where the trends are and what we need in the field to win the next war. We are now in the process of identifying all the trends, both enemy and friendly, from the Central Battle as it exists today and projecting them forward to a future Central Battle. That will give us an idea of what we have to do.

To give you some idea of the things that are appearing in these trends I mentioned, take a moment to think about these ideas. It is increasingly apparent in the hostile electronic battlefield of modern war that we can no longer afford the luxury and emotional reassurance found in hearing the voices of everyone with whom we want to communicate. The enemy won't allow it, nor can the limited channels and time available bear the strain. Instead we must pursue the capabilities offered by data or digital transmission means, transmissions made in multisecond bursts that are impossible to jam or to pinpoint sending locations. Communicators have been urging us to go that way for a long time, but we didn't listen very closely. I should correct this by telling you it won't be easy to wean the steely-eyed fighters away from their microphones. It is a great pacifier, but we've got to do it.

The tyranny of the message centers is no stranger to you, but it is another of our communications problems. We somehow must break it. It is stronger than the First Amendment now. If the forms aren't right and priorities just so, the message doesn't get sent. In V Corps in Europe, we broke that by getting the messages put right on tape and sent, then giving them to the message center to log in and fill in forms and all the other wonderful things they do.

As a corollary to this, you communicators can't afford to allow the data-digital transmission method to fall into the deadening bureaucracy of the message centers. If you do, and set up a bureaucratic system for preparing data, filling forms, etc., the steely-eyed ones will go right back to their pacifiers and demand to talk personally to everyone.

We're also wrestling with the problem of communicating quickly over long distances. Our own hierarchy of command often stifles this—you know, company to battalion, battalion to brigade, brigade to division, and so on. By the time information arrives, it's too old to be reliable. We're experimenting with systems that use satellite transmissions right from the battalion and broadcast simultaneously to all the levels. Coupled with digital transmissions in a burst mode, it can be a real timesaver.

Recently, both in TRADOC and in DARCOM, we've started a review of all those alphabet soup systems we've got in the field or on the drawing boards with an eye toward eliminating those that are redundant. We're getting the single-system zealots to talk to each other in the hope they may borrow from each other. If that doesn't work, we may have to just bang all their heads together until they get the message that the systems have to be interrelated and as common as possible. We're making some progress.

In the area of electronic equipment development, I hope you realize that modular construction is the way to go. Both from the ease of maintenance and common design, it has some exciting prospects. It may also provide a way to beat the obsolescence problem in the electronic field. Hopefully, too, it will help to rein in the single-system design fanatics.

As a last point, we're starting to work on an integrated communications plan that will dovetail with the Central Battle today and provide a roadmap to the future Central Battle. If we can get everyone to agree, including the steely-eyed fighters, we may be able to provide more than just promises to the field. Admittedly it is something akin to grabbing a moving train, but it must be done.

Communication is so vital to our success on the battlefield that we can't afford to accept anything but the best and most reliable, even if it means going back to hand and arm signals. In the recent *Infantry Magazine*, there is an article by a battalion commander who forced his unit to undergo some silent exercises. He turned the radios off. The first time, confusion was the order of the day, but they learned. Messengers became the most used means, but they proved they could operate. I'm not advocating that harsh a solution, but we could afford to do with less if what we had is the very best. That's our goal in TRADOC.

Position Location Reporting System

Message for Lieutenant General Donald R. Keith
Deputy Chief of Staff for Research, Development, and Acquisition
14 April 1978

A. Reference Roy Thurman's LVW 0621, 101452Z APR 78, SAB.

I'd just like to endorse Roy's recommendation. I know we've a lot of priority items and not enough money to go around. But if we are indeed to move the Army into the twentieth century in command-control, this one issue alone is at the moment the key to that effort. We just must go ahead with it.

Combat Mission Failure Criteria

Message to Major General James H. Merryman
7 June 1978

1. Monday, in the briefing leading to development of a TRADOC position on what to do next in the ITV program, the perennial argument developed over what should be charged as a failure and what should not. One chart toted up mission failures in several subsystems as charged by OTEA, and compared those with TRADOC and/or DARCOM views as to what should be counted as a failure.

2. This is not a new problem, but it is one which continues to cost us inordinate time and effort arguing it out. It is one which also encourages considerable ill will between tester, user, and developer. Therefore I'd like to put a stop to it.

3. In operational testing we should first arrive at some agreed-upon mission failure definitions with the tester—OTEA in the case of major systems. Then we should count failures. There should be no equivocating about these failures by the developer or the trainer. Where failures are high, the developer, anxious to show his system meets the documented need, immediately begins to manipulate the test numbers, trying to show that test outcomes were really in the required performance band, even though they weren't. To some extent the trainer tries the same act with regard to human errors. The result is always a split between the three—tester, developer, trainer: the former trying to be fairly objective, the latter two trying to prove that their special areas of interest didn't really cause the thing to fail all that much.

4. What should really happen, once mission failures are toted up, is developer, trainer, and tester together carefully review failures, determine the causes therefor, and set about to fix the faults—each in his own field of responsibility.

5. When failure rates exceed allowable performance bands set forth in requirements documents, the interested parties must jointly arrive at a judgment about what effect the failure rate should have on decisions to proceed with the next phase of development. Alternative decision possibilities can then be drawn up, costed and risk estimated, and presented to the decisionmaking person or body for decision. This procedure implies that failure rates above stated bands, or performance rates below stated thresholds, are not necessarily causes for terminating the development. Rather a judgment must be made in each case based on analysis of the failures by type and cause.

6. It is also important, following this line of reasoning, that we begin work to develop more realistic failure criteria for operational testing. These criteria should derive from combat mission failure definition agreed upon by tester, developer, and trainer prior to test. Use of criteria in this context can and should lead us away from the mean miles between failure-type criteria so characteristic of our current need statements, test designs, and test outcomes.

7. I'd like for you to explore the ideas set forth above with OTEA and others as appropriate and, when you're ready, let's talk about it.

Rationalization, Standardization, and Interoperability

Defense Science Board
Newport, Rhode Island
31 July 1978

The Undersecretary asked me to talk about rationalization, standardization, and interoperability. I will do that. He also asked me to be candid. That I will do also, so what you hear may be a different perspective on this matter than you have heard before or will hear later in this meeting.

Defense Department and Army efforts to improve NATO effectiveness through more effective use of defense resources committed to the Alliance are governed by two key documents—DOD Directive 2010.6 and Army Chief of Staff Memorandum77-34. These documents share an imposing title, "Standardization and Interoperability of Weapon Systems and Equipment Within the North Atlantic Treaty Organization (NATO)." While both are masterpieces of bureaucratic perfection, their interpretation varies so widely that there is confusion about what they mean and what the Army should be doing in response to them, so let's first define the terms "rationalization," "standardization," and "interoperability" in an effort to develop a perspective about how the Army approaches its RSI work.

First, rationalization. Its definition includes the other two key words—standardization and interoperability. Rationalization is a difficult idea with which to work. Normally, those working on RSI matters give the "R" lip service, turning quickly to hardware developments or to services or supply matters. Yet rationalization is a cardinal concept; it implies a logical reasoning process, the result of which is a goal or goals for standardization and interoperability.

To the Army, rationalization produces common operational concepts, concepts that become the basis for joint development of tactics, hardware, services, procedures, organization, or training systems. This is a very important idea. If the concepts are not well thought out, then any attempt to achieve commonality is sure to fail.

Now this is not at all a new problem. The Training and Doctrine Command's primary business is to establish, for the US Army, operational concepts to be used as a basis for developing tactics, equipment, organization, and training systems. We do this by analyzing the Central Battle and from our analysis deriving what systems, tactics, organizations, and training we need to fight the battle successfully. Rationalization simply says we must now do that in the context of the NATO coalition and that we must now add those considerations of services, supply, and operational cooperation brought on by the nature of coalition warfare.

Since the Army considers true rationalization flows from carefully thought-out operational concepts, we also believe the RSI business is considerably more than a series of agreements to buy something from the other fellow. The Chief of Staff of the Army has reiterated that position in writing.

Now, standardization. Again there is redundancy, and so some confusion. To illustrate the confusion, the major Army command most concerned with RSI—US Army, Europe—tends to define interoperability as we have defined standardization. It takes a more narrow definition of standardization by applying it strictly to hardware or to procedural software that is integral to a hardware system. Maybe it really isn't important whether it's called rationalization, standardization, or interoperability just as long as it works. But let's define standardization as the process of developing and using the same or common weapons or hardware and the software procedures related thereto.

Now, interoperability. By definition it deals with services that can be provided or accepted by members of the coalition. As I have already noted, USAREUR tends to use the word to include not just services but supply and operational matters such as fire support and the working together of maneuver forces. In either definitional context, interoperability is a micro rather than macro approach to cooperation in coalition warfare. In the opinion of many observers, it is the most promising part of RSI, for often it can be had at little or no cost.

So for Army-working definitions for RSI—it has but one goal, to improve the capability of the NATO Alliance to deter war and, if war comes, to surely win the first and then succeeding battles of that war.

Earlier we saw that commonality of operational concepts provided the means to achieve standardization and interoperability. It's instructive to note that the Soviets also subscribe to this idea and have indeed achieved their own brand of RSI within the Warsaw Pact. Although one suspects Soviet RSI to be perhaps more one-sided and dictative than ours, there is no question as to its success.

Soviet operational concepts for modern warfare have deep roots. In the 1930s, with help from the Germans, the Soviets developed concepts they embrace today. They were, and are, mass, momentum, and continuous land combat. Mass means numbers and concentration of forces. Momentum means sustaining depths of advance of 40 to 50 kilometers a day. Continuous combat means echelonment of forces—when one unit is consumed in battle, the next one is committed without pause. Night-vision equipment enables the fight to continue around the clock.

For many years Soviet weapons, organizations, and training systems were not a perfectly matching support for the concepts. However, they did not abandon the concepts. Instead, by requiring training, weapons, and organizational systems to develop capabilities required by the concepts, they have gradually built the forces—the Warsaw Pact AirLand Forces—that NATO faces today, forces fully capable in every way of carrying out the operational concepts in battle. It's a good process; it works; it's continuing.

It has resulted in integrated weapons, organization, and training systems designed to work together as a combined arms team; designed to be redundant, with a variety and quantity of systems whose capabilities overlap one another; designed to be complementary, mutually supportive of one another; and designed to be multicapable, used for a variety of purposes.

Soviet rationalization is reflected in common Warsaw Pact operational concepts. Standardization in equipment has been achieved by manufacture in many countries of common designed—Soviet-designed—major weapon systems. Interoperability is paramount in an integrated logistics system. It is an admirable achievement, one that NATO must work hard to match.

Now Soviet-style rationalization in a coalition of truly sovereign nations is clearly not possible. Simply put, NATO will have a hard time matching the Pact. However, we have made some beginnings; let's look at those.

Our most important achievement is the series of staff talks between the US Army and the German Bundeswehr. Although contacts between the German and US Armies have been working at the colonel level since 1961, they were for a long time informal, had no real clout, and there was no structure to them. Then, in 1974, the Vice Chief of Staff of the German Army, Lieutenant General Schulz, recommended to General Abrams, the US Army Chief of Staff, that the talks be formalized and expanded. The format suggested by General Schulz was adopted as the management system for the talks. Notice that General Schulz rightly placed first emphasis on common concepts. General Weyand, succeeding as US Army Chief of Staff on General Abrams' death, responded and directed General DePuy, my predecessor, to be his executive agent. TRADOC thus became the operative Army element for US/German RSI. General DePuy emphasized conceptual agreement as the first consideration.

On 23 June 1978 we concluded Staff Talks V. I won't recite the details of the talks, but I'd like to describe how we manage the talks and report some highlights.

We adopted a three-phased approach to cooperation. First was development of a harmonized concept. The idea was that this phase would produce a series of joint concept papers dealing with topics such as threat, terrain, antiarmor operations, fire support, airmobile operations, and so on. We now have more than 16 subject area papers in varying states of agreement and discussion. One country or the other was assigned the lead in each subject area. As the papers are agreed to by Chiefs of Delegation, the respective Army Chiefs of Staff sign the papers.

From joint concept papers we proceed to Phase II, the definition of equipment requirements. The device agreed upon to enable us to progress from tactical concepts to equipment development is a Military Equipment Characteristics Document (MECD). It is a joint user expression of military requirements. It is a fairly detailed document that sets forth the requirement, numbers needed and when, feasibility analysis, cost parameters, and procedures for follow-on actions. Five specific equipment developments are now in varying stages of MECD development.

When Phases I and II are done, we enter the final phase—cooperative fulfillment of the requirement. This stage involves materiel, training, logistics, and analysis. Several concepts and equipment developments have reached this stage and are now being evaluated. The criteria for evaluation are that the item or concept must increase combat effectiveness, enhance NATO RSI, reduce duplication in research and development, have broad application in both numbers and cost, synchronize the initial operational capability, and have a possibility for quick success.

Obviously, at each Staff Talk, there are items to be considered in each phase, and since the process is dynamic, items are added, modified, or deleted. Possibly our most important achievement is harmonization of the basic operations manuals of each Army—US FM 100-5 and German HDv 100/100. These two books describe how each Army fights. Each deals with tactics, weapons, and command-control. Each has been translated into the other language. There are no contradictions. Considering that the US and Germany represent the two largest military contributors to NATO, this is a significant achievement indeed. The two Chiefs of Staff have signed nine concept papers and approved the management system.

The next round of US/German Staff Talks is scheduled for February 1979. Obviously the important work of the effort is not done at the conferences. Personnel from TRADOC and other commands and agencies of the Army are dedicated to working on papers, concepts, or tests—either full-time or part-time. At the Army Staff and OSD levels, there is also much RSI work in progress. However, most of that work is centered around hardware exchange—mostly weapons systems—and does not proceed from or indeed consider operational concepts as a first order of business. In fact, so many independent agents are operating in this arena that it is most difficult to even keep track of who all has agreed to what all.

We have also started talks similar to the US/German series with the United Kingdom. Since the British Army has no comparable field manual to our FM 100-5, *Operations*, we must attack the conceptual problem piecemeal rather than as a coherent whole. Also, the British tend to think more in terms of operations in northern Germany, since First British Corps will fight there. While this effort represents a problem set somewhat different from that of the US/German talks, we believe there is mutual benefit to be had. Further, it is obviously necessary to somehow rationalize all of NATO's military operations in order to achieve the maximum benefit from the obvious potential of RSI. We hope to use our experience with and the products of the US/German talks as a basis for achievement in this important next step.

Now let's talk a little about standardization—myth and reality. Since the most visible part of RSI is that which deals with hardware systems, a realistic perception on the matter is not only appropriate but essential. Standardization has three essential components. First is a common resolve—yes, a determination—to achieve standardization. Second is the technical ability to accomplish standardization. Of these two, the first is the more important, for without it the obstacles presented by technical problems associated with achieving standardization will never be overcome. Even with the necessary common resolve, some technical difficulties may well prove nigh insolvable. Third is the reconciliation of the materiel acquisition process in each country with demands imposed by standardization programs. Here again, despite considerable common resolve, bureaucratic and political encumbrances may intrude to make standardization well-nigh impossible.

While it would be convenient to assume away the problem of common resolve, it's never wise to do that. Let me be specific. You may recall that, several years ago, there was a common resolve to standardize the NATO rifle cartridge. After considerable study, it was agreed that 7.62 millimeters was the correct caliber. That was agreed to.

Then two things happened. First, several nations chambered their weapons differently so that, while the bore diameter was common, the cartridges were not, in fact, interchangeable. Second, the United States unilaterally adopted the M16 rifle at 5.56 millimeters. While reasons for these decisions are now buried in obscurity, the problems they created, including considerable logistical implications, plague NATO today. So now we're about to participate in a NATO weapons shootoff that hopefully will resolve the problem, if only we can muster and sustain the requisite common resolve.

The Main Battle Tank Program in the 1960s was an ambitious effort to provide NATO a common main battle tank. It was backed by considerable common resolve, a common resolve that was dissolved as the program dragged on and costs rose. Most recently, the tank gun standardization saga is a live issue that may outlast all our resolve and patience on both sides of the Atlantic.

The technical ability to achieve standardization is properly the subject of a separate dissertation. But let's consider five key problems that must be confronted head on if standardization is to succeed.

First is the technical generation gap. Generally it's fair to say that the Europeans are a generation behind us in technical developments related to electronics, communications, computers, and certain imagery systems. Therefore, if the Alliance is to take advantage of opportunities afforded by most modern technology, many stipulations applicable to competitive development must be set aside or modified. For example, in communications the opportunities offered by US advanced technology in digital data distribution systems and in microprocessing systems must be adopted over less technically advanced, but probably less costly, analog systems.

Second is the problem of design philosophy—essentially in automotive systems. The Germans, who are our major design competitors in this field, tend to produce big, heavy, well-designed systems and components, overdesigned by our standards. We would reject many such systems as being too much for too much money. In addition, the Germans design essentially for systems employment in Europe and so do not include in their design the extremes of climate, terrain, and other operational/environmental features on which we have traditionally insisted. So, while their gear performs better in a narrower environmental range than does ours, it tends to perform less well across a broader environmental range. German tank engine cooling systems, for example, tend to overheat in the desert, a fact to be expected. They didn't design their systems for desert operations.

Third is the problem of production plant methods and procedures. As a general observation, the European heavy equipment production plant is much more modern than that in the United States. With old, manpower-intensive production equipment, we are competing with a new, equipment-intensive, computer-driven production plant, especially in Germany. On the other hand, our aviation-electronics and computer production plants in the United States tend to be better than those in Europe. All this leads to an impossible imbroglio in production, one for which the only solution is to insist on form, fit, and function of componentry—not asking what is in or goes on in the components.

Fourth is the problem of technology transfer. This sounds simple, but in practice it's so difficult that, once the transfer is made, one must begin testing as if a new system were being dealt with. This simply adds to development time and cost and plant and problems. So long as we in the United States continue to insist on a US-based production capability, this problem cannot be solved. The MAG 58 machinegun is a good example. We could not beat it with a US-made machinegun. But we insisted on a US-based production capability, the product of which must be subjected to the same rigorous testing as was the original product.

Fifth, common resolve to standardize must eventually be translated into a very strongly disciplined configuration control program. This must be joint. It has to be ruthless, and it will defy effectiveness parameters from mechanical factors to cost factors. But it must be, or the materiel developers in each country will tinker away whatever chance at standardization the program may have had.

Now a few brief words about the materiel acquisition process. There are several things to remember. First, in each nation it's quite different, and many of the differences are so significant that they require reconciliation. Second, in most nations it is much more a product of nationally owned or nationally subsidized plants than it is in the United States. In the US, since the advent of our present management system in the early 1960s, our totally competition-based least-cost system has not, from concept to fielding, produced a single major weapon system for the Army. That's 16 years.

So, to a system already so encumbered that we don't know if we can produce a new system, we have added a requirement to enter foreign competitors to each development under consideration. The basic ideas of selecting the least cost alternative and competitive development at each step of the cycle prevent us from the kind of logical product improvement so characteristic of the

Soviet hardware development system and inevitably mean a more costly system, despite our least-cost rule.

By now it should be apparent that we believe the interoperability of NATO forces offers the greatest potential for improvement. Interoperability, as noted, can be assisted by microstandardization, but it can also be deliberately designed into completely different national weapons systems. For example, the present use of the 105mm tank gun by most NATO forces allows common ammunition servicing. Some interoperability can be achieved quite cheaply by training practices alone—the use of other nations' rations for troop messes, for example. Interoperable design of automotive engines to accept a range of types and grades of fuel from any source is achievable; some multifuel trucks are in use in several forces.

Host-nation support—preplanned and prepracticed—can serve interoperability. This support can range from equipment (vehicles, pumps, commercial construction equipment) to manpower for logistical installations (supply clerks, warehousemen, dock workers) to logistical facilities (docks, pipelines, communication networks) to specialists (mechanics, recovery experts, and building tradesmen). The possibilities are enormous, but the agreements and the organization must be developed and exercised.

We should also recognize that interoperability of this kind defies the longstanding NATO dictum that logistics is a national responsibility. Apparently we have abandoned that idea without ever really saying so. It's long past the time when we should address ourselves to that problem head on. Thus we see a NATO land force in which not all necessarily use the same equipment, but in which a US, German, or Dutch tank soldier could drive his tank into any supply area and get the part he needs and ammunition, fuel, and rations to quickly return him to the battle. That is real interoperability!

Operationally, interoperability means the ability to use the firepower—including aerial fires and the maneuver forces of other nations—just as one would use those of one's own forces. As in past coalition wars, this capability comes largely from the exchange of liaison parties equipped with the necessary communications equipment and procedural documentation to conduct their business. While this is, for the most part, a matter to be worked out within the operating forces of the Alliance—that is, USAREUR and its Alliance counterparts—we can help by developing common procedures, common lexicons of operational terminology describing the procedures, and common training systems to facilitate their use.

So RSI is a serious, complicated business. It can provide the industrial and military backbone for coalition warfare. We believe the future of NATO, both economically and militarily, depends on it. But we must be very realistic about its possibilities as well as its problems.

May I set forth in conclusion a summary that might serve to guide us in our further efforts at RSI?

First, declarations alone don't produce rationalization of anything. They must be backed by a firm and active resolve.

Second, rationalization, in its full definitional context, calls for early agreement on military operational concepts.

Third, rationalization cannot be had unless we improve our technical capability to achieve rationalization. This includes:

- An early collaborative research and development effort designed to close the technical generation gap between us and European technical capabilities. This should include efforts to solve the technology transfer problem in the R&D phase insofar as possible.
- Comprehensive rationalization of defense industrial procedures and methods, possibly to include elimination of redundant or duplicative capabilities.
- Comprehensive rationalization of differing design philosophies between the industrial plants of the Alliance countries.
- Procedures for configuration control of common items produced in several countries.

Fourth, rationalization in NATO must be based on common long-range equipment plans. These plans must include:

- Time-phased binding schedules for succeeding generations of equipment.
- Reconciliation of the materiel acquisition systems in the several countries with requirements of the NATO long-range plan. This will mean in many cases going away from the least cost, totally competitive development pattern that has characterized the US system since the early 1960s.

Fifth, there is the matter of money and the need to develop the many ways in which common development of materiel can be funded. These include:

- Licensed production.
- A common NATO fund.
- Legitimate compensation balancing between the partners.

Finally, there is a great deal more to be gained in the immediate future through interoperability as we've defined it—working together to develop common operating procedures and the necessary supporting software or hardware. This includes interoperability of operational forces, supporting forces, and supply services. While less exotic than most weapon system developments, it is nonetheless the area of greatest potential payoff.

While full standardization can be acknowledged as a goal, it should not be allowed to delay the achievement of an interoperable NATO force. We cannot afford the luxury of delay, for the stakes of the game—national and world survival—are too high.

Position Location Reporting System

Message to Multiple Addressees
11 August 1978

1. For years, monitoring our own radio transmissions has revealed that as much as 65 percent of our voice traffic involves telling each other where we are on the battlefield and where we are going. We have under development the Position/Location Reporting System (PLRS), which is our only near-term system to move US Army command and control into the twentieth century.

2. Emerging results of the POS/NAV/COEA indicate not enough attention was paid to identifying requirements for the C3 capabilities offered by PLRS. Hindsight tells us now that this should have been done.

3. We need an intensive effort to correct this earlier oversight on the part of user reps at schools/centers. As the TRADOC proponent for PLRS, Bill Hilsman will take charge of this effort through provision of information on PLRS to the user reps and pulling together their requirements for review and incorporation into the PLRS/C3 programs.

4. Full cooperation is required from all school commandants in this effort. As seats of doctrine within your functional areas, your requirements for C3 support are critical to the integration of the battlefield.

Acquisition Cycle Compression

Message to General Frederick Kroesen
Army Vice Chief of Staff
26 September 1978

1. On 14 September I had a conversation with General Dr. Karl Schnell, Deputy MOD Bonn for Armaments. He reported to me that the Germans have from the British a letter at the prime minister-presidential level which proposes tripartite development, between the UK-GE-US, of the next generation tank—that is, after Leo II, XM1, Chieftain. I understood him to say that the letter had been presented simultaneously to both GE and US governments. To date I can't find anyone in the US who has any knowledge, or at least who will admit to having that knowledge.

2. Subsequently in the UK press it was announced that the Brits were undertaking a new tank development to replace Chieftain. The announcement apparently made no mention of a tripartite arrangement. Can't tell at this writing if this is the same thing as in 1 above.

3. In addition, General Dr. Schnell and I agreed that the US and GE should get together soon to work out requirements for the next generation tank. This is a subject I had already spoken with Bob Baer about. If the length of our materiel acquisition process is of concern to us, and it continues to be cited as an inhibitor to modernization of the force, then we must begin to look for ways to speed things up. Considerable study of the Soviet system, often cited as a contrast to our own in timing, reveals that their acquisition cycles are really not all that much shorter than ours. What does characterize their system is a dovetailing process by which a Mark I can be in production, a Mark II in advanced development, and a Mark III in engineering development, all at the same time. All too frequently, in fact almost without fail, we wait until a system is fielded before considering even PIPs to the fielded materiel. The reason most frequently cited for this is that to consider a PIP or a new model any sooner in the cycle would jeopardize the development of the original system. While I believe this is true, telescoping the cycle seems to me to be the only way we can hope to shorten up the acquisition process. If that be the case, we need to figure out how we can do that without jeopardizing the project. This will involve some consensus building in OSD, OMB, and the Congress—considerable consensus building, I would judge. But if this course of action appears to be the best to pursue, and indeed at the moment it seems to be the only one, then we'd better get on with it. I would suggest that we need to deliberately select two or three major systems and try this approach. I will come back in a few weeks with a proposal to do this; the proposal will incorporate a plan to pick up on an offer made me by Bill Perry in a recent visit here, and to try to move at least some of the dovetailed developments off line in an attempt to get a quick turnaround on some technology which is at the moment controlled largely by ARPA.

4. With regard to the Schnell conversation, he and I agreed to exchange letters in October or early November, then get together as soon as possible to follow up. I will pursue this. I would also urge that, if any of you can ferret out whether or not there has been an offer from the government of the UK to the GE and US (one or both), then we'd better get hold of that offer and get cracking quickly lest some instant field marshal in OSD or some other layer above us get in the act and become the Army's de facto tank developer.

Technocrat Mentality

Letter to Major General Hillman Dickinson
US Army Communications Research and Development Command
26 September 1978

Thanks for sending me the Technology Trends Colloquium (whatever that is) report. I've read it with some interest . . . there's still a bit of the technocrat mentality reflected in the report.

First, it's a long damn way from what's on the bench in the lab to what's in the field. In some of those technologies things are moving so fast that there is no practical way I know of, within the confines of 5000.1 and A109, to field anything reflecting the advanced state of the art in a timely way. I don't say we should give up trying, but as a practical fellow with lots of frustrating experience in the field, I have to be the professional skeptic once again.

Second, in the very vague proposals about small units with high technology gear doing the same thing larger units with lower technology levels can do, there is, once again, the pure technocrat. This idea is a myth. Conceptually it follows the oft-repeated strain that in the last thirty years or so technology has allowed us to substitute equipment for people on the battlefield. This is by way of being a self-fulfilling prophecy. The fact that the Soviets have fielded all the technology we have, and more and faster besides, belies the argument. What we're saying is that we can't, or won't, afford the trauma of facing up to a judgment of what's necessary and sufficient to meet the Soviet threat, given our current 40 percent disadvantage in military investment vis-a-vis the Sovs. It's an intellectual cop-out for lack of resolve, and for our unwillingness to get tough in something besides talk. In the 1967–1968 period I became involved in an ill-fated project known at the time as SIAF—small independent action forces. These, it was alleged, would be small units, highly trained, equipped with the latest technology, which would accomplish marvelous things on the battlefield against the wily NVA. It was a follow-on to the McNamara Wall, and in fact the product of some of the same minds that conceived the Wall. We found, despite considerable euphoria by its authors, the idea foundered because of our inability to convert technological concepts to practical hardware that us dumb guys could use in the field at night, in the rain, when tired, 24 hours a day, ad infinitum.

Information Explosion

Royal Canadian Armored Corps Association
Shilo, Manitoba, Canada
28 September 1978

It is no great secret that military organizations practice war in peacetime. This is done, hopefully, on the premise that preparedness will prevent the very war for which they practice. The US Army Training and Doctrine Command, which I represent, has the specific job of preparing the Army for the next war. That future war has become the central focus of a great deal of study, and today I'd like to share with you a problem we've uncovered that applies to us and to all our NATO allies.

Like most wars, the next war will likely begin with some sort of explosion. Unlike past wars, however, the first explosion of the next war will not be the sound of an enemy shell, bomb, or missile. Indeed, the first explosion of the next war has already occurred; its reverberations grow louder daily. That explosion is information—information in such quantity and detail and from such a multitude of sources that it literally overwhelms the receiver; information that clutters and confuses communications nets, message centers, headquarters, staffs, and even commanders; information that provides knowledge of nearly everything on, over, and under the battlefield; information that holds the key to victory if it can only be organized, cataloged, and used correctly and on time.

The information explosion is a new experience. In the US Army, it compounds some already serious shortcomings. The US Army has never enjoyed high marks for its ability at information collection and processing in war. As war grew more complex after the turn of this century, our Army approached the problem of battlefield information in a most gentlemanly way. Information collection about the enemy or our own forces was left largely to scouts—to the cavalry. A long and acrimonious debate arose as to whether the cavalry was to accomplish its mission by stealth, or if indeed it could be expected to fight in order to get the job done. Fortunately, our leisurely operational attitude was supplemented by great amounts of sheer enthusiasm that usually sufficed to overcome the inherent deficiencies of our system.

It was, however, an inefficient system. When the US Army joined the Allies in World War I, the strain on our information gathering and processing system, even in that static environment, was overwhelming. Commanders often found they knew as little about the location and activities of their own troops as they did of those of the enemy. Information they did have lagged hours—sometimes days—behind the real situation. Orders issued under these conditions often served only to confuse. The most reliable method of communication of the time, the telephone, depended on vulnerable wire laid on the ground. In an attack, when troops moved, telephones became impractical and information flow slowed or dried up completely.

. . .

Between the World Wars, the US Army information and intelligence systems were largely the property of the Signal Corps. Here almost every effort to bring about a sustained development program was thwarted by the hypochondria over ethics and morality that stunted the US intelligence program in the late 1920s. The national attitude on the matter is perhaps best illustrated by then-Secretary of State Henry L. Stimson who, upon discovering that US intelligence services were collecting information on other nations, abruptly terminated State Department support of these programs, saying, "Gentlemen do not read each other's mail."

During World War II battlefield information procurement and processing posed a difficult challenge for both sides.

. . .

Today, all armies have or are developing a host of the most sophisticated and marvelous information systems; the mind is boggled with their capabilities. There are systems to locate enemy units to 150 kilometers ahead of the battle line; to seek out enemy radars, radio waves, even artillery rounds, and within milliseconds determine the location of the originating equipment; to direct a multitude of information systems to singly or simultaneously provide a flood of information; to correlate, integrate, and disseminate data from many sources, hopefully providing a coordinated battlefield picture; and to afford increasingly responsive, accurate, and secure command and control at all levels. The list of systems seems endless; their acronyms—SOTAS, TOS, RAWS, RPV, LBSD—make the mind spin.

However, it is quite apparent that all the old problems of information collection and dissemination are still with us. Today there is more information. Sometimes it moves faster, and it's somewhat more secure, but humans who must use it have the same old problems. Message centers, headquarters, and staffs are still constructed according to the shopworn, hierarchical, sequential principles of long ago. So much information is coming in that the system is in a state of overload, and we still can't find out what we need to know. Communications systems, designed by communicators not users, still don't bring us what we need to know. Users, unable to wean themselves from the sweet sound of their own voices, cling to outmoded voice nets when digital data transmission technology promises so much more.

With an information explosion in progress and all developments promising even more, there is an urgent need to recognize what information overload means—there is a limit to what an organization or system can handle. What happens to living systems or organizations as they approach overload is that output increases with input for a time, but levels off, then quickly drops to near zero. The result is breakdown. Overload of information systems is a today problem—it's here, not in the future. Indeed, it is the very problem faced by commanders in the historical examples I mentioned.

. . .

Historically, as I have indicated, military commanders have sought multiple channels—bypass channels—to both speed and filter the flow.

None of tomorrow's systems, in design or development today, take this multiple-channel phenomenon into account. Should they? Can we afford not to have them do so? Have we the resources to provide enough redundancy to permit multiple channels? Will some advance in technology obviate the need for multiple channeling? Or, better yet, shouldn't we be smart and design systems now to provide us the specific vital information we need to control the battle? Information, funneled through sure and accurate discriminating processes, using discriminators provided by the commander, would get to the commander when it happens—in real time—not hours later after it's been administratively fumbled over by a myriad of well-meaning but essentially clumsy hands.

In large measure the faults alluded to above must be laid at the feet of tactical commanders. This is so because, as a general rule, the tactical types have not been forthright with information systems designers and developers as to what information systems must do. As a result, the systems are being designed by men who imperfectly perceive what commanders want to know. Because the tactical fellows haven't thought it out very carefully, they tend to be unhappy with the result. Therefore it's of cardinal importance that the users set forth clearly just what and how much information of what kinds is really required to win the war.

Let me be specific. Directing the Central Battle today can be reduced to a rather simple cycle. The commander must get information, information that has gone through a conscious and deliberate discrimination process and that is available on time—real time data. Then he must decide what to do about it. If the discrimination process works, then he has time to decide, since he is not busy sorting out the information himself. He must tell someone to do something, and finally keep track of doing it. It's a continuous cycle.

Now let's look at why the battle can't be directed in today's environment. First, there is the sheer mass of information we've talked about. There are the multiple sources that provide information from everywhere and often duplicate each other. There is no priority discipline, no discrimination process, so routine reports flow in along with important ones. Those important ones are lost in the shuffle or held up for processing. Then there is the sequential reporting—company to battalion to brigade to division and so on. By the time information arrives, it is out of date, useless, good only for a history book.

Now these are external problems. But we've got some real internal problems too. The first is what I call our self-constipating staff system, where everybody has to massage and mull over and otherwise fondle every bit of information that comes in before it gets to the commander. The result is everything that is unusual is submerged into a normative point of view, lost in the routine of background noise. The other problem is the tyranny of the message center. If the individual in the message center doesn't get to stamp, spindle, fold, and otherwise mutilate the message in accordance with the rules of bureaucracy, then it doesn't go. The same applies to incoming messages. The result is the same—there is no real time data.

To date we've relied on a lot of new technology to solve our problems. Each new gadget or technical innovation is embraced with the enthusiasm of a child for a new toy. "Yes, that's it! My, don't we need that!" Then, when we get it in the field, the search begins to fit it into a framework crowded with many other systems. The result is a hodgepodge, a network that doesn't work very well and really never has. System overload is an inevitable outcome. The background noise of the overload is so strong that critical information can't be identified, much less isolated. It is important to decide "how much is enough" and then limit the addition of more information from more sources. Commanders' essential information requirements must dictate what information goes to what level and when. Specific requirements must direct the developers to tailor the system to user needs.

If today's problem is significant, tomorrow's is even more so, for lurking just over the information technology horizon are truly remarkable and necessary opportunities for improvement. Today we have the potential to replace 1940–1950 voice transmission technology with digital data flow networks, coupled with distribution techniques that send information to many users in one quick burst. There are multiple frequency switching procedures for both low- and high-speed transmissions that can make enemy electronic jamming and direction finding impossible. There is a potential to both send and receive from communications satellites directly to frontline units, providing commanders access to worldwide information. The list goes on.

But we can't use the present or future technology if we don't "set our house in order" by solving the overriding human interface problems. This demands a fresh and objective review of how requirements for systems are set forth, but even more critical is the demand for a critical analysis of how we intend to process, distribute, and assimilate information. The battlefields of the next war are not the place to do this—indeed, when battle is joined, it will be too late.

Our potential foes have already solved some of these problems. They have invested heavily in the technology of radio-electronic combat and are prepared to wage it now. Jamming, counterjamming, electronic direction finding, and many capabilities are common to the liturgy of their battle planning. This is so despite the fact that, technically, their national technologies are collectively a generation behind those of the West. They are a generation ahead in fielding their "old" technology. The result is that the military forces of NATO are about to be relegated to the backwaters in a field in which the West leads the world. The information explosion is the first battle of the next war. It's here today!

Tank Development

Letter to Richard P. Hunnicutt
Berkeley, California
6 November 1978

Thanks so much for sending me a copy of your story of the Sherman tank. It's obviously a classic. I wish we could today produce one model of a tank in the time we managed several models in the Sherman's life. I guess at this point I'd settle for just one model of a new tank and hope we could get it out while I'm still on active duty!

Perry Initiatives

Message to Lieutenant General Donald Keith
Deputy Chief of Staff for Research, Development, and Acquisition
27 November 1978

1. Some weeks ago Bill Perry visited me at my invitation. During the course of our day together he offered to respond to some technical challenges I laid out for him, and at the same time invited us to participate with him in some initiatives he has already started. This was the beginning of what is now called the "Perry Initiatives" on which we have been working for some weeks now. We have been fully coordinating with Bill Richardson from the beginning; some of your people have helped us, as have some of Jack Guthrie's people. We are about ready to go back to Bill Perry and lay out what we believe should be done. Before doing that we want to lay out our proposals for you and for Guthrie.

2. The purpose of this message is to respond to some statements you are reported to have made about this whole matter. I would first like to allay your fears that we are charging off on some wild goose chase which will in the end cost you money or cause some ongoing program to lose out in favor of some new and exotic enterprise. Since we've not talked personally I have to assume that my informants are exaggerating whatever you may have said, and that yours haven't properly informed you as to what we are about. I hope we can talk about this definitively in the near future.

3. I am responding to Bill Perry in this wise, first because he alleged that we are all dragging our feet and that he has not yet found anyone in the Army who would rise to his challenge. Secondly, it is quite clear that the RDTE strategy of the DOD is marching to his tune—pursuing a course of action that I'm not at all certain is the way to go, without a single military voice being raised in assent or dissent. I object to that—so should you. Third, he has lots of money; he will spend it for whatever suits his personal convictions regarding our RDTE strategy—no matter that it's the way to go or not. Therefore, given that the money will be spent anyway, why shouldn't we try to steer at least some of it into programs of our own designing? In addition why can't we, by joining up with him, try to steer what he already has started into useful and productive channels? If we don't, we'll just get another box of complicated junque that we can't train anyone to use.

4. We'll get with you soon. Meanwhile let's not let the midgets who surround us get our intentions and our work all out of perspective.

XM1 Tank Issues

Message to Major General Thomas Lynch
Fort Knox, Kentucky
27 November 1978

This responds to your KNX 0613: XM1 OT II. I agree fully with your stated position. I would also add to your list of things that have to be fixed: the hydraulic system, the fuel system, the tank commanders' station, and the fire control system. Some aspects of this machine are strongly reminiscent of M114s, Sheridans, and other DARCOM spectaculars. The ASARC should be an interesting meeting.

Operational Testing

Message to General Frederick Kroesen
Army Vice Chief of Staff
29 December 1978

1. This follows our telephone conversation in which I promised to provide you some further observations on operational testing after my message about Guthrie's recommendation concerning development testing.

2. When we created OTEA out of the bowels of ACSFOR, shortly after I succeeded you in MF, the concept was that it was to be an independent evaluator of tests conducted by other agencies. There was a lot of controversy over that; there were pressures to give it the role it now has from the outset, and there were those of us who objected to the entire concept from the beginning. I was one of those latter. My reasons then are my reasons now. Nothing that I have seen or that has happened since has caused me to change my mind. I'm not saying I told you so, for I know full well why it was created and why it was eventually given the mission it now has. But let me review my position.

3. Any agency at the DA Staff level which is given independent evaluation authority over the work of major subordinate commands or agencies of the Army is in an untenable position. This was the consistent problem with the Weapons Systems Analysis Directorate of the A/VICE office. We were a bunch of smart analysts who took the work of CDC (primarily) and set out to prove quite independently that that agency didn't know its business. And so we were continually in the mode of telling the Chief of Staff that one of his lieutenant generals had just wasted several million dollars and considerable time studying, testing, or evaluating some system, submitting at the end a report in which we had no confidence statistically, analytically, or from several other standpoints. This antagonist relationship boded no good for the Army Staff, the A/VICE office, subordinate commands, or almost anything else in the Army. At one point its manifestations caused Jack Stockfisch to lose his job as the director of Weapons Systems Analysis. It was, in sum, an impossible situation. Creation of OTEA as an independent evaluator of tests could only serve to aggravate a problem which we were already having trouble coping with.

4. From the beginning of OTEA's existence as a tester of major systems TRADOC's boards and schools have had considerable difficulty with OTEA's analysts and staff. Some of this reflected OTEA's growing pains and has to some extent gone away as the agency has matured. But the major problem persists, and it is that problem which causes me the greatest concern. Let's take the XM1 test for example. Driven by the statistician's desire for a clean test design, and the need for statistical confidence, correct sample sizes, and all the things that excite analysts, we were forced into a test design in which the tactics used for a considerable part of the test were quite unreal and uncharacteristic, and the data therefrom are therefore suspect because of the mode in which they were gathered. For example, the test design required M60 baseline tanks in the moving tank sequences to stop, range, start up, and fire on the move. The excuse for this procedure was that the tank optical rangefinder can't range on the move and so, to get a ranging, the tank had to stop. We know that with an AOS tank we can range on the move, or fire battlesights, and that we must have a speed of 8 to 11 or 12 miles an hour to get a stable firing platform. The firing tanks never attained that speed; the crews knew time was important, and so they fired the minute the tank began to move after starting up from the ranging exercise. I saw

one gun go off towards the moon as the tank rocked back when the driver tramped on it, trying to get the thing moving. Hit performances were miserable, and understandably so. This part of the test design reflected the need for statistical consistency in the test data. But in the end I have trouble understanding what was tested—certainly not the fire control performance of the M60 tank in its proper tactical employment mode. So that part of the test was in fact a nontest. There are many other examples, but that is sufficiently illustrative. The end result is that here we have an operational test in which we don't test an operational mode at all, at least not consistently, and in the end we have a test outcome which could well be all screwed up simply because some parts of it, at least, were not conducted using sound tactics or intelligent variations thereof.

5. Now what I've just described is the kind of thing we used to love to get our hands on in the Weapons Systems Analysis Directorate—we'd eat that one up; we'd wind up in the Chief of Staff's office telling him that the major commander responsible for the test had really done it again, and that we had no confidence in the test outcome. So who today is it that calls OTEA's hand? Who is it that performs the independent evaluation by means of which we make the judgment that the test was right or not? No one, that's who. All we have is the completely premature and out of bounds judgment by the hand of OTEA that the tank isn't ready for an ASARC decision. Is that his job? Has he that authority? Or is that the purpose of the ASARC, based on the best test advice of OTEA, and on the advice of the user, MACOM commanders and others? You see, the testing czar concept simply allows for no reasonable checks and balances on the test agency, and some are needed—especially for major systems. Therefore, if one elects to have a super-test agency which answers only to the Chief of Staff, then one must also create an independent evaluator of some sort, somewhere, who in some mode performs that unpleasant business I described earlier.

6. You will conclude that I've just come down in favor of the very thing I spoke out against in the beginning. Not true. An independent test evaluation is a necessary evil, but the independent evaluator must be made a party to the crime from the beginning, and must follow the test design and evaluation procedures from the time the paper is blank until the computer has stopped running. In that way you avoid the dismal situation we encountered in WSA days. My concern is that OTEA overrides everyone in the pursuit of proper test design—never mind what it is we're trying to test, never mind the tactics aren't right. And that's a situation I just don't believe we should put up with. Please don't misunderstand me—I'm not trying to pick a fight with OTEA, nor am I criticizing the people involved. They have a mission carved out for them and, as we all do, are trying their best to fulfill their charter. It is the system with which I take exception. I have a solution—but that can be the subject of another conversation if you care to pursue this any further.

Computer Systems Needs

Message to Major General Clay Buckingham
Fort Belvoir, Virginia
29 January 1979

. . .

2. It is difficult to address your request that we assess the support effectiveness of standard [computer] systems without some in-depth analysis. However, pending results of an in-depth study by my staff, the following general comments seem appropriate. The concept of standard systems was a sound one and, in its time, was the way to go to get the Army ADP program underway, but events have overtaken us. In the 1960s, the concept of standard systems was a logical approach to the Army's ADP needs. This was a time of large-scale, high-cost, batch-processing computers. For these large computers we wrote large, unwieldy standard systems which addressed primarily the bookkeeping aspect of the Army's requirements. MOBEX 76 and MOBEX 78 have proven conclusively that these large, bulky computers and associated systems cannot support us for mobilization. As you are pursuing "viable," I realize that you recognize that we are in a new age. Computers are smaller, cheaper, and less constrained by the environment. Database management systems make it possible to have selected data available to senior managers via interactive terminals on an almost instant basis. We are pursuing such a concept at HQ TRADOC in a system called STARS. STARS will be designed to afford instant access by the user to a large interactive database. We feel, in view of today's technology, that this is a management tool we need to manage. With the requirement placed upon us by mobilization and funding restraints, we must have such access to data to make the necessary decisions in these critical times.

3. Some of my immediate concerns are that disk operating system-enhancement (DOS-E) has not provided the saturation relief promised and, in some cases, has actually increased run times. DOS-E still has not been completely fielded and problems in installation exist at many of our posts. As stated earlier, field validation test (FVT) quality and scheduling have improved markedly; however, we must work together to ensure that incompletely tested systems changes are not forced into the field by overzealous functional groups. Our functional areas are inundated with output, making management by exception a dream rather than a reality. We are far from the present state of the art, that is a capability for key decisionmakers to have the information at their fingertips necessary to make logical and instant decisions necessary in times of crisis.

4. The bottom line is, for the kind of effective support we need, let's get with technology of the times. The days of the large number cruncher computer for management information systems are past, and small, interactive systems in a decentralized mode are here. This is the road to the kind of support we at TRADOC need.

5. I welcome this opportunity to share with you some of the immediate needs of TRADOC. If you require more detailed information, my staff will be happy to provide any information available.

Telescoping Materiel Acquisition

Message to Multiple Addressees
30 January 1979

1. A major Army initiative in this year's POM will be introduction of the concept of telescoping the overall materiel acquisition process. While little can be done to shorten the acquisition cycle of an individual system, we can reduce the time required to field successive generations of systems. The need for a new approach became evident in the evolution of the Battlefield Development Plan. It pointed out that the extended length of the current heel-to-toe materiel acquisition process, when coupled with rapidly expanding technology, was severely limiting our ability to modernize the force in the face of an expanding threat.

2. The concept of telescoping involves the idea of either product improvement ongoing developments or initiating a milestone zero cycle for a second generation system while the first system is being developed. It is an approach similar to that practiced in the business world—marketing "good" products while continuing to develop "better" follow-on models. In our case it will involve a conscious decision to begin product improvement on equipment still in engineering development, or alternatively to initiate a follow-on, or second generation, system, including a MENS if necessary.

3. DCSCD will work out the details for developing the list of candidate systems which we will then coordinate with DARCOM and forward to DA. Since this concept is a departure from established policy, a concentrated effort will have to be made to change the current attitudes in OSD, OMB, and Congress prior to including such programs in the POM.

4. I cannot overemphasize the necessity for developing sound telescoping candidates that are firmly justified by operational needs. These programs will be the forerunners of a new, quicker, and, I believe, better way of providing updated weapons capability to soldiers in the field.

Improved TOW Vehicles

Message to Colonel Robert Sunell
11th Armored Cavalry Regiment
6 March 1979

1. I've been told you're to get 18 ITVs [Improved TOW Vehicles] in September for evaluation. Without trying to bias that evaluation, may I offer some suggestions.

2. It is important that your MOEs are well and properly chosen at the outset. Unless you spell out pretty clearly what you're trying to find out, you'll quite likely fall afoul of a host of very emotional arguments about everything from scouting to maintenance.

3. I'd suggest first that you review the 8th Mech evaluation of the new cavalry organization. In it I believe you'll find words to the effect that the TOWs tend to detract scouts from their first and most important business—scouting. In addition there was a finding that the ATGM is not exactly a quick-firing system, and so not really in tune with the cavalry mode of operation. Both are valid observations—you'll recognize them from way back. They were true then and remain so today. So don't try to reinvent that wheel.

4. It's also true that the ITV was to be part of a pair which included cannon vehicles and ITVs. We'll not get the cannon until we get IFV, so the question you've got to sort out is of what use is the ITV if you don't have its companion vehicle, the cannon-mounting scout. Related is the question of whether or not one should keep the TOW vehicles just so that the scouts will have some firepower better than the Mark 2 50-caliber. And that's a hard one. If you don't give the scouts some combat capability, they will have to hang close to the tanks or sneak and peek to do their job.

5. Clearly the TOW is an overwatch system. Therefore its utility is at the longer ranges. So what part of your target array is available to be engaged beyond tank gun high Ph ranges (1,500–2,000 meters)? That number, as I recall, is on the order of 15 percent. The threat in that array can be recon or advance guard against a movement to contact, or first echelon regiments if they are coming in breakthrough formation. The utility to you of opening fire and hitting at those ranges is far greater than the 15 percent . . . probably about 25 on a scale of 100 in some circumstances. It could be especially important for the cavalry to be able to reach out beyond gun range. Added to that is the utility of having some long-range armament in the scouts so they don't have to always seek shelter of the tanks or sneak and peek.

6. If one considers that the ITV and the M2-mounting M113 are but interim systems pending the cavalry fighting vehicle, then the question becomes one of how best to use the ITV as part of the scout team and with the tanks. It seems to me that's what you ought to concentrate on.

7. So I'd suggest your MOE concentrate on finding out if the thing works—can it be manned, fired, and maintained by a scout crew of the size and with the skills we've got in there, and can we devise tactics that are satisfactory? I'm confident those questions can be answered—one way or t'other. But don't let the thing get bogged down in some grand eval scheme. We've done all that. If you want help from me, from Dick Coffman, or anyone else in old TRADOC, just let me know.

Weapons System Development

Letter to General John R. Guthrie
US Army Materiel Development and Readiness Command
22 March 1979

I appreciated your response to my . . . letter addressing our mutual concerns on total systems development. I must admit I share your views on the impact of the proliferation of new weapons systems in a resource-poorer future. We must work together in the Army to pull this together through our various efforts such as the TRADOC Priorities and the constrained programming efforts now ongoing at DA. However, within this larger problem of affordability still lies the necessity to field those systems we do select with whatever is necessary to ensure the systems are used to their full design capability. Specifically, I am concerned how our two commands should jointly approach weapons systems development to ensure success of the total system.

The training development interface between DARCOM and TRADOC is not working as smoothly as it should. There have been marked improvements in the recent past, but I think we can move ahead more rapidly in this area if our two headquarters can agree jointly on the appropriate approach. We have fielded—and are developing—some of the best equipment in the world. The problem is that we are not able, in the field, to employ this equipment up to its design capability.

Training can become a force multiplier, but only if our soldiers can use equipment at or near its design capability. Moreover the possibility exists that, if we are able to do the training job better, we may realize savings that can be applied to your major concerns.

I suggest that we must look carefully at the areas of training devices and research and development for training. Training devices—both systems and non-systems, including such items as targets, training ammunition, and battle simulations—are presently handled in the same procedural fashion as the development of major systems. Does this add unnecessary delays to the fielding of such devices? Does it permit us to exploit the best technology that industry has to offer?

Training research and development—communicative technology is but one example—appears to have insufficient emphasis and focus. There are laboratories and commands looking at all aspects of hardware, but it is difficult to pinpoint who is looking systematically at the future for improved training. If my view is correct that training is the real lever for substantial improvements in force readiness, we may be able to make a sizable contribution by addressing this problem.

Perhaps, by DARCOM and TRADOC working together on the training interface of our respective missions, we can get more out of the limited resources that the Army has. With that in mind, I accept your suggestion that in-house capability may be the way to go.

Acquisition Cycle Compression

Message to Multiple Addressees
23 April 1979

1. One of the major conclusions of the Battlefield Development Plan I was that, if we are to pursue force modernization, the acquisition cycle must be compressed. Funding constraints do not permit the compression for all systems. As a result of this funding constraint, efforts have been undertaken to identify and pursue the accelerated acquisition of a few systems with high tactical leverage. One of these efforts is a program which has been termed "Perry Initiatives."

2. In August of last year, Dr. Perry (USDRE) offered his assistance in accelerating the acquisition of a few systems with high tactical leverage. On 11 April 1979 three candidates were briefed to Dr. Perry. The first system is an Anti-Standoff Jammer (ASOJ); it is a surface-to-air missile to destroy or push back standoff jammers. The second system is a multipurpose lightweight missile system (MLMS) which is a heliborne system composed of two types of missiles, one to suppress enemy air defense close to the FEBA and the other to provide self-protection from threat helicopters. The last system, PLRS/JTIDS hybrid-DACC, is a communications system and concept to solve the data distribution problem and to integrate all communications systems on the battlefield.

3. At the conclusion of the briefing, Dr. Perry directed that three work groups be formed with participation from OSD, DCSRDA, TRADOC, DARCOM and others as required. The work groups are charged with developing a complete expedited acquisition strategy and to provide recommendations to Dr. Perry early in May. The acquisition strategy will include a plan for going to Congress, a schedule with a funding profile, and the addressing of multiservice and NATO issues. Actual NATO involvement in the programs will be addressed at a later date.

4. This is our opportunity to see if in fact we can shorten the acquisition cycle. If we are successful, it will lay the groundwork for the acquisition of future systems.

5. You will be periodically informed on the progress of the initiatives.

Top Attack Systems

Message to Dr. Walter LaBerge
Under Secretary of the Army
10 July 1979

. . .

2. For some time now I've followed the growing technical world interest in top attack systems. So, as the evil old user with a grim view of technocrats and a realistic view of what we are spending our meager funds for, may I presume to sound an alarm.

3. I understand top attack is a feasible concept technically. I am also quite certain that, given the way things work, the Sovs can fix their top armor against top attack and do it in less time than it will take us to develop systems for top attack. If you're talking about it, rest assured they know about it. If they believe you're spending any money on it, rest assured they're at work on countermeasures. If they are at work on countermeasures, rest assured that in five or six years they will have fixed whatever you're trying to attack so that your investment is obsolete before it's fielded. That's not necessarily to be critical of anyone. It is, however, just the nature of the way things work in our respective systems.

4. May I suggest, therefore, that we concentrate on an equally important opportunity, and on a problem for which we ourselves do not now have adequate countermeasures—bottom attack. You've heard me say I believe the randomly scattered mine is the biggest challenge on the battlefield. I have been saying that since 1969, but haven't been persuasive enough to get anyone to invest in systems that reflect what I believe to be the operational concepts we must embrace if we're to cope with this problem. On the offense we have the means for bottom attack—FASCAM. What we should do is improve that capability so that we can deliver them from a variety of artillery systems—tube and rocket, from helicopters, and from a variety of offensive support aircraft—all NATO attack aircraft. We should combine them with DPICM in some effective mix that destroys some targets by other than mine techniques, but which depends on the bottom attack concept for its effectiveness. I tell you that is the most insidious threat on the battlefield, and despite their extensive mine clearing equipment the Soviets are but little better prepared to handle it than are we. As we defend, the Soviets' growing capability to deliver mines against us in the same fashion tells me that we must shore up our countermine capability along the lines you and I have talked of.

5. My concern, and the reason for this message, is that we appear to be starting off in a new direction—top attack, spending a fortune to develop the technology and systems, when in fact we have bottom attack systems in hand or in development. All we really need is to finish up some developments and provide ourselves with a sufficient inventory of weapons and some more delivery systems, and we have a potent bottom attack capability for something less than the Shah's ransom it will cost us to proceed along the primrose path to top attack systems. The fact that detection of scatter mines is a terribly difficult problem technically should not deter us from trying to find a solution. I've written Bob Fossum asking him to take this one on with our help; so far I've not had a response. The way I read the FYDP and the POM, if we start a top attack program it will be at the expense of a bottom attack program; we'll wind up spending a lot of money and not having enough of either to do us any good in battle. If I recall correctly, the main purpose of this outfit is to fight battles—but then I could be mistaken.

Command-Control and Communications

Message to Dr. Gerald P. Dinneen
Assistant Secretary of Defense for C3I
25 January 1980

1. For the past two years TRADOC has studied intensively the Army's organization, command-control and communications for combat. We have concentrated on concepts to give a clearer roadmap for balancing the realities of threat, economy, and public mood. We have recognized that mobility, speed, and precision in applying combat power are essential, not just a bonus effect as implied in the cliché "force multiplier."

2. Such recognition led us to examine the most expeditious route to an integrated communications, navigation, and identification system to provide the position reporting and identification required for command-control and the rapid data transmission paths required for our sophisticated weaponry. Hence the PLRS/JTIDS test bed.

3. We looked further in review of our entire integrated tactical communications system (INTACS) and have found a continued need for high frequency radio, improved in transmission characteristics and frequency selection, to complement tactical satellite and cover those intermediate ranges too great for FM radio. While the mobile subscriber equipment remains a desired goal, it appears it will be some time before engineering complexities and ECCM can be resolved to the satisfaction of most of us. Hence we have a continuing need for multichannel terrestrial relays in the brigade area with a heavy reliance on net radio interface. Tactical facsimile appears to me to be the biggest boon to record coordination between units (particularly of different nationalities) while avoiding what I call "the tyranny of message center."

4. Recently we gathered a cast of thousands to view the TRI-TAC equipment at Huachuca and review the program. I'm convinced the central logic of the TRI-TAC architecture is the only approach to integrating the many devices and opportunities that technology has made possible for us to achieve that mobility, speed, and precision of which I spoke earlier. However, I am concerned about the size and weight of some of the configurations and their interconnecting cables, our apparent inability to field switches and network control simultaneously, and our abysmal acquisition strategy which guarantees obsolescence, perpetual transition, and token buys at high cost.

5. The Army desperately needs modern equipment now, and it appears that we can improve our situation only by continuing to buy what's available now until a comparable TRI-TAC item is in production. Noting the sad state of our active forces and reserves, I doubt that we run much risk of duplication. For instance, I understand that, after 20 years of purchasing the AN/VRC-12 net radio, we are at only 54 percent of the Army's acquisition objective. Meanwhile we continue fruitless arguments, based on gut feelings, as to which piece of unfielded equipment or whose technical design plan is better.

6. I would like to discuss these matters with you soon to see if we can agree on a program that will enable us to proceed with solving the field Army's problems instead of offering solace and promises. If you agree, we can ask a few key people to join us and our people can set up the meeting.

Army of the Future

US Army Materiel Development and Readiness Command Executive Seminar
Atlanta, Georgia
14 February 1980

It appears that, if you had some idea as to where our Army is headed and the requirements associated with that direction, you might be able to get in front of the power curve and help us get there. So for the next few minutes, let me lay out for you what kind of an Army we're designing in the next five or so years.

You, in DARCOM and in industry, generally deal with the physical equipment side of the house. In TRADOC we must deal with the ideas, the concepts, that eventually result in equipment requirements. How well we do our work has a direct bearing on how well you can do yours. We visualize the battlefield—today, tomorrow, and in the longer term—and then attempt to portray how our Army will fight on that battlefield. We try to do that in a realistic atmosphere, but generally without the constraints of present technology.

Once we've decided on how we'll fight, we attempt to fit in the constraints of manpower, budgets, and present equipment. It's an iterative process, obviously. For instance, if we can't have the overwhelming manpower strength we need to defeat every enemy, then we go back, change how we'll fight, how we're organized, and what we'll use, and then we try again. To replace manpower we might require more mobility, more firepower, or different tactics as a substitute. It's a constant process of balancing the realistic with the ideal. From this process, eventually we come to equipment requirements—better guns or an improved communications system. Sometimes we find that more sophistication is not the answer, but that better training on what we have is.

Now I tell you this so you'll understand that we're not just dreaming up requirements to find ways to spend money or keep you all busy and frustrated. It may seem like that at times, but I assure you that it's not the case. Nor are we looking for scapegoats on whom we can lay the blame for the loss of a battle or a war. The stakes are too high for that, and we're all in this together. There's no room or time for pointing fingers.

This process I've described is not formally laid out, but it's our everyday work. But we have some large-scale projects that illustrate how we go about it. The first one is a document called the Battlefield Development Plan (BDP), first published in November 1978. It was designed to focus, prioritize, and integrate TRADOC efforts in training and materiel developments, force structure, concepts, and doctrine. Some have called it a roadmap to the future, but it's really a very candid assessment of where we are today and where we're going. The BDP does state priorities for what we need, and the results are realistic and eye opening. We are now completing BDP II and starting on BDP III.

I must admit that the BDP is classified, so I cannot go into much detail on it. From the BDP, however, a series of studies have started that are designed to reorganize the Army for the future. What happened was that we realized we were about to introduce some 40-plus major systems into the Army with no overall plan to maximize their potential. The studies are now gathered under the title "Army 86" and include everything from Army divisions, light and heavy, through corps to the echelons above corps. All these studies and the new organizations that will result will require equipment development and production in both the near term and midterm.

They are not pie-in-the-sky–Buck Rogers requirements; we focused on 1986 to prevent that approach. They are hard, realistic and include a transition plan to get from today to 1986. Most of the organizations are based on the equipment that is in development or coming on board. There is, however, always room for something better, but frankly, to get it, we've got to identify tradeoffs. The bow wave of procurement spending is already too large.

There is an exception, however, that has some interesting potential—the light division. We are looking for equipment in that organization that will make it tactically mobile with a high intensity of firepower that allows it to be less manpower intensive. The equipment needed for this division must be capable of being *produced*—not just developed, but *produced*—in the next five years. Obviously all the equipment must be strategically mobile, and that means more than one or two vehicles per plane. That's the requirement.

The other Army 86 studies—Corps 86 and Echelons Above Corps—are still in their infancy, so I can't say specifically what will be needed. But you can judge that we are looking hard at command and control systems and a much different approach to combat service support.

One of the concepts we embraced in the Army 86 effort was redundancy, robustness, and resiliency—R3. The words are awkward, but they mean that we must build, in our organizations and our equipment, alternate and supplementary ways of doing things. It means the ability to take losses and still achieve our mission. It means that centralization, although cost-effective, may not be effective on the battlefield. One computer controlling all of a theater army's supply system may be effective, but it's also very risky without an alternative, maybe even a manual system, as a backup.

There is another aspect of TRADOC work that impacts on requirements. We are now beginning to rewrite our Army doctrine—how we fight—to reflect that the next battlefield will be an integrated one—with nuclear and chemical weapons. We believe that this may occur early in the battle, just as our potential enemies have been writing, saying, and training for the past 20 years. They have the capability; they train with it; they deploy it even into fellow Communist countries. We must believe their words and actions. Now rewriting the doctrine won't be easy, for somewhere in those same 20 years past our Army turned away from the integrated battlefield. In the 1950s we had doctrine for and trained on an integrated battlefield. But when Vietnam came along and our focus on contingency operations became all-consuming, we enjoyed a great strategic and tactical nuclear advantage, so our doctrine talked less and less about an integrated battlefield. Our units, facing the reality of a war of insurgency, trained for that war and not some future event.

Vietnam is over, and I don't have to make the obvious point that the world conditions have changed. We no longer enjoy any advantages, conventional or integrated. Facing that, we must train for an integrated battlefield that may become a reality. The Catch-22 is that, if we don't plan and train for that integrated battlefield, we make it more likely to occur. Our potential enemy forces—equipped and trained as they are—are watching, and our only protection is in being prepared.

What that means to you is that this new doctrinal approach will result in more requirements in both new equipment and improvements to what we already have. For instance, we must look now at improving the NBC protection of our vehicles, developing and producing better individual decontamination kits, and resurrecting our almost moribund 1950 vehicle decontamination programs with 1980 technology and longer range and more versatile delivery means. The list goes on and on. If we really need some ideas, I suggest we all go back and read our manuals and professional magazines of the 1950s—*Infantry, Armor, Artillery, Army R&D*. Many of the ideas on those pages died at birth, overtaken by the events of the 1960s.

Our new doctrine will also include a strong emphasis on deepening the battlefield by finding and attacking the follow-on enemy echelons. If you look at the enemy battle array, you will note that the enemy depends on an echelonment of forces to achieve mass, momentum, and continuous combat. Much of our effort in the past has focused on destroying the enemy first echelons—on target servicing, to use the analyst's term. We've done a lot in this area, but the simple fact is that, if we let the enemy echelons move into the battle unhindered, then they will overwhelm the first-echelon target-servicing capability.

The problem is twofold. We must see deep and attack deep. Today we can't do either very well from the ground. Our doctrine to fight the battle is that brigades fight first-echelon regiments and must see out to 12 hours, while divisions fight first-echelon divisions but must see out to 24 hours. Corps fight the second-echelon divisions and must see out to the second-echelon army. So the requirement is that we need some systems, both air and ground, that will do these jobs. And those systems must fit with the battlefield systems we have today.

There's another aspect to the fight in the follow-on echelons that generates requirements. We've looked at the composition of the enemy array and found that it isn't just a simple case of finding and killing tanks. When you examine the array, you find that the follow-on echelons are much heavier in combat support and combat service support elements than tanks. There are quantities of targets—command posts, radar installations, logistics vehicles, ammunition, and fuel depots—that are lucrative and hard to replace.

Now two competing theories have evolved. One says kill tanks, infantry fighting vehicles, and artillery in the deep echelons and they won't be around to fight later. Good, we need all the help we can get. But those combat targets—tanks, etc.—are hard to find and hard to kill. Only one or two losses from those elements won't make much of a dent. They're easily replaced.

The other theory says destroy the command posts, communications means, and logistical sinews and you really disrupt the enemy. It makes sense when you consider that one or two fuel tankers destroyed will immobilize 10 or more tanks. One or two command or communications centers may immobilize a whole brigade—90 tanks. That's the requirement we must achieve now. We need intelligence and weapons systems that will find these "soft" targets and put them away. Strip the combat services support from the tanks and the tanks will stop. It happened to us in World War II in Third Army. It happened to the Japanese in the Pacific in that same war.

I've talked about organizational changes and doctrinal changes. Let me just touch briefly on an equally important area that affects our Army requirements—training. Later in this seminar, General Bob Shoemaker from Forces Command will discuss the environment and problems of training in today's Army, so I don't want to steal his thunder. But this environment and the problems impact on the requirements for equipment.

We don't have many college graduates in the Army as enlisted men, so it is no good if you engineers develop a system that works perfectly for them in a laboratory-like environment. It's got to work for our troops in a less-than-ideal environment, where it's wet and cold and one of the crew is on detail and another just reported in and a third one never saw this type of equipment before, an environment where the system's manual has pages missing and the stress level is very high because the risk is great and the only law is Murphy's.

Soldiers aren't dumb; they're our own sons and daughters, but they have a great many things to do and time is always short. The average infantryman has more than 140 tasks he must do to perform his job, and sophisticated equipment that doesn't have a very high payoff only adds to his burden. He takes shortcuts and does those things well that he feels really enhance his fighting ability. You can't fool him with nice-to-have eyewash, and he's quick to spot things that are designed wrong. His only question is, "When can you correct it?" So we can't build into our equipment sophisticated problems. He doesn't have time for them, and I don't blame him. There are examples of this—the M551 Sheridan assault vehicle, sometimes called a tank; our antiquated mine detection equipment; the old coaxial tank machinegun; even our basic M16 rifle when it first appeared.

One more aspect of sophistication and complexity is important to keep in mind. When you're in the lab trying to develop that last little bit of sophistication, stop and think. Do we really need it? There are studies—one on tank fire control, I remember—that indicate the last 10 percent of complexity usually has a very low percentage payoff. This is where the maxim "the better is the enemy of the good enough" often comes into play. Programs are delayed for years just to achieve that last 10 percent.

I've talked about doctrine, organizations, training, and equipment prospects for our Army. John Blanchard asked me to be as specific as possible on requirements, so let me give you a partial list of our chief concerns over a broad spectrum, something all of you can take back and on which you can turn your smart guys loose.

In command and control, C2, we need today a system that will tie together—jury-rig, if you prefer—all our command and control systems. We had a system called the Tactical Operations System, TOS, that was going to do this, but it's been killed, overtaken by time and events. I don't want to be the father to the "son of TOS," but we need something and we need it now, not 10 years away, something that works with reasonable use of the state of the art and no need to wait for a technological breakthrough.

In regard to our fight with the follow-on echelons, we need the ability to see deep, one that discriminates targets. We need to search out the enemy command, control, and communications and the other soft targets I mentioned earlier. This capability has to be in real time, and it must be accurate enough for targeting purposes. We don't need more 8x10, 3-day-old glossies—"Army happy pix," the Air Force calls them. This integrated system could be sensors, radar, TV, RPVs or a combination, but it must be redundant so all our eggs aren't in one basket.

Along with C2 and see deep, we expect to have an awesome amount of information flowing back to the commander, so much that he is in danger of going into overload. General MacArthur, among others, suffered from this in Manila at the time of Pearl Harbor. It was costly. We can't afford it on the next battlefield. We need a system or an arrangement that will get the critical information needs to the commander, but only the critical ones. We need some sort of preprogrammed discriminating system tied into all our information-gathering devices, both friendly and enemy. Our studies show us that this information engineering, the sorting of information, is a great effectiveness multiplier. TRADOC has a group involved in this, and DARCOM and industry should too. The requirement is there.

In the forgotten and neglected area of barrier warfare, we must have lots of everything. We need multiple technology detectors that combine metal, nonmetal, influence, infrared, and whatever signatures, and do it at high speed. They must also do it over a wide area, in real time, and with a reasonable degree of accuracy. The technology must be simple and easy to operate and maintain and rugged enough to survive.

Our means of elimination of barriers must also be part of a family of alternatives. Plows, rollers, flails, explosive devices, and so on must be developed and fielded. The same technology requirements are necessary here. Above all, they must be easy and simple to operate and maintain. Complexity has been the problem with most of our efforts in the past.

I have mentioned the requirements for the integrated battlefield already, so I won't repeat them. They are urgent and compelling. The shame is that, at one time, we led the world in this. I daresay we may have forgotten more than many have yet learned. We must relearn it quickly.

Our tank development program illustrates another problem. Our XM1 being fielded this month is super; we need it badly. But now we've got to look at it and start reducing the weight. We've been striving for protection and firepower so hard we've come up to around 60 tons combat loaded. We can't go on or we'll have a 200-ton invulnerable tank that won't move. We used to call that a castle. So the requirement now is to reduce the weight.

This list could go on and on, but I'm going to stop now so you can ask me questions if you have any. Let me summarize by saying there is an Army master plan. In TRADOC, we call it the BDP, and there are many studies and requirements that flow from it. Anyone with a good idea is welcome. There is no prize for solving any of these problems—unless you can place a value on liberty and freedom. Whether you can or can't, they are what are at stake here. We cannot afford to delay any longer.

Modernization

AUSA Spring Banquet
Fort Sill, Oklahoma
10 April 1980

As most of the country is aware, we are in the throes of the budget approval process. This process has generated, and rightly so, a national debate on modernization of all the services. It is very popular right now to conjure up lists of the various weapons systems needed by our military establishment. Anyone with a little knowledge can quote extensive figures on airplanes, aircraft carriers, ICBMs, nuclear warheads, warships, and so on—both ours and those of our potential enemies. Much of this equipment counting is glamorous and attention-getting. Most of it has strategic implications and applies, it seems, to our sister services.

When you stop to think there hasn't been much debate or even listings of what the Army is doing, it makes you wonder. Maybe we're not getting our story told, or maybe the Army is so complicated no one understands, or maybe no one cares. Yet I can't accept the last hypothesis—that no one cares. Over the past 20 years, more men and women have served in the US Army than any other service. More families in our nation have had sons and daughters in Army green than in the uniform of any other service. I don't believe there is a rational US citizen who doesn't agree that a strong Army is the key to our survival as a nation.

So, deep down, we all care, we all want a strong Army. Well, maybe the Army is so big and complicated that it's hard to understand. But every day you can hear and read dissertations by self-appointed field marshals on what the Army should do about a myriad of problems. Everybody has a solution, and that's because everybody feels they can relate to the Army. It's about people, and we all are experts on that—"we are one."

No, it's not that no one cares or no one feels overawed by the Army. It's so simple we all can relate to it. Well, if that's so, if citizens care and are capable of understanding, then the problem must be elsewhere. I think it is; I think the Army has failed to explain its story to our citizens so they'd be knowledgeable enough to discuss the kind of Army we need. That's, of course, where the AUSA comes in. It's your job and mine to get the facts on the Army spread around to our fellow citizens.

Let me share some thoughts with you on what the Army is trying to do and where we have some problems. Recently you had the Chief of Staff, General "Shy" Meyer, out here. I'm sure somewhere during that visit he mentioned he had two goals—modernizing the force and manning the force. Now if he has those two goals, you can be sure that those are the goals for everyone in the Army. But, everything being equal, modernizing the force—talking about equipment—is much easier than discussing manning the force. So everyone concentrates on the former and slides over the latter.

But it's not that easy. The two problems are really interrelated and you can't concentrate on one at the expense of the other. There is no doubt we have a modernization problem. No nation can spend its national treasure as we did in Vietnam at the expense of the future without having problems. In the case of modernization, the future has arrived. For almost 10 years, 1964–1974, we did little reinvesting in new equipment. Our focus was on the battlefields in the far-off Indochina Peninsula.

Not so our potential enemy. Like a squirrel in fall—before a heavy winter—he was busy inventing, building, and modernizing his armed forces—particularly his army. When we painfully wrenched our national attention away from that forlorn conflict, we found that we had lost the lead in modernization and technology that allowed us a margin of safety. In terms of quantity, we were and are behind; in terms of quality in the hands of troops, we are no longer ahead.

Collectively the Army set to work to right the equation, but in 1974 to 1976 there were no large sums available to achieve the goal. And, I might add, there was no national interest in doing so. It seemed more popular to look back than ahead, so we stalled and publicly whipped ourselves for alleged sins of commission. All the time we, as a nation, committed the dangerous sin of omission—we failed to remain prepared for the future. Everyone agreed on no more Vietnams, and that quickly became no more wars.

No one, particularly soldiers, wants wars. But, human history being what it is, they are a reality. Struggles continue. The race doesn't go to the fastest or to the one best prepared but, to paraphrase Damon Runyon, that's the way to bet. That's the way our potential enemies are betting. They don't seem to share our view of the future. They have built and produced much more than is needed merely for defense.

We won't review what woke us all up—one event or a seeming kaleidoscope of crises upon crises—but it appears we are awake, finally, as a nation. And all the programs for new equipment that were planned and developed since 1974 are beginning to come into fruition. The Army is in the throes of it. It is perhaps the most dramatic reequipping since the early years of the 1940s.

Tanks, guns, computers, lasers, radars—some 40 major systems alone—are available to come into our Army. We need them all, and a lot of them. Many of them seem to be small unglamorous items—like a new boot or helmet—when compared to the aircraft carriers, fighter planes, ICBMs, and warheads that grab the headlines. They aren't even political items, so they have a tough time getting support. We need help with that support; it's that simple. We're not in competition with the other services, but we're woefully behind them all in modernization. We support any and all of their efforts because we realize that ultimately they will help the Army. We also need help because the threat we face is formidable. But there is a deeper problem involved in modernizing the force.

Modernization is not solely the buying of new and more technically advanced equipment to be issued to the force. Rather, modernization begins with doctrine—concepts about how the Army should conduct its business. These concepts lead in turn to new tactical schemes for battle fighting, new equipment developments, new organizations in which equipment and tactics are combined, and revised training—the cement that bonds tactics, organizations, equipment, and soldiers into successful battle teams. If that training is successful, readiness is the result. If the whole system succeeds, preparedness is the outcome and deterrence is the perceived product.

The main ingredient of modernizing the force is indeed training, and training is what manning the force is all about. Now there's a very heated argument being bandied about about the quality of the soldiers we have in the Army today. I'm not going to get into that. No one seems to be able to prove much on one side or the other. I will tell you that today's soldiers are trainable. We need more time to do that training and more trainers to accomplish it, but the soldier can be trained.

There are three facts you need to keep in mind when people start talking quality. One has to do with today's weapons, one with the training environment, and one with quantity. Now today's weapons systems are variously described as complicated, complex, modular, state of the art, space age, and a whole host of awesome tag lines. To the engineer who designs a piece of equipment, it is complex. All those wires have to go to the correct places, the laws of physics and chemistry have to be obeyed, and so on.

We need an engineer for all that. But the soldiers couldn't care less about it. They merely need to know how to aim and fire it or read information from it. Our new XM1 tank is an example. It has a laser rangefinder—very complex to the engineer. The soldier who puts crosshairs on a target, punches a button, and reads the range doesn't see the complexity, doesn't want to know about it. To the tankers, it either works or doesn't. Complex to them, no, merely sophisticated. We don't teach them the laws of laser light to use the laser. So when you talk about complexity, you've got to make a distinction between it and sophistication—between the engineers' views and the soldiers'.

The next fact deals with the training environment today. That environment is most easily described as being hostile. It's an environment where sound training gives way to many so-called higher priority programs. Training is something that is done *after* the priority tasks have been accomplished. Time is the most precious resource. Trainability depends on time or the lack thereof. We pay lip service to training time and then wonder why the soldiers don't learn.

The hostile training environment has another ingredient—turbulence. That's where soldiers move around so much—from unit to unit or job to job or position to position—that they don't get time to learn individual skills, let alone to act as members of a team. Half that turbulence is created by headquarters or actions outside a unit; the other half is created by the unit commander trying to cross-level what's left in a desperate attempt to get the job done. The result is confusion, continued changes, and insecurity. No one, high or low quality, can learn anything under those circumstances.

This turbulence is partly a creation of the third fact I talked about—quantity. We just don't have enough soldiers to go around, and we're trying to do everything with what we have. We aren't assigning soldiers; we're parceling out shortages.

The most pressing issue facing the Army has to do with numbers. It is no secret that the available pool of 17- to 21-year-old males will decrease by more than 15 percent over the next 10 years. Considering physical and mental qualifications, prior service, and educational and military commitments, only one out of four of today's young men between 17 to 21 years of age is qualified and eligible for active military duty. The Army must enlist 1 out of every 16 of these young men. Last year the Army was unable to recruit sufficient numbers. In succeeding years this situation will likely worsen. You can see how a lack of numbers can contribute to turbulence and a hostile training environment.

Now all of this gets us back to the central issue. How can we, and especially you as concerned citizens, help tell the Army story? How can we, by telling this story, attract young people who want to serve? First of all we've got to be truthful—the Army is service and it's hard and it's unique. It doesn't translate to future jobs easily, but it does translate to future security. It's not glamorous and we can't treat it as another marketing job, selling a product. We're selling only one product—the liberty and freedom of our country.

Second, we've got to give time and resources to the training—meaningful training—of our soldiers. Soldiers aren't dumb, they're our own sons and daughters, but they have a great many things to learn and do and time is very short. The average infantry soldier has more than 140 tasks he must do to perform his job. He takes shortcuts and does those things well that he feels really enhance his fighting ability. So we've got to keep equipment sophisticated but simple. Things that don't have a high combat payoff only add to his learning burden.

Third, we've got to remember that soldiers have—indeed must have—some simple but different values than the society as a whole. That's natural. Their orientation is toward a battlefield where they may have to make the ultimate sacrifice—their lives. Soldiers on a battlefield have a way of quickly weeding out all that's worthless and unimportant and getting down to basics. On the battlefield, there are only four important values—candor, commitment, courage, and competence.

Candor means openness, plus honesty, plus simplicity. The stakes are too high, and the time is too short, to fool around with hidden meanings, little white lies, and status games. The candor of the battlefield is why lies told there are punished, not with gossip, but with action. The battlefield is the most honest place in the world.

Commitment starts with candor. It is a pledge to a person or an idea. Soldiers make few commitments because their world is small—first to their buddy, then their crew or squad and maybe to their small unit, their platoon. It's hard to see or commit beyond that. Commitment is the real message of the Congressional Medal of Honor.

Courage is important too, not because there is no fear, but because courage controls fear. Courage is the choosing to take a risk despite fear. Courage is contagious and spreads, just as fear can. Soldiers live and die by courage.

Competence is the oldest value on the battlefield. It is the underpinning for all the other values. You cannot fake competence. On the battlefield, the leaders and the led respect competence more than any other value—except courage.

Our task is to encourage the growth of these values, even before a young person joins the services. They are not so different from society as a whole, but they are most important. We will not win any conflict unless they are present.

Finally, we must admit that all the modernization we can do depends in the final analysis on the soldiers who must shoulder the burdens of modernization. Their environment is less than ideal—because it's wet and cold, and one of the crew is on detail and another just reported in, and a third is only half-trained. It's an environment where the system's manual has pages missing and the stress level is very high because the risk is great and the only law is Murphy's.

In that environment, only the well-trained, dedicated soldiers will survive. We must help them tell their story. The time is getting short.

User Influence on Laboratories

Message to Multiple Addressees
11 April 1980

1. In my visit to TARADCOM, I observed many programs which I believe we could stand to monitor a bit closer. For example, ACVT is paying for experiments with a jiggling gunner's seat that only examines conventional optics and video screen displays along with isometric and conventional gunner controls. We possibly should also be looking at fiber optics, helmet sight, and other areas to improve the man-machine interface in fire controls. In the engine arena, the major approach being used is to increase output from old engines. This approach fails to address RAM, reduced size, application of new fuels, improved fuel consumption, and what we want to power our next family of fighting vehicles.

2. It is important that we continue to influence where the labs are spending their dollars. Now that DARCOM has accepted expansion of the tank science and base development plan to include infantry and tactical support vehicles, we must press the assault. Your people must continue to participate actively through the action committees, influence the projects, then keep track of how the money is being spent. Your guys are also going to have to ferret out those pet projects that are inherently hidden in each lab, and either state our level of interest or our desire to see them killed.

3. Coordination between your staff and mine must be close and continuous. We cannot afford to have resources wasted on programs that are going nowhere.

Modernization Issues

Message to General E. C. Meyer
Army Chief of Staff
13 May 1980

1. We are in receipt of the POM alternatives for FY82–86. The choices available to you are not easy ones. As your officer-in-charge of determining the requirements to modernize the future Army, I strongly support your efforts to secure more TOA (Case C), but must express my grave concern over the alternatives (Cases A and B).

2. As you know, the first edition of our Battlefield Development Plan (BDP I) reported that the US was losing the race with the Soviets to modernize its ground forces and that the Soviets would be difficult to match and impossible to surpass with the current level (FY81–85) of funding. It concluded that the US must increase its pace of modernization and maximize its combat readiness. Our net assessment was later confirmed by the Army Science Board.

3. We are currently in the final stages of completing the second edition of the BDP. While it will not be available to you until after the POM is submitted, it will show that Soviet capabilities are growing. They continue to outspend us, outproduce us, and are introducing the latest technology into their inventory and into surrogate states in Eastern Europe and the Middle East. They have also demonstrated a willingness to accept greater risk in power projection. BDP II restates that we will seriously lag behind the Soviet modernized capability, even if we fund all of our current programs. Case C is essential merely to avoid widening the gap. When we lay BDP II alongside POM 82–86, we find that Cases A and B actually slow modernization, thereby increasing the disparity to be portrayed in BDP II.

4. In our Army, and with our allies, we have great acceptance for the conceptual idea that to win against Soviet-style forces we need to fight concurrent battles against assault echelons and follow-on echelons. Cases A and B effectively foreclose our ability to implement the second echelon concept. Moreover, Cases A and B preclude initiation of the BDP I recommendation to implement "telescoping" (preplanned product improvements) as a way to accelerate modernization.

5. If we do not get all the TOA proposed in Case C, or if we get no additional TOA proposed in Case C, or if we get no additional TOA, we still must modernize. Three major sources of funding appear available to the Army: kill some major systems; defer force structure actions such as additional POMCUS, 10 heavy battalions, and an additional brigade; and, finally, reduce end strength.

6. As regards systems, it is patently clear that we cannot defer a large number of systems one year as POM A and B would have us do. This kind of discontinuity in the program adds to program costs and makes the affordability problem worse next year. Thus, instead of deferring a large number of systems, we should kill some of them. While our prime candidate to kill is Roland, we clearly cannot concentrate on one mission area and kill Roland, DIVAD, Improved Chaparral, and air-to-air Stinger as POM A and B would do.

7. As regards force structure, it is clear from early work that there will be a bill to pay for Army 86. This will occur even if Case C is approved. Much needs to be done with our force structure to accommodate the emerging light and heavy divisions and the support structure that they require. We discussed several of the ways to pay this bill at your commander's conference

last October, such as incorporating the separate and special mission brigades into our division force structure. Lacking approval of Case C, certainly we should not proceed with the heavying up proposals in Cases A and B (10 heavy battalions and the brigade for the 24th). We should also consider diverting the money for some POMCUS into modernization. We might even be forced into eliminating end strength to afford modernization. This is a bitter pill—one that would be tough to take and would further complicate our force structure modernization.

8. In summary, POM Cases A and B are clearly unsatisfactory in terms of Army modernization. If Case C is not forthcoming from OSD, we must make some very difficult choices other than Cases A and B in order to provide for Army equipment and force structure modernization. This may require use of all three major sources of funding described above: killing systems, deferring force structure actions, and reducing end strength. As you and the Secretary stated in your FY81 posture statement: "The alternative to expediting force modernization is increasing obsolescence. That alternative is unacceptable."

Tank Cooperative Development

Message to General E. C. Meyer
Army Chief of Staff
4 August 1980

Whatever forum we decide on, it will be necessary in my judgment for us to have members of that forum sit with the AFV working party during its deliberations and vice versa. If we do not take that step, the risk of having this project go astray is compounded. In addition, it will be necessary for someone from the DCSRDA staff to be present in both groups. Lacking that, the iron majors in DCSRDA and DAMO-RQ will go to work to subvert whatever the rest of us decided that they've not blessed from the beginning.

Future Close Combat Vehicle Concepts

Message to Multiple Addressees
13 August 1980

. . .

2. After some study of the matter I believe I understand what has been done.

3. After further study of the matter I do not understand why it was done.

4. From the background of more experience than I need, I tell you that this course of action is exactly the wrong way to go about beginning development of anything.

5. In order for there to be a meaningful briefing for industry on the RFP, which there always must be, it is necessary for some military fellow to stand up and recite fairly precisely what it is we want the machine to do when it gets built. If that is done correctly, it is necessary for us to think through in some detail the concept for employment of the equipment. If that is done we have in fact created an operational concept. If that has been done, then there is no reason why we should not proceed from the basis of an operational concept we provide to industry at the outset. If we've done all that in the beginning, why do we ask industry to go back to the lower left-hand corner and begin all over again? Doing that simply drags out the development process, and it's already unconscionably long.

6. In order for there to be a meaningful evaluation of what industry produces in response to the RFP, it is necessary for some military fellows to sit down and score the industry proposals against their idea of what they wanted in the first place. If they have thought the matter out sufficiently in advance, then there's no reason why we can't shorten the development cycle by providing the operational concept with the RFP. If they have not thought it out in detail at the outset, then I fail to understand how they are able to score what industry submits.

7. Almost any logic train one uses to think this problem through leads to the same conclusion—we have to think out the operational concept at the beginning. On the face of it I would conclude that we have not done that; we have asked industry to do that; soon or late we will try to score responses to the RFP on some flimsy basis which we throw together at the last minute—when we remember that we should have figured out something in advance.

8. Whenever we let one of the DARCOM commands go out with an RFP for anything without some user controls on it in terms of an operational concept, it amounts to a license to steal. Developments undertaken on this basis almost always produce the wrong machine which usually does not do what the user had in mind in the first place. There are those who say this is all DARCOM's fault. I say it is all the user's fault—we simply dare not let those self-serving little work teams in DARCOM perpetuate their existence at our expense and, unless we get into it with both feet from the beginning, we are bound to be taken to the cleaners again. We have a boneyard full of equipment withdrawn from service and parked—M-114s, M-551s, and so on, all because the development process we're describing did not have a strong user voice from the beginning. The days when we can just invite DARCOM to make something for us, then hope it turns out well, are gone forever. Neither side will come out a winner, and the Army will lose again in the longer run. We cannot afford any more mistakes like that.

9. Someone will try to assure me that all is well. I know that's not true. My long and very, very sorry experience is that what you've done is just exactly the wrong way to go about development of any kind.

10. Having said all that, I would like for you to provide me as soon as convenient with the conceptual notions against which you intend to score the proposals that come back from industry with regard to this RFP.

Man-Machine Interface

Letter to General Walter T. Kerwin Jr.
Alexandria, Virginia
16 September 1980

Thanks for your man-machine interface paper. . . . No question, we've a problem. My considerable frustration is that we've not been able to harness in an organized way the great amounts of money we spend annually in human research. All too many times we conclude, as did the classic study on night operations with tanks done at Knox several years ago, that it is likely to be more difficult to operate with tanks at night since generally one can't see quite as well at night as in daylight. It only cost us $2 million plus to learn that from the researchers. I've started some work along these lines, but our lack of general and specific knowledge is nothing short of abysmal.

The only part of your report that I take exception with is the notion that TRADOC should not be the sole voice representing the user. I'm not guarding turf here—God knows I've enough problems without protesting the migration of some of them to someone else. The point I make is that we've worked mightily these past few years to ensure that we are the user voice. I don't go to Washington, or anywhere else, on any topic involving the Army as a whole without at least having touched base with my fellow major commanders. This is a self-assumed requirement. I have done it because, from my former vantage point in Ops and ACSFOR, I saw the divisiveness generated when the Army staff solicits the views of major commands direct. The disparateness of their responses reflects regional and other biases, and their isolation from the context of how whatever proposal being considered relates to the Army as a whole. Without an ACSFOR to do that for you, TRADOC has to do it—in my judgment. If it isn't done that way the iron majors on the Army staff will decide the issue their own way—usually on parameters unrelated or only marginally related to how the Army in the field views the problem.

It may be fair to assert that TRADOC has not been doing enough in this specific area. To that I plead guilt. One of my problems has been that, in cases wherein some research or evaluation is required, we have experienced some difficulty getting into units to do the evaluation. Bob Shoemaker has been especially turf conscious in this regard. It has taken me two years of accommodation and working around to get him to agree to let us come along in a joint FORSCOM-TRADOC evaluation visit schedule. In part he's right to defend his turf—we can quickly overload units in the field with well-intentioned but ill-timed and ill-devised evaluations. In that regard I support him fully. At the moment we have the most extensive system of field visit, evaluation and coordination that I've seen in recent years—at least since I became aware that it was necessary. From TRADOC's standpoint I would prefer that we be given regulatory authority for standardization, evaluation, and the kinds of investigation necessary to do better at man-machine interface. Then I need to be tasked to do it. At the moment I have only the tasking and am forced to negotiate at excruciating length with my fellow commanders to avoid giving the impression that I'm out telling them how to run their units, or out finding out just how bad—or good—they really are. It's not a happy world at all. But it is necessary under the current system.

In short, I am convinced that there has to be a single user representative with the authority and responsibility to do whatever is necessary to provide the Chief of Staff, and others, with the best worked out scheme for solving whatever problems we have in the man-machine area and many others related to materiel development.

Proposed Test Command

Message to General John W. Vessey Jr.
Army Vice Chief of Staff
14 October 1980

The proposal to organize a test command is one to which I object—I am unalterably, unqualifiedly, unrelentingly, uncompromisingly, and violently opposed to such a course of action. Besides that I am against it.

Mobile Protected Gun

Message to General John W. Vessey Jr.
Army Vice Chief of Staff
22 October 1980

1. Thursday you will hear the mobile protected gun (MPG) briefing from Fort Benning and Fort Knox. It will not completely scratch your itch, nor does it scratch mine.

2. The study identifies a need to improve light forces' armor-defeating capability and provide direct fire to support the assault. The need applies to both Army and Marine forces. Models available can't play dismounted infantry; therefore, conclusions as to capabilities to support the assault by fire are based on military judgment. What could be measured was the capability to defeat armor. Questions of survivability based on mobility/agility are addressed, but not completely resolved. These questions have been around for years without answers because we still don't know how to measure the payoff.

3. Threat was defined as T72 export for the study. The analysis compared a 90mm automatic cannon with a variety of other systems. The 90mm won on rate of fire. It was second to the 105mm on lethality if you believe the projections. Conclusions were the 90mm is marginal in lethality and we can't have it before 1989. By that time the threat will be much harder than the T72 export. Even if we buy the 90mm gun and it works we will still be 10 years behind.

4. If we are looking for a gun to defeat tanks then we are asking for a new tank. If it is to accompany light forces, we probably are talking about a souped-up 120mm mounted on a light, unarmored chassis. I just don't think that will work. I would be delighted with a smaller caliber, but it must do the job.

5. The near-term solution provides for a six-inch TOW on HMMWV combined with a 25mm cannon on a wheeled armor vehicle. We can do this fairly quickly (FY84–85). It provides an answer we can live with for a while.

6. We need to use the time the near-term solution buys to get the entire development community together and lay out what we think the 90s threat will be and what type weapon will be needed to defeat it. We may not be able to get a suitable tank cannon within weight constraints peculiar to light forces. We may find we are stuck with a missile for the tank threat, with a gun to strip out light armor and provide assault fires, or we may decide we have the technology to build a gun to do the antitank job. Right now we just don't have the answer to this dilemma, and the APAS study could not provide it.

7. Whatever weapons we decide we need must be carried on a family of vehicles using a common chassis. This family must be strategically and tactically transportable if we are to meet all contingencies. Weapon systems would be tailored to address different threats so that all targets are serviced.

8. I think we need for light infantry:

   a. A vehicle with a long-range antitank capability (probably a guided missile).

   b. Another vehicle mounting a cannon to provide assault fires for the infantry and to defeat light armor.

c. Other combat and combat support systems that I have not completely defined in my mind yet, but which need careful study (for an example, an AD gun). All must be strategically and tactically transportable and logistically supportable. They also have [to have] more survivability than we can presently provide.

9. The family concept will give us a logistical advantage we have not before had through commonality of equipment, spare parts, and soldier training. This will be a distinct advantage in situations with a severely constrained logistics tail.

10. Bottom line is the study won't provide you an answer to the 75/90mm question. The near-term solution is the best fix we can get right now. It will probably be with us for the next 10–15 years. What we have got to do now is get to work on what we want in the long term.

Mobile Protected Gun

Message to General John W. Vessey Jr.
Army Vice Chief of Staff
8 November 1980

. . .

2. While I recognize the benefits of attacking deep, I also believe the force we put on the ground must be survivable and able to cope with the threat it faces. I also believe that we can attack deep with light forces against an unsophisticated threat. I don't believe they will make it against a sophisticated enemy except in selected terrain and scenarios. In most future contingencies, we will probably be against either Soviet or Soviet surrogate forces who already are equipped with large quantities of sophisticated armored systems. Therein we are developing high-technology systems to see and shoot deep without unnecessarily exposing our soldiers, as all agree that trained soldiers are much harder to replace than equipment. I do recognize why the questions are being asked about the priority of the system capability. We will look at it again.

3. We considered all the options you want us to look at in our initial drill. I will tell you why they fell out the first time, and why I suspect they probably will again.

a. The combination of the M113 w/25mm and ITV is a good solution with respect to doing the job. Problem is that we are short about 13, 000 M113/M113-derivative and IFV/CFV chassis in the inventory, with no funded program to fill this gap other than slow acquisition of IFV/CFV. As you know, we have tried to get more M113s since FY79 and have been unsuccessful except for a few ITV bottoms. With only 1,756 ITVs to be produced, we must choose to either reprioritize its distribution or eliminate it from POMCUS. (The 371 targeted for POMCUS would meet about half the light division requirement.) We perceive that no more M113s will be bought, therefore this combination was not recommended. If budget conditions change, this combination is certainly attractive.

b. The IFV was eliminated based on not being C-130 transportable (tactical mobility) and not being C-141 transportable in a "ready-to-fight" condition. Selecting the IFV also requires reprioritization of assets and increasing the buy. It generates a significant increase in the logistic support tail, which would be detrimental to the light division's transportability and supportability.

c. M551 w/25mm was considered but eliminated in the near term because three other M551 options were on the list. We felt it would be bracketed in cost between the M551 w/105 and the M113 w/25. Its effectiveness would approximate that of the M113 w/25. It offers no potential for commonality with the troop carrier.

4. The infusing of two battalions of M551 into the 9ID testbed program will be costly in terms of dollars for refurbishment, shipping, and support. It will also create turmoil in the personnel stream, as we would have to significantly increase MOS production at Knox and people in the force with the correct MOSC will have to be identified, reassigned, and trained as a unit. In addition, for the next two years there is already a full plate agreed to. If we add this, something planned will have to come out. These are only the surface problems. I am sure there will be more discovered. If you like, we will get together with FORSCOM and DARCOM and lay out the whole cost and what options are available.

5. I agree that we need to reexamine our philosophy on lightweight guns killing threat tanks at their toughest point. That is why I said in my earlier message we may be stuck with a gun/missile combination. Bottom line is something has to kill the threat and I believe that, once that threat tank is fired on, the second round will address the toughest aspect.

6. I propose that we come back quickly to you to answer the near-term question, but that we establish a TRADOC/DARCOM/USMC joint working group to cope with the long-term answer. The group would have your best threat guy and do the following:

a. Sort out the threat.

b. Sort out which technologies offer the most potential.

c. Recommend to you what technologies be pursued to achieve the best system to do the job. This may be either a gun or a gun/missile combination.

7. I emphasize this should be a working group, not another study. Once you make your decision, we ought to get moving to get the long-term in the force. If it matches up with the Marines, fine. If not, they will have to decide what they want to do. The reason they have been getting Congressional support is that they appear to know where they are going and that we do not. Once we articulate our plan, we will get support too.

8. The working group recommendation could significantly affect the 90mm program. Given the narrow safety margin of the 90mm projected lethality, and our past track record of success in achieving projected capabilities, I remain to be convinced that the 90mm will make it.

9. I still have difficulty in understanding our logic. We project the 90mm as having capabilities about equal to the 105mm. We say this is good enough, yet we have demonstrated to ourselves and the Congress the 105mm is not good enough and we need the 120mm. Once we drop below having to kill tanks, the 25mm will take care of everything else. It is here.

10. . . . the near-term solution provides the best opportunity to get with the Marines. Once we have finished our relook, I hope to get on with it.

Soldier-Machine Interface

Letter to Lieutenant General William R. Richardson
Combined Arms Center
13 April 1981

This letter is a follow-up to my . . . letter concerning [Major General Sinclair L.] Melner's Soldier-Machine Interface paper. Hopefully it will provide direction for further deliberations which I see as being centered around the complexity issue you are gearing up to address.

Soldier Support Center has done yeoman work in preparing this comprehensive compilation of our people-related problems. All are problems which must eventually be solved, but resource restraints preclude our attacking them all simultaneously. I still feel that the best way to start is by selecting the doables and, in turn, getting the doables done.

We are now finalizing the formal directive for the Soldier-Machine Interface Requirements (Complexity) study I mentioned above. CAC (as lead), and with support from the TRADOC community and from DARCOM, will be tasked to look at the impact of complexity of new weapon systems on the manpower, personnel, training and logistics aspects of force modernization. A related study, Human Capability to System Performance to Combat Effectiveness, is currently being conducted by SSC and is referred to in the SMI paper.

. . .

Back to the few doables. Our immediate and most pressing people problem is manning the force of today and of the immediate future with soldiers who can (are trained) and will fight and win. Our highest priority must be given to achieving that goal. The complexity study effort aims at satisfying the "can fight" element of this goal. We hope the effort will lead to developing better ways to articulate our personnel requirements and also better ways to impact future systems design based on projected personnel availability. I feel that we are now moving in the right direction in the "can fight" or "skill" aspect and do not need to start any major new undertakings until we see what we get from the current efforts. We just need to tie them together. . . .

Even more critical than a "can fight" soldier is a "will fight" soldier. It is in this area that we must seek new initiatives and exploit everything we know and can learn about group dynamics of soldiers. . . . The current Regimental System Study and ARCOST initiatives are only the beginning. While tying current "skill" efforts to the complexity effort, our priorities should now be development of new ways to enhance the will of our soldiers, institutionalism of what we learn concerning both skill and will, and follow-on efforts as required.

Modernization Timing

Message to General E. C. Meyer
Army Chief of Staff
4 June 1981

1. Re Bob Shoemaker's OMCP 1221 EO this subject.

2. With all due respect to Bob's point of view, which I believe I understand as well as he does, I must take complete, violent, vigorous, unqualified, unrelenting, stubborn exception. Aside from that, I'm against his notion set forth in ref.

3. The first thing we all must understand is that, if we wait for everything to be just right to move ahead with whatever aspect of modernization we are considering, we will never, ever modernize anything. It has always been thus, and it will ever continue to be that way. While that's unfortunate, and I share Bob's deep concern about it, and I decry it as does he, it is and will continue to be unfortunate, but not a situation that should prevent us from moving ahead as quickly as possible to modernize, even though everything is not just exactly right.

4. Secondly, we must also recognize that DARCOM is simply not organized to issue to the field new systems, to include all their components. I know of no more difficult case than that of the TACFIRE; the same thing is happening in the development of requirements for the corps support weapon system—Bill Burns is having to coordinate between four DARCOM commands because there is no central coordinating mechanism in the command—for R&D or for procurement and issue. This is said not in criticism of DARCOM at all, but simply to highlight a fact that all the rest of us must recognize. In part we have sought to solve pieces of that problem with the TRADOC systems managers. While they have the responsibility to signal what's not going right, they have not the authority to order anything done about it. Nor were they intended to do so, nor could they do so; they belong to TRADOC, not DARCOM.

5. Thirdly, no mechanism we could create could foresee all the problems we can expect to encounter when we start fielding these new systems. Therefore, in our demand-supported supply world we must create a demand on the system to produce. That can be done only by issuing the item and forcing the system to respond. I just finished going over with the 8th Mech their plan for fielding TACFIRE and for the Firefinder fielding due to come upon them this fall. All the problems in Bob's message they have encountered, continued to demand, insist, pressure, cajole, and be otherwise persuasive; as a result, they will field the system about on time. Had they waited until they had everything in hand, TACFIRE would never have seen the light of day. The system didn't know what all they needed; it could not have been expected to know that; demand produced the desired results.

6. While some instances of temporary unreadiness may result from pressing on with issue of new systems, with reorganizations, with other aspects of modernization, and while we should take every reasonable precaution to avoid doing something dumb, we simply cannot, must not, hold up waiting for things to get better. They won't get better until we put that gear, organization, tactic, or whatever modernization aspect we're considering out here in the hands of our soldiers and create the experience and so the demand to get the thing up and fully operational. For God's sake don't slow down; if anything, let's speed up. The guys out in this world are ready for modernization. Let's give them a chance to make it work!!

Current Imperatives

Association of the United States Army
MacDill Air Force Base, Florida
11 September 1981

It might be useful if I started our association together with one person's perspective on what seem to be today's imperatives for our Army. First, we're embarked on the largest Army modernization program since World War II. While modernization happens to military forces all the time, today it's made more acute by several facts. One is the pace at which technology is advancing. In the electronics field alone, technology is said to turn over every 3½ years; other technical fields are not far behind. That being the case, it is difficult, if not impossible, to keep truly modern equipment in the hands of troops. This rapid and constant changing technology encourages a tendency to wait just a little longer to field something so that the latest exciting developments can be incorporated. These two things combine to stretch out the research, development, and acquisition cycle to the end that, probably for the first time in history, it now takes longer to develop and field the equipment itself than it does to produce the soldier who will operate it.

Second, it's important to remember that, while modernizing forces does include the infusion of new and modern equipment, equally important is the requirement to modernize the way we fight—tactics, the way we organize for battle—organization, and the way we get soldiers and units ready for battle—training. Studies show that new equipment can add some 30 percent to our battle-fighting capability; new organizations about 10 percent; sound tactics and lots of the right kind of training can add as much as 60 percent. Today we provide our soldiers no more than 25 percent of the training other armies of the world give their soldiers at the same grade levels and in the same skills. Some of those armies even have American equipment. One might say the soldiers of those other armies are not as smart as ours or that we know something about training the other armies do not know. Neither is the case. We just haven't been willing in the last few years to provide the resources to train our soldiers and units as they should be trained.

Third, we are, as a nation, an impatient people with a very short memory. Perhaps it's our television mentality—if we can't wind the whole thing up in a one-hour program that includes 40 minutes of action and 12 commercials, it's no good. Modernizing today's military forces is a long-term project, for we are striving to overcome the materiel and morale effects of Vietnam and the decade since Vietnam. We can't expect to do that overnight. Patience, in large measures, will be required.

So, while we applaud increases in defense spending, which portend, in time, the advent of modern equipment, we must be very careful to show that, while we wait for that gear to come along, we are doing other things to improve our fighting capabilities, things that may also cost. So it is that a very careful balance must be struck between dollars spent for materiel and dollars spent to ensure our capability to use the equipment correctly. We must avoid giving the impression that we are the guys with insatiable appetites, for if ever that tag is hung on us, surely our national impatience will overtake our credibility, and should that happen, very real requirements will be set aside and we will be in great trouble.

So often in these discussions I am reminded of one of the very first things I learned in the Army. Thirty-eight years ago this month, the one-eyed corporal who was my drill sergeant in basic training said to me, as we completed that very difficult phase of my military career, "Young man, if you can ever get your feet straightened out and your head screwed on straight, you may be an officer in this Army. Who knows? If ever that happens, you just remember one thing—battles are won by the blood of the soldiers, the guts of their leaders, and the disciplined way their units fight in battle." Thirty-eight years later, I have to echo the words of that wise and very tough old soldier. For on that complex, most dangerous, and difficult battlefield of the next war, the determined, prompt, correct, and calculated battle actions of trained commanders, soldiers, and units are all that will win. Commanders trained to train and trained to lead the soldiers and units entrusted to them. Soldiers trained to follow, trained to fight determinedly and fiercely in units whose survival and success are their very life's blood. Equipment we need, force structure we need, but above all, we need trained soldiers, leaders, and units, and we must invest in that training far more than we've been willing to do in the last few years.

For in the tough times still facing us, sometime, somewhere, once again, the fate of the nation's aims, goals, programs, ambitions, and perhaps even survival will surely rest on the determined actions of a few good soldiers, a few good leaders, and a few good units who trained well in time of peace in order that they might be ready to fight well in time of war.

That is our challenge. I am confident that if we can band together, determined, strong, and with one voice, we too can win. But it will take all the skill, dedication, and ability at our command to do it.

Joint Tactical Cruise Missile System Rationale

Letter to Donald C. Latham
Deputy Under Secretary of Defense for
Communications, Command, Control, and Intelligence
27 May 1983

Thank you for sending me the "Value of JTACM" paper. I'm glad someone is at work trying to show the worth of the system. Unfortunately, I have some serious misgivings about the way the analysts have gone about this one.

To begin with, the purpose, and indeed the value to the supported unit commander, of JTACMS is not necessarily in the attrition it causes in following echelons. Rather, it is in the combination of breaking up the mass and slowing down the velocity—destroying the momentum—that the true value lies. Attrition there may be, but it is secondary to the need to destroy or disrupt momentum. Therefore, the paper uses the wrong measure of worth and in this case it is, in my judgment, a fatal error, for it puts the analysis squarely on the horns of the old firepower score WEI/WUV formula dilemma.

Second, looking at the analysis from the standpoint of Central Duel calculus, if we take the direct fire systems—tanks and ATGMs—as the overstressed systems and introduce something to relieve the stress, then it does not necessarily follow that we can remove from the battlefield any or all of the overstressed systems. What is true is that the overloaded systems at the FLOT simply cannot do their job if there is not some control over the enemy systems presentation rate at the FLOT. It is control over this systems presentation rate that JTACMS, or any other system aimed at the same targets, seeks to establish. Unfortunately, the calculus we're using is not powerful enough to calculate and overcome this fatal flaw in Central Duel logic. We encountered the same poblem several years ago when trying to rationalize ATGMs, M1s and AH64s in combination—the first of the Central Duel studies.

Following the logic of this study to an absurd conclusion, one could show that it is possible to replace all divisions with JTACMS. While the missile enthusiasts might embrace that notion, I doubt that it would get wide acclaim elsewhere. In the paper enemy forces are attrited, friendlies are not—at least as nearly as I could determine. Neither is reconstitution of damaged vehicles calculated. JTACMS has no residual value in the calculus; fighting vehicles do. All of these things lead to a grossly misleading portrayal of force ratios.

Cost figures for JTACMS appear optimistic and are misleading. PGMs on the warhead will produce missile costs in the $460–500K range at least. Security forces, support personnel, targeting people, communications, and other systems are not included in the costs and they must be in order to produce relevant comparisons. Launchers are part of JTACMS costs; they are not included.

Finally, the ability of JTACMS to do the job we have spelled out for it is related to accuracy of acquisition systems and functionality of the supporting fusion system. A missile with a 100 meter CEP [circular error probable] driven by an acquisition system with a 200 meter CEP is not exactly the most accurate of weapons. Therefore the probability of success of the missile must be considered in the full context of all supporting elements required to make it do its job.

In short, one kind of unit can't be traded for another, and particularly is it not possible to trade personnel costs around as the paper does without a full and complete audit trail of manpower authorizations required for full fielding. It's a seriously flawed paper.

My recommendation would be that we seek a better measure of merit, then lay down a much more thoroughgoing, comprehensive and relevant analysis.

16. Planning

Page

Contingency Planning Intelligence 696
Short-Term Goals 697

Contingency Planning Intelligence

Letter to Charles W. Snodgrass
Electronic Data Systems Corporation
26 August 1981

I do hope that we made some inroads into the problems of tactical intelligence; we've a whole host of similar and overlapping problems in the business which I have taken up as the world's contingency planner for our side. If we can just draw on our experience in the tactical intelligence field perhaps it won't take quite so long for us to get this one defined and under development. Let's hope not.

Short-Term Goals

Memorandum to Readiness Command Chief of Staff
24 September 1981

There are several actions I'd like the staff to begin or continue to work on, or perhaps change directions. Since I'll be away for most of October, this memorandum will serve to get everyone working on at least these following subjects which are of immediate concern to me.

1. Goals. With regard to the goals and taskings exercise; since we went over the list of tentative goals last time, I've discovered a 1 October 1979 document entitled *Mission Book*. I am told that it has not been brought up to date in some time. Review that book with a view to determining what we could do to revise it to meet the need for a goals and tasks document. Reading through, it appears to me to be very nearly what we want. Accordingly, suspend work on other goals and missions activities until we can decide whether to use the Mission Book with some amendments or to begin anew.

2. In addition to assigned tasks, several areas of endeavor need to be reoriented, opened up, reviewed. Principal among these are:

- A capability for simulations or war games by which we could examine employment, deployment and sustainment plans; examining the adequacy of plans by iterating them frequently enough to eliminate aberrant patterns before, not after the fact. That would also give us the capability to offer to the supported commands alternative schemes of employment, deployment, sustainment; schemes which might make better use of limited resources than might be the case with the present system.
- An exercise evaluation system for quantitative evaluation of joint exercises with emphasis on aspects of joint exercises which are of primary concern to USREDCOM and JDA. Those elements are: Indications and Warning Systems, Intelligence Collecting and Reporting Systems, Command and Control Systems, the Joint Deployment System, interoperability amongst elements of joint forces, Tactics Techniques and Procedures for Joint Forces, Structuring of Forces for Contingency Operations.
- A staff operations analysis capability in USREDCOM and JDA to carry on the work indicated above.
- A foreign area specialist capability on the USREDCOM staff to provide specialized and continuous analysis and evaluation coverage of areas of specific interest to USREDCOM.
- A change in the primary thrust of USREDCOM's exercise program to one designed to focus on the matters set forth above; to conduct more CPXs rather than FTXs; to extract from all of the joint exercise program useful data for revising plans and joint tactics, techniques and procedures.
- A staff element whose primary field of endeavor is the development and marketing of joint tactics, techniques and procedures.
- USREDCOM and JDA liaison with and representation at headquarters of the supported commands and other appropriate agencies.
- The perceived need to bring all USREDCOM resource management activities under a single office.

With that general description as background, some specific tasks for staff sections are set forth in the following paragraphs.

3. J1. . . . Based on my expressed need for ops analysts, area specialists, joint tactics, techniques and procedures development, force structure development, and liaison, draw up a revised set of requirements which will include specialists in those skills in appropriate numbers. Develop a revised plan for transition to the separate stand alone RDJTF. Complete and present to me your current work to determine at what level we cannot or should not accept any more people in USREDCOM/JDA/JCSE/RDJTF because of overcrowded facilities at MacDill.

4. J2. Determine what would be required to provide an analysis and evaluation capability for at least the following areas: the Persian Gulf, the Eastern Littoral of the Mediterranean, the Horn of Africa/Red Sea, Africa, Central America, South America, Korea, and Southeast Asia. This should include country by country or area by area expertise sufficient to give background/area coverage and situational coverage daily, if required. Review concepts for I&W, intelligence collection and reporting, intel fusion analysis—evaluation and dissemination for joint forces in order to describe the state of the art.

5. J3. Draft concepts for a revised exercise program, along the lines suggested in 2 above, for an exercise evaluation program, for more direct evaluation of functions in which USREDCOM has specific interest and responsibility. Review concepts for command and control of joint forces, for ensuring interoperability of joint forces, for evaluating tactics, techniques and procedures for joint forces; purpose—to describe the state of the art.

6. J4. Draft a concept for a war game/simulation capability to test and evaluate plans, a program to review all plans for which USREDCOM is responsible or is responsible for supporting; purpose of the review is to determine if the plans are suppportable logistically. Draft concepts for force structuring combat service support forces to support combat and combat support forces with emphasis on force package tailoring and the tailoring of alternative support force packages with visibility of the effects of the variances on capabilities of the force as a whole—combat, combat support and combat service support.

7. J5. Draft concepts for a planning system which includes a simulation feedback loop from J3-conducted simulations to test adequacy of plans. Draft proposals for a staff agency responsibility for joint tactics, techniques and procedures, their development, test, evaluation, coordinating, monitoring and marketing. Draft proposal for a TAC-TRADOC-USREDCOM "troika" which would together develop and implement joint tactics, techniques and procedures for the airland battle.

8. J6. Review communications linkages in the I&W system, intel collection and reporting system, intel fusion—evaluation—analysis and dissemination system, operational command and control system to describe current status with regard to interoperability, standardization, equipment and provisions for the linkages themselves. Review the status of joint techniques and procedures for command and control to describe status and what is needed next.

9. JDA. Assist J3, J4, J5 in tasks above as required. Draft proposals for quantitative evaluation of the Joint Deployment System. Review current transportation allocation analytical models to determine their adequacy as means of optimizing deployment transportation modes and for quickly developing alternative deployment programs as contingencies and unforeseen circumstances arise in the process of deployment.

10. Chief of Staff.

- Finish up work on proposal to bring all USREDCOM resource management activities under one office.
- Review the USREDCOM Liaison Officer situation both to and from other headquarters. Draw up proposals that would provide better first-hand representation for USREDCOM Headquarters at headquarters of supported commands, lateral commands or agencies, components.
- Especial attention to new functions such as the ops analysis and tactics/techniques/procedures function to determine if this should be imbedded in the current staff, or if they require separate staff agencies.
- Ensure that we bring together into a single analysis all aspects of the RDJTF "stand alone" problem—manpower, facilities, office space, et al., taking into account the new missions already assigned, and best estimates of requirements stemming from new directions set forth above.
- With the exec, lay out a calendar to bring the results of the above to me as time permits during the month of November.

GPO U.S. GOVERNMENT PRINTING OFFICE: 2009—553-445 R1